MODERN CONTROL SYSTEMS THEORY

UNIVERSITY OF CALIFORNIA ENGINEERING AND SCIENCES EXTENSION SERIES

Balakrishnan · Space Communications
Beckenbach · Modern Mathematics for the Engineer, First Series
Beckenbach · Modern Mathematics for the Engineer, Second Series
Brown and Weiser · Ground Support Systems for Missiles and Space Vehicles
Dorn · Mechanical Behavior of Materials at Elevated Temperatures
Huberty and Flock · Natural Resources
Langmuir and Hershberger · Foundations of Future Electronics
Leondes · Computer Control Systems Technology
Leondes · Modern Control Systems Theory
Parker · Materials for Missiles and Spacecraft
Puckett and Ramo · Guided Missile Engineering
Ridenour and Nierenberg · Modern Physics for the Engineer, Second Series
Robertson · Modern Chemistry for the Engineer and Scientist
Sines and Waisman · Metal Fatigue
Zarem and Erway · Introduction to the Utilization of Solar Energy
Le Galley and McKee · Space Exploration
Garvin · Natural Language and the Computer

MODERN CONTROL SYSTEMS THEORY

MASANAO AOKI

GEORGE A. BEKEY

DALE D. DONALSON

H. C. HSIEH

FRANCIS H. KISHI

JAMES S. MEDITCH

RICHARD A. NESBIT

PETER R. SCHULTZ

EDWIN B. STEAR

ALLEN R. STUBBERUD

Edited by

CORNELIUS T. LEONDES

Professor of Engineering
University of California,
Los Angeles

McGraw-Hill Book Company, Inc.

NEW YORK ST. LOUIS TORONTO LONDON SAN FRANCISCO SYDNEY

MODERN CONTROL SYSTEMS THEORY

Library of Congress Catalog Card Number 64-66024

37192

To Emerson

March 1, 1930 – September 13, 1964

The Authors

Masanao Aoki, Associate Professor, Department of Engineering, University of California, Los Angeles

George A. Bekey, Associate Professor, Electrical Engineering Department, University of Southern California, Los Angeles

Dale D. Donalson, Head, Functional Design Section, Hughes Aircraft Company, Culver City, California

H. C. Hsieh, Deceased, Formerly Assistant Professor, Department of Electrical Engineering, Northwestern University, Evanston, Illinois

Francis H. Kishi, Member of Technical Staff, TRW Space Technology Laboratories, Redondo Beach, California

Cornelius T. Leondes, Professor, Department of Engineering, University of California, Los Angeles

James S. Meditch, Member of the Technical Staff, Aerospace Corporation, Los Angeles, California

Richard A. Nesbit, Beckman Corporation, Santa Monica, California

Peter R. Schultz, Head, Injection Guidance Section, Systems Research and Planning Division, Aerospace Corporation, Los Angeles, California

Edwin B. Stear, Assistant Professor, Department of Engineering, University of California, Los Angeles

Allen R. Stubberud, Assistant Professor, Department of Engineering, University of California, Los Angeles

Preface

At the beginning of the 1940's techniques for the analysis and synthesis of control systems were virtually nonexistent, certainly as we know them today. The development of techniques for simple lead or/and lag compensation design for desirable transient response characteristics for relatively simple linear systems was just beginning. The root locus method with the insight it would provide was still almost ten years away. Describing function techniques so useful in the design and understanding of many nonlinear control systems was also almost ten years away. Applied needs were ahead of the tools available to meet these needs in many cases.

The last decade has seen a very significant increase in the number of applications of control systems theory. It has also witnessed a substantial increase in the body of the theory itself. Among the various reasons for this is, of course, the pressures which the increasingly complex nature and variety of applicational needs place on the theory itself. Another factor has been that the field holds an enormous amount of appeal in terms of challenging problems, real or with a slight bent to the imaginary, for such individuals as mathematicians and physicists as well as theoretically oriented engineers. Some of the prime examples of this, of course, are mathematicians R. Bellman and his work on dynamic programming and Lev S. Pontryagin and his contribution of the maximum principle to the control systems theory field.

It is probably fair to say that, whereas in the early forties the available theoretical techniques were in many cases not adequate for dealing with numerous applied situations, the last several years have seen a situation wherein many of the theoretical techniques developed or under development may be years before they are really applied, if ever. However, even in these instances there are examples where such results have led eventually to other results with fruitful implications from an applied point of view. Besides this there is the greater understanding and insight that inevitably results in either event.

This book provides a rather good example of some of the above observations. Most of the techniques presented in this book are less than five years old. The power of these techniques from an applied point of view is quite clear. For instance, consideration might be given to the very first chapter, "Linear Time Variable System Synthesis Techniques," by Dr. A.R. Stubberud. The general area of synthesis techniques for linear time variable systems had, for the most part, been in a relatively unsatisfactory state until a few years ago. Dr. Stubberud's research results over the past several years have done much to remedy this situation with numerous practical implications. For instance, some aircraft landing control problems, air-to-air missile problems, and many others could well use the availability of such techniques as are presented in Chapter 1. In addition to this it might be noted that in some cases nonlinear systems can to a good degree of accuracy be replaced, for analysis and synthesis purposes, by a linear time variable system. Comments of a somewhat similar nature can be made of much of the rest of the book.

Criteria for the selection of the contents of this book involved several considerations. First of all, emphasis was given to those techniques which it appears will stand the test of time. Secondly, those important techniques which have already been very well documented elsewhere were for the most part omitted where feasible and appropriate. For instance, the important method of dynamic programming is discussed to a certain extent in Chapter 9 by Schultz, but from a comparative rather than a developmental point of view. The reason for this is that there are already a number of books and reports of wide circulation which develop the subject nicely. On the other hand, those who have seen "The Mathematical Theory of Optimal Processes" by Pontryagin et al., Interscience, 1962, will find Chapter 7 by Meditch most helpful. In addition, Meditch's chapter may be quite adequate for the user in many cases.

A word discussing each of the various chapters in the book and their role in the current state of control theory is in order at this time. The area of the synthesis of linear time variable feedback systems as discussed in Chapter 1 is an extremely important one in the general area of applied control systems theory, and some comments on the nature of this chapter have been presented earlier in this preface.

The synthesis of systems with random inputs for both the multivariable or multipole systems as well as for systems with nonstationary stochastic inputs has seen some rather recent developments. In addition, the area of suitable mathematical descriptions of the random processes serving as inputs to control systems has been rather lightly treated in the literature particularly for the more complex processes encountered in the design of engineering systems. Chapter 2 takes up these various important topics.

It has been said by some that the next few decades will see a fairly large body of practicing engineers have a fairly good working familiarity with the techniques of functional analysis and its importance and utility in the solution of certain classes of engineering problems. Chapter 3 serves as an introduction to the area of functional analysis and provides some illustrations of the utility of these techniques in the solution of engineering problems in the control systems theory area. There have been some rather excellent texts appearing recently in the field of functional analysis aimed to a certain extent at an engineering audience. The Russians have made particularly good textbook contributions in this area.

In simulation techniques for the determination of the mean square error performance of linear time variable systems with nonstationary stochastic inputs, such as the method of adjoint simulation developed over ten years ago by Laning and Battin, it is important to be able to generate a nonstationary stochastic signal from stationary white noise. In various of the techniques developed recently for the optimum synthesis of linear time variable filters with nonstationary stochastic inputs the requirement for a mathematical description of the shaping filter to generate a nonstationary stochastic signal from a white noise source is essential. Because of the great importance of shaping filters to a fairly wide variety of control system problems Chapter 4 has been devoted to a treatment of this subject.

Recently there have appeared some rather excellent textbooks on the subject of Lyapunov's direct method for the analysis of nonlinear control systems. These texts include the book by W. Hahn, recently translated into English and published in this country, as well as the well-known book by LaSalle and Lefschetz. However, both books are lacking in a presentation of what is certainly one of the most important techniques yet devised for the generation of Lyapunov functions, namely, the variable gradient method developed recently by D. G. Schultz. The reason for this, of course, is the fact that Schultz's technique was announced after the publication of these texts. In addition, several other results have occurred since the publication of these texts. Because of the great importance of Lyapunov's direct method for applied problems and because it was desired to develop a treatment with these recent significant developments more readily accessible and perhaps more readily useful to engineers, Chapter 5 on Lyapunov's direct method is included here. A particularly notable feature of this chapter is the presentation of Lyapunov techniques for nonautonomous systems.

With the trend toward more sophisticated control problems, newer techniques have been developed to deal with such problems. One of the most important of these technique areas has been adaptive control systems theory. Chapter 6 is devoted to a development of one approach to discussing the various methods and techniques for synthesizing adaptive control systems. Rather than get into a discussion of what adaptive control systems are at this point, further comments will be deferred until Chapter 6.

The role of Chapter 7 in the book was discussed earlier in this preface.

Chapter 8 is devoted to a development of another approach to system optimization problems, viz., Krein's problem. Techniques here rely rather heavily on functional analysis metheds discussed earlier. Krein's L method is introduced and described in this chapter. It is pointed out how the solution of the 2 point boundary value problem, which arises in so many optimization techniques and which is discussed further in Chapter 9, can be avoided, although another computational problem of similar complexity frequently arises. Numerous, rather powerful applications of these techniques are presented in this chapter.

Chapter 9 is devoted to a development and comparison of various analytical techniques for the optimum synthesis of linear control systems. The techniques studied and compared here include dynamic programming and the Pontryagin maximum principle. It should prove quite helpful in clarifying the relative merits of these techniques for the class of problems discussed there.

Stochastic approximation theory affords one of the few approaches to the examination of the question of the stochastic stability of control systems. Unsatisfactory though the present status of these techniques are, a review and examination of where the field stands on this rather important question is quite pertinent and is discussed in Chapter 10.

Computers are, of course, assuming an ever greater role in control systems to result in what might be referred to as a computer control systems technology of increasing power and utility. Fundamental to the development of such systems are techniques for the analysis and synthesis of discrete time systems. The great importance of this area is recognized by material presented in Chapter 11.

Human operators occur in many control systems either as an integral part or else as an emergency backup element in case of failure of key system elements. In this latter situation a control system design which takes into account the operator's performance characteristic or human transfer function can be quite vital. For instance, such analytical techniques can conceivably be used to establish bounds on control system design parameters which will assure that the human operator can indeed control a complex system such as an aerospace vehicle when suddenly placed in an emergency situation because of the failure of a control element. Chapter 12 is devoted to the general question of a mathematical description of the human operator in control systems.

The book closes with a final chapter on some comments relative to the application of modern control methods to aerospace vehicle control systems.

It is a pleasure at this time to note a number of acknowledgments. First of all, the editor would like to acknowledge the AFOSR and the individuals in this organization who have helped to make the programs of research in control system theory at UCLA and other institutions active in the field possible. Many of the results presented here came directly from AFOSR and other military-sponsored research projects either at UCLA or at other institutions. Without such support these results, many of them with important applied implications, would not have been possible.

This book grew out of a national two-week summer course that has been held at UCLA over the past several years. Comments by the students who took this course, who were for the most part rather senior practicing engineers, have been most helpful in the evolution of the course as well as in the development of the manuscript for this book. The contributions of these students are gratefully acknowledged.

Finally, it is a great pleasure to acknowledge the coauthors of this book, who in spite of schedules which were already top-heavy with commitments and responsibilities accepted the opportunity to contribute in this manner. The association with them in this venture has been for me a most gratifying experience.

CORNELIUS T. LEONDES

Contents

1

Linear Time-Variable System Synthesis Techniques

A. R. STUBBERUD
ASSISTANT PROFESSOR OF ENGINEERING
UNIVERSITY OF CALIFORNIA, LOS ANGELES, CALIFORNIA

A number of general techniques for analyzing and synthesizing linear, time-invariant systems have been developed by control systems engineers. On the other hand, general techniques for the analysis and synthesis of linear, time-variable, and nonlinear systems have received much less attention. The reasons for the predominance of linear, time-invariant systems are, first, the general solutions for linear, time-invariant equations are more simply obtained than general solutions for time-variable and nonlinear equations; and second, many time-variable and nonlinear equations can be adequately approximated by linear, time-invariant systems.

In this chapter some techniques for the synthesis of linear, time-variable systems are presented. These techniques are quite general and parallel some of the classical synthesis techniques for linear, time-invariant systems.

1.1 Some Characteristics of Linear Systems

A general linear system is shown in Fig. 1-1. The input to this system is a function of time, denoted $x(t)$, and the corresponding output is $y(t)$. In general, $x(t)$ and $y(t)$ may be vectors; however, in this chapter, they shall be assumed scalars.

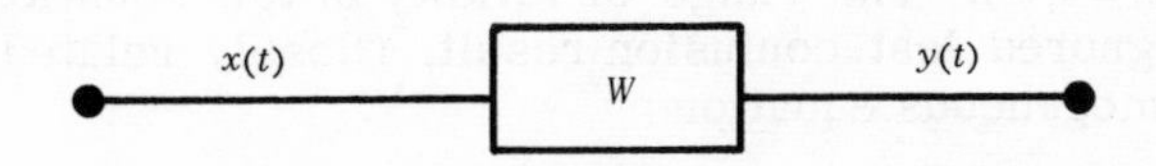

FIG. 1-1. General linear system W.

A linear system is defined as follows.

DEFINITION:*

A system is called linear if

(1) an input x_1 produces an output y_1;

(2) an input x_2 produces an output y_2; and

(3) an input $c_1x_1 + c_2x_2$ produces an output $c_1y_1 + c_2y_2$ (where c_1 and c_2 are arbitrary constants).

This definition assumes that at the time of application of $x(t)$ to the system, $y(t)$ and its derivatives are zero.

A wide variety of systems are covered by this definition; for example, systems which can be described by linear algebraic equations, linear ordinary differential equations, and linear partial differential equations. In this chapter, only those described by linear ordinary differential equations will be considered. In particular, those considered can be described by a differential equation of the form

$$\sum_{i=0}^{n} a_i(t) \frac{d^i y}{dt^i} = \sum_{i=0}^{n} b_i(t) \frac{d^i x}{dt^i} \quad ; \quad t_0 \leq t < +\infty \tag{1.1}$$

or in operator form

$$L(y) = M(x) \tag{1.1$'$}$$

where x is the input and y the output. Three further assumptions will be made in the sequel: (1) the $a_i(t)$ and $b_i(t)$ are continuous and have enough continuous derivatives to carry out the indicated operations; (2) $a_n(t) \equiv 1$ (without loss of generality); and (3) the initial value of y $[y(t_0)]$ is zero, i.e., only the transfer characteristics of Eq. (1.1) are of interest. These systems will also be referred to as linear, nonstationary systems.

1.2 Some Properties of Eq. (1.1)

Let Eq. (1.1$'$) be written in the form

$$L(y) = r(t) \tag{1.2}$$

where $r(t) \equiv M(x)$. The range of validity of this equation $(t_0 \leq t < +\infty)$ will be ignored lest confusion result. Closely related to Eq. (1.2) is the homogeneous equation

$$L(u) = 0 \tag{1.3}$$

*A system satisfying this definition is sometimes called *zero state linear*, but in this chapter this added distinction will not be made.

If n sets $(j = 1, 2, \ldots n)$ of initial conditions of the form

$$\left.\frac{d^{i-1}u_j}{dt^{i-1}}\right|_{t = t_0} \begin{array}{ll} = 1 & j = i \\ = 0 & j \neq i \end{array} \Bigg\} \quad i = 1,2,\ldots, n \tag{1.4}$$

are applied to Eq. (1.3), n linearly independent solutions $u_j(t)$ result. The linearity of Eq. (1.3) then allows the solution $u(t)$ resulting from arbitrary initial conditions

$$\frac{d^i u}{dt^i} = u_0^{(i)} \qquad i = 0,1,\ldots, n-1 \tag{1.5}$$

to be written

$$u(t) = \sum_{j=1}^{n} u_0^{(j-1)} u_j(t) \tag{1.6}$$

The functions $u_j(t)$ are called a fundamental set of Eq. (1.3). This is not the only fundamental set, since another set of linearly independent solutions $v_1(t) \ldots v_n(t)$ can always be determined by the relation

$$\begin{bmatrix} v_1(t) \\ \cdot \\ \cdot \\ \cdot \\ \cdot \\ v_n(t) \end{bmatrix} = \begin{bmatrix} a_{11} & \cdots & a_{1n} \\ \cdot & & \cdot \\ \cdot & & \cdot \\ \cdot & & \cdot \\ \cdot & & \cdot \\ a_{n1} & & a_{nn} \end{bmatrix} \begin{bmatrix} u_1(t) \\ \cdot \\ \cdot \\ \cdot \\ \cdot \\ u_n(t) \end{bmatrix} \tag{1.7}$$

or in matrix form

$$\mathbf{v} = [a]\mathbf{u} \tag{1.8}$$

where $[a]$ is any nonsingular, constant matrix.

On the other hand, all sets of n solutions of Eq. (1.3) are not fundamental sets, since they may not be linearly independent. A sufficient condition for a set of solutions $u_1(t) \ldots u_n(t)$ to be linearly independent is that the Wronskian determinant

$$\Delta = \begin{vmatrix} u_1 & u_2 & \cdots & u_n \\ \frac{du_1}{dt} & \frac{du_2}{dt} & \cdots & \frac{du_n}{dt} \\ \cdot & \cdot & & \cdot \\ \cdot & \cdot & & \cdot \\ \cdot & \cdot & & \cdot \\ \frac{d^{n-1}u_1}{dt^{n-1}} & \frac{d^{n-1}u_2}{dt^{n-1}} & \cdots & \frac{d^{n-1}u_n}{dt^{n-1}} \end{vmatrix} \tag{1.9}$$

be unequal to zero.

A particularly important solution of Eq. (1.3) is the weighting function. The weighting function is defined as the solution of Eq. (1.3) subject to the initial conditions

$$\left.\frac{d^i u}{dt^i}\right|_{t=\tau} = 0 \qquad i = 0,1,2,\ldots,n-2$$
$$\left.\frac{d^{n-1} u}{dt^{n-1}}\right|_{t=\tau} = 1 \tag{1.10}$$

where

$$t_0 \leq \tau < +\infty$$

This solution $u(t)$ can be written in terms of an arbitrary fundamental set $u_1(t)..u_n(t)$, i.e.,

$$u(t) = \sum_{i=1}^{n} \beta_i u_i(t) \tag{1.11}$$

where the β_i are determined by substituting Eq. (1.11) into Eqs. (1.10), thus forming the n equations (in matrix form)

$$\begin{bmatrix} u_1(\tau) & \cdots & u_n(\tau) \\ \left.\frac{du_1}{dt}\right|_{t=\tau} & \cdots & \left.\frac{du_n}{dt}\right|_{t=\tau} \\ \vdots & & \\ \left.\frac{d^{n-1}u_1}{dt^{n-1}}\right|_{t=\tau} & \cdots & \left.\frac{d^{n-1}u_n}{dt^{n-1}}\right|_{t=\tau} \end{bmatrix} \begin{bmatrix} \beta_1(\tau) \\ \beta_2(\tau) \\ \vdots \\ \beta_n(\tau) \end{bmatrix} = \begin{bmatrix} 0 \\ 0 \\ \vdots \\ 1 \end{bmatrix} \tag{1.12}$$

Since the determinant of the matrix is the Wronskian, this set of equations always has a solution. The resultant solution, the weighting function, will be written

$$u(t) = G(t, \tau) = \sum_{i=1}^{n} \beta_i(\tau) u_i(t) \tag{1.13}$$

The general solution of Eq. (1.2) can now be written in terms of the weighting function by using the convolution integral, i.e.,

$$y(t) = \int_{t_0}^{t} G(t, \tau) r(\tau)\, d\tau \tag{1.14}$$

That $y(t)$ satisfies Eq. (1.2) can be easily shown by direct substitution if it is recalled that for

$$z(x) = \int_{b(x)}^{a(x)} F(x, t)\, dt \tag{1.15}$$

then

$$\frac{dz}{dx} = \int_{b(x)}^{a(x)} \frac{\partial F(x, t)}{\partial x}\, dt + F(x, a)\frac{da}{dx} - F(x, b)\frac{db}{dx} \tag{1.16}$$

The definition of the weighting function can be generalized if $r(t)$ is written in the form

$$r(t) = \sum_{i=0}^{n} b_i(t)\frac{d^i x}{dt^i} \tag{1.17}$$

If Eqs. (1.13) and (1.17) are substituted in Eq. (1.14), which is then integrated by parts, it can be shown that

$$y(t) = \int_{t_0}^{t} \sum_{i=1}^{n} u_i(t)\left\{\sum_{j=0}^{n} (-1)^j \frac{d^j[b_j(\tau)\beta_i(\tau)]}{d\tau^j}\right\} x(\tau)\, d\tau + b_n(t)x(t) \tag{1.18}$$

A weighting function $W(t, \tau)$ can now be defined for Eq. (1.1), which can be convolved directly with $x(t)$ to form the output $y(t)$, i.e.,

$$y(t) = \int_{t_0}^{t} W(t, \tau)x(\tau)\, d\tau \tag{1.19}$$

Comparing Eqs. (1.18) and (1.19), it is apparent that $W(t, \tau)$ is given by

$$W(t, \tau) = \sum_{i=1}^{n} u_i(t)\alpha_i(\tau) + b_n(t)\delta(t-\tau) \tag{1.20a}$$

$$\equiv W_1(t, \tau) + b_n(t)\delta(t-\tau) \tag{1.20b}$$

where

$$\alpha_i(\tau) = \sum_{j=0}^{n} (-1)^j \frac{d^j[b_j(\tau)\beta_i(\tau)]}{d\tau^j} \tag{1.21}$$

and $\delta(t-\tau)$ is the Dirac delta function.

1.3 Determination of a Differential Equation From a Weighting Function

It has been shown that to each differential equation of the form of Eq. (1.1) there corresponds a weighting function of the form of Eqs. (1.20). The actual process of forming $W(t, \tau)$ from Eq. (1.1) is usually quite difficult. In particular, finding the fundamental set of a general linear time-variable differential equation is very difficult, and the functions usually cannot be described in closed form. On the other hand, forming a differential equation from a given weighting function is much easier, as we shall now show.

Assume that a weighting function of the form of Eq. (1.20a) is given and that the differential equation to which it corresponds is to be formed. The $a_i(\tau)$, as given in Eq. (1.21), are not known. Since there are n independent solutions $u_i(t)$ in $W(t, \tau)$, the order of the differential equation must be n. That equation has the form of Eq. (1.1), and the $a_i(t)$ and $b_i(t)$ are to be determined.

Since the $u_i(t)$ are solutions of the homogeneous Eq. (1.3), then

$$\sum_{j=0}^{n} a_j(t) \frac{d^j u_i}{dt^j} = 0 \qquad i = 1, 2, .. n \tag{1.22}$$

Equations (1.22) are n simultaneous equations in the $n + 1$ unknown $a_j(t)$. If $a_n(t)$ is chosen to be unity, these equations can be solved for the remaining $a_j(t)$.

Once the $a_j(t)$ are known, the $b_j(t)$ can be determined. Suppose the weighting function $G(t, \tau)$ corresponding to the homogeneous Eq. (1.3) is known. Then the solution of Eq. (1.1) is given either by Eq. (1.19) or by

$$y(t) = \int_{t_0}^{t} G(t, \theta) \left[\sum_{j=0}^{n} b_j(\theta) \frac{d^j x(\theta)}{d\theta^j} \right] d\theta \tag{1.23}$$

Equating these two results forms

$$\int_{t_0}^{t} G(t, \theta) \left[\sum_{j=0}^{n} b_j(\theta) \frac{d^j x(\theta)}{d\theta^j} \right] d\theta = \int_{t_0}^{t} W(t, \theta)\, x(\theta)\, d\theta \tag{1.24}$$

Now if the differential operator $\sum_{p=0}^{n} a_p(t) \frac{d^p}{dt^p}$ is applied to both sides of Eq. (1.24), Eq. (1.25) is formed

$$\sum_{j=0}^{n} b_j(t) \frac{d^j x}{dt^j} = \sum_{p=0}^{n} a_p(t) \frac{d^p}{dt^p} \left[\int_{t_0}^{t} W(t, \theta)\, x(\theta)\, d\theta \right] \tag{1.25}$$

Now let Eq. (1.20b) be substituted in Eq. (1.25), and for simplification let

$$\begin{aligned} F_0(t) &= b_n(t) \\ F_i(t) &= \left.\frac{\partial^{i-1} W_1(t,\tau)}{\partial t^{i-1}}\right|_{\tau\to t^-} = \left.\frac{\partial^{i-1} W(t,\tau)}{\partial t^{i-1}}\right|_{\tau\to t^-} \end{aligned} \tag{1.26}$$

Then the derivatives on the right-hand side of Eq. (1.25) are given by

$$\begin{aligned} \frac{d^p}{dt^p}\left[\int_{t_0}^{t} W(t,\theta)\,x(\theta)\,d\theta\right] &= \int_{t_0}^{t} \frac{\partial^p}{\partial t^p} W_1(t,\theta)\,x(\theta)\,d\theta \\ &+ \sum_{k=0}^{p} \frac{d^k}{dt^k}\left[F_{p-k}(t)\,x(t)\right] \qquad p = 0, 1, \ldots n \end{aligned} \tag{1.27}$$

The right-hand side of Eq. (1.25) can, consequently, be given by

$$\sum_{p=0}^{n} a_p(t) \int_{t_0}^{t} \frac{\partial^p}{\partial t^p}\left[W_1(t,\theta)\right] x(\theta)\,d\theta + \sum_{p=0}^{n} a_p(t) \sum_{k=0}^{p} \frac{d^k}{dt^k}\left[F_{p-k}(t)\,x(t)\right] \tag{1.28}$$

Since

$$\sum_{p=0}^{n} \int_{t_0}^{t} a_p(t) \frac{\partial^p}{\partial t^p}\left[W_1(t,\theta)\right] x(\theta)\,d\theta = 0 \tag{1.29}$$

and

$$\frac{d^k}{dt^k}\left[F_{p-k}(t)\,x(t)\right] = \sum_{j=0}^{k} \binom{k}{j} \frac{d^{k-j}}{dt^{k-j}}\left[F_{p-k}(t)\right] \frac{d^j x}{dt^j} \tag{1.30}$$

where $\binom{k}{j}$ is the binomial coefficient, then

$$\sum_{j=0}^{n} b_j(t) \frac{d^j x}{dt^j} = \sum_{p=0}^{n} \sum_{k=0}^{p} \sum_{j=0}^{k} \binom{k}{j} a_p(t) \frac{d^{k-j}}{dt^{k-j}}\left[F_{p-k}(t)\right] \frac{d^j x}{dt^j} \tag{1.31}$$

Now, if the summations on the right-hand side of Eq. (1.31) are rearranged as follows

$$\sum_{p=0}^{n} \sum_{k=0}^{p} \sum_{j=0}^{k} = \sum_{p=0}^{n} \sum_{j=0}^{p} \sum_{k=j}^{p} = \sum_{j=0}^{n} \sum_{p=j}^{n} \sum_{k=j}^{p} \tag{1.32}$$

it then follows that

$$b_j(t) = \sum_{p=j}^{n} \sum_{k=j}^{p} \binom{k}{j} a_p(t) \frac{d^{k-j}\left[F_{p-k}(t)\right]}{dt^{k-j}} \qquad j = 0,1,2,\ldots,n \tag{1.33}$$

Equations **(1.33)** then allow the $b_j(t)$ to be determined in terms of the $a_p(t)$ and the $F_i(t)$.

1-4 An Operator Algebra for Differential Equations

In Sections 1.5 through 1.12, an operator algebra for linear differential equations will be presented. This algebra will be used in later sections to develop some synthesis techniques for linear time-variable systems.

1-5 The Necessary Operations

Since a linear differential equation of the form of Eq. (1.1) is a linear transformation of x into y, the algebra is one of linear transformations. In the following development, capital letters (A, B, C . . .) will represent differential equations of the form of Eq. (1.1).

The algebra of differential equations involves three operations:

(1) Addition of two differential equations, i.e.,

$$A + B = C \tag{1.34}$$

(2) Multiplication of a differential equation by a scalar, i.e.,

$$Ap(t) = B \tag{1.35}$$

or

$$p(t)A = C \tag{1.36}$$

(If $p(t)$ is a constant, $B = C$.)

(3) Multiplication of two differential equations, i.e.,

$$BA = C \tag{1.37}$$

These operations are indicated in block diagram form in Fig. 1-2. Obviously, in order for this albegra to be useful, each of these

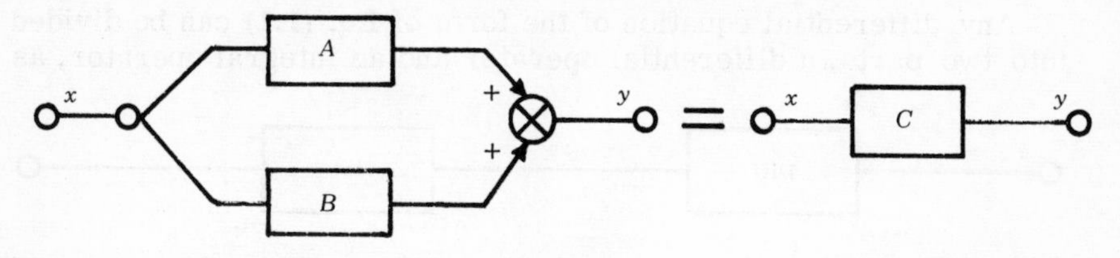

(a) Addition of two differential equations

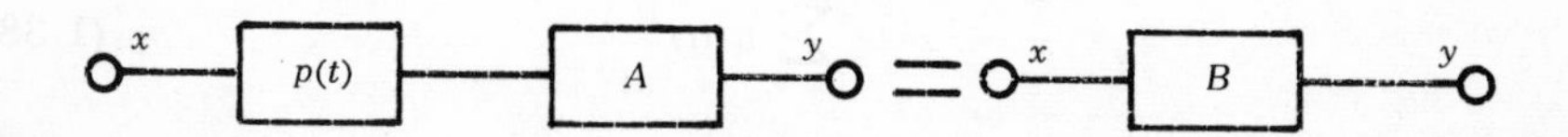

(b) Post-multiplication of a differential equation by a scalar

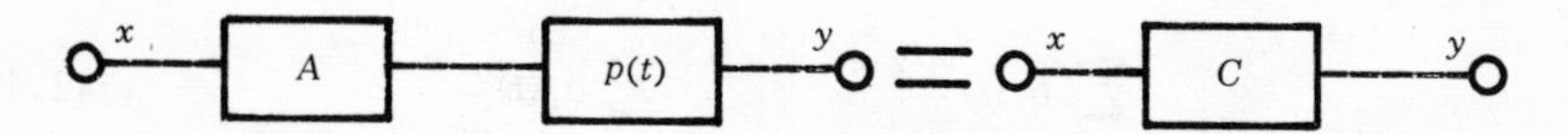

(c) Pre-multiplication of a differential equation by a scalar

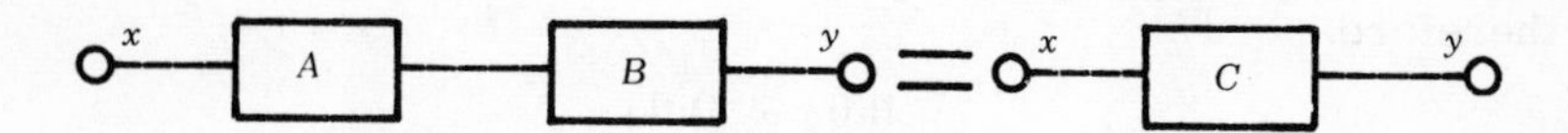

(d) Multiplication of two differential equations

FIG. 1-2. Operations of transformation algebra.

operations must be defined. In addition, the following useful properties of this algebra will be defined:

1) A unity element
2) A zero element
3) An additive inverse
4) A multiplicative inverse.

1.6 Multiplication of Two Differential Equations

It is convenient to define multiplication first since it is used in defining addition. It is obvious from Fig. 1-2d that multiplication is identical to convolution if A and B represent weighting functions.

Any differential equation of the form of Eq. (1.1) can be divided into two parts, a differential operator and an integral operator, as

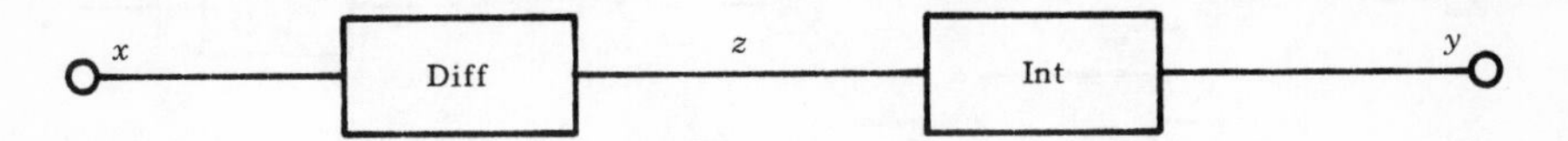

FIG. 1-3. Block diagram of differential equation terms of the notation of Eq. (1-1).

indicated in Fig. 1-3. The relationships between the variables x, y, and z are, in terms of the notation of Eq. (1.1):

$$z = \sum_{j=0}^{n} b_j(t) \frac{d^j x}{dt^j} \tag{1.38}$$

$$\sum_{i=0}^{n} a_i(t) \frac{d^i y}{dt^i} = z \tag{1.39}$$

Now, if two differential equations are multiplied together, this operation can be represented by Fig. 1-4a. The equations which define the relationships between the variables in this system are:

$$\sum_{j=0}^{n} b_j(t) \frac{d^j x}{dt^j} = x_1 = \sum_{i=0}^{n} a_i(t) \frac{d^i y}{dt^i} \tag{1.40}$$

$$\sum_{s=0}^{m} f_s(t) \frac{d^s y}{dt^s} = y_1 = \sum_{k=0}^{m} c_k(t) \frac{d^k z}{dt^k} \tag{1.41}$$

In Fig. 1-4b, the two inner operators of Fig. 1-4a have been "interchanged." In general, these operators are not commutative; therefore

$$\text{Diff}_2 \neq \text{Diff}_3$$

$$\text{Int}_1 \neq \text{Int}_3$$

It is necessary at this point to determine the relationships between x_1, w, and y_1 in terms of the parameters of Eqs. (1.40) and (1.41). Assume that these relationships can be written

$$\sum_{\alpha=0}^{m} g_\alpha(t) \frac{d^\alpha x_1}{dt^\alpha} = w \tag{1.42}$$

and

$$\sum_{\beta=0}^{n} h_\beta(t) \frac{d^\beta y_1}{dt^\beta} = w \tag{1.43}$$

where $g_\alpha(t)$ and $h_\beta(t)$ are as yet unknown coefficients. Substituting the values for x_1 and y_1, each in terms of y, in Eqs. (1.42) and (1.43) produces the relationship

$$\sum_{\alpha=0}^{m} \sum_{i=0}^{n} \sum_{c=0}^{\alpha} \binom{\alpha}{c} g_\alpha(t) \frac{d^{(\alpha-c)} a_i(t)}{dt^{(\alpha-c)}} \frac{d^{(i+c)} y}{dt^{(i+c)}}$$

$$= \sum_{\beta=0}^{n} \sum_{s=0}^{m} \sum_{d=0}^{\beta} \binom{\beta}{d} h_\beta(t) \frac{d^{(\beta-d)} f_s(t)}{dt^{(\beta-d)}} \frac{d^{(\ell+d)} y}{dt^{(\ell+d)}} \tag{1.44}$$

If the coefficients of like derivatives of y are equated, a system of $m + n + 1$ simultaneous equations in the $m + n + 2$ unknown $g_\alpha(t)$ and $h_\beta(t)$ are formed. Arbitrarily choosing $h_n(t) = 1$ (without loss of generality) allows these equations to be solved for the remaining $m + n + 1$ unknowns. This operation shows the equivalence of Figs.

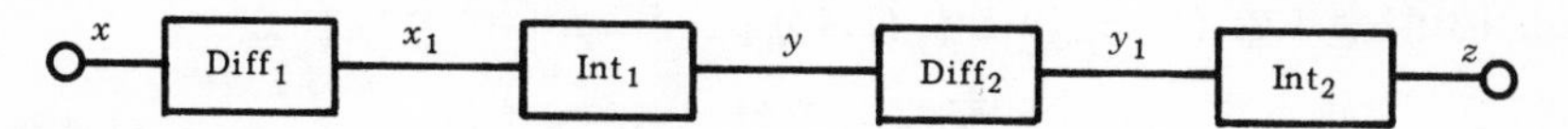

(a) Product of two differential equations

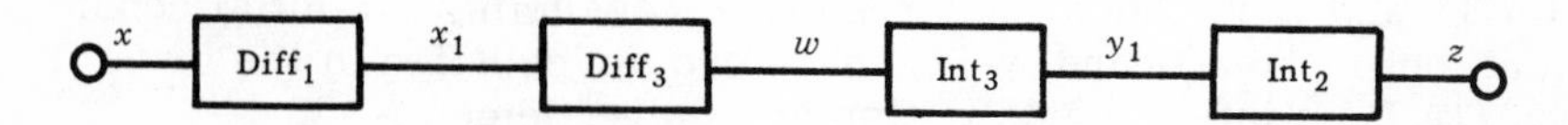

(b) Equivalent block diagram of Figure 1-4a

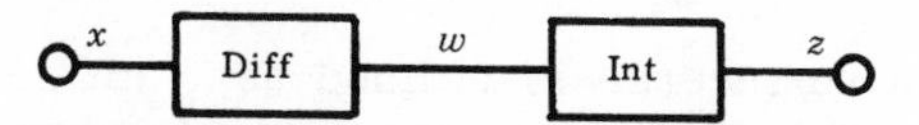

(c) Equivalent block diagram of Figure 1-4b

FIG. 1-4. Multiplication of two differential equations.

1-4a and 1-4b. Figure 1-4b can be further reduced to Fig. 1-4c by combining the differential and integral operators. The resulting equation is

$$\sum_{\beta=0}^{n} \sum_{k=0}^{m} \sum_{f=0}^{\beta} \binom{\beta}{f} h_\beta(t) \frac{d^{(\beta-f)} c_k(t)}{dt^{(\beta-f)}} \frac{d^{(k+f)} z}{dt^{(k+f)}}$$

$$= \sum_{\alpha=0}^{m} \sum_{j=0}^{n} \sum_{r=0}^{\alpha} \binom{\alpha}{r} g_\alpha(t) \frac{d^{(\alpha-r)} b_j(t)}{dt^{(\alpha-r)}} \frac{d^{(j+r)} x}{dt^{(j+r)}} \tag{1.45}$$

where x is the system input and z the output. Multiplication of two differential equations defined by Eqs. (1.40) and (1.41) is then carried out by the steps indicated in Eqs. (1.42), (1.43), (1.44), and (1.45), and is denoted symbolically by

$$AB = C \tag{1.46}$$

where A corresponds to (1.41), B to (1.40), and C to (1.45).

1.7 The Unity Element

The unity element is defined as that element which, when applied to a function, leaves the function unchanged. In the differential equation algebra, the unity element is any differential equation of the form

$$\sum_{i=0}^{n} a_i(t) \frac{d^i x}{dt^i} = \sum_{i=0}^{n} a_i(t) \frac{d^i y}{dt^i} \tag{1.47}$$

This can be seen if the output y is written

$$y = x + y_1 \tag{1.48}$$

Substituting Eq. (1.48) in Eq. (1.47) produces

$$\sum_{i=0}^{n} a_i(t) \frac{d^i y_1}{dt^i} = 0 \tag{1.49}$$

Thus, y_1 is the unforced response [complementary solution of Eq. (1.47)] and x is the forced response. Assuming no initial conditions on y, $y_1 = 0$ and $y = x$; therefore, an equation of the form of Eq. (1.47) satisfies the definition of a unity element.

1.8 The Multiplicative Inverse

A multiplicative inverse is defined as a differential equation (which will be denoted A^{-1}) with the property of producing a unity element when multiplied by the differential equation A. Symbolically this is denoted

$$AA^{-1} = I \tag{1.50}$$

The differential equation of the multiplicative inverse can be found as follows. Consider the cascade transmittances of Fig. 1-5a, where the transmittances of Fig. 1-2d have been separated into integral and differential operators. The operator Diff_1 has the form

$$\sum_{j=0}^{n} b_j \frac{d^j x}{dt^j} = z \tag{1.51}$$

and Int_1 has the form

$$z = \sum_{i=0}^{n} a_i \frac{d^i y}{dt^i} \tag{1.52}$$

If the weighting function corresponding to Int_1 is called $W_1(t, \tau)$, then

$$y(t) = \int_{-\infty}^{t} W_1(t, \tau) z(\tau)\, d\tau \tag{1.53}$$

where, by the definition of a weighting function,

$$\sum_{i=0}^{n} a_i \frac{\partial^i W_1(t, \tau)}{\partial t^i} = 0 \tag{1.54}$$

Then, operating on $y(t)$ with a differential operator of the form

$$\sum_{i=0}^{n} a_i \frac{d^i}{dt^i} \tag{1.55}$$

produces $z(t)$; thus the differential operator in Eq. (1.55) is the inverse of Int_1.

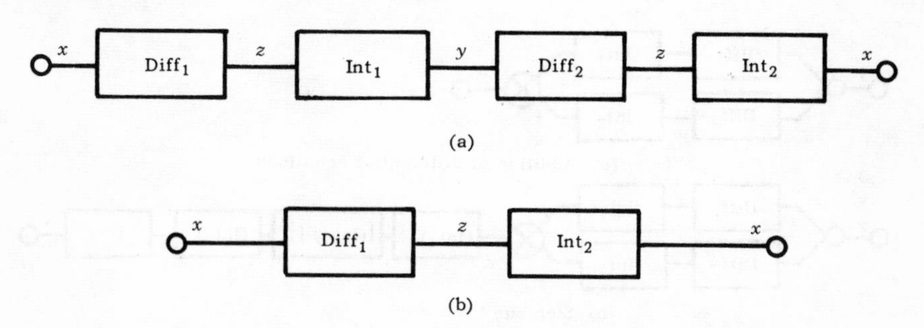

FIG. 1-5. Cascade of a transmittance and its inverse.

If Diff_2 is chosen with this differential form, Fig. 1-5a can then be reduced to Fig. 1-5b.

If Int_2 is now given the form

$$z = \sum_{j=0}^{n} b_j \frac{d^j x}{dt^j} \tag{1.56}$$

Fig. 1-5b takes the form of a unity element, as defined by Eq. (1.47), which is necessary if both input and output are to be x. Thus if A is represented by the differential equation

$$\sum_{i=0}^{n} a_i(t) \frac{d^i y}{dt^j} = \sum_{j=0}^{n} b_j(t) \frac{d^j x}{dt^j} \tag{1.57}$$

where x is the input and y is the output, then A^{-1} is represented by the differential equation

$$\sum_{j=0}^{n} b_j(t) \frac{d^j x}{dt^j} = \sum_{i=0}^{n} a_i(t) \frac{d^i y}{dt^i} \tag{1.58}$$

where x is the output and y is the input.

1.9 Addition of Two Differential Equations

Addition of two differential equations is represented symbolically in Fig. 1-6a. As in the case of multiplication, the differential equations have been divided into differential and integral operators. The addition is performed step by step as indicated by the block diagrams of Fig. 1-6. In step 1, the original system is multiplied by the cascade combination of two differential operators denoted $(\mathrm{Int}_1)^{-1}$ and $[(\mathrm{Int}_2)']^{-1}$, and two integral operators Int_1 and $(\mathrm{Int}_2)'$, their respective multiplicative inverses. This combination then represents a unity element. Since the system is linear, $(\mathrm{Int}_1)^{-1}$ can be moved to the left of the summing junction, thus eliminating Int_1

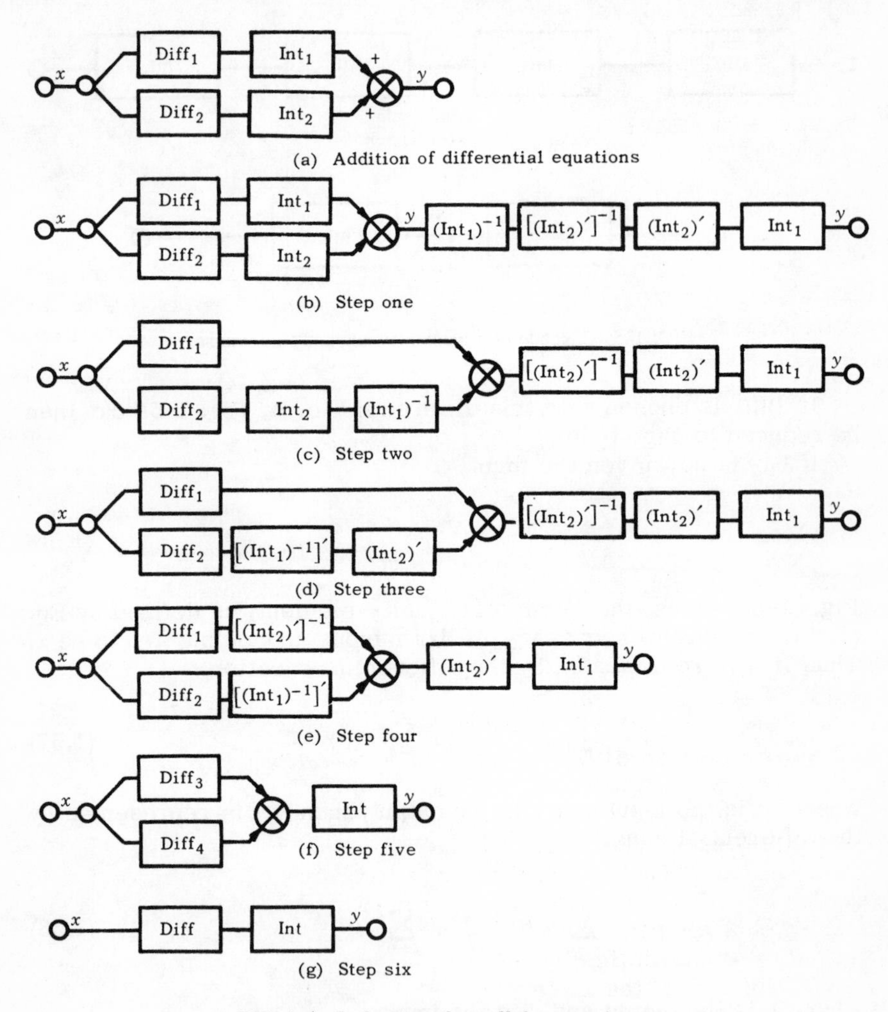

FIG. 1-6. Reduction of parallel transmittances.

from the upper element, as shown in Fig. 1-6c. Now, the integral operator Int_2 and the differential operator $(\mathrm{Int}_2)^{-1}$ are multiplied and the product separated into its new differential and integral operators, as indicated in step 3, where the new differential operator is denoted $[(\mathrm{Int}_1)^{-1}]'$ and the new integral operator is denoted $(\mathrm{Int}_2)'$. $(\mathrm{Int}_2')^{-1}$ is now moved to the left of the summing junction, thus cancelling $(\mathrm{Int}_2)'$. To the left of the summing junction, only differential operators now appear, and to the right only integral operators. These can then be combined as indicated in steps 5 and 6. Thus addition of differential equations has been accomplished.

1.10 The Zero Element

The zero element in differential-equation algebra is defined as any differential equation whose output is always identically zero. The zero element is denoted 0.

1.11 The Additive Inverse

The additive inverse of a differential equation A is defined as that differential equation B which, when added to A, produces a zero element, i.e.,

$$A + B = 0 \tag{1.59}$$

It is quite obvious that if A is the differential equation

$$\sum_{i=0}^{n} a_i(t) \frac{d^i y}{dt^i} = \sum_{j=0}^{n} b_j(t) \frac{d^j x}{dt^j} \tag{1.60}$$

where x is the input and y is the output, then B has the form

$$\sum_{i=0}^{n} a_i(t) \frac{d^i y}{dt^i} = -\sum_{j=0}^{n} b_j(t) \frac{d^j x}{dt^j} \tag{1.61}$$

1.12 Multiplication of Differential Equation by a Scalar

Multiplication of a differential equation by a scalar can be considered a degenerate case of multiplication of two differential equations, where one of the differential equations is the degenerate equation $y = p(t)x$, in which x is the input and y is the output. The techniques developed for multiplying differential equations are then applicable to multiplication of a differential equation by a scalar.

In addition to the properties of differential-equation algebra mentioned above, the following laws hold:

1. Addition is commutative, i.e., $A + B = B + A$.
2. Addition is associative, i.e., $A + (B+C) = (A+B) + C$.
3. Multiplication is not commutative, i.e., $AB \neq BA$. (In the stationary case multiplication is commutative.)
4. Multiplication is associative, i.e., $A(BC) = (AB)C$.
5. Distributivity is valid, i.e., $A(B+C) = AB + AC$.

1-13 Synthesis by Plant Cancellation

In this and the following sections, two techniques for synthesizing linear, time-variable feedback systems are presented. These techniques are based on the differential-equation algebra previously developed.

The first synthesis technique is a plant-cancellation technique. The steps to be followed in this technique are:

a) Determine the closed-loop weighting function, or closed-loop differential equation, from the specifications. (It is desirable to specify the system by its weighting function since the system output for any input can then be determined by convolution.)

b) If the closed-loop weighting function has been specified, determine the corresponding closed-loop differential equation from the weighting function.

c) Determine the open-loop differential equation from the closed-loop differential equation.

d) Determine the differential equations for the appropriate compensation networks and synthesize these networks by appropriate analog computer systems.

Implementation of this procedure requires techniques for (1), determining a weighting function, or differential equation, which meets the given specifications; and (2), manipulating this weighting function in such a way that the appropriate compensation networks can be developed.

The application of the differential-equation algebra to the synthesis problem is as straightforward conceptually as the application of Laplace-transform algebra to the synthesis of linear stationary systems. The advantage of such an algebra is that all of the manipulative operations can be performed symbolically and the numerical details carried through only at the end of the process.

To describe the technique of synthesis, the feedback configuration in Fig. 1-7 will be used. In this figure, r is the input, c is the output, K, G, and H are differential equations, and m and ϵ are intermediate variables in the system. In the following, the operation (.) represents the operation a differential equation performs on a variable to produce a new variable. Let the relationship between the input $r(t)$ and the output $c(t)$ be

$$c = W \cdot r \tag{1.62}$$

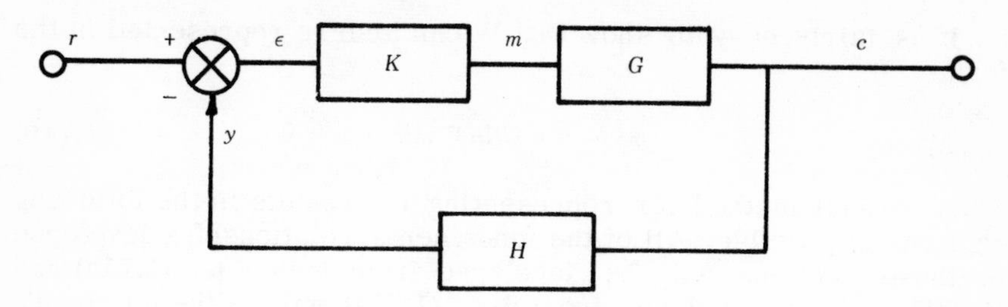

FIG. 1-7. General feedback configuration.

where W is the desired over-all differential equation. From the figure it is seen that

$$y = H \cdot c \tag{1.63}$$

and

$$\epsilon = r - y = r - H \cdot c \tag{1.64}$$

Since

$$m = K \cdot \epsilon \qquad \text{and} \qquad c = G \cdot m \tag{1.65}$$

then

$$r = I \cdot \epsilon + HGK \cdot \epsilon = (I + HGK) \cdot \epsilon \tag{1.66}$$

by virtue of the fact that

$$I \cdot \epsilon = \epsilon \tag{1.67}$$

Applying the multiplicative inverse of $(I + HGK)$ to both sides of Eq. (1.66), we have

$$\epsilon = (I + HGK)^{-1} \cdot r \tag{1.68}$$

Then, since

$$c = GK \cdot \epsilon \tag{1.69}$$

subsitution of Eq. (1.68) in Eq. (1.69) produces the equation

$$c = GK(I + HGK)^{-1} \cdot r \tag{1.70}$$

Comparison of Eqs. (1.62) and (1.70) reveals that

$$W = GK(I + HGK)^{-1} \tag{1.71a}$$

Equation (1.71a) may be considered a fundamental relationship for Fig. 1-7.

It is fairly easy to show that W can also be represented in the form

$$W = (I + GHK)^{-1} GK \tag{1.71b}$$

Thus, a dual method for representing the results in the following sections is possible. All of the fundamental relationships developed in those sections will be developed from both Eqs. (1.71a) and (1.71b). Those developed from Eq. (1.71a) will be the (a) result, and those developed from Eq. (1.71b) will be the (b) result. Two special cases of the configuration in Fig. 1-7 will now be considered.

1-14 Synthesis of a Feedback System Unconstrained by a Fixed Plant

Suppose that a given over-all differential equation W is to be realized as a feedback system with a unity feedback, i.e., it is to have the configuration of Fig. 1-8. The problem then is to determine G in terms of W. Since in this case

$$K = H = I \tag{1.72}$$

Equation (1.71) reduces to

$$W = G(I + G)^{-1} \tag{1.73a}$$

$$W = (I + G)^{-1} G \tag{1.73b}$$

Solving Eqs. (1.73) for G, the relationships

$$G = (I - W)^{-1} W \tag{1.74a}$$

$$G = W(I - W)^{-1} \tag{1.74b}$$

are obtained. Therefore, a given differential equation W can be synthesized as a unity feedback system with G as the feedforward element through Eqs. (1.74). The actual differential equation of G is obtained by performing the operations indicated.

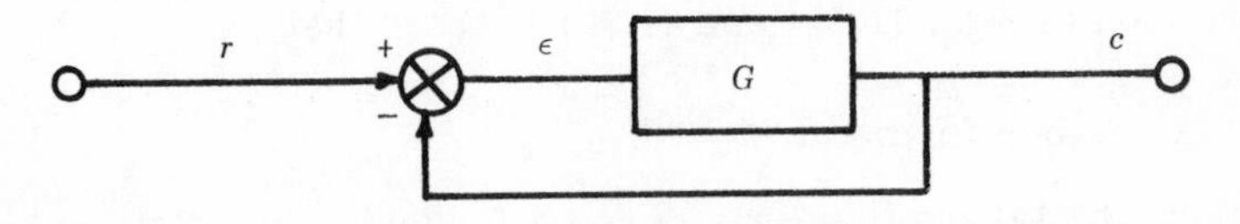

FIG. 1-8. Unity feedback system.

1.15 Synthesis of a System Constrained by a Fixed Plant

In this case, let W be the differential equation of the over-all system of the configuration of Fig. 1-9. The differential equation of the known fixed plant is G. The problem is to find the differential equation of a suitable compensation network K such that this system has the desired over-all response indicated by W. Setting $H = I$ in Eqs. (1.71) and solving for K, the relationships

$$K = G^{-1}(I - W)^{-1}W \tag{1.75a}$$

$$K = G^{-1}W(I - W)^{-1} \tag{1.75b}$$

are formed. Performing the operations indicated symbolically in Eqs. (1.75), the differential equation for K is obtained.

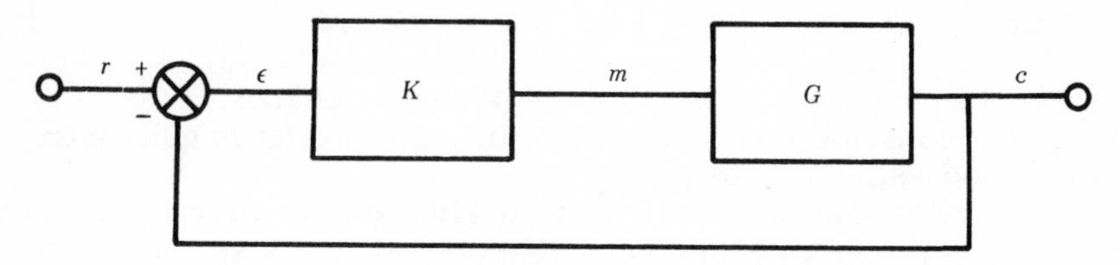

FIG. 1-9. Feedback system with fixed plant.

1.16 Example

As an example of the process described above, consider the following problem. Suppose that G in Fig. 1-9 is a fixed plant described by the differential equation

$$\frac{d^2c}{dt^2} + \frac{dc}{dt} + e^{-t}c = m(t) \tag{1.76}$$

Let the desired over-all differential equation W be

$$\frac{d^2c}{dt^2} + 2\frac{dc}{dt} + c = r \tag{1.77}$$

$I - W$ is then formed by addition and is given by

$$\frac{d^2z}{dt^2} + 2\frac{dz}{dt} = \frac{d^2c}{dt^2} + 2\frac{dc}{dt} + c \tag{1.78}$$

where z is the input and c the output. The differential equation $(I - W)^{-1}W$ is then formed as

$$\frac{d^2c}{dt^2} + 2\frac{dc}{dt} = \epsilon(t) \tag{1.79}$$

The differential equation K is then produced from

$$K = G^{-1}(I - W)^{-1}W \tag{1.80}$$

and the equation for this compensation network is found to be

$$\begin{aligned} &\frac{d^2\epsilon}{dt^2} + \frac{2 + 3e^{-t}}{1 + e^{-t}}\frac{d\epsilon}{dt} + (1 + e^{-t})\epsilon \\ &= \frac{d^2m}{dt^2} + \frac{3 + 4e^{-t}}{1 + e^{-t}}\frac{dm}{dt} + \frac{2 + 3e^{-t}}{1 + e^{-t}}m \end{aligned} \tag{1.81}$$

Thus, the compensation network is completely specified by Eq. (1.81). This network can be synthesized by analog computer elements through the technique of Section 1.27.

1.17 Comment

The disadvantages of this method of synthesis are:

1) Cancellation compensation ordinarily results in quite complex compensation networks.

2) Perfect cancellation is not possible due to practical considerations (in the case of an unstable plant this is a serious problem).

3) By merely cancelling the plant and putting in the desired open-loop transfer function, one cannot radically change the transfer characteristics of a system. Physical considerations will dictate the amount of change possible.

Even with these disadvantages, the method is both straightforward and feasible from a practical viewpoint, and is therefore a valuable synthesis tool.

1.18 Constraints on the Choice of an Over-all Differential Equation

In choosing an over-all differential equation W, certain practical constraints on the form of the plant, the compensation network, and the over-all system will restrict the relationship which may exist between the orders of the differential and integral operators of the compensation and the over-all system. In general, it can be said that physical systems will have a smoothing effect on an input signal, i.e., the system will usually contain at least one over-all integration (the order of the integral operator is one greater than the order of the differential operator). At the very least, the orders of the integral and differential operators will be the same, and the order of the differential operator will never be greater than the order of the integral operator.

Examine Eq. (1.71) with $H = I$, i.e., the equation

$$W = GK(I + GK)^{-1} \tag{1.82}$$

By virtue of the previous discussion, the following restrictions will be placed on G and K. If G has the form

$$\sum_{i=0}^{N} a_i(t)\frac{d^i c}{dt^i} = \sum_{i=0}^{M} b_i(t)\frac{d^i m}{dt^i} \tag{1.83}$$

where c is the output and m is the input, then $N > M$. If K has the form

$$\sum_{i=0}^{R} f_i(t)\frac{d^i m}{dt^i} = \sum_{i=0}^{P} d_i(t)\frac{d^i \epsilon}{dt^i} \tag{1.84}$$

where m is the output and ϵ is the input, then $R \geq P$. The product GK then has the form

$$\sum_{i=0}^{N+R} g_i(t)\frac{d^i c}{dt^i} = \sum_{i=0}^{M+P} h_i(t)\frac{d^i \epsilon}{dt^i} \tag{1.85}$$

where c is the output and ϵ is the input, and apparently $N+R > M+P$. Actually the orders of the operators in Eq. (1.85) will be given by $N+R-n$ and $M+P-n$ where n is the order of any terms common to both operators; however, since the difference in these orders is unchanged by this reduction it can be ignored in what follows. From Eq. (1.85), the equation for $(I + GK)^{-1}$ can be written

$$\sum_{i=0}^{N+R} \left[g_i(t) + h_i(t)\right] \frac{d^i y}{dt^i} = \sum_{i=0}^{N+R} g_i(t)\frac{d^i x}{dt^i} \tag{1.86}$$

where y is the output and x is the input. Finally, W can be formed by substituting Eqs. (1.85) and (1.86) in Eq. (1.82). Thus W will have the form

$$\sum_{i=0}^{2(N+R)} p_i(t)\frac{d^i c}{dt^i} = \sum_{i=0}^{(N+R+M+P)} k_i(t)\frac{d^i r}{dt^i} \tag{1.87}$$

and from the constraints on N, M, R, and P

$$2(N+R) - (N+R) - (M+P) \geq N - M \tag{1.88}$$

The relationship (1.88) then indicates that in choosing an over-all differential equation W, it is necessary that the difference in orders of the integral and differential operators be equal to or greater than the difference in orders of the integral and differential operators of the fixed plant.

1.19 An Algebraic Synthesis Technique

The technique developed in this section closely resembles that developed in Ref. 10, page 238. It allows an over-all system of the type in Fig. 1-9 to be synthesized approximately, and has the advantage that the compensation network does not cancel the fixed plant. The basic equation for the technique is Eq. (1.75a).

To aid in the development of this technique, each differential equation will be "factored" into a differential operator premultiplied by an integral operator, i.e.,

$$\begin{aligned} K &= I_k D_k \\ G &= I_g D_g \\ W &= I_w D_w \end{aligned} \tag{1.89}$$

Now if I (the unity element) is given the form

$$I = I_w I_w^{-1} \tag{1.90}$$

then $(I - W)$ is given by

$$(I - W) = I_w (I_w^{-1} - D_w) \tag{1.91}$$

and

$$(I - W)^{-1} = (I_w^{-1} - D_w)^{-1} I_w^{-1} \tag{1.92}$$

Substituting Eqs. (1.89) and (1.92) in Eq. (1.75a), we find that K is given by

$$K = I_k D_k = D_g^{-1} I_g^{-1} (I_w^{-1} - D_w)^{-1} I_w^{-1} I_w D_w \tag{1.93}$$

or

$$I_k D_k = D_g^{-1} I_g^{-1} (I_w^{-1} - D_w)^{-1} D_w \tag{1.94}$$

At this point in the development, a restriction is placed on K requiring that K must not cancel the integral operation I_g of the plant. This requirement prevents "cancellation" of the linearly independent solutions (of the differential equation) of the plant, and substitution of a new set of dynamics. The restriction then requires that $(I_w^{-1} - D_w)^{-1}$ must satisfy the relationship

$$(I_w^{-1} - D_w)^{-1} = I_g I_c \tag{1.95}$$

where I_c is an integral operator as yet unknown. If we substitute Eq. (1.95) in Eq. (1.94), Eq. (1.94) becomes

$$I_k D_k = D_g^{-1} I_c D_w \tag{1.96}$$

On equating integral and differential operators, the following relationships are formed

$$I_k = D_g^{-1} I_c \tag{1.97}$$

$$D_k = D_w \tag{1.98}$$

The equations used to determine the compensation are Eq. (1.95), rewritten in the form

$$I_w^{-1} = D_w + I_c^{-1} I_g^{-1} \tag{1.99}$$

and Eqs. (1.97) and (1.98). In examining these equations, it must be remembered that I_g and D_g are known and fixed. In addition, in determining an acceptable over-all differential equation to meet the system specifications, it would be desirable to be able to fix both I_w and D_w. However, this is not possible using this technique. Only I_w can be fixed from the specifications, which is equivalent to being able to fix the linearly independent solutions of the over-all system weighting function but not being able to fix their multiplying constants. Thus, in Eq. (1.99), I_w^{-1} and I_g^{-1} are known while D_w and I_c^{-1} are unknown differential operators to be determined by equating the coefficients of like derivatives on both sides of Eq. (1.99). Before this is done, however, certain constraints must be placed on the orders of the various differential and integral operators. To aid in the determination of these constraints, the following representations of various transmittances are defined. Let W be the differential equation

$$\sum_{i=0}^{P_w} b_i(t) \frac{d^i c}{dt^i} = \sum_{i=0}^{Z_w} a_i(t) \frac{d^i r}{dt^i} \qquad b_{P_w}(t) = 1 \tag{1.100}$$

Let K be the differential equation

$$\sum_{i=0}^{P_k} h_i(t) \frac{d^i m}{dt^i} = \sum_{i=0}^{Z_k} g_i(t) \frac{d^i \epsilon}{dt^i} \tag{1.101}$$

Let G be the differential equation

$$\sum_{i=0}^{P_g} \beta_i(t) \frac{d^i c}{dt^i} = \sum_{i=0}^{Z_g} \beta_i(t) \frac{d^i m}{dt^i} \qquad \beta_{P_w}(t) = 1 \tag{1.102}$$

Finally, let I_c be the differential equation

$$\sum_{i=0}^{P_c} f_i(t) \frac{d^i y}{dt^i} = x \tag{1.103}$$

As indicated previously, at this point the known I_w^{-1} is substituted in Eq. (1.99) from Eq. (1.100), the known I_g^{-1} is substituted in Eq. (1.99) from Eq. (1.102), and the representations for the unknown D_w and I_c are substituted in Eq. (1.99) from Eqs. (1.100) and (1.103). The coefficients of like orders of derivatives are equated and a set of simultaneous linear algebraic equations in the unknown $a_i(t)$ and $f_i(t)$ results. The solution of this set then produces the D_w and the I_c. For a solution to the equation to exist, the orders of the various operators are constrained as follows. On the basis of physical reasoning it is required that

$$P_w \geq Z_w \qquad P_g \geq Z_g \qquad P_k \geq Z_k \tag{1.104}$$

From Eq. (1.99) it is seen that

$$P_c + P_g = P_w \quad \text{or} \quad P_c = P_w - P_g \tag{1.105}$$

Upon equating like orders of derivatives in Eq. (1.99), a total of $P_w + 1$ equations in $Z_w + P_c + 2$ unknowns is obtained. For a solution to exist, it is necessary that

$$P_w + 1 \leq Z_w + P_c + 2$$

or

$$P_w \leq Z_w + P_c + 1 \tag{1.106}$$

Substituting Eq. (1.105) in (1.106), it is seen that

$$Z_w \geq P_g - 1 \tag{1.107}$$

From condition (1.104) and Eq. (1.96)

$$Z_g + P_c = P_k \geq Z_k = Z_w$$

or

$$P_c \geq Z_w - Z_g \tag{1.108}$$

Comparing Eq. (1.105) and expression (1.108)

$$P_w - P_g \geq Z_w - Z_g$$

or

$$P_w \geq P_g + Z_w - Z_g \tag{1.109}$$

Combining expressions (1.107) and (1.109)

$$P_w \geq 2P_g - Z_g - 1 \tag{1.110}$$

Conditions (1.105), (1.107), and (1.110) then restrict the orders of the operators P_w, Z_w and P_c. The restriction on P_w must be taken into account when P_w is chosen to satisfy the system specifications. The conditions for Z_w and P_c determine suitable forms for D_w and I_c in Eq. (1.99).

If, in expressions (1.107) and (1.110), the inequality is chosen, more unknowns than equations will be obtained. Then, it is possible to choose some of the unknowns in an arbitrary manner, e.g., a manner allowing optimization of some criterion function.

1.20 Example

Suppose that a fixed plant is given by

$$\frac{d^2c}{dt^2} + \frac{dc}{dt} = m \tag{1.111}$$

Thus

$$I_g \cdot c = \frac{d^2c}{dt^2} + \frac{dc}{dt} \qquad \text{and} \qquad D_g = 1 \tag{1.112}$$

Let the desired over-all integral operator have the form

$$I_w \cdot c = \frac{d^3c}{dt^3} + e^{-t}\frac{d^2c}{dt^2} + \frac{dc}{dt} + e^{-2t}c \tag{1.113}$$

which satisfies condition (1.110). Choosing

$$D_w = a_1(t)\frac{d}{dt} + a_0(t) \tag{1.114}$$

which satisifes condition (1.107), I_c must have the form

$$I_c = b_1(t)\frac{d}{dt} + b_0(t) \tag{1.115}$$

Substituting Eqs. (1.112), (1.113), (1.114), and (1.115) in Eq. (1.99), the relationship

$$\begin{aligned}\frac{d^3c}{dt^3} + e^{-t}\frac{d^2c}{dt^2} + \frac{dc}{dt} + e^{-2t}c &= b_1(t)\frac{d^3c}{dt^3} + \left[b_1(t) + b_0(t)\right]\frac{d^2c}{dt^2} \\ &\quad + \left[a_1(t) + b_0(t)\right]\frac{dc}{dt} + a_0(t)c\end{aligned} \tag{1.116}$$

is formed. Equating coefficients, we form the simultaneous equations

$$\begin{aligned} b_1(t) &= 1 \\ b_1(t) + b_0(t) &= e^{-t} \\ a_1(t) + b_0(t) &= 1 \\ a_0(t) &= e^{-2t} \end{aligned} \tag{1.117}$$

The solutions of Eq. (1.117) are

$$\begin{aligned} b_1(t) &= 1 \\ b_0(t) &= e^{-t} - 1 \\ a_1(t) &= 2 - e^{-t} \\ a_0(t) &= e^{-2t} \end{aligned}$$

The compensation network K is then given by

$$\frac{dm}{dt} + (e^{-t} - 1)m = (2 - e^{-t})\frac{d\epsilon}{dt} + (e^{-2t})\epsilon \tag{1.118}$$

1.21 Comment

The foregoing synthesis technique has the advantages of:

1) Simplicity of solution.

2) Formation of a noncancellation compensation network.

The disadvantages of this technique are:

1) The over-all system response cannot be specified completely.

2) The form and the stability of the compensation network are not under the designer's control.

1.22 The Approximation Problem

Before the synthesis techniques can be used, an over-all system weighting function which satisfies the system specifications must be determined. The problem of determining such a weighting function will be called the approximation problem. Actually, for any of these synthesis techniques, it would be adequate to determine a system differential equation; however, it is more desirable to know the system weighting function since the system output can then be determined for any input by application of the convolution integral.

There are many methods by which a system weighting function or a system differential equation might be obtained; for instance, the function or equation might be given directly in the system specifications. In any event, once one or the other has been established, the techniques of the preceding sections can be followed.

1-23 Approximation Method for Polynomial Inputs

In this section, a method for determining weighting functions for a particular class of synthesis problems is presented. The method is restricted to systems which receive inputs expressed as polynomials in time, and whose outputs can be approximated as a separable function (see below).

The approximation method can be formulated as follows. In Fig. 1-1, let $x(t)$ be a polynomial in time given by

$$\begin{aligned} x(t) = x(t-\tau) &= \sum_{n=0}^{N} c_n (t-\tau)^n \qquad t \geq \tau \\ &= 0 \qquad t < \tau \end{aligned} \tag{1.119}$$

where c_n are constants, t is time, and τ is the time of application of $x(t)$ to the linear system W. The function $y(t)$ is the output of the linear system W produced as a result of the application of $x(t)$. It is assumed that an analytic expression for $y(t)$ can be determined from the specifications. W is the unknown linear system whose weighting function $W(t, \tau)$ is to be specified in a form capable of synthesis as a feedback system. Output $y(t)$ can then be written in the form of Eq. (1.19). Substituting Eq. (1.119) in Eq. (1.19), can be written

$$\begin{aligned} y(t) = y_x(t,\tau) &= \sum_{n=0}^{N} c_n \int_{\tau}^{t} W(t,\theta)\,[\theta-\tau]^n\,d\theta\,; \qquad t \geq \tau \\ &= 0 \qquad t < \tau \end{aligned} \tag{1.120}$$

The problem now reduces to solving the integral equation (1.120) for $W(t, \tau)$. Fortunately, this equation reduces readily to a linear, stationary differential equation easily solved. The class of permissible $W(t, \tau)$ is the class defined in Eqs. (1.20) and (1.21). Substituting Eq. (1.20) in Eq. (1.120), the relationship

$$y_x(t,\tau) = \sum_{n=0}^{N} c_n \int_{\tau}^{t} W_1(t,\theta)\,[\theta-\tau]^n\,d\theta + b_m(t) \sum_{n=0}^{N} c_n (t-\tau)^n \tag{1.121}$$

results. The solution of this equation will be broken into two parts: (1) determination of $b_m(t)$; (2) determination of $W_1(t, \tau)$.

First, a technique for determining $b_m(t)$ will be presented. Assume that c_k $(0 \leq k \leq N)$ is the particular nonzero c_n with the lowest valued subscript, i.e.,

$$y_x(t,\tau) = \sum_{n=k}^{N} c_n \int_{\tau}^{t} W_1(t,\theta)\,(\theta-\tau)^n\,d\theta + b_m(t) \sum_{n=k}^{N} c_n (t-\,)^n \tag{1.122}$$

If, now, the kth partial derivative of $y_x(t, \tau)$ with respect to τ is formed, the relationship

$$\frac{\partial^k y_x(t, \tau)}{\partial \tau^k} = \sum_{n=k}^{N} c_n \int_{\tau}^{t} W_1(t, \theta)(-1)^k (n)(n-1)\ldots(n-k+1)(\theta-\tau)^{n-k}\, d\theta$$
$$+ b_m(t) \sum_{n=k}^{N} c_n (-1)^k (n)(n-1)\ldots(n-k+1)(t-\tau)^{n-k} \tag{1.123}$$

results. Now if the limit of

$$\frac{\partial^k y_x(t, \tau)}{\partial \tau^k}$$

as $\tau \to t$ is formed, the equation

$$\left.\frac{\partial^k y_x(t, \tau)}{\partial \tau^k}\right|_{\tau \to t} = b_m(t)\, c_k (-1)^k (k!) \tag{1.124}$$

is formed. Solving Eq. (1.124) for $b_m(t)$, the desired relationship for $b_m(t)$ is obtained, i.e.,

$$b_m(t) = \frac{(-1)^k \left.\dfrac{\partial^k y_x(t, \tau)}{\partial \tau^k}\right|_{\tau \to t}}{(k!) c_k} \tag{1.125}$$

Once $b_m(t)$ is known, the expression

$$b_m(t) \sum_{n=0}^{N} c_n (t-\tau)^n \tag{1.126}$$

can be formed, and a new function $y_x'(t, \tau)$ can be obtained from Eq. (1.122) as

$$\begin{aligned} y_x'(t, \tau) &= y_x(t, \tau) - b_m(t) \sum_{n=0}^{N} c_n (t-\tau)^n \\ &= \sum_{n=0}^{N} c_n \int^{t} W_1(t, \theta)(\theta-\tau)^n\, d\theta \end{aligned} \tag{1.127}$$

The second problem now is to determine $W_1(t, \tau)$ from Eq. (1.127). First, derivatives of $y_x'(t, \tau)$ with respect to τ are taken. The first derivative is:

$$\frac{\partial y_x'(t,\tau)}{\partial\tau} = \sum_{n=0}^{N} c_n \int_\tau^t W_1(t,\theta)(-1)n(\theta-\tau)^{n-1}\,d\theta - c_0 W_1(t,\tau) \quad \textbf{(1.128)}$$

The kth derivative $(k < N+1)$ is:

$$\frac{\partial^k y_x'(t,\tau)}{\partial\tau^k} = \sum_{n=0}^{N} c_n \int_\tau^t W_1(t,\theta)(-1)^k(n)(n-1)\ldots(n-k+1)(\theta-\tau)^{n-k}\,d\theta$$

$$\sum_{j=0}^{k-1} c_j \frac{\partial^{k-1-j}W_1(t,\tau)}{\partial\tau^{k-1-j}}(-1)^j(j!) \qquad k = 1,2,\ldots,N$$

(1.129)

and the $(N+1)$st is:

$$\frac{\partial^{N+1} y_x'(t,\tau)}{\partial\tau^{N+1}} = \sum_{j=0}^{N}(-1)^{j+1}(j!)c_j\frac{\partial^{N-j}W_1(t,\tau)}{\partial\tau^{N-j}} \quad \textbf{(1.130)}$$

Equation **(1.130)** is seen to be a linear, stationary differential equation whose solution is $W_1(t,\tau)$. To solve this differential equation uniquely, N initial conditions must be obtained. Equations (1.129) produce these initial conditions if τ is made to approach t, i.e.,

$$\left.\frac{\partial^k y_x'(t,\tau)}{\partial\tau^k}\right|_{\tau\to t} = \sum_{j=0}^{k-1}(-1)^{j+1}(j!)\,c_j\left.\frac{\partial^{k-1-j}W_1(t,\tau)}{\partial\tau^{k-1-j}}\right|_{\tau\to t} \quad \textbf{(1.131)}$$

or

$$\left.\frac{\partial^{k-1}W_1(t,\tau)}{\partial\tau^{k-1}}\right|_{\tau\to t} = -\frac{1}{c_0}\left\{\left.\frac{\partial^k y_x'(t,\tau)}{\partial\tau^k}\right|_{\tau\to t} - \sum_{j=1}^{k-1}(-1)^{j+1}(j!)\,c_j\left.\frac{\partial^{k-1-j}W_1(t,\tau)}{\partial\tau^{k-1-j}}\right|_{\tau\to t}\right\} \qquad k = 1,2,\ldots N$$

(1.132)

Equation **(1.132)** represents a set of recursive relationships which provides the N necessary "initial conditons." Actually, since the derivatives of $W_1(t,\tau)$ are made with respect to τ, Eq. **(1.132)** represents final conditions. This inconvenience is easily remedied by making the substitution

$$\tau = t - z \quad \textbf{(1.133)}$$

in Eqs. **(1.130)** and **(1.132)**. The resulting equations can then be easily solved, and by making the inverse substitution

$$z = t - \tau \quad \textbf{(1.134)}$$

in the solution obtained, $W_1(t,\tau)$ is recovered.

In a special case of this development, Eqs. (1.125) and (1.130) are examined for the degenerate case of $x(t)$, i.e.,

$$x(t-\tau) = c_N (t-\tau)^N \tag{1.135}$$

Equation (1.125) becomes

$$b_m(t) = \frac{(-1)^N \left. \dfrac{\partial^N y_x(t,\tau)}{\partial \tau^N} \right|_{\tau \to t}}{(N!)\, c_N} \tag{1.136}$$

Equation (1.130) becomes:

$$\frac{\partial^{N+1} y_x'(t,\tau)}{\partial \tau^{N+1}} = (-1)^{N+1} (N!)\, c_N W_1(t,\tau) \tag{1.137}$$

or

$$W_1(t,\tau) = \frac{(-1)^{N+1} \dfrac{\partial^{N+1} y_x'(t,\tau)}{\partial \tau^{N+1}}}{(N!)\, c_N} \tag{1.138}$$

Equations (1.137) and (1.138) should then be used if the input is a polynomial with but one nonzero coefficient. This case apparently includes step-function inputs.

1.24 Method for Nonpolynomial Inputs

Consider again Fig. 1-1. Suppose that now $x(t)$ is not a polynomial but a separable function of t and τ, i.e., it can be represented by

$$\begin{aligned} x(t) &= \sum_{n=0}^{N} c_n(\tau)\, x_n(t) \qquad && t \geq \tau \\ &= 0 && t < \tau \end{aligned} \tag{1.139}$$

In addition, assume that $y(t)$ can be expressed as a separable function of t and τ in accordance with the given system specifications, i.e.,

$$\begin{aligned} y(t) &= \sum_{m=0}^{M} a_m(\tau)\, y_m(t) \qquad && t \geq \tau \\ &= 0 && t < \tau \end{aligned} \tag{1.140}$$

Assuming enough differentiability of the $c_n(\tau)$, $x_n(t)$, $a_m(\tau)$, and $y_m(t)$, both Eqs. (1.139) and (1.140) can be assumed to be weighting functions of linear systems. Consequently, by the method of Sec. 1-3, the differential equations which correspond to each can be determined. Figure 1-1 could be redrawn as in Fig. 1-10. In Fig. 1-10, X represents the differential equation corresponding to Eq. (1.139),

FIG. 1-10. System equivalent to Fig. 1-1.

and W the differential equation corresponding to the unknown system. If Y represents the differential equation corresponding to Eq. (1.140), then, according to the notation of differential-equation algebra,

$$Y = WX \tag{1.141}$$

or

$$W = YX^{-1} \tag{1.142}$$

Equation (1.142) then indicates that if the input and output of an unknown linear system are known separable functions of t and τ, the over-all differential equation of the system can be determined by the product of Y and X^{-1}.

While this technique is more straightforward than that for polynomial inputs, the solution contains less readily available information since it is in the form of a differential equation and not a weighting function.

Since a polynomial of the form of Eq. (1.119) is a separable function of t and τ, it is apparent that this technique can be used also for polynomial inputs. It is also apparent that the $y_x(t, \tau)$ defined by Eq. (1.120) must have the properties of a weighting function of a linear system described by an ordinary differential equation of the form of Eq. (1.1). However, due to physical constraints, $y_x(t, \tau)$ will not contain delta functions.

1.25 Example

As an example of the technique developed in Sec. 1-23, consider the following problem. The input to a linear system has the form:

$$x(t - \tau) = 2u(t - \tau) + (t - \tau) - 2(t - \tau)^2 \tag{1.143}$$

From the system specifications, the output of the system is required to have the form:

$$y_x(t,\tau) = -\frac{7}{6}t^5 + t^4\left[\frac{5}{6} + \frac{10}{3}\tau\right] + t^3\left[1 - \frac{3}{2}\tau - 3\tau^2\right]$$
$$+ t^2\left[2\tau + \frac{1}{2}\tau^2 + \frac{2}{3}\tau^3 + 1\right] + t\left[-3\tau^2 + \frac{1}{6}\tau^3 + \frac{1}{6}\tau^4 - \tau + 2\right] \quad (1.144)$$

From the given input $x(t-\tau)$ and the desired output, the weighting function $W(t,\tau)$ which represents the desired linear system is to be determined. The output $y_x(t,\tau)$ can then be represented by Eq. (1.122), in which $N = 2$, $c_0 = 2$, $c_1 = 1$, and $c_2 = -2$.

The first step in determining $W(t,\tau)$ is to find $b_m(t)$. Taking the limit of $y_x(t,\tau)$ as $t \to \tau^+$, we obtain

$$\lim_{t\to\tau} y_x(t,\tau) = 2\tau = b_m(\tau)c_0 \quad (1.145)$$

Therefore

$$b_m(t) = \frac{2t}{c_0} = t \quad (1\text{-}146)$$

The second term of $y_x(t,\tau)$ in Eq. (1.122) is then

$$b_m(t)\sum_{n=0}^{N} c_n(t-\tau)^n = 2t + t^2 - t\tau - 2t^3 + 4t^2\tau - 2t\tau^2 \quad (1.147)$$

Inserting Eq. (1.147) in Eq. (1.122) and solving for $y'_x(t,\tau)$, we obtain

$$y'_x(t,\tau) = 3t^3 - \frac{3}{2}t^3\tau - 3t^3\tau^2 - 2t^2\tau + \frac{1}{2}t^2\tau^2 + \frac{2}{3}t^2\tau^3$$
$$+ \frac{5}{6}t^4 - \frac{10}{3}t^4\tau - t\tau^2 + \frac{1}{6}t\tau^3 + \frac{1}{6}t\tau^4 - \frac{7}{6}t^5 \quad (1.148)$$

Next, the derivatives of $y'_x(t,\tau)$ with respect to τ are formed

$$\frac{\partial y'_x}{\partial\tau} = -\frac{3}{2}t^3 - 6t^3\tau - 2t^2 + t^2\tau + 2t^2\tau^2 + \frac{10}{3}t^4 - 2t\tau + \frac{1}{2}t\tau^2 + \frac{2}{3}t\tau^3 \quad (1.149)$$

$$\frac{\partial^2 y'_x}{\partial\tau^2} = -6t^3 + t^2 + 4t^2\tau - 2t + t\tau + 2t\tau^2 \quad (1.150)$$

$$\frac{\partial^2 y'_x}{\partial\tau^3} = 4t^2 + t + 4t\tau \quad (1.151)$$

Referring to Eqs. (1.130) and (1.132), we see that

$$\frac{\partial^3 y'_x(t,\tau)}{\partial\tau^3} = 4t^2 + t + 4t\tau$$
$$= -2c_2W_1(t,\tau) + c_1\frac{\partial W_1(t,\tau)}{\partial\tau} = c_0\frac{\partial^2 W_1(t,\tau)}{\partial\tau^2} \quad (1.152)$$

$$W_1(t, t^-) = -\frac{1}{c_0}\left\{\left.\frac{\partial y_x'(t, \tau)}{\partial \tau}\right|_{\tau \to t^-}\right\} = 2t^2 \tag{1.153}$$

$$\left.\frac{\partial W_1(t, \tau)}{\partial \tau}\right|_{\tau \to t} = -\frac{1}{c_0}\left\{\left.\frac{\partial^2 y_x'}{\partial \tau^2}\right|_{\tau \to t^-} - c_1 W_1(t, t^-)\right\} = t \tag{1.154}$$

The solution of the set of Eqs. (1.152), (1.153), and (1.154) is then the desired weighting function. To simplify the algebra involved and make the solution easier, the following substitutions are made

$$\begin{aligned} z &= t - \tau \\ k_2 &= 2t \\ k_1 &= 4t^2 + \frac{t}{2} \\ f(z) &= W_1(t, t - z) \end{aligned} \tag{1.155}$$

Then, since $c_0 = 2$, $c_1 = 1$ and $c_2 = -2$, Eqs. (1.152), (1.153), and (1.154) can be rewritten as:

$$\frac{d^2 f}{dz^2} + \frac{1}{2}\frac{df}{dz} - 2f = k_2 z - k_1 \tag{1.152$'$}$$

$$f(0) = 2t^2 \tag{1.153$'$}$$

$$f'(0) = -t \tag{1.154$'$}$$

The complementary solution of Eq. (1.152′) has the form

$$f_c(z) = Ae^{-\gamma_1 z} + Be^{-\gamma_2 z} \tag{1.156}$$

where A and B are functions of the initial conditions and γ_1 and γ_2 are roots of the polynomial

$$\gamma^2 + \frac{1}{2}\gamma - 2 = 0 \tag{1.157}$$

The particular solution has the form

$$f_p = \frac{k_1}{2} - \frac{k_2}{8} - \frac{k_2}{2}z \tag{1.158}$$

The complete solution then has the form

$$f(z) = Ae^{-\gamma_1 z} + Be^{-\gamma_2 z} + \frac{k_1}{2} - \frac{k_2}{8} - \frac{k_2}{2}z \tag{1.159}$$

Applying the initial conditions of Eqs. (1.153′) and (1.154′), the constants A and B can be determined as

$$A = \frac{\left[f(0) + \frac{k_2}{8} - \frac{k_1}{2}\right]\left[-\gamma_2\right] - f'(0) - \frac{k_2}{2}}{\gamma_1 - \gamma_2} \tag{1.160}$$

$$B = \frac{f'(0) + \frac{k_2}{2} + \gamma_1\left[f(0) + \frac{k_2}{8} - \frac{k_1}{2}\right]}{\gamma_1 - \gamma_2} \tag{1.161}$$

Substituting the values of Eqs. (1.153′), (1.154′), and (1.155), it can be shown that

$$\begin{aligned} A &= 0 \\ B &= 0 \end{aligned} \tag{1.162}$$

and $f(z)$ therefore equals the particular solution of the differential equation, i.e.,

$$f(z) = \frac{k_1}{2} - \frac{k_2}{8} - \frac{k_2}{2}z \tag{1.163}$$

Substituting the values of Eq. (1.155) in Eq. (1.163), it can be shown that

$$W_1(t, \tau) = t^2 + t\tau \tag{1.164}$$

The over-all weighting function is then

$$W(t, \tau) = t^2 + t\tau + t\delta(t - \tau) \tag{1.165}$$

and the problem is solved. Convolution of (1.165) and (1.143) will indeed produce $y_x(t, \tau)$ as given in Eq. (1.144).

1.26 Approximation of Separable Functions

From the discussion in the previous sections, it is obvious that for a large class of synthesis problems the response function must have the form of a separable function. It is then necessary to the synthesis techniques to have a method of approximating a function of two variables as a separable function of the two variables. Some work has been done on this problem; for instance, Cruz [8] proposed an impulse-train approximation of weighting functions; and Cruz and Van Valkenberg [4] proposed a method of approximating a weighting function as a separable function. This section describes a simple method of approximating a function of two variables as a separable function.

As a tool to help explain this approximation method, a type of final-value controller will be considered. Suppose that a system is to be designed according to the following specifications. The system will receive a step input at some unknown time between $t = 0$ and $t = T$, and must settle to within n percent of its final steady-state value (unity) by time T_s where $T_s > T$. On the other hand, the system should always be kept as sluggish as possible. Thus, if the system is designed as a nonstationary linear system, a system step response $y(t, \tau)$ which satisfies the specifications should first be determined. From this $y(t, \tau)$, the appropriate weighting function can be found and the design completed.

Figure 1-11 represents a two-dimensional view of $y(t, \tau)$ in which the τ axis is perpendicular to the plane of the paper. The planes $y(t, 0)$ and $y(t, T)$ are the two planes which bound the region of interest of $y(t, \tau)$; i.e., if the function $y(t, \tau)$ is suitable in this region, the specifications can be met.

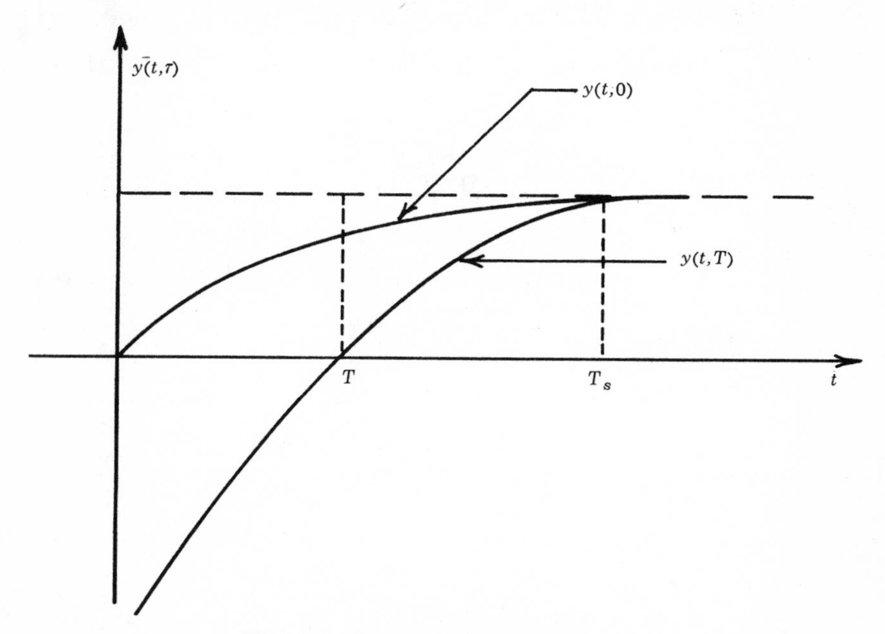

FIG. 1-11. Two-dimensional view of $y(t, \tau)$.

Suppose that $y(t, \tau)$ and $y(t, T)$ are chosen in the form

$$
\begin{aligned}
y(t, 0) &= 1 - e^{\alpha_1 t} \\
y(t, T) &= 1 - e^{\alpha_2 (t-T)}
\end{aligned}
\tag{1.166}
$$

where a_1 and a_2 are chosen so that

$$y(T_s, T) = y(T_s, T) = 1 - \frac{n}{100} \quad \textbf{(1.167)}$$

It is seen then that $y(t, \tau)$ meets the specifications for two values of τ, $\tau = 0$ and $\tau = T$. In addition, $y(t, \tau)$ for $0 \le \tau \le T$ might be fixed for several additional values of τ; however, for the purpose of this exposition, two will suffice. Obviously, if $y(t, \tau)$ could be represented by relationships similar to Eq. **(1.166)** for all $0 \le \tau \le T$, the specifications might be met exactly. The number of values of τ for which $y(t, \tau)$ is fixed, by Eq. **(1.166)**, for example, will also be the maximum number of terms in the final separable function; thus, the better the desired approximation, the higher the order of the resultant system.

Suppose Fig. 1-11 is redrawn as in Fig. 1-12, with the t axis normal to the plane of the paper. Outside the region $0 \le \tau \le T$, $y(t, \tau)$ is arbitrary. At the boundaries of the region, $y(t, \tau)$ satisfies Eq. (1.166).

If the only constraints on $y(t, \tau)$ are at $\tau = 0$ and $\tau = T$, $y(t, \tau)$ can be approximated in several ways. For instance, let

$$y(t, \tau) = y(t, 0) + [y(t, T) - y(t, 0)]\, u(\tau - T) \quad \textbf{(1.168)}$$

where $u(\tau - T)$ is a step function occurring at $\tau = T$. Then $y(t, \tau)$ corresponds to the full line function in Fig. 1-12. If Eq. **(1.168)** is Laplace-transformed with respect to τ, the resulting transform $Y(t, s)$ is given by

$$Y(t, s) = \frac{y(t, 0)}{s} + \frac{y(t, T) - y(t, 0)}{s} e^{-sT} \quad \textbf{(1.169)}$$

Now if $\dfrac{e^{-sT}}{s}$ is approximated by a Padé approximation [1], i.e.,

$$\frac{e^{-sT}}{s} \cong -\frac{2}{T} \frac{(s - 3/T)}{s\; s^2 + 4s/T + 6/T^2} \quad \textbf{(1.170)}$$

then

$$Y(t, s) \cong Y^*(t, s)$$

$$= \frac{y(t, 0)}{s} + [y(t, T) - y(t, 0)] \left\{ -\frac{2}{T} \frac{(s - 3/T)}{s\left[s^2 + 4s/T + 6/T^2\right]} \right\} \quad \textbf{(1.171)}$$

The inverse transform of $y^*(t, s)$ will be denoted $y^*(t, \tau)$. Other Padé approximations could be used in Eq. **(1.169)**. Obviously, a

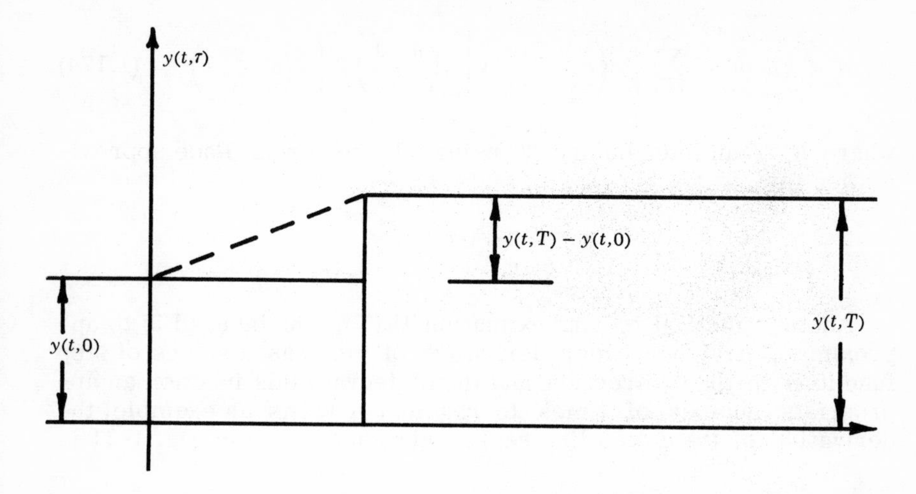

FIG. 1-12. Approximation of $y(t, \tau)$.

better approximation would be obtained if a higher order polynomial were used. This particular approximation was chosen because it is relatively simple and because the numerator is of lower degree than the denominator. This latter condition forces the second term of $y^*(t, \tau)$ to be zero at $\tau = 0$, and, hence, $y^*(t, 0) = y(t, 0)$. Taking the inverse transform of $Y^*(t, s)$, the approximation $y^*(t, \tau)$ is

$$
\begin{aligned}
y^*(t, \tau) &= y(t, 0)\left[e^{-2\tau/T} \sin\left(\frac{\sqrt{2}\,\tau}{T} + \psi\right)\right] \\
&\quad + y(t, T)\left[1 - 3e^{-2\tau/T} \sin\left(\frac{\sqrt{2}\,\tau}{T} + \psi\right)\right] \qquad t \geq \tau
\end{aligned}
\tag{1.172}
$$

where $\psi = \sin^{-1}(1/3)$, and $y(t, 0)$ and $y(t, T)$ are given in Eq. (1.166).

As a very rough check on the accuracy of the approximation $y^*(t, \tau)$, it is seen that

$$
\begin{aligned}
y^*(t, 0) &= y(t, 0) \\
y^*(t, T) &= 0.398y(t, 0) + 0.602y(t, T) \\
y^*(t, \infty) &= y(t, T)
\end{aligned}
\tag{1.173}
$$

The accuracy of this type of approximation can be improved by: (1) approximating $y(t, \tau)$ by several steps in the τ direction (see Fig. 1-11), i.e., letting

$$y(t, \tau) = y(t, 0) + \sum_{n=1}^{N} \left\{ y\left(t, \frac{n}{N} T\right) - y\left[t, \left(\frac{n-1}{N}\right) T\right] \right\} u\left(\tau - \frac{n}{N} T\right) \qquad (1.174)$$

where N is an integer; and (2) using a higher order Padé approximation for

$$e^{-\frac{n}{N} sT}$$

Another method of approximation that might be used is to approximate a higher order derivative of $y(t, \tau)$ as a series of step functions in the τ direction and then integrate this function an appropriate number of times to regain $y(t, \tau)$. As an example, the derivative of the dotted line representation of $y(t, \tau)$ in Fig. 1-11 is

$$y^{(0,1)}(t, \tau) = \frac{\partial}{\partial \tau}[y(t, \tau)] = \frac{y(t, T) - y(t, 0)}{T} [u(\tau) - u(\tau - T)] \qquad (1.175)$$

If this equation is transformed, if the Padé approximation in Eq. (1.170) is used, and an inverse transform is then taken, the approximation of $y^{(0,1)}(t, \tau)$, denoted $y^{*(0,1)}(t, \tau)$, is given by

$$y^{*(0,1)}(t, \tau) = \frac{y(t, T) - y(t, 0)}{T} \cdot 3e^{-2\tau/T} \sin\left[\frac{\sqrt{2}\tau}{T} + \psi\right] \qquad (1.176)$$

where $\psi = \sin^{-1}(1/3)$. Then, since

$$y^*(t, \tau) = y(t, 0) + \int_0^{\tau} y^{*(0,1)}(t, \theta)\, d\theta \qquad (1.177)$$

the final approximation is:

$$y^*(t, \tau) = y(t, 0)^{-2\tau/T} \left\{ \frac{1}{2} \sin \frac{\sqrt{2}\tau}{T} + \cos \frac{\sqrt{2}\tau}{T} \right\}$$
$$+ y(t, T) \left\{ 1 - e^{-2\tau/T} \left[\frac{1}{2} \sin \frac{\sqrt{2}\tau}{T} + \cos \frac{\sqrt{2}\tau}{T} \right] \right\}; \qquad t \geq \tau$$
$$y^*(t, \tau) = 0 \qquad t < \tau$$
$$(1.178)$$

From a comparison with the approximation in Eq. (1.172), it is seen that

$$\begin{aligned} y^*(t, 0) &= y(t, 0) \\ y^*(t, T) &= 0.116y(t, 0) + 0.884y(t, T) \\ y^*(t, \tau) &= y(t, T) \end{aligned} \tag{1.179}$$

This procedure then gives a more accurate approximation to $y(t, \tau)$ than the first procedure. Again, a better approximation can be found if $\frac{\partial}{\partial \tau}[y(t, \tau)]$ is approximated by more terms than for $y(t, \tau)$ [see Eq. (1.174)].

1.27 Synthesis of Time-Variable Differential Equations with Analog Computer Elements

One of the advantages of putting the compensation network of a linear control in the form of a differential equation is that then the compensation network can be readily built. In particular, if the networks are time-variable, they can be built with analog computer elements [6].

The types of differential equations which must be synthesized will have the form of Eq. (1.1). The only problem which arises in synthesizing Eq. (1.1) is how to handle the right hand side, which contains derivatives of the input x. This problem is circumvented by writing Eq. (1.1) in an equivalent vector matrix form

$$\dot{\mathbf{y}} = \mathbf{A}\mathbf{y} + \mathbf{f}x \tag{1.180}$$

where $\mathbf{y}$ is an n-vector. The solution of Eq. (1.1) is then given by

$$y(t) = \mathbf{c}^T\mathbf{y} + rx \tag{1.181}$$

where $\mathbf{c}$ is an n-vector and r is a scalar determined by the relationship between Eqs. (1.1) and (1.180).

The problem, then, is to find a vector equation, in the form of Eq. (1.180), equivalent to Eq. (1.1). There are several forms which this vector equation might take (see Refs. [15, 18]); however, only the form which the author has found easiest to use will be presented in this section. The development is that of Matyash [6] with some slight changes for convenience.

Using the notation of Eq. (1.2), let

$$L = \sum_{i=0}^{n} a_i(t)\frac{d^i}{dt^i} \qquad M = \sum_{i=0}^{n} b_i(t)\frac{d^i}{dt^i} \tag{1-182}$$

Now let us define two subsidiary operators

$$L_1 = \sum_{i=0}^{n} a_i(t)\frac{d^i}{dt^i} \tag{1-183}$$

and

$$M_1 = \sum_{i=0}^{n} \beta_i(t) \frac{d^i}{dt^i} \tag{1.184}$$

where $\alpha_i(t)$ and $\beta_i(t)$ are to be defined subsequently.

Now assume that the following set of equations is valid

$$y_0 = -\alpha_n y - \beta_n x \tag{1.185}$$

$$-\dot{y}_{k-1} = y_k + \alpha_{n-k} y + \beta_{n-k} x \qquad k = 1,2,\ldots,n-1 \tag{1.186}$$

$$-\dot{y}_{n-1} = \alpha_0 y + \beta_0 x \tag{1.187}$$

Now if y_0 from Eq. (1.185) is substituted in Eq. (1.186) for $k = 1$, and in the resulting equation, y_1 is determined and substituted into Eq. (1.186) for $k = 2$, etc., all of the y_i are eliminated and Eq. (1.187) becomes an nth order differential equation in y

$$\sum_{i=0}^{n} (-1)^i \frac{d^i[\alpha_i y]}{dt^i} = -\sum_{i=0}^{n} (-1)^i \frac{d^i[\beta_i x]}{dt^i} \tag{1.188}$$

Now if the adjoint operators of L_1 and M_1 are denoted L_1^* and M_1^* respectively, i.e.,

$$L_1^*(y) = \sum_{i=0}^{n} (-1)^i (\alpha_i y)^{(i)} \tag{1.189}$$

and

$$M_1^*(x) = \sum_{i=0}^{n} (-1)^i (\beta_i x)^{(i)} \tag{1.190}$$

then Eq. (1.188) can be written

$$L_1^*(y) = -M_1^*(x) \tag{1.191}$$

and if, now, Eqs. (1.1) and (1.188) are made identical

$$\begin{aligned} L_1^*(y) &= L(y) \\ -M_1^*(x) &= M(x) \end{aligned} \tag{1.192}$$

Then since, for K a linear differential operator,

$$(K^*)^* = K$$

Equations (1.192) can be written

$$\begin{aligned} L_1(y) &= L^*(y) \\ M_1(x) &= -M^*(x) \end{aligned} \tag{1.193}$$

In other words, the α's and β's of the n first-order differential equations in Eqs. (1.185), (1.186), and (1.187) are determined from the adjoint operators L^* and M^*.

A further refinement which can be made is the following. Make $a_n(t) \equiv 1$ (without loss of generality). Then

$$\alpha_n \equiv (-1)^n$$

and, form Eq. (1.185)

$$y = (-1)^{n+1}[y_0 + \beta_n x] \tag{1.194}$$

and substituting this relationship for y in Eqs. (1.186), the final form of the n first-order equations becomes

$$\begin{aligned} \dot{y}_0 &= (-1)^n \alpha_{n-1} y_0 - y_1 - \left[\beta_{n-1} + (-1)^{n+1} \beta_n \alpha_{n-1}\right] x \\ \dot{y}_1 &= (-1)^n \alpha_{n-2} y_0 - y_2 - \left[\beta_{n-2} + (-1)^{n+1} \beta_n \alpha_{n-2}\right] x \\ &\text{-------------------------} \\ \dot{y}_{n-2} &= (-1)^n \alpha_1 y_0 - y_{n-1} - \left[\beta_1 + (-1)^{n+1} \beta_n \alpha_1\right] x \\ \dot{y}_{n-1} &= (-1)^n \alpha_0 y_0 \qquad - \left[\beta_0 + (-1)^{n+1} \beta_n \alpha_0\right] x \end{aligned} \tag{1.195}$$

Equation (1.194) corresponds to Eq. (1.181), and Eq. (1.195) to Eq. (1.180).

Equations (1.194) and (1.195) can be synthesized by analog computer elements as shown in Fig. 1-13a. The symbols used in Fig. 1-13a are defined in Figs. 1-13b, 1-13c, and 1-13d. In Fig. 1-13b, the output of an adder e_0 is given by

$$e_0 = -(e_1 + e_2 + e_3) \tag{1.196}$$

In Fig. 1-13c, the output of an integrator e_0 is given by

$$e_0 = -\int^t (e_1 + e_2 + e_3)\, dt \tag{1.197}$$

In Fig. 1-13d, the output of a multiplier is given by

$$e_0 = a(t)\, e_1 \tag{1.198}$$

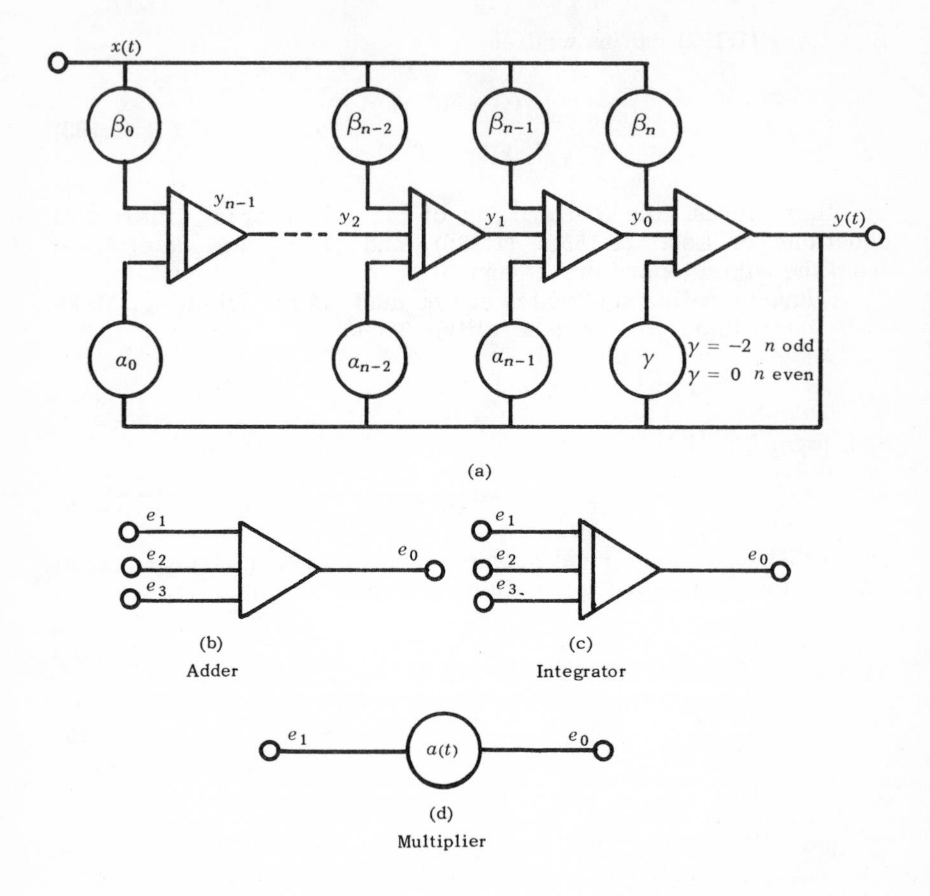

FIG. 1-13. Analog computer schematic of Eqs. (1-194) and (1-195).

The advantage of synthesizing Eqs. (1.194) and (1.195) instead of Eq. (1.1) is that no differentiation is required in the first two equations.

1.28 Summary

In this chapter, we have developed some techniques which can be used to synthesize linear feedback systems containing fixed elements which are either time-variable or must have an over-all response characteristic varying with time. The development has been reasonably complete; however, certain problems which arise have not been explored. A more complete development can be found in Ref. [17].

REFERENCES

1. Truxal, J. G., Control Systems Synthesis, McGraw-Hill Book Company, Inc., New York, 1955.
2. Borskiy, V., "On the Properties of Impulsive Response Functions of Systems with Variable Parameters," Automation and Remote Control, Vol. 20, July, 1959, pp. 822-830.
3. Control Systems Engineering, edited by W. W. Siefert and C. W. Steeg, Jr., McGraw-Hill Book Company, Inc., New York, 1960.
4. Cruz, J. B. and Van Valkenberg, M. E., "The Synthesis of Models for Time-Varying Linear Systems," Proceedings of the Symposium on Active Networks and Feedback Systems, Polytechnic Press of the Polytechnic Institute of Brooklyn, New York, 1960, pp. 527-544.
5. Batkov, A. M., "The Problem of Synthesizing Linear Dynamic Systems with Variable Parameters," Automation and Remote Control, Vol. 19, January, 1958, pp. 42-48.
6. Matyash, I., "Methods of Analog Computer Solution of Linear Differential Equations with Variable Coefficients," Automation and Remote Control, Vol. 20, July, 1959, pp. 813-821.
7. Malchikov, S. V., "On the Synthesis of Linear Automatic Control Systems with Variable Parameters," Automation and Remote Control, Vol. 20, December, 1959, pp. 1543-1549.
8. Cruz, J. B., "A Generalization of the Impulse Train Approximation for Time-Varying Linear System Synthesis in the Time Domain," IRE Transactions on Circuit Theory, Vol. CT-6, December, 1959, pp. 393-394.
9. Laning, J. H. and Battin, R. H., Random Processes in Automatic Control, McGraw-Hill Book Company, Inc., New York, 1956.
10. Computer Control Systems Technology, edited by C. T. Leondes, McGraw-Hill Book Company, Inc., New York, 1961.
11. Birkhoff, G. and MacLane, S., A Survey of Modern Algebra, MacMillan Inc., New York, 1960.
12. Stear, E. B. and Stubberud, A. R., "Signal Flow Graph Theory for Linear Time-Variable Systems," Transactions of the AIEE (Communications and Electrons), Vol. 58, January, 1962, pp. 695-701.
13. Stubberud, A. R., "A Technique for the Synthesis of Linear, Nonstationary Feedback Systems, Part I: The Approximation Problem," accepted for publication in AIEE Transactions.
14. Stubberud, A. R., "A Technique for the Synthesis of Linear, Nonstationary Feedback Systems, Part II: The Synthesis Problem," accepted for publication in AIEE Transactions.
15. Gladkov, D. I., "On the Synthesis of Linear Automatic Control Systems," Avtomatika i Telemekhanika, Vol. 22, 1961, pp. 306-313.

16. Miller, K. S., "Properties of Impulsive Responses and Green's Functions," IRE Transactions on Circuit Theory, Vol. CT-2, March, 1955, pp. 26-33.
17. Stubberud, A. R., The Analysis and Synthesis of Linear, Time-Variable Systems, University of California Press (in press).

2

System Synthesis With Random Inputs

H. C. HSIEH*
DECEASED, FORMERLY ASSISTANT PROFESSOR
ELECTRICAL ENGINEERING
NORTHWESTERN UNIVERSITY, EVANSTON, ILLINOIS

C. T. LEONDES
PROFESSOR OF ENGINEERING
UNIVERSITY OF CALIFORNIA, LOS ANGELES, CALIFORNIA

PART I Optimum Synthesis of Multivariable Systems With Stationary Random Inputs

2.1 Introduction

Modern control theory is entering an era of emphasis on the problems of multivariable systems. Although this development has been inspired considerably by previous work on single-variable systems, there have nevertheless been problems in multivariable systems intrinsically different from those of the corresponding single-variable case.

The nature of the problem considered in this part of the chapter can be stated simply: A set of stationary random processes, taken to be a mixture of useful signals and perturbing noise, is available. These input processes are to be operated on by a linear device to obtain a set of outputs optimum in the least square sense. This is the multivariable optimum linear prediction and filtering problem.

The pioneer work of the optimum linear prediction and filtering problem was done by N. Wiener [1] about two decades ago. After that, various extensions of his basic theory to other single-variable problems have been made. These synthesis techniques are well known now and can be found in several textbooks [2, 3, 4]. However, the problems associated with multivariable cases have not

*Previously with the Department of Engineering, University of California, Los Angeles.

been investigated until recently. These problems are extremely important to modern technology including aerospace control and communication problems.

The early work in the synthesis of optimum multivariable filters and control systems in the least square sense with stationary inputs was done by Hsieh and Leondes [5, 6] and also by Amara [7]. The vector Wiener-Hopf equation was solved by using the transform method and the method of undetermined coefficients. A more direct approach for solving this problem is made through factorization of the input rational spectral density matrix. However, the procedures thus far developed for carrying out this factorization are rather complicated. Wiener and Masani [8, 9] considered the synthesis of a discrete multivariable filter and presented a method of factorization in terms of an infinite series of matrices. This method is unsuited for any practical purpose. Youla [10] developed an algorithm for solving this factorization problem with rational spectral matrix. Unfortunately, his method is quite complicated, and a simple rational matrix requires a great deal of computation. More recently, Davis [11] gave another method for factoring the spectral density matrix. The basic idea behind his approach is intuitive, and the procedure for carrying out the factorization is somewhat simpler.

In the following sections, the least-square filtering problem will be formulated and its associated Wiener-Hopf equation will be derived in very general terms by using some matrix and operator theories. The method of undetermined coefficients, proposed by Hsieh and Leondes and also by Amara, will be presented in detail. An example will be given for amplifying the procedure. Following that, the technique developed by Davis for factoring the spectral density matrix factorization will be shown. Finally, the extension of these methods for solving a general control system problem will be illustrated.

2.2 Statement of the Continuous Multivariable Filtering Problems in the Wiener Sense

A multivariable system is defined as a system with n inputs and m outputs. For a linear system, the inputs and outputs are related by a weighting function matrix. Let $\mathbf{I}(t)$ be an n-dimensional input vector and $\mathbf{C}(t)$ be an m-dimensional output vector. For simplicity, a vector always implies a column vector. Denote the weighting function between the kth input and the jth output by $\mathbf{W}_{jk}(t, \tau)$. Then $\mathbf{W}(t, \tau)$ will be an $\mathrm{m} \times \mathrm{n}$ weighting function matrix. The output vector can now be expressed by

$$\mathbf{C}(t) = \int_{-\infty}^{t} \mathbf{W}(t, \tau)\, \mathbf{I}(\tau)\, d\tau \tag{2.1}$$

The purpose of this section is to formulate the Wiener filtering theory for the multivariable case [5]. To facilitate the formulation and the solution of the problem, some definitions and terminology will be introduced. Let $X(t)$, $-\infty < t < \infty$, be an n-dimensional real vector random process. Then the covariance matrix of the process is given by

$$R(s,t) = E\{X(s)X'(t)\} \tag{2.2}$$

where $X'(t)$ is the transpose of $X(t)$. Thus $R(s,t)$ is an $n \times n$ matrix. It is obvious that, for each s and t, $R(s,t)$ is nonnegative definite, and $R'(s,t) = R(t,s)$. We shall denote the norm of a matrix M (square or not) by $\|M\|$ and define it as

$$\|M\| = \sqrt{\text{Trace }(MM')} \tag{2.3}$$

For the problem considered, each input will be a corrupted signal, the sum of a wanted signal $s_k(t)$ and an unwanted noise $n_k(t)$. Thus the input vector can be written as

$$I(t) = S(t) + N(t) \tag{2.4}$$

where $S(t)$ and $N(t)$ are n-dimensional vectors. It is assumed that all the signals and noises are stationary processes and that they are all real.

The filter under investigation is defined in the sense that the ideal outputs can be the result of any linear operation on the signal portions of the inputs. Let $D(t,\tau)$ be the $m \times n$ ideal weighting function matrix. Then the ideal output vector $C_d(t)$ is

$$C_d(t) = \int_{-\infty}^{\infty} D(t,\tau)S(\tau)\,d\tau \tag{2.5}$$

It should be noted that the ideal weighting function may not be physically realizable to include the case of prediction. The system error vector is defined as the difference between the actual output vector and the ideal output vector

$$\epsilon(t) = C(t) - C_d(t) \tag{2.6}$$

The formulation of this problem is best understood by referring to Fig. 2-1. The part of the diagram below the dotted line represents the actual system, while the part above the dotted line represents the hypothetical ideal system together with the comparator for generating the error signals.

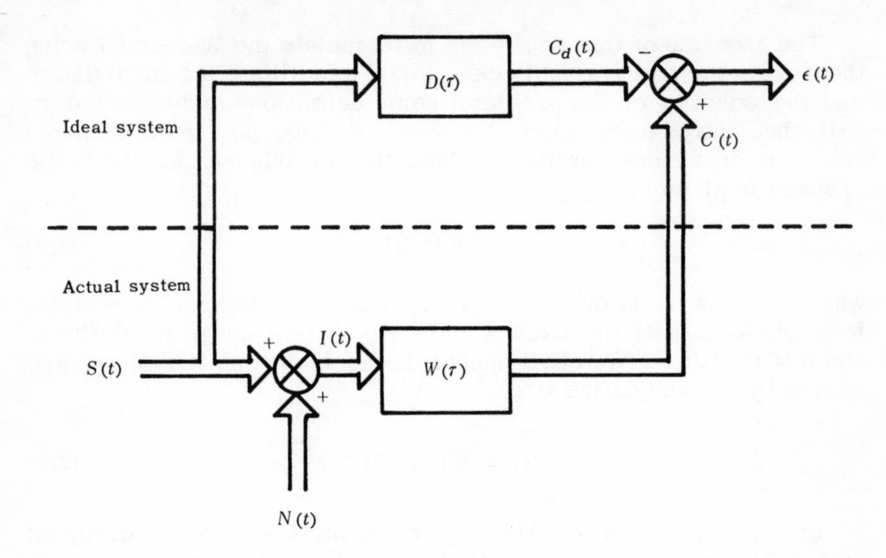

FIG. 2-1. Error generation diagram.

There is no loss of generality in assuming that the signal and noise processes have zero means. This assumption implies that the expected value of the error vector is automatically zero. The problem now is to choose the physically realizable (also stable) weighting function matrix such that the expected value of the square of the norm of the error vector is minimized.

$$E\left\{\| \epsilon(t) \|^2\right\} = \text{min w.r.t.} \quad \mathbf{W}(t, \tau) \tag{2.7}$$

where $\mathbf{W}(t, \tau) = 0$ for $t < \tau$.

2.3 The Minimization of the System Error

Let us now evaluate the performance criterion. First we observe that for stationary inputs and the time-invariant ideal system, the optimum actual system will also be time-invariant. Secondly, we notice that the minimization of $E\left\{\| \epsilon(t) \|^2\right\}$ is equivalent to the minimization of the expected value of the square of each of its components. Let us now consider the jth component.

$$\epsilon_j(t) = \int_0^\infty \mathbf{W}_j'(\tau)\, \mathbf{I}(t-\tau)\, d\tau - C_{dj}(t) \tag{2.8}$$

where $\mathbf{W}_j'$ is a $1 \times n$ row vector, the jth row of the weighting function matrix. Then we have

$$
\begin{aligned}
E\left\{\epsilon_j^2(t)\right\} &= E\left\{\int_0^\infty \mathbf{W}_j'(\tau)\int_0^\infty \mathbf{I}(t-\tau_1)\mathbf{I}'(t-\tau_2)\mathbf{W}_j(\tau_2)\,d\tau_2\right. \\
&\qquad \left. - 2\int_0^\infty \mathbf{W}_j'(\tau_1)\mathbf{I}(t-\tau_1)C_{dj}(t)\,d\tau_1 + C_{dj}^2(t)\right\} \\
&= \int_0^\infty \mathbf{W}_j'(\tau_1)\,d\tau_1\int_0^\infty \mathbf{R}(\tau_1-\tau_2)\mathbf{W}_j(\tau_2)\,d\tau_2 \\
&\qquad - 2\int_0^\infty \mathbf{W}_j'(\tau_1)\mathbf{B}_j(\tau_1)\,d\tau_1 + E\left\{C_{dj}^2(t)\right\}
\end{aligned}
\tag{2.9}
$$

where $\mathbf{R}(\tau_1-\tau_2) = E\left\{\mathbf{I}(t-\tau_1)\mathbf{I}'(t-\tau_2)\right\}$ ($n \times n$ input covariance matrix)

$\mathbf{B}_j(\tau_1) = E\left\{\mathbf{I}(t-\tau_1)C_{dj}(t)\right\}$ ($n \times 1$ input and ideal output covariance vector).

Since the last term in Eq. (2.9) does not depend on $\mathbf{W}_j$, then the problem becomes a minimization of

$$
\begin{aligned}
Q(\mathbf{W}_j) &= \int_0^\infty \mathbf{W}_j'(\tau_1)\,d\tau_1\int_0^\infty \mathbf{R}(\tau_1-\tau_2)\mathbf{W}_j(\tau_2)\,d\tau_2 \\
&\qquad - 2\int_0^\infty \mathbf{W}_j^T(\tau_1)\mathbf{B}_j(\tau_1)\,d\tau_1
\end{aligned}
\tag{2.10}
$$

Let us define the inner product of two vectors $\mathbf{x}(t)$ and $\mathbf{y}(t)$ in the square integrable vector function space, or the L_2 space, by*

$$[\mathbf{x}, \mathbf{y}] \triangleq \int_a^b \mathbf{x}'(t)\mathbf{y}(t)\,dt \tag{2.11}$$

and also introduce the notation

$$R\mathbf{W}_j \triangleq \int_0^\infty \mathbf{R}(\tau_1-\tau_2)\mathbf{W}_j(\tau_2)\,d\tau_2 \tag{2.12}$$

Here R is evidently a linear operator in n-dimensional L_2 space. Then Eq. (2.10) can be written as

$$Q(\mathbf{W}_j) = [R\mathbf{W}_j, \mathbf{W}_j] - 2[\mathbf{B}_j, \mathbf{W}_j] \tag{2.13}$$

where the integration limits for the inner product are from 0 to ∞.

*Functions which are square integrable are defined as functions in L_2 space.

The problem now is to choose an optimum weighting function vector which will minimize Q. Toward this end, let us assume a vector $\mathbf{V}_j$ such that

$$R\mathbf{V}_j = \mathbf{B}_j \qquad \text{for} \quad \tau_1 \geqq 0 \tag{2.14}$$

By substituting Eq. (2.14) in Eq. (2.13) and adding and subtracting a term $[R\mathbf{V}_j, \mathbf{V}_j]$, we have

$$Q(\mathbf{W}_j) = [R\mathbf{W}_j, \mathbf{W}_j] - 2[R\mathbf{V}_j, \mathbf{W}_j] + [R\mathbf{V}_j, \mathbf{V}_j] - [R\mathbf{V}_j, \mathbf{V}_j] \tag{2.15}$$

Since the linear operator R is self-adjoint, i.e.,

$$[Rx, y] = [Ry, x]$$

then Eq. (2.15) can be written as

$$\begin{aligned} Q(\mathbf{W}_j) &= [R\mathbf{W}_j, \mathbf{W}_j] - [R\mathbf{W}_j, \mathbf{V}_j] - [R\mathbf{V}_j, \mathbf{W}_j] + [R\mathbf{V}_j, \mathbf{V}_j] - [R\mathbf{V}_j, \mathbf{V}] \\ &= [R(\mathbf{W}_j - \mathbf{V}_j), \mathbf{W}_j - \mathbf{V}_j] - [R\mathbf{V}_j, \mathbf{V}_j] \end{aligned} \tag{2.16}$$

Now we observe that, although the covariance matrix is in general nonnegative definite, it is also positive definite in practice. Thus we are only considering nonsingular problems. This means, then, that for any $\mathbf{Z}_j$,

$$[R\mathbf{Z}_j, \mathbf{Z}_j] > 0$$

and

$$[R\mathbf{Z}_j, \mathbf{Z}_j] = 0$$

if and only if, $\mathbf{Z}_j = 0$. (Except for a set of measure zero.*) Thus Q will be a minimum if we choose the optimum $\hat{\mathbf{W}}_j$ as

$$\hat{\mathbf{W}}_j - \mathbf{V}_j = 0$$

or

$$\hat{\mathbf{W}}_j = \mathbf{V}_j$$

Equation (2.14) then can be rewritten as

$$R\hat{\mathbf{W}}_j = \mathbf{B}_j \qquad \text{for} \quad \tau_1 \geqq 0 \tag{2.17}$$

The minimum Q is then

$$Q_{\min} = -[\mathbf{B}_j, \mathbf{W}_j] \tag{2.18}$$

*Those not familiar with the terminology of real variable theory need only replace this parenthetical statement by the note that consideration is given here only to functions which make sense from a physical point of view.

Hence Eq. (2.17) is the equation which the optimum weighting function vector must satisfy. This is the generalized Wiener-Hopf equation for multivariable systems. It is the necessary and sufficient condition, as has already been shown.

2.4 Optimization Equation in the Transform Domain

It is well known in the filtering theory of single-variable systems that, with stationary input processes, the transform method can be used to solve the Wiener-Hopf integral equation. For multivariable systems with stationary processes, transform methods can also be applied to the vector integral equation such that an algebraic equation in terms of the system transfer functions will result [5]. However, in order to determine these physically realizable system transfer functions, the techniques used must be quite different from those employed in the single variable system.

First we notice that, due to the physical realizability requirement of the system weighting function matrix, Eq. (2.17) is necessarily true only for $\tau_1 \geqq 0$. Since the covariance matrix $\mathrm{R}(\tau)$ is not zero for its negative argument, Eq. (2.17) must be modified before the transform method can be formally used. Let $\mathrm{f}_j(\tau_1)$ be defined as

$$\mathrm{f}_j(\tau_1) = 0 \qquad \text{for} \quad \tau_1 \geqq 0$$

Then Eq. (2.17) can be written as

$$R\mathbf{W}_j = \mathrm{B}_j + \mathrm{f}_j \qquad \text{for all value of } \tau_1 \tag{2.19}$$

The legitimacy of using the Fourier transform in Eq. (2.19) will first depend upon whether every function in this equation is Fourier-transformable. With the requirement of a physically realizable and stable system, the transfer function matrix can only have singularities or poles in the left half of the s-plane. (Here we define $s = j\omega$.) The power spectral densities of stationary processes, assumed to be rational functions in s, will have singularities in the entire plane and will be analytic in a strip around the imaginary axis. In particular, the auto-power spectral densities will have poles and zeros placed symmetrically with respect to both imaginary and real axes. The transform of the arbitrarily defined vector function f_j, which vanishes for $\tau_1 \geqq 0$, can only possess singularities in the right half of the s-plane. Thus, for an appropriate specification of the function of the ideal system, the input and ideal output covariance vector B_j is also Fourier-transformable. With all the foregoing stated properties about the known functions in Eq. (2.19) and the requirement of a physically realizable and stable system transfer function matrix, the transform method can certainly be applied. In order to solve for the optimum transfer function matrix, one additional requirement should be imposed on the spectral density matrix. This will be given detailed discussion in the next section.

Let the Fourier transform now be taken on both sides of Eq. (2.19). Then we have

$$\mathbf{G}\mathbf{Y}_j = \mathbf{K}_j + \mathbf{F}_j^- \tag{2.20}$$

where $\mathbf{G} = \int_{-\infty}^{\infty} \mathbf{R}(t) e^{-st} dt$ (input spectral density matrix)

$\mathbf{Y}_j = \int_{0}^{\infty} \mathbf{W}_j(t) e^{-st} dt$ (System transfer function vector)

$\mathbf{K}_j = \int_{-\infty}^{\infty} \mathbf{B}_j(t) e^{-st} dt$ (Input and ideal output spectral density vector)

$\mathbf{F}_j^- = \int_{-\infty}^{\infty} \mathbf{f}_j(t) e^{-st} dt$

Here, for simplicity, the argument s in Eq. (2.20) has been omitted. The vector $\mathbf{F}_j^-$, still unknown at this moment, can only have singularities in the right half-plane. The vector $\mathbf{K}_j$ and the matrix $\mathbf{G}$ will have singularities over the entire plane. Hence the vector $\mathbf{G}\mathbf{Y}_j$ will also have singularities over the entire plane. Since the singularities of $\mathbf{F}_j^-$ are restricted entirely to the right half-plane, it must be true then that

$$\{\mathbf{G}\mathbf{Y}_j\}_+ = \{\mathbf{K}_j\}_+ \tag{2.21}$$

Here the symbol $\{\quad\}_+$ denotes that part of the function which has poles in the left half-plane only. To be specific, this operation on any arbitrary function $\mathbf{H}(s)$ in the s-domain means that

$$\{\mathbf{H}(s)\}_+ = \int_0^{\infty} \mathbf{h}(t) e^{-st} dt \tag{2.22}$$

where

$$\mathbf{h}(t) = \frac{1}{2\pi j} \int_{-j\infty}^{j\infty} \mathbf{H}(s) e^{ts} ds$$

It should be noted that, when $\mathbf{H}(s)$ is a rational matrix in s, the operation $\{\quad\}_+$ is equivalent to expanding each of its components into partial fractions and then keeping those terms which have poles in the left half-plane only. However, when $\mathbf{H}(s)$ is a more general function in s, Eq. (2.22) must be used to carry out the operation $\{\quad\}_+$. Equation (2.21) can be considered the Wiener-Hopf equation in the transform domain for the multivariable filtering problem. When the transfer function matrix satisfies this equation, the minimum mean-square error becomes

$$E\left\{\epsilon_j^2\right\}_{\min} = \frac{1}{2\pi j}\int_{-j\infty}^{j\infty}\left[\left(\mathbf{Y}_d\right)_j^* \mathbf{G}_{ss}\left(\mathbf{Y}_d\right)_j - \mathbf{Y}_j^* \mathbf{G}_{IS}\left(\mathbf{Y}_d\right)_j\right] ds \qquad (2.23)$$

Here $\mathbf{G}_{ss}$ = Signal spectral density matrix
$\mathbf{G}_{IS}$ = Input and signal cross-spectral density matrix
$\left(\mathbf{Y}_d\right)_j$ = Ideal transfer function vector
and the asterisk denotes the transpose of the transfer function vector and the change of its argument from s to $-s$.

2.5 Solution of the Optimum Transfer Function Matrix by the Method of Undetermined Coefficients

For all the multivariable filtering problems considered in this chapter, the random processes are assumed to be stationary, and their power spectral densities assumed to be rational functions. The rational power spectral density matrix will then have the following properties:

P1. $\bar{\mathbf{G}}(s) = \mathbf{G}(\bar{s})$, i.e., $\mathbf{G}(s)$ is real. Here the upper bar denotes the complex conjugate.

P2. $\mathbf{G}'(-s) = \mathbf{G}(s)$, i.e., $\mathbf{G}(s)$ is a para-Hermitian matrix. Thus, if we let $s = j\omega$, then $\mathbf{G}(j\omega)$ is an ordinary Hermitian matrix for real ω.

P3. $\mathbf{G}(j\omega)$ is a nonnegative definite matrix. In other words, $\mathbf{b}'\,\mathbf{G}(j\omega)\,\mathbf{b} \geqq 0$ for any arbitrary vector $\mathbf{b}$ and every real finite ω.

For detailed discussions of the properties of covariance matrices and power spectral density matrices of multiple stationary random processes, the reader is referred to Cramer's paper [12].

Let us now consider the solution to Eq. (2.20). It can be inverted to give

$$\begin{aligned}\mathbf{Y}_j &= \mathbf{G}^{-1}\left(\mathbf{K}_j + \mathbf{F}_j^-\right) \\ &= \frac{\mathbf{A}}{|\mathbf{G}|}\left(\mathbf{K}_j + \mathbf{F}_j^-\right)\end{aligned} \qquad (2.24)$$

Here the elements of the matrix $\mathbf{A}$, the cofactors of the spectrum density matrix, will not have any pole on the imaginary axis and thus are analytic in a strip around the imaginary axis. So that the transform method will be legitimate for solving the problem, the spectral density matrix must have the additional property:

P4. The inverse of the spectral density matrix $\mathbf{G}^{-1}(s)$ must be analytic along the imaginary axis.

We conclude then, that the determinant $|\mathbf{G}|$ must not have any finite zero along the imaginary axis. This is the additional requirement which must be imposed on the power spectral density matrix. Thus Eq. (2.24) will be a legitimate expression. Every term in this equation has an analytical strip around the imaginary axis. Hence the path for its inverse transform can be taken along the imaginary axis.

We notice then, that the determinant of the power spectrum matrix $|\mathbf{G}|$ will always be a rational function in s^2. If it now satisfies the foregoing requirement that it have no finite zeros along the imaginary axis, this determinant can be decomposed such that

$$|\mathbf{G}| = G^+(s)\, G^-(s) \tag{2.25}$$

where all the poles and zeros of $G^+(s)$ must lie in the left half-plane, and $G^-(s)$, which is equal to $G^+(-s)$, can only have poles and zeros in the right half-plane.

The feasibility of using the transform method and the method of undetermined coefficients for solving the vector integral equation depends on the factorization of $|\mathbf{G}|$ as shown in Eq. (2.25). Thus Eq. (2.21) can be rewritten as

$$\mathbf{Y}_j = \frac{\mathbf{A}}{G^+ G^-}\mathbf{K}_j + \mathbf{F}_j^-$$

or

$$G^+\mathbf{Y}_j = \frac{1}{G^-}\mathbf{A}\mathbf{K}_j + \frac{1}{G^-}\mathbf{A}\mathbf{F}_j^- \tag{2.26}$$

Let us now investigate each term in the above equation. With the requirement of a physically realizable transfer function matrix for the filter, it is apparent that the vector on the left-hand side of Eq. (2.26) can only have poles in the left half-plane. It must be true then, that

$$G^+\mathbf{Y}_j = \left\{\frac{1}{G^-}\mathbf{A}\mathbf{K}_j\right\}_+ + \left\{\frac{1}{G^-}\mathbf{A}\mathbf{F}_j^-\right\}_+ \tag{2.27}$$

The vector $\frac{1}{G^-}\mathbf{A}\mathbf{K}_j$ is completely known. Therefore we shall have

$$\left\{\frac{1}{G^-}\mathbf{A}\mathbf{K}_j\right\}_+ = \int_0^\infty \mathbf{g}(t)e^{-st}\,dt$$

where

$$g(t) = \frac{1}{2\pi j} \int_{-j\infty}^{j\infty} \frac{1}{G^-} \Lambda K_j e^{ts} ds$$

Now the left half-plane poles in $\frac{1}{G^-} \Lambda F_j^-$ can only come from the matrix Λ. Although F_j^- is still unknown at this stage, it is constrained to have poles in the right half-plane only and thus contributes no left half-plane poles to the optimum transfer functions. Therefore the kth element of the vector $\left\{\frac{1}{G^-} \Lambda F_j^-\right\}_+$ will contain left half-plane poles from the kth row elements of matrix Λ, and can be expressed as

$$\frac{\sum_{p=0}^{m_k} a_p s^p}{\prod_{i_k} (s - r_{i_k})} = \frac{P_{jk}(s)}{\prod_{i_k} (s - r_{i_k})} \tag{2.28}$$

Here the coefficients of the polynomial $P_{jk}(s)$ which gives the zeros in the above expression are still unknown since F_j^- is unknown. The determination of the order of this polynomial as well as its unknown coefficients will be our main concern now. It should be noted that, for multiple poles r_{i_k}, the term $(s - r_{i_k})$ in Eq. (2.28) will be counted as many times as its multiplicity.

Let us now combine Eqs. (2.27) and (2.28). A typical kth component of the transfer function vector can then be expressed as

$$Y_{jk} = \frac{1}{G^+} \left[\left\{ \frac{1}{G^-} \sum_{k'=1}^{n} A_{k'k} K_{jk'} \right\}_+ + \frac{\sum_{p=0}^{m_k} a_p s^p}{\prod_{i_k} (s - r_{i_k})} \right] \tag{2.29}$$

Here $A_{k'k}$ is the element at the kth column and kth row of the matrix Λ, and is a cofactor of the original power spectral density matrix. Since the left half-plane poles from the terms $A_{k'k}$ will eventually be cancelled by the poles of G^+, the optimal transfer functions are given by Eq. (2.29) and can be rewritten as

$$Y_{jk}(s) = \frac{\left[\prod_u (s - a_u)\right] \left[\sum_{p=0}^{n_k} C_p s^p\right]}{\left[\prod_v (s - b_v)\right] \left[\prod_\ell (s - d_\ell)\right]} \tag{2.30}$$

where b_v = the zeros of G^+

d_l = left half-plane poles of the ideal transfer function vector $(\mathbf{Y}_d)_j$ if they are different from those of the spectral densities or have not been cancelled by the poles of G^+

a_u = poles of G^+ which have not been cancelled by the poles of the kth row element of the vector $\left\{\frac{1}{G^-}\mathbf{A}\ \mathbf{K}_j + \mathbf{F}_j^-\right\}_+$

and $\{C_p\}$ is a new set of unknown coefficients which must be determined. In the actual solution of a problem, Eq. (2.30) is a more convenient form for the optimum transfer functions.

It has been stated in the previous section that the optimum transfer function vector must satisfy the transformed Wiener-Hopf equation as given by Eq. (2.21). If we now substitute Eq. (2.30) in Eq. (2.21) and equate the residues associated with the same left half-plane poles on both sides of the equation, we obtain a set of linear algebraic equations in terms of these coefficients $\{C_p\}$. The number of linear independent equations thus derived will determine the number of the total unknown coefficients to be used in the transfer function vector. The distribution of these coefficients among the components of the transfer function vector can be determined from the form of the spectral density matrix $\mathbf{G}$ such that the inverse transform of $\{\mathbf{G}\mathbf{Y}_j\}_+$ will be a well-behaved time-function vector. It should be noted that the optimum transfer function vector determined must give a finite mean-square error as expressed by Eq. (2.23). This will give us a general guide as to the maximum number of unknown coefficients that can be included for each transfer function. This procedure will be amplified by the example given in the next section.

In conclusion, the major steps for determining the optimum transfer function vector $\mathbf{Y}_j$ of the multivariable filtering problem using the method of undetermining coefficients are as follows:

1. Factor the spectral density determinant $|\mathbf{G}|$ into $G^+(s)$ and $G^-(s)$.

2. Express the system transfer functions in the form given in Eq. (2.30).

3. Equate the residues associated with the same left half-plane poles on both sides of Eq. (2.21) and count the number of independent linear equations thus obtained. This determines the total number of unknown polynomial coefficients to be used in the numerators of the transfer functions.

4. Make a proper distribution of these coefficients among the components of the transfer function vector and then solve the resultant linear algebraic equation.

From Eq. (2.30), it is seen that the poles of the transfer functions will consist of two parts. First, all the system transfer functions will have poles which are zeros of $G^+(s)$. These poles, in general, are completely different from the left half-plane poles of all the spectral densities. Second, the transfer functions associated with a particular output terminal may contain the left half-plane poles from the ideal transfer functions vector $(\mathbf{Y}_d)_j$.

It should be noted that for the single-variable filtering problem, the matrix Λ is equal to 1. Hence, Eq. (2.27) is reduced to

$$G^+ Y = \left\{ \frac{K}{G^-} \right\}_+$$

Note that $\left\{ \frac{F^-}{G^-} \right\}_+ \equiv 0.$

2.6 An Illustrative Example for the Synthesis of Multivariable Filters

As a simple illustration of the procedure thus far presented, let a system with two inputs and one output be chosen. The signal portions of the inputs are assumed to be the same. However, they are corrupted by different white noises. Let

$$G_{SS}(s) = \frac{1}{-s^2 + 1}$$

$$G_{N_1 N_1}(s) = \frac{1}{2}$$

$$G_{N_2 N_2}(s) = \frac{1}{4}$$

The function of the filter is to extract the signal from these two different channels (Fig. 2.2).

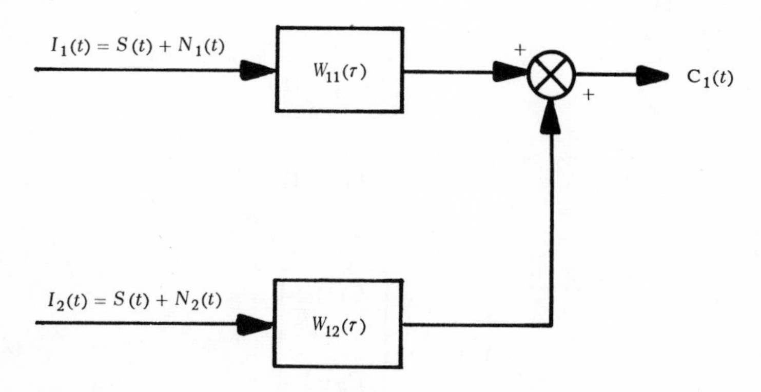

FIG. 2-2. Variable 2 x 1 filter for the example.

For this problem, the following matrices are obtained

$$\mathbf{G} = \begin{bmatrix} \dfrac{-s^2+3}{2(-s^2+1)} & \dfrac{1}{-s^2+1} \\ \dfrac{1}{-s^2+1} & \dfrac{-s^2+5}{4(-s^2+1)} \end{bmatrix} \tag{2.31}$$

$$\mathbf{K}_1 = \begin{bmatrix} \dfrac{1}{-s^2+1} \\ \dfrac{1}{-s^2+1} \end{bmatrix} \tag{2.32}$$

$$\mathbf{A} = \begin{bmatrix} \dfrac{-s^2+5}{4(-s^2+1)} & \dfrac{-1}{-s^2+1} \\ \dfrac{-1}{-s^2+1} & \dfrac{-s^2+3}{2(-s^2+1)} \end{bmatrix} \tag{2.33}$$

and

$$\mathbf{Y}_{d\,1} = \begin{bmatrix} \dfrac{1}{2} \\ \dfrac{1}{2} \end{bmatrix}$$

Thus

$$G^+(s) = \frac{s+\sqrt{7}}{\sqrt{8}\,(s+1)}$$

$$G^-(s) = \frac{s-\sqrt{7}}{\sqrt{8}\,(s-1)}$$

Let

$$\mathbf{Y}_1(s) = \begin{bmatrix} \dfrac{P_1(s)}{s+\sqrt{7}} \\ \dfrac{P_2(s)}{s+\sqrt{7}} \end{bmatrix} \tag{2.34}$$

Here $P_1(s)$ and $P_2(s)$ are two polynomials whose degrees have to be determined. By substituting Eqs. (2.31), (2.32), and (2.34) in Eq. (2.21), we have

$$\begin{Bmatrix} \dfrac{-s^2+3}{2(-s^2+1)} \dfrac{P_1(s)}{s+\sqrt{7}} + \dfrac{1}{-s^2+1} \dfrac{P_2(s)}{s+\sqrt{7}} \\[2ex] \dfrac{1}{-s^2+1} \dfrac{P_1(s)}{s+\sqrt{7}} + \dfrac{-s^2+5}{4(-s^2+1)} \dfrac{P_2(s)}{s+\sqrt{7}} \end{Bmatrix}_+ = \begin{Bmatrix} \dfrac{1}{-s^2+1} \\[2ex] \dfrac{1}{-s^2+1} \end{Bmatrix}_+$$

Evaluating and equating the residues at the poles $s = -1$ and $s = -\sqrt{7}$ from the above vector equation, we have four equations in terms of $P_1(s)$ and $P_2(s)$.

$$\frac{2}{2 \times 2} \frac{P_1(-1)}{-1+\sqrt{7}} + \frac{1}{2} \frac{P_2(-1)}{-1+\sqrt{7}} = \frac{1}{2} \tag{2.35}$$

$$\frac{1}{2} \frac{P_1(-1)}{-1+\sqrt{7}} + \frac{4}{4 \times 2} \frac{P_2(-1)}{-1+\sqrt{7}} = \frac{1}{2} \tag{2.36}$$

$$\frac{4}{2 \times 6} P_1(-\sqrt{7}) - \frac{1}{6} P_2(-\sqrt{7}) = 0 \tag{2.37}$$

$$-\frac{1}{6} P_1(-\sqrt{7}) + \frac{2}{4 \times 6} P_2(-\sqrt{7}) = 0 \tag{2.38}$$

It is easily seen that Eqs. (2.35) and (2.36), and Eqs. (2.37) and (2.38) are dependent. Hence there are only two independent equations, and $P_1(s)$ and $P_2(s)$ will be chosen simply as constants. Thus we have the simultaneous equations

$$\frac{C_1}{-1+\sqrt{7}} + \frac{C_2}{-1+\sqrt{7}} = 1$$

$$2C_1 - C_2 = 0$$

They give the results

$$C_1 = \frac{-1+\sqrt{7}}{3} = \frac{2}{1+\sqrt{7}} = P_1(s)$$

$$C_2 = \frac{4}{1+\sqrt{7}} = P_2(s)$$

Hence these two components $Y_{11}(s)$ and $Y_{12}(s)$ of the filter have the same cutoff frequency. However, their gains are inversely proportional to the noise power of their channels. This result is what we would expect. The mean-square error can be evaluated by using Eq. (2.23) as

$$E\{\epsilon_1^2\}_{min} = \frac{1}{2}\left[1 - \frac{6}{(1+\sqrt{7})^2}\right] = 0.2743$$

It will be interesting to compare the performance of the multivariable filter considered above with the performances of the single-variable filters optimal with respect to the two channels separately. For the first channel alone, the transfer function of the optimal filter is

$$Y_1(s) = \frac{2}{1+\sqrt{3}} \frac{1}{s+\sqrt{3}}$$

and its corresponding mean-square error is

$$E\{\epsilon^2\}_{min} = \frac{1}{2}\left[1 - \frac{2}{(1+\sqrt{3})^2}\right] = 0.3660$$

For the second channel alone, the transfer function of the optimal filter is

$$Y_2(s) = \frac{4}{1+\sqrt{5}} \frac{1}{s+\sqrt{5}}$$

and its corresponding mean-square error is

$$E\{\epsilon^2\}_{min} = \frac{1}{2}\left[1 - \frac{4}{(1+\sqrt{5})^2}\right] = 0.3090$$

It is evident, now, that the performance of the multivariable filter, where two input channels are available simultaneously, is better than the performances of the two single-variable filters.

2.7 Solution of the Optimum Transfer Function Matrix by the Spectral Density Matrix Factorization Method

It has been shown in the previous section that in the transform domain, the Wiener-Hopf equation takes the form given by Eq. (2.20)

$$\mathbf{G}\mathbf{Y}_j = \mathbf{K}_j + \mathbf{F}_j^-$$

To find a physically realizable transfer function vector which satisfies the above equation, another approach can be taken. We shall show that this approach essentially reduces the factorization of the input spectral density matrix G to the product of two matrices with some specific properties.

For a rational matrix to be a legitimate power spectral density matrix, it must satisfy the three basic properties: P1, P2, and P3, as stated in Sec. 2.4. We have also shown that, to obtain the solution of the Wiener-Hopf integral vector equation by using the transform method, an additional requirement must be imposed on the power spectral density matrix. This is property P4, specifying that the inverse of the power spectral density matrix must also be analytic along the imaginary axis of the s-plane. Suppose now that the spectral density matrix can be factored in the form

$$\mathrm{G}(s) = \mathrm{H}^*(s)\,\mathrm{H}(s) \tag{2.39}$$

where the asterisk denotes, as usual, the transpose of a matrix and the change of its argument from s to $-s$. Here we stipulate that the matrix $\mathrm{H}(s)$ must have the properties:

T1. $\mathrm{H}(s)$ is rational and analytic together with its inverse $\mathrm{H}^{-1}(s)$ in the right half of the s-plane.

T2. $\mathrm{H}(s)$ is real.

It is evident, than, that $\mathrm{H}^*(s)$ as well as its inverse $(\mathrm{H}^*)^{-1}(s)$ will be analytic in the left half-plane, and thus can have only poles in the right half-plane.

Let us now substitute Eq. (2.39) in Eq. (2.20). We have

$$\mathrm{H}^*\mathrm{H}\,\mathrm{Y}_j = \mathrm{K}_j + \mathrm{F}_j^-$$

Thus

$$\mathrm{H}\,\mathrm{Y}_j = (\mathrm{H}^*)^{-1}\mathrm{K}_j + (\mathrm{H}^*)^{-1}\mathrm{F}_j^- \tag{2.40}$$

Since both $(\mathrm{H}^*)^{-1}$ and F_j^- are analytic in the left half-plane, then

$$\left\{(\mathrm{H}^*)^{-1}\mathrm{F}_j^-\right\}_+ \equiv 0 \tag{2.41}$$

Therefore we obtain, from Eq. (2.40), that

$$\mathrm{H}\,\mathrm{Y}_j = \left\{(\mathrm{H}^*)^{-1}\mathrm{K}_j\right\}_+$$

or

$$\mathrm{Y}_j = \mathrm{H}^{-1}\left\{(\mathrm{H}^*)^{-1}\mathrm{K}_j\right\}_+ \tag{2.42}$$

Equations (2.41) and (2.42) indicate clearly the importance of the requirement that $\mathrm{H}(s)$ as well as $\mathrm{H}^{-1}(s)$ be analytic in the right half-plane. The filter thus obtained will be physically realizable and stable. We have shown, then, that by using the second approach for solving a multivariable filtering problem, it suffices to exhibit a factorization of the rational spectral density matrix $G(s)$ as given by Eq. (2.39). Once this factorization is completed, the transfer function matrix of the optimum filter can readily be obtained by using Eq. (2.42).

2.8 Factorization of the Rational Spectral Density Matrix

The problem of factoring a rational spectral density matrix has recently been the concern of many investigators. The first workable algorithm for affecting such a decomposition in a closed form solution was developed by Youla [10]. His method is centered around the Smith canonic form of a polynomial matrix. A rigorous proof has been given that the factorization of such a rational spectral density matrix is always possible. The factorization procedure is strictly algebraic in nature. Unfortunately, the steps involved are quite complicated.

It should be noted that the decomposition of the spectral density matrix as given by Eq. (2.39) can be physically interpreted as the equivalent of synthesizing a multivariable shaping filter $\Phi(s)$, with the additional requirement that $\Phi^{-1}(s)$ is also analytic in the right-half plane. When this system is excited by a vector white noise source, the output spectral density matrix is

$$G(s) = \Phi(-s)\, I\, \Phi'(s) = \Phi(-s)\, \Phi'(s) \tag{2.43}$$

Here the identity matrix is the input spectrum of the white noise and $G(s)$ is the output spectrum. By comparing Eqs. (2.39) and (2.43), we have

$$H(s) = \Phi'(s)$$

Thus we can take the spectral density matrix factorization problem as the "Strengthen Shaping Filter Problem." However, it should be noted that in solving a filtering problem, the synthesis of the shaping filter is just an artifice. The result will not be explicitly used. Through this observation, Davis [11] recently proposed a factorization method using a series of matrix transformations. His method is intuitive in nature and involves very little of advanced matrix theory. The following development adheres very closely to his original paper.

Let us assume that a stable system $\Phi(s)$ is excited by a vector white noise source. If now a series of linear systems $T_1(s), \ldots,$ $T_n(s)$ are cascaded with $\Phi(s)$, and if the resulting spectral density matrix is a unity matrix, then the whole cascaded system must be the inverse of the desired system $\Phi(s)$. This situation is shown in Fig. 2-3, and the argument given above constitutes the basic idea for factoring the spectral matrix. Thus we have

$$\Phi^{-1}(s) = I_n(s)\, T_{n-1}(s) \cdots T_1(s)$$

and

$$\Phi(s) = I_1^{-1}(s) \cdots T_{n-1}^{-1}(s)\, T_n^{-1}(s) \tag{2.44}$$

Since $\Phi^{-1}(s)$ is the inverse shaping filter, then

$$\left[T_n(-s)\, T_{n-1}(-s) \cdots T_1(-s)\right] G(s) \left[T_1'(s) \cdots T_{n-1}'(s)\, T_n'(s)\right] = I \tag{2.45}$$

Hence we have

$$G(s) = \left[T_1^{-1}(-s) \cdots T_{n-1}^{-1}(-s)\, T_n^{-1}(-s)\right] \left[T_1^{-1}(s) \cdots T_{n-1}^{-1}(s)\, T_n^{-1}(s)\right]'$$
$$= \Phi(-s)\, \Phi'(s)$$

Due to the requirement that both $\Phi(s)$ and $\Phi^{-1}(s)$ must be stable, care should be taken in choosing each component of the inverse shaping filter. This will obviously be satisfied if each component

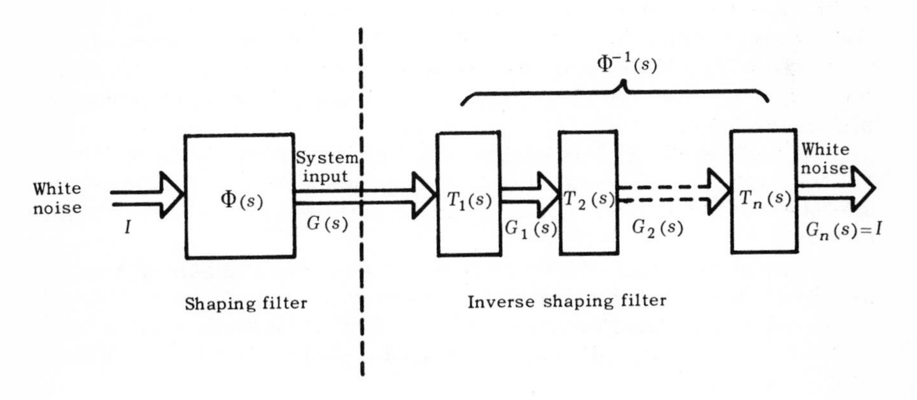

FIG. 2-3. The shaping and inverse shaping filters.

$T_i(s)$ and its inverse $T_i^{-1}(s)$ are individually stable. Thus we shall restrict the structure for each of the linear systems $T_i(s)$ to one of two simple types:

1. $T_i(s)$ is an identity matrix except for one or more jth diagonal elements $t_{jj}(s)$. The inverse of this matrix is another diagonal matrix with jth element $1/t_{jj}(s)$. In this case, the stability of $T_i(s)$ as well as $T_i^{-1}(s)$ is assured if all the elements $t_{jj}(s)$ are both stable and of minimum phase.

2. $T_i(s)$ is an identity matrix, except for the off-diagonal elements of the nth row, $t_{nj}(s)$. The inverse of this matrix is identical with $T_i(s)$ except for a possible reversal of the sign of $t_{nj}(s)$. For stable $T_i(s)$ and $T_i^{-1}(s)$, all elements $t_{nj}(s)$ must be stable but not necessarily of minimum phase.

By referring to Fig. 2-3 and Eq. (2.45), it is evident that each component of the inverse shaping filter will make the transformation from its input to its output spectrum according to the relationship

$$G_i(s) = T_i(s)\,G_{i-1}(s)\,T_i'(s) \qquad i = 1, 2, \ldots, n \tag{2.46}$$

where

$$G_0(s) \triangleq G(s)$$
$$G_n(s) = I$$

Using Eq. (2.46) repeatedly, it will yield the result given in Eq. (2.45). Now let us first consider the effect of the transformation by using the type 1 component. The premultiplication of $T_i(-s)$ is equivalent to multiplying the jth row of $G_{i-1}(s)$ by $t_{jj}(-s)$, and the post-multiplication of $T_i(s)$ is equivalent to multiplying its jth column by $t_{jj}(s)$. For the type 2 component, the transformation is seen to consist of multiplying the jth row of $G_{i-1}(s)$ by $t_{nj}(-s)$ and then forming the nth row elements of $G_i(s)$ by adding these terms. A similar operation will be performed on the columns by using $t_{nj}(s)$. Note that the determinant of this type 2 transformation matrix is unity.

With all these specified properties concerning the components of the inverse shaping filter, the procedure for factoring the spectral density matrix can then be carried out in three major phases:

1. Pole-Removal Phase: all poles in every element of the input spectral density matrix will be removed.
2. Determinant Reduction Phase: the resultant matrix with polynomial elements in s will be transformed until its determinant is a constant.
3. Elements Order Reduction Phase: this resultant matrix will be further reduced to a constant matrix which can be readily factored.

These three steps for carrying out the factorization will be given in detail accompanied by an example.

The first phase of the solution is to remove all the poles of each element in the input spectral density matrix $G(s)$ and to leave this matrix with polynomial elements only. By using the type 1 transformation, the rows and columns of $G(s)$ can be multiplied by polynomial factors which will cancel all the poles in the particular row and column. Since $T_1(s)$ and $T_1^{-1}(s)$ are required to be stable, the row multiplication performed by $T_1(-s)$ must remove the right half-plane poles, and a column multiplication performed by $T_1'(s)$ must remove the left half-plane poles. Since the spectral matrix is para-Hermitian, it is enough to consider only the row operations.

As an illustrative example, let us consider a spectral matrix given by

$$G(s) = \begin{bmatrix} \dfrac{-2s^2 + 5}{(-s+1)(-s+2)(s+1)(s+2)} & \dfrac{-2s^2 - 5s + 13}{(-s+1)(-s+2)(s+3)(s+5)} \\ \\ \dfrac{-2s^2 + 5s + 13}{(-s+3)(-s+5)(s+1)(s+2)} & \dfrac{-2s^2 + 34}{(-s+3)(-s+5)(s+3)(s+5)} \end{bmatrix} \tag{2.47}$$

Thus, in the first phase of our synthesis procedure, the first row of the spectral matrix must be multiplied by $(-s + 1)(-s + 2)$ and the second row, by $(-s + 3)(-s + 5)$. This means that we shall have for $T_1(-s)$

$$T_1(-s) = \begin{bmatrix} (-s+1)(-s+2) & 0 \\ \\ 0 & (-s+3)(-s+5) \end{bmatrix}$$

The resultant spectral matrix is

$$\begin{aligned} G_1(s) &= T_1(-s)\,G(s)\,T_1'(s) \\ &= \begin{bmatrix} -2s^2 + 5 & -2s^2 - 5s + 13 \\ \\ -2s^2 + 5s + 13 & -2s^2 + 34 \end{bmatrix} \end{aligned} \tag{2.48}$$

The purpose of the second phase of our procedure is to transform the matrix of polynomial elements $G_1(s)$ into another matrix of polynomial elements with a constant determinant independent of s. Suppose now that the determinant of $G_1(s)$ is factored in the general form

$$|G_1(s)| = k\prod_j (s + a_j)(-s + a_j)$$

Here a_j are positive numbers or complex-conjugate pairs with positive real parts and they can be repeated. If a multiplying transformation $T_2(s)$ of type 1 with an nth diagonal element of $1/s+a_i$ is used, then the determinant of the transformed spectrum $G_2(s)$ will be given by

$$|G_2(s)| = k\prod_{j \neq i} (s + a_j)(-s + a_j)$$

Thus a factor $(s + a_i)(-s + a_i)$ has been removed from $|G_1(s)|$. However, the resultant matrix $G_2(s)$ is no longer a polynomial matrix because of the pole terms $1/s+a_i$ in the nth row and $1/-s+a_i$ in the nth column. The next step now is to remove this pole from each term without altering the determinant of the matrix. We also notice that it is enough to consider the row operation only.

Let us now consider a typical term in the nth row of $G_2(s)$. It can be expressed in the form

$$\frac{t_{nj}(s)}{-s+a_i} = P_{nj}(s) + \frac{a_{nj}}{-s+a_i} \tag{2.49}$$

where

$$a_{nj} = t_{nj}(a_i)$$

Our objective is then to remove these terms $\frac{a_{nj}}{-s+a_i}$ without affecting the determinant. A proper choice for the transformation $T_3(s)$ will be

$$T_3(-s) = \begin{bmatrix} 1 & & & & & \\ & 1 & & & 0 & \\ & & \ddots & & & \\ & 0 & & \ddots & & \\ & & & & \ddots & \\ \frac{k_1}{-s+a_i} & \frac{k_2}{-s+a_i} & \cdots\cdots & \frac{k_{n-1}}{-s+a_i} & & 1 \end{bmatrix}$$

It has a determinant of 1. By investigating the nth row terms of the matrix $G_3(s) = T_3(-s)\,G_2(s)\,T_3'(s)$, it is obvious that to eliminate the pole terms the coefficients $k_1, k_2, \ldots, k_{n-1}$ must satisfy the algebraic equations

$$\sum_{r-1}^{n-1} k_r t_{rj}(a_i) + t_{nj}(a_i) = 0 \qquad j = 1, 2, \ldots, n \tag{2.50}$$

It should be noted that in these n simultaneous equations, only $(n-1)$ equations are independent since

$$|G_2(a_i)| = 0$$

Hence the set of coefficients $\{k_r\}$ can be uniquely determined. Obviously, this procedure can be repeated until the resultant matrix has constant determinant and polynomial elements. Thus, for the example considered

$$|G_1(s)| = (-s+1)(s+1)$$

Hence $T_2(s)$ is chosen as

$$T_2(-s) = \begin{bmatrix} 1 & 0 \\ 0 & \frac{1}{-s+1} \end{bmatrix}$$

which multiplies the second row of $G_1(s)$ by $\frac{1}{-s+1}$. The second row and first column element of $G_2(s)$ is

$$\frac{-2s^2 + 5s + 13}{-s+1} = 2s - 3 + \frac{16}{-s+1}$$

The transformation $T_3(-s)$ will then be taken as the form

$$T_3(-s) = \begin{bmatrix} 1 & 0 \\ \frac{k}{-s+1} & 1 \end{bmatrix}$$

which must remove the unwanted pole. Thus we have

$$k\left[-2s^2 + 5\right]_{s=1} + 16 = 0$$

or

$$k = -\frac{16}{3}$$

Note that this number k can also be determined from the equation

$$k \quad -2s^2 - 5s + 13 \quad _{s=1} + \quad -2s^2 + 34 \quad _{s=1} = 0$$

After these two successive transformations, we shall have

$$G_3(s) = T_3(-s)\,T_2(-s)\,G_1(s)\,T_2'(s)\,T_3'(s)$$

$$= \begin{bmatrix} -2s^2 + 5 & \frac{26}{3}s - \frac{41}{3} \\ -\frac{26}{3}s - \frac{41}{3} & \frac{(26)^2}{18} \end{bmatrix} \tag{2.51}$$

and

$$|G_3(s)| = 1$$

In the final phase of our procedure, the matrix of polynomial elements with a determinant independent of s will undergo a series of transformations to remove all the powers of s from each element and obtain a numerical matrix. This constant matrix can then be factored very easily. Since the determinant is just a constant, it is evident that expansion of the determinant for the matrix must

must have zero coefficients for all the terms with nonzero exponents of s. Let us consider the terms of the highest powers of s in each element of the matrix. We form an array by using these terms and replacing all the noncontributing terms by zero. As an example, let these highest terms be

$$\begin{matrix} -2s^6 & 2s^5 & 7s^2 \\ -2s^5 & 2s^4 & 3s^2 \\ 7s^2 & 3s^2 & -2s^2 \end{matrix}$$

The form of this array is proper since the original matrix is para-Hermitian. The highest possible power of s for the determinant of this array, and consequently, also for the original matrix, is s^{12}. Replacing terms which do not contribute to the coefficient of s^{12} by zero yields a new array

$$\begin{matrix} -2s^6 & 2s^5 & 0 \\ -2s^5 & 2s^4 & 0 \\ 0 & 0 & -2s^2 \end{matrix}$$

Since the coefficient for the term s^{12} must be zero, the determinant of the above array will accordingly be zero. This means that at least two rows are linearly dependent. In the example given by the above array, the first row can be removed by multiplying the second row by $-s$ and adding it to the first row. This can be accomplished by a premultiplication of the type 2 transformation $T(-s)$

$$T(-s) = \begin{bmatrix} 1 & -s & 0 \\ 0 & 1 & 0 \\ 0 & 0 & 1 \end{bmatrix}$$

Thus, after the completion of premultiplication and post-multiplication by $T(-s)$ and $T'(s)$ respectively, the highest possible power for s in the original matrix will be reduced by two. This procedure will be repeated until each element becomes a constant.

Let us now return to the main illustrative example considered. Referring to Eq. (2.51), the desired transformation matrix is obviously given by

$$T_4(-s) = \begin{bmatrix} 1 & -\frac{3}{13}s \\ 0 & 1 \end{bmatrix}$$

Thus we have

$$G_4(s) = T_4(-s)\,G_3(s)\,T_4'(s) = \begin{bmatrix} 5 & -\frac{41}{3} \\ -\frac{41}{3} & \frac{(26)^2}{18} \end{bmatrix} \triangleq C$$

The final step will then be the factorization of a constant matrix. Since this matrix is symmetric and positive definite, it is always possible to factor it into the product of two nonsingular triangular matrices as

$$C = N N'$$

where N is a lower-triangular constant matrix. This factorization is unique up to the post-multiplication of N by a unitary constant matrix U since

$$NU \cdot (NU)' = NN'$$

where

$$UU' = I$$

Hence for our example,

$$\begin{bmatrix} 5 & -\frac{41}{3} \\ -\frac{41}{3} & \frac{(26)^2}{18} \end{bmatrix} = \begin{bmatrix} n_{11} & 0 \\ n_{21} & n_{22} \end{bmatrix} \begin{bmatrix} n_{11} & n_{21} \\ 0 & n_{22} \end{bmatrix}$$

Solving the above equation successively yields

$$N = \begin{bmatrix} \sqrt{5} & 0 \\ -\frac{41}{3\sqrt{5}} & \frac{1}{\sqrt{5}} \end{bmatrix}$$

Since

$$G_5(s) = I = T_5 G_4 T_4'$$

we have

$$T_5 = N^{-1}$$

Hence the overall solution for $\Phi(s)$ as given by Eq. (2.44) is

$$\Phi(s) = T_1^{-1}(s), \ldots.. T_4^{-1}(s) T_5^{-1}$$

$$= \frac{1}{13\sqrt{5}} \begin{bmatrix} \dfrac{41s + 65}{(s+1)(s+2)} & \dfrac{-3s}{(s+1)(s+2)} \\ \\ \dfrac{41s + 169}{(s+3)(s+5)} & \dfrac{-3s + 13}{(s+3)(s+5)} \end{bmatrix} \tag{2.52}$$

The procedure for factoring a given rational spectral density matrix is now complete. It should be noted that, post-multiplying $\Phi(s)$ by any arbitrary unitary matrix will still give the right answer to the problem since Eq. (2.43) remains unchanged. It can also be easily shown that Eq. (2.42) is not affected by this unitary transform. Thus the optimum transfer function matrix obtained is certainly unique.

2.9 Discussion of the Solutions for Optimum Multivariable Filters

The synthesis of optimum multivariable filters with stationary vector input has been presented in two different approaches. In the method of undetermined coefficients, the procedure is rather simple. A form for the optimum transfer function matrix is assumed and the unknown coefficients associated with the polynomials in the numerators of its elements are then determined by solving a set of linear algebraic equations. However, this method is somewhat implicit in nature. The distribution of the unknown coefficients among each element of the transfer function matrix demands keen observation.

In the second method, based on the factorization of the input spectral density matrix, the solution obtained is explicit in nature. Nevertheless, the steps taken for carrying out the factorization are somewhat involved. It should be noted that when the requirement of stable $H^{-1}(s)$ is removed, and we thus consider strictly a shaping filter problem, the factorization of the input spectral matrix can be carried out in a different and simpler way. Youla [10] has shown that the matrix $\mathbf{H}(s)$ in this case will be an upper-triangular matrix. Maytyash and Shilkhanek [13] have given a simple way for accomplishing this factorization by introducing an all-pass factor in each element of $\Phi(s)$.

2.10 Optimum Synthesis of Multivariable Semi-Free Configuration Control Systems

The essential difference between a filtering problem and a control problem lies in the degree of freedom in choosing the configuration of the systems. In filtering problems, the optimum transfer functions of the system are completely determined by the input processes through the minimization procedure. However, in the control problem, a fixed plant has to be included in the system to perform certain tasks. Thus compensation must be introduced to the system in conjunction with the fixed plant such that the overall performance of the system is optimal in some sense. The system transfer function now depends not only on the input processes but also on certain properties of the fixed plant, such as whether or not it is a nonminimum phase. In multivariable systems, a nonminimum phase plant is a system whose transfer function determinant has zeros in the right half of the s-plane. This kind of control problem is usually referred to as the problem of semifree configuration.

The multivariable plants under consideration are linear time-invariant systems and are also stable [6]. The inputs to the system will consist of additive stationary random signals and noise. Thus the overall systems are time-invariant. It is often necessary that the systems should contain feedback loops between inputs and outputs. However, if an optimum system with cascade combination of controller and fixed plant is obtained, its feedback configuration can easily be obtained. Thus the main problem is now to synthesize this cascade controller.

The cascade configuration of the system is shown in Fig. 2-4. Here $\mathbf{Q}(\tau)$ is the $n \times n$ weighting function matrix of the controller, and $\mathbf{P}(\tau)$ is the $m \times n$ weighting function matrix of the fixed plant. $\mathbf{I}(t)$ is an n-dimensional system input vector, $\mathbf{r}(t)$ is an n-dimensional control input vector, and $\mathbf{C}(t)$ is an m-dimensional system output vector. For most of the practical problems, the number of system inputs is usually equal to or greater than the number of outputs; i.e., $n \geqq m$. Thus we have

$$\mathbf{r}(t) = \int_0^\infty \mathbf{Q}(\tau)\mathbf{I}(t-\tau)\,d\tau \tag{2.53}$$

and

$$C(t) = \int_0^{\infty} P(\tau) r(t - \tau) d\tau \tag{2.54}$$

Combination of Eqs. (2.53) and (2.54) gives

$$C(t) = \int_0^{\infty} P(\tau_2) d\tau_2 \int_0^{\infty} Q(\tau_1) I(t - \tau_1 - \tau_2) d\tau_1 \tag{2.55}$$

It is quite evident from Fig. 2-4 that each system output is closely related to the entire cascade controller. Hence we have to minimize the square of the norm of the error vector.

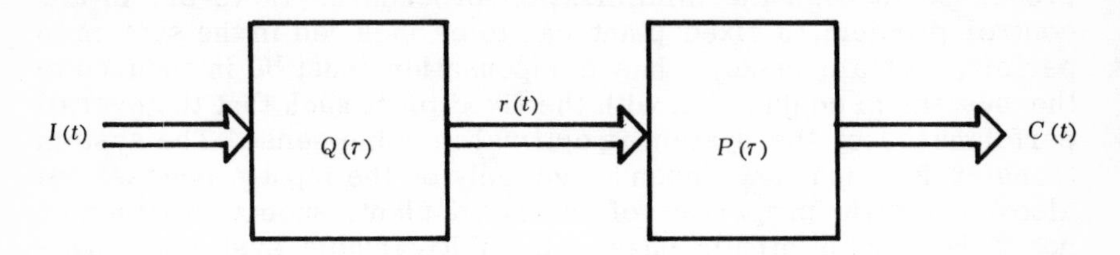

FIG. 2-4. System diagram with cascade controller.

It can easily be shown [6] that, in the frequency domain, the generalized Wiener-Hopf equation for this semifree configuration control problem is

$$\left\{\sum_{j=1}^{m} \sum_{f=1}^{n} \sum_{k=1}^{n} P_{jf'}{}^{*} P_{jf} G_{I_{k'} I_k} Q_{fk}\right\}_{+} = \left\{\sum_{j=1}^{m} \sum_{k=1}^{n} P_{jf'}{}^{*} (Y_d)_{jk} G_{I_{k'} S_k}\right\}_{+}$$

$$f' = 1, 2, \ldots n$$
$$k' = 1, 2, \ldots n \tag{2.56}$$

Here $P_{jf}(s)$ = plant transfer function
$Q_{fk}(s)$ = controller transfer function
$G_{I_{k'} I_k}(s)$ = input spectral density
$G_{I_{k'} S_k}(s)$ = input and signal spectral density
$(Y_d)_{jk}(s)$ = ideal system transfer function
$P_{jf}{}^{*}(s) = P_{jf}(-s)$

It is evident that the $n^2 \times n^2$ matrix N whose elements are of the form

$$N_{f'k'fk} = \left\{\sum_{j=1}^{m} P_{jf'}{}^{*} P_{jf}\right\} G_{I_k I_{k'}} \tag{2.57}$$

is also para-Hermitian. Here the fk are used to number the columns of the N matrix, and the $f'k'$ are used to number its rows.

It is quite evident now that the techniques developed in the previous sections can well be applied to synthesize the controller. For most control problems, the number of inputs is equal to or greater than that of the outputs ($n \geq m$). To convert the cascade configuration to the feedback configuration, the canonical form of the multipole system suggested by Freeman [14] can be used. The synthesis of the forward and feedback controllers can best be carried out by considering the sensitivity problem for the overall system as investigated by Horowitz [15].

For a system with an equal number of inputs and outputs, if the plant is of minimum phase such that the determinant of its transfer function matrix has no zeros in the right half-plane, the controller matrix can easily be obtained as

$$\mathbf{Q} = \mathbf{P}^{-1}\mathbf{Y}$$

where Y is the optimum system transfer function matrix with the plant absence, i.e., the solution of the corresponding filtering problem. For a single-variable system, this corresponding fact is well known [16].

PART II Additional Topics

2.11 Techniques for the Description of Random Processes

Fundamental to the analysis and synthesis of any control system with random or stochastic inputs is the development of a suitable mathematical description of the random process. The techniques available for the determination of such a description fall naturally into two principle categories: experimental, or mathematical analysis. In the first category, various experimental techniques are applied to determine the power spectral densities or the appropriate correlation functions of the particular random process involved. Because, in general, it is possible to measure only samples of the random processes involved, and because of limitations in the experimental measuring devices themselves, certain data-processing techniques have to be developed and applied together with experimental approaches to the determination of power spectral densities or correlation functions [4, 17].

With regard to the second category of the determination, mathematical description of a random process through mathematical analysis, this method of approach means establishing an appropriate mathematical model of the physical random process involved. This model may have to be simplified in a meaningful manner to become more amenable to analysis. Thus, in the open literature at least, this approach has been principally presented for rather basic physical situations such as thermal, shot, and impulse noise [18]. The analyses of rather complex physical systems available in the open literature have tended to concentrate mainly on radar systems [19, 20] and guidance systems [21, 22, 23, 24]. The reason for this, of course, is that these are classes of systems of wide applicability, and the development of such analyses is dictated by necessity. References [25] through [48] provide additional sources of experimental and analytical techniques for the determination of random processes. In addition, reference [38] as well as the references listed there presents analytical techniques for determining the effect of rather complex feedback systems with nonlinear processing elements on the mathematical description of input random processes.

The question naturally arises as to the relative merties of the experimental and mathematical analysis approaches to the determination of the descriptive parameters of a random process, and what connection exists between the two. Some physical systems may be so complex or indeterminate as to preclude the possibility of meaningful mathematical analysis for determination of the random processes. A possible example here is the determination of windspeed correlation functions useful in the analysis and synthesis of velocity-damped inertial navigation systems [49]. Another is the determination of the power spectral density of atmospheric turbulence for use in the structural design of aircraft [50]. Yet another is the determination of power spectra of ocean waves for possible use in boat stabilization system design [51]. On the other hand, in some instances it might not be practical to conduct experimental measurements. For instance, in the determination of radar noise associated with tracking targets, experimental measurements are readily possible. However, if it is desired to vary the target shape and dimensions to study their effect, economy and the need for clarity dictate the development of the mathematical analysis of an appropriate mathematical model of the system [20]. Thus, it is clear that experimental and mathematical techniques can either complement one another or be used individually when only one is feasible.

This section concentrates on the particular area of the experimental techniques for correlation functions because of their fundamental importance to the synthesis techniques presented in Part I as well as in the remainder of this chapter and the next. Nothing more will be said here of the technique of determining descriptive parameters of random processes through the analysis of mathematical models [19, 20]. The important question of experimental techniques for power spectral density measurements [27] for individual random processes or cross-spectral densities [53], for pairs of random processes and their various subtopics such as aliasing, prewhitening, resolution problems, uncertainty principles, sampling theorems [53], error estimates, etc., will also be ignored here although these are also quite important for the synthesis techniques to be presented later.

One point must be made with respect to the questions of stationarity and ergodicity [2]. Resting on the assumption of ergodicity, the techniques for processing experimental data for determining the parameters of a random process have tended to emphasize the use of time averages of members of an ensemble rather than ensemble averages [2]. Needless to say, not all random processes are approximately stationary and ergodic. Strictly speaking, there is no such thing in the physical world as a stationary ergodic process. Such processes are mathematical artifices which are nevertheless reasonably good approximations in many physical situations. The question of determining when a real physical process can, to a good degree of approximation, be assumed stationary and erogic rests on a combination of experimental evidence and physical knowledge of the situation involved. For example, without conducting any experimental measurements, one can readily see that the input noise to a satellite tracking radar is essentially stationary and ergodic for a synchronous satellite. On the other hand, the radar or IR (infrared) noise input to an active, self-contained guidance system for an air-to-air or ground-to-air missile can be readily visualized as nonstationary [2] in character. There are certain mathematical conditions which can be examined to determine whether or not a random process is ergodic, as indicated by the Ergodic Theorem [54]. But this may require a considerable amount of work demanding an explicit mathematical definition of the random process involved. Thus, it is in general difficult to use this theorem for any complex physical system. Its use would most naturally occur in the analysis of a mathematical artifice reasonably approximating physical situations. Experimental techniques can be utilized with regard to nonstationary random processes. However, the fact remains that now the practical difficulty exists of evaluating ensemble averages rather than time averages. Thus, where it can be feasibly carried out, mathematical analysis of an appropriate physical model appears that much more attractive for nonstationary random processes.

Considering now the experimental determination of correlation functions, the basic definition for the correlation function of two random processes whose ensemble members will be designated as $x(t)$ and $y(t)$ is given as the ensemble average or [2]

$$\emptyset_{xy}(t_1, t_2) = \int_{-\infty}^{\infty}\int_{-\infty}^{\infty} x(t_1)y(t_1)f(x, t_1; y, t_2)\,dxdy \qquad (2.58)$$

where $f(x, t_1; y, t_2)$ is the joint probability density function for $x(t_1)$ and $y(t_2)$. When the random processes involved are stationary

$$\phi_{xy}(t_1, t_2) = \phi_{xy}(t_2 - t_1) = \phi_{xy}(\tau) \tag{2.59}$$

When they are both stationary and ergodic

$$\phi_{xy}(\tau) = \lim_{T \to \infty} \frac{1}{2T} \int_{-T}^{T} x(t)x(t+\tau)\,dt \tag{2.60}$$

that is, the ensemble and time averages are equal. Any experimental approach to the measurement of correlation functions would involve averages carried out over finite time periods or

$$\phi_{xy}(\tau, T) = \frac{1}{T} \int_{0}^{T} x(t)y(t+\tau)\,dt \tag{2.61}$$

where the symbol T has been introduced into the argument of the correlation function on the left-hand side of this equation to display the dependence on T, the time over which the correlation function measurement is made. If the ensemble average of this last equation is taken, there follows

$$\phi_{xy}(\tau) = \int_{-\infty}^{\infty}\int_{-\infty}^{\infty} \phi_{xy}(\tau, T) f(x, y, \tau)\,dx\,dy \tag{2.62}$$

This equation indicates that, on the average, the measurement $\phi_{xy}(\tau, T)$ is equal to $\phi_{xy}(\tau)$, the desired correlation function. Needless to say, this is a rather fortunate but not unexpected result.

Thus, if we can establish the expressions for the variance about this average and the conditions under which this variance is sufficiently small, we will have established conditions under which our correlation function measurements, by the process of Eq. (2.61), are, in general, sufficiently close to the actual correlation function. As a result, it is now necessary to develop such expressions. Using now the symbol $E[\quad]$ to designate ensemble averages, the variance of a correlation function measurement is defined as

$$\sigma_{xy}{}^2(\tau, T) = E\left[\left\{\phi_{xy}(\tau, T) - \phi_{xy}(\tau)\right\}^2\right] \tag{2.63}$$

$$= E\left[\phi^2_{xy}(\tau, T)\right] - \phi^2_{xy}(\tau) \tag{2.64}$$

where

$$E\left[\phi^2_{xy}(\tau, T)\right] = \frac{1}{T^2} \int_{0}^{T}\int_{0}^{T} E\left[x(t_1)y(t_1+\tau)x(t_2)y(t_2+\tau)\right] dt_1\,dt_2 \tag{2.65}$$

Using the symbol

$$\gamma_{xy}^2(\tau, t_1, t_2) = E\left[x(t_1)y(t_1+\tau)x(t_2)y(t_2+\tau)\right] \tag{2.66}$$

we note that, because of stationarity,

$$\gamma_{xy}^2(\tau, t_1, t_2) = \gamma_{xy}^2(\tau, 0, t_2 - t_1) \tag{2.67}$$

$$= \gamma_{xy}^2(\tau, \nu) \qquad \nu = t_2 - t_1 \tag{2.68}$$

Thus

$$E\left[\phi_{xy}^2(\tau, T)\right] = \frac{1}{T^2}\int_0^T \int_{t_2-T}^{t_2} \gamma_{xy}^2(\tau, \nu)\, d\nu\, dt_2 \tag{2.69}$$

Interchanging the order of integration in these two equations there follows

$$E\left[\phi_{xy}^2(\tau, T)\right] = \frac{1}{T^2}\int_{-T}^{0} \int_0^{T+\nu} \gamma_{xy}^2(\tau, \nu)\, dt_2\, d\nu + \frac{1}{T^2}\int_0^T \int_\nu^T \gamma_{xy}^2(\tau, \nu)\, dt_2\, d\nu \tag{2.70}$$

But the identity $\gamma_{xy}^2(\tau, \nu) = \gamma_{xy}^2(\tau, -\nu)$ holds. Substituting this into the first integral on the right-hand side of Eq. (2.70) we have, in straightforward manner, since the integrands in Eq. (2.70) do not depend on t_2,

$$E\left[\phi_{xy}^2(\tau, T)\right] = \frac{2}{T^2}\int_0^T (T - \nu)\, \gamma_{xy}^2(\tau, \nu)\, d\nu \tag{2.71}$$

But

$$\frac{2}{T^2}\int_0^T (T - \nu)\, d\nu = 1 \tag{2.72}$$

Thus the variance is given as

$$\sigma_{xy}^2(\tau, T) = \frac{2}{T^2}\int_0^T (T - \nu)\left[\gamma_{xy}^2(\tau, \nu) - \phi_{xy}^2(\tau)\right] d\nu \tag{2.73}$$

As a result, if it is desired that any given measurement of the correlation function lie within some percent of the true or average measurement of the correlation function with some particular degree of certainty, parameters such as T in Eq. (2.73) have to be established accordingly. For instance, if it is assumed that the measurements of the correlation function have a normal distribution, and if this can be established by a number of measurements as it is in more elementary situations, then, from elementary probability theory, for a 95% certainty that an arbitrary measured value of the correlation function lies within "n" percent of the mean or true correlation function, it is necessary that

$$.001\, n\, \phi_{xy}(\tau) = 2\sigma_{xy}(\tau, T) \tag{2.74}$$

or in other words

$$\frac{\phi_{xy}(\tau)}{\sigma_{xy}(\tau, T)} = \frac{200}{n} \tag{2.75}$$

If the processes $x(t)$ and $y(t)$ belong to a stationary, ergodic, and Gaussian random process, Eq. (2.75) can in some cases be evaluated as a function of parameters such as T. Bendat [4] does this for a number of situations and presents fairly extensive numerical results and graphs. However, there must be certain knowledge of the forms of the correlation functions involved.

Actually, this is not unreasonable. For instance, Blasingame [49] established the exponential cosine form of the correlation function of windspeed. The results presented in Bendat [4] or Laning and Battin [2] can then be used in any experimental measurements of this particular random process once the general form is established. It might also be mentioned that Bendat [4] considers such other problems as the effect of noise in the measurement of correlation functions.

2.12 Synthesis With Nonstationary Inputs

In the original work [1] on the optimum synthesis of the systems with random inputs, Wiener assumed that the random processes involved were stationary and ergodic. However, examples occur in practice where the random processes involved are nonstationary [55, 61] and for these cases an extension of synthesis techniques is necessary. Such extensions were carried out and briefly summarized as of July 1961 in the resume article of Zadeh [56]. Since that time, there have been a few other contributions [57].

The various approaches to solution of the problem of determination of the optimum weighting function with nonstationary random inputs with a mean-square error criterion involve either a direct attack on the integral equation whose solution defines the optimum weighting function, or utilization of the knowledge of the shaping filter which generates the input nonstationary random process from a white noise process. This shaping filter approach is discussed in Chapter 4 of this book as well as in Zadeh's [56] article. We will review here two fairly significant approaches to the solution of the problem of optimum system synthesis in the mean-square error sense. Other approaches are referred to in Zadeh's paper [56].

Thus, consider first the work of Kalman and Bucy [57] with regard to the problem of determining the system whose N outputs, $\hat{x}_1(t), \hat{x}_2(t), \ldots, \hat{x}_N(t)$, are the best linear estimates based on minimizing the mean squared error $[\mathrm{x} - \mathrm{x}]^2$ of the corresponding N components of the difference between the desired system output components, $x_1(t), x_2(t), \ldots, x_N(t)$, and the actual outputs, $\hat{x}_1(t), \hat{x}_2(t), \ldots, \hat{x}_N(t)$. The inputs to the optimum filter or system will be written as $z_1(t), z_2(t), \ldots, z_M(t)$, and these are taken as linear combinations of the system input signal components and corrupting white noise as will be explicitly shown below. The random signal is assumed to be generated by white noise introduced into a linear, in general, time varying system, viz.,

$$\frac{d\mathbf{x}(t)}{dt} = F(t)\mathbf{x}(t) + G(t)\mathbf{u}(t) \tag{2.76}$$

where $\mathbf{x}(t)$ is the column vector of the signal and $\mathbf{u}(t)$ is a column vector having zero mean white noise components and a covariance matrix

$$\operatorname{cov}[\mathbf{u}(t), u(\tau)] = E[\mathbf{u}(t), u(\tau)] = Q(t)\,\delta(t - \tau) \tag{2.77}$$

In these last two equations the capital letters signify matrices, and as before $E[\quad]$ designates the expected value.

The actual input to the system, $\mathrm{z}(t)$, can be expressed as

$$\mathrm{z}(t) = H(t)\mathbf{x}(t) + \mathrm{v}(t) \tag{2.78}$$

where $\mathrm{v}(t)$ is a zero mean white noise vector whose components enter into each component of $\mathrm{z}(t)$ and whose covariance matrix may be written as

$$\operatorname{cov}[v(t), v(\tau)] = R(t)\,\delta(t - \tau) \tag{2.79}$$

The optimum system is then shown, by Kalman and Bucy, to be given by

$$\frac{d\hat{\mathbf{x}}(t)}{dt} = [F(t) - \kappa(t)H(t)]\hat{\mathbf{x}}(t) + \kappa(t)\,\mathrm{z}(t) \tag{2.80}$$

Thus the form of this optimum system is partially specified by what may be referred to as the signal processing as defined in

Eq. (2.78). The matrix of time variable gain elements $\kappa(t)$ of Eq. (2.80) is given from (the prime denotes the transpose)

$$\kappa(t) = P(t)H'(t)R^{-1}(t) \tag{2.81}$$

The matrix $P(t)$ is the symmetric covariance matrix of the vector $x(t) - \hat{x}(t)$ and is obtained by solving the nonlinear equation

$$\frac{dP(t)}{dt} = F(t)P(t) + P(t)F'(t) - P(t)H'(t)R^{-1}(t)H(t)P(t) + G(t)Q(t)G'(t) \tag{2.82}$$

This concludes the development of the work of Kalman and Bucy. In an earlier paper, Kalman [58] presented the results for the case of sampled data systems. Results for this case were also developed by Hsieh and Leondes [59]. The second approach we will consider is that of Shinbrot's.

The response of a linear time variable system whose weighting function $W(t, \tau)$ we wish to optimize in the mean-square error sense is given as

$$x(t) = \int_{-\infty}^{t} W(t, \tau)\, i(\tau)\, d\tau \tag{2.83}$$

The input $i(t)$ consists of signal pulse noise, i.e.,

$$i(t) = s(t) + n(t) \tag{2.84}$$

Since the input is applied at some finite time, which we may arbitrarily take as zero,

$$x(t) = \int_{0}^{t} W(t, \tau)\, i(\tau)\, d\tau \tag{2.85}$$

We represent the desired output as $\mu(t)$. Thus, the mean-square error which we wish to minimize is given as

$$\overline{\epsilon^2} = E\left[\left\{\mu(t) - \int_{0}^{t} W(t, \tau)\, i(\tau)\, d\tau\right\}^2\right] \tag{2.86}$$

Proceeding in the usual manner we obtain as the necessary and sufficient condition that $W(t, \tau)$ minimize $\overline{\epsilon^2}$

$$\phi_{\mu i}(t, \tau) = \int_{0}^{t} W(t, \sigma)\, \phi_{ii}(\tau, \sigma)\, d\sigma \qquad \text{for} \quad 0 \le \tau \le t \tag{2.87}$$

where, in the notation of the previous section, ϕ_{ii} and $\phi_{\mu i}$ are the correlation functions for the nonstationary processes involved. By noting the definition of the correlation functions then

$$\phi_{ss}(t, \tau) = \phi_{ss}(\tau, t) \tag{2.88}$$

As is explained in Chapter 4, nonstationary correlation functions can be represented as separable with a degree of accuracy depending on the order of the representation, viz.,

$$\left.\begin{aligned} \phi_{ss}(t, \tau) &= \sum_{\rho=1}^{a} a_\rho(t)\, b_\rho(\tau) \\ \phi_{\mu s}(t, \tau) &= \sum_{\rho=1}^{a} c_\rho(t)\, b_\rho(\tau) \end{aligned}\right\} \quad \text{for } t \geq \tau \tag{2.89}$$

From Eq. (2.88) we note that we may write

$$\phi_{ss}(t, \tau) = \sum_{\rho=1}^{a} a_\rho(\tau)\, b_\rho(t) \qquad \text{for } \tau > t \tag{2.90}$$

Now letting $\mathbf{a}$, $\mathbf{b}$, and $\mathbf{c}$ denote the vectors whose elements are a_ρ, b_ρ, and c_ρ, respectively, we can write more briefly

$$\begin{aligned} \phi_{\mu s}(t, \tau) &= \begin{cases} \mathbf{a}'(t) \cdot \mathbf{b}(\tau) & \text{for } \tau \leq t \\ \mathbf{a}'(\tau) \cdot \mathbf{b}(t) & \text{for } \tau > t \end{cases} \\ \phi_{\mu s}(t, \tau) &= \mathbf{c}'(t) \cdot \mathbf{b}(\tau) \qquad \text{for } \tau \leq t \end{aligned} \tag{2.91}$$

If the input noise $n(t)$ is taken to be white noise, then Eq. (2.87) becomes

$$\phi_{\mu s}(t, \tau) = \int_0^t W(t, \sigma)\, \phi_{ss}(\tau, \sigma)\, d\sigma + \lambda g(t, \tau) \qquad \text{for } 0 \leq \tau \leq t \tag{2.92}$$

where we have used

$$\begin{aligned} \phi_{\mu i}(t, \tau) &= \phi_{\mu m}(t, \tau) \\ \phi_{ii}(t, \tau) &= \phi_{mm}(t, \tau) + \lambda \delta(t - \tau) \end{aligned} \tag{2.93}$$

Substituting Eq. (2.91) in Eq. (2.92) there follows

$$\begin{aligned} \mathrm{c}'(t) \cdot \mathrm{b}(\tau) &= \mathrm{a}'(\tau) \cdot \int_0^{\tau} \mathrm{b}(\sigma) W(t, \sigma)\, d\sigma \\ &\quad \mathrm{b}'(\tau) \cdot \int_{\tau}^{t} \mathrm{a}(\sigma) W(t, \sigma)\, d\sigma + \lambda g(t, \tau) \qquad \text{for } 0 \le \tau \le t \end{aligned} \tag{2.94}$$

Now, setting

$$v(t, \tau) = \mathrm{a}'(t) \cdot \mathrm{b}(\tau) - \mathrm{a}(\tau) \cdot \mathrm{b}(t) \tag{2.95}$$

then, writing $\int_{\tau}^{t}$ in Eq. (2.94) as $\int_0^t - \int_0^{\tau}$, there results

$$\begin{aligned} &\left[\mathrm{c}'(t) - \int_0^t \mathrm{a}'(\sigma) W(t, \sigma)\, d\sigma\right] \cdot \mathrm{b}(\tau) \\ &\quad = \int_0^{\tau} W(t, \sigma) v(\tau, \sigma)\, d\sigma + \lambda g(t, \tau) \qquad \text{for } 0 \le \tau \le t \end{aligned} \tag{2.96}$$

A separable solution for $W(t, \tau)$ of this equation is now sought. In particular, we seek

$$W(t, \tau) = [\mathrm{h}'(t) \cdot \gamma(\tau)]\, u(t - \tau) \tag{2.97}$$

where $u(t - \tau)$ is the unit step function introduced for reasons of physical realizability. Substituting Eq. (2.97) in Eq. (2.96) gives

$$\begin{aligned} &\mathrm{c}'(t) - \int_0^t \mathrm{a}'(\sigma) [\mathrm{h}'(t) \cdot \gamma(\sigma)]\, d\sigma \cdot \mathrm{b}(\tau) \\ &\quad = \mathrm{h}'(t) \cdot \left[\int_0^{\tau} \gamma(\sigma) v(\tau, \sigma)\, d\sigma + \lambda \gamma(\tau)\right] \qquad \text{for } 0 \le \tau \le t \end{aligned} \tag{2.98}$$

This equation is certainly satisfied if

$$\begin{aligned} &\text{(a)} \quad \mathrm{b}(\tau) = \int_0^{\tau} \gamma(\sigma) v(\tau, \sigma)\, d\sigma + \lambda \gamma(\tau) \qquad 0 \le \tau \le t \\ &\text{(b)} \quad \mathrm{c}(t) - \int_0^t \mathrm{a}(\sigma) [\mathrm{h}'(t) \cdot \gamma(\sigma)]\, d\sigma = \mathrm{h}(t) \qquad t \ge \tau \end{aligned} \tag{2.99}$$

In the derivation of this equation, the integral Eq. (2.92), in which $g(t, \tau)$ depends on the two independent variables t and τ, has been replaced by two equations for the two components of the separable representation of $g(t, \tau)$. Defining

$$\begin{aligned} e'(\tau) &= \left[a_1(\tau), \ldots, a_\alpha(\tau), b_1(\tau), \ldots, b_\alpha(\tau)\right] \\ f'(\tau) &= \left[b_1(\tau), \ldots, b_\alpha(\tau), -a_1(\tau), \ldots, -a_\alpha(\tau)\right] \end{aligned} \tag{2.100}$$

Then Eq. (2.95) becomes

$$v(t, \tau) = e'(t) \cdot f(\tau) \tag{2.101}$$

and thus Eq. (2.99a) becomes

$$b_p(\tau) = e'(\tau) \cdot \int_0^\tau f(\sigma)\gamma_p(\sigma)\,d\sigma + \lambda\gamma_p(\tau) \qquad \begin{array}{l} \text{for } 0 \le \tau \le t \\ p = 1, \ldots, a \end{array} \tag{2.102}$$

where the vectors $b(t)$ and $\gamma(t)$ are written in terms of their components. Thus

$$b_p(\tau) = \sum_{q=1}^{2\alpha} e_q(\tau) \int_0^\tau f_q(\sigma)\gamma_p(\sigma)\,d\sigma + \lambda\gamma_p(\tau) \qquad \text{for } 0 \le \tau \le t,\ p = 1, \ldots, a \tag{2.103}$$

If the components $e_q(\tau)$ of $e(\tau)$ are not all linearly independent, the appropriate terms of the right-hand side of Eq. (2.103) can be collected to obtain equations of the form

$$b_p(\tau) = \sum_{q=1}^{\beta} e_q(\tau) \int_0^\tau \omega_q(\sigma)\gamma_p(\sigma)\,d\sigma + \lambda\gamma_p(\tau) \qquad \text{for } 0 \le \tau \le t,\ p = 1, \ldots, a \tag{2.104}$$

where the functions $e_q(\tau)$ are linearly independent and the $\omega_q(\sigma)$ functions result from the combination of functions made in going from Eq. (2.103) to Eq. (2.104). Now, Eq. (2.104) can be immediately reduced to a system of differential equations in the sought-for functions $\gamma_p(\tau)$ by differentiating Eq. (2.104) r times. The result is

$$\begin{aligned} \sum_{q=1}^{\beta} e_q^r(\tau) \int_0^\tau \omega_q(\sigma)\gamma_p(\sigma)\,d\sigma &= b_q^r(\tau) - \lambda\gamma_p^r(\tau) \\ &- \sum_{s=1}^{r} \binom{r}{s} \sum_{q=1}^{\beta} e_q^{(r-s)}(\tau) \frac{d^{s-1}}{d\tau^{s-1}}\left[\omega_q(\tau)\gamma_p(\tau)\right] \\ &\text{for } 0 \le \tau \le t,\ p = 1, \ldots, a \end{aligned} \tag{2.105}$$

where $\binom{r}{s}$ denotes the binomial coefficient

$$\binom{r}{s} = \frac{r!}{s!(r-s)!}$$

Equations (2.105) with $r = 0, 1, \ldots, \beta - 1$ represent β simultaneous equations in the β unknowns

$$\int_0^T \omega_q(\sigma)\,\gamma_p(\sigma)\,d\sigma \qquad q = 1, \ldots, \beta \text{ (p fixed)} \tag{2.106}$$

In Eq. (2.106) it is the γ_p which must be determined. Rewriting Eq. (2.105) in matrix form we have

$$EW = F \tag{2.107}$$

where E_{rq}, the element in the rth column and qth row of the E matrix is given as $e_q^r(\tau)$. Thus the E matrix is the Wronskian of the linear independent $e_q(\tau)$ functions. The elements of the W matrix are given by Eq. (2.106). In particular,

$$W_{qp} = \int_0^T \omega_q(\sigma)\,\gamma_p(\sigma)\,d\sigma \tag{2.108}$$

The elements of the F vector are given by the right-hand side of Eq. (2.105). In particular,

$$F_{rp} = b_p^r(\tau) - \lambda\gamma_p^r(\tau) - \sum_{s=1}^{r}\binom{r}{s}\sum_{q=1}^{\beta} e_q^{r-s}(\tau)\,\frac{d^{s-1}}{d\tau^{s-1}}\left[\omega_q(\tau)\,\gamma_p(\tau)\right] \tag{2.109}$$

Thus the solution for the elements of the W matrix is given by

$$W = E^{-1}F \tag{2.110}$$

Differentiating Eq. (2.110) once results in a set of linear differential equations for the sought-for unknowns γ_p. These equations may be solved on an analog or digital computer, or analytically, as the particular problem warrants.

One observation might be made here. Examination of Eq. (2.110) shows that the W matrix has as its elements W_{qp}. This means, since $q = 1, 2, \ldots,$ and $p = 1, 2, \ldots, \alpha$, that we have a total of $\beta\alpha$ equations to determine the α unknowns γ_p, or more equations than unknowns. It might seem, as a result, that these various equations are different and do not possess a common solution. This cannot be so since Eq. (2.99a) is a Volterra integral equation [62], which always possesses a unique solution. There are any number of ways in which the set of α equations to determine the γ_p can be selected. One is to select a particular row in the set of matrix differential equations resulting from the differentiation of Eq. (2.110).

Now that the $\gamma_p(\tau)$ are determined, it remains to determine $h(t)$ in Eq. (2.97) in order to determine $W(t, \tau)$. This can be readily done from Eq. (2.99b). Each component of this equation is given as

$$h_p(t) + \sum_{q=1}^{a} h_q(t) \int_0^t a_p(\sigma)\gamma_q(\sigma)\,d\sigma = c_p(t) \qquad p = 1, \ldots, a \quad (2.111)$$

Everything is known in Eq. (2.111) except $h_p(t)$ or $h_q(t)$. Thus these quantities and hence $W(t, \tau)$ may be determined.

The development of the explanation of Shinbrot's method has been somewhat more detailed than that of the method of Kalman and Bucy. The reason for this is the somewhat more complex nature of Shinbrot's technique. In any event, the practical realization of the optimum system of both methods has to be balanced against practical engineering considerations. In particular, the question of approximating the optimum system by a system that can be realized in a somewhat simpler manner may have to be examined in some cases where the resultant optimum weighting function is entirely too complex to be realized practically. This, then, makes it necessary to develop methods for analyzing systems involving nonstationary random processes. This question will be taken up in the next section.

2.13 The Method of Adjoint Stimulation

For simplicity, consideration is first given to a linear time-varying system having a single input and output. Assume that the system under consideration receives a random input $x(t)$ defined over $-\infty < t < \infty$, and let $e(t)$ denote the corresponding inaccuracy or error which $x(t)$ produces. Then if $W(t, \tau)$ is the weighting function relating $x(t)$ and $e(t)$, we have

$$e(t) = \int_{-\infty}^{t} W(t, \tau)\,x(\tau)\,d\tau \quad (2.112)$$

To calculate the correlation function for $e(t)$, we note that

$$e(t_1)e(t_2) = \int_{-\infty}^{t_1} W(t_1, \tau_1)\,d\tau_1 \int_{-\infty}^{t_2} W(t_2, \tau_2)\,x(\tau_1)\,x(\tau_2)\,d\tau_2 \quad (2.113)$$

Averaging over the ensemble, there results

$$\phi_{ee}(t_1, t_2) = \int_{-\infty}^{t_1} W(t_1, \tau_1)\,d\tau_1 \int_{-\infty}^{t_2} W(t_2, \tau_2)\,\phi_{xx}(\tau_1, \tau_2)\,d\tau_2 \quad (2.114)$$

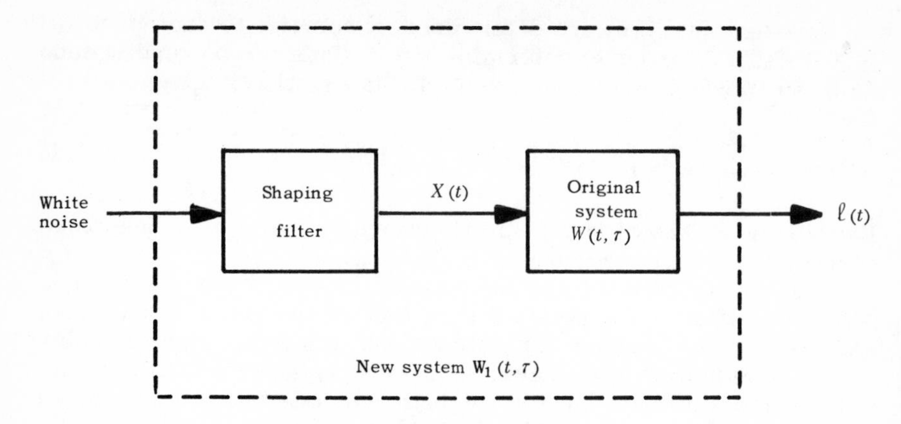

FIG. 2-5. The use of shaping filter.

Referring to Fig. 2-5, the use of a shaping filter whose input is white noise and whose output is the nonstationary random process $x(t)$ makes it possible to rewrite Eq. (2.114) as

$$
\begin{aligned}
\phi_{ee}(t,t) = \overline{e^2(t)} &= \int_{-\infty}^{t} W(t,\tau_1)\,d\tau_1 \int_{-\infty}^{t} W(t,\tau_2)\,\phi_{xx}(\tau_1,\tau_2)\,d\tau_2 \\
&= \int_{-\infty}^{t} W_1(t,\tau_1)\,d\tau_1 \int_{-\infty}^{t} W_1(t,\tau_2)\,(\tau_2-\tau_1)\,d\tau_2 \qquad (2.115) \\
&= \int_{-\infty}^{t} W_1(t,\tau)^2\,d\tau
\end{aligned}
$$

The question of the synthesis of the shaping filter in Fig. 2-5 is dealt with in Chapter 4. We will assume here that this shaping filter has been synthesized according to the techniques of Chapter 4. Thus the question of the determination of the system's mean-square error rests on the evaluation of $W_1(t,\tau)$. The straightforward but highly tedious way of doing this is to apply impulse functions at the input to the system $W_1(t,\tau)$. From the practical point of view of analog computer equipment, it is better to apply a step function at the output of the integrator to which an impulse function would normally be applied. The impulse function is applied at τ and the response observed at t. This process is repeated for many values of τ in order to determine $W_1(t,\tau)$ as a function of τ for use in Eq. (2.115). Fortunately, in many problems it is unnecessary to do this for many values of t. For instance, in the design of fire control systems it is necessary to determine $\overline{e^2(t)}$ at the time of target approach and not for the entire time of flight. However, even here, where it is necessary to determine $\overline{e^2(t)}$ only at one or a few instances of time, a great deal of work may be required.

It would therefore be desirable to develop a method of analog computer simulation which would make it possible to determine $\overline{e^2(t)}$ in one computer run. Or, to put this another way, to determine $W_1(t, \tau)$ as a function of in a single analog computer run. The method of adjoint simulation makes it possible to do just this. It will not be developed.

First, we set up the weighting function of a system describable by a linear differential equation

$$\dot{\mathbf{x}}(t) = A(t)\mathbf{x}(t) + \mathbf{f}(t) \tag{2.116}$$

where

$$\mathbf{x}(t) = \begin{bmatrix} x_1(t) \\ x_2(t) \\ \cdot \\ \cdot \\ \cdot \\ \cdot \\ x_n(t) \end{bmatrix}; \quad f(t) = \begin{bmatrix} f_1(t) \\ f_2(t) \\ \cdot \\ \cdot \\ \cdot \\ \cdot \\ f_n(t) \end{bmatrix}$$

and where $A(t)$ is a matrix whose elements are $a_{ij}(t)$. Also $x(0) = 0$. Let M denote the solution of the matrix equation

$$\frac{dM(t)}{dt} = A(t)\mathrm{M}(t); \qquad \mathrm{M}(0) = I \tag{2.117}$$

where I is the identity matrix. To solve Eq. (2.116), we employ Lagrange's method of variation of parameters. Let $\mathbf{x} = \mathrm{M}\mathbf{u}$. Substituting this in Eq. (2.116) there results

$$\begin{aligned} \frac{d\mathbf{x}(t)}{dt} &= \mathrm{M}(t)\frac{d\mathbf{u}(t)}{dt} + \frac{dM(t)}{dt}\mathbf{u}(t) \\ &= \mathrm{M}(t)\frac{d\mathbf{u}(t)}{dt} + A(t)\mathrm{M}(t)\mathbf{u}(t) \\ &= A(t)\mathrm{M}(t)\mathbf{u}(t) + \mathbf{f}(t) \end{aligned} \tag{2.118}$$

Hence

$$\mathrm{M}(t)\frac{d\mathbf{u}}{dt} = \mathbf{f}(t) \tag{2.119}$$

As a result

$$\mathbf{u}(t) = \int_0^t \mathrm{M}^{-1}(s)\mathbf{f}(s)\,ds \tag{2.120}$$

and

$$\mathbf{x}(t) = \int_0^t \mathrm{M}(t)\mathrm{M}^{-1}(s)\mathbf{f}(s)\,ds \tag{2.121}$$

Then, if for physical realizability we define the weighting function $\kappa(t, s)$ as

$$\begin{aligned} \kappa(t, s) &= \mathrm{M}(t)\mathrm{M}^{-1}(s) && s \leq t \\ \kappa(t, s) &= 0 && s > t \end{aligned} \tag{2.122}$$

the solution may be written as

$$\mathbf{x}(t) = \int_0^\infty \kappa(t, s)\mathbf{f}(s)\,ds \tag{2.123}$$

provided we are interested only in the case $t \geq 0$. The function $\kappa(t, s)$ is called the weighting function, or Green's function, of the system.

Recalling the remarks made at the beginning of this section, we see that for the problem under consideration it is desirable that the function $\kappa(t, s)$ be a function of s for some fixed value of t. That is, given T, we wish to know $\kappa(T, s)$ as a function of s. We will now show how the adjoint system makes this possible. Let us begin by introducing a pair of variables t' and s' related to the variables t and s by

$$\begin{aligned} t' &= T - t \\ s' &= T - s \end{aligned} \tag{2.124}$$

where T is any fixed real number. Let us define a function $L(t', s')$ by the relation (the primes on capital letters indicate a transposed matrix and the primes on lower-case letters indicate different variables)

$$L(t', s') = \kappa'(s, t) \tag{2.125}$$

where $\kappa'(s, t)$ is the transposed weighting function matrix of the original system. Note that in Eq. (2.125), in the argument for $\kappa'(s, t)$, the variables s and t assume the positions and roles assumed by t and s, respectively, in the Eq. (2.122). The reason for this rather abrupt reversal of notation is that, in the expression for $L(t', s')$, we wish arbitrarily to use t' and s' as shown. In any event, to examine Eq. (2.125) in a little more detail, suppose $t' < s'$. Then $t > s$, hence $\kappa(s, t) = 0$, and thus $L(t', s') = 0$, as it should be from physical realizability considerations. For $t' \geq s'$ we have $t \leq s$, and hence

$$L(t', s') = \kappa'(s, t) = \left[\mathrm{M}(s)\mathrm{M}^{-1}(t)\right]' = \left[\mathrm{M}^{-1}(t)\right]' \cdot \mathrm{M}'(s) \tag{2.126}$$

We now define the function $z(t')$ by the formula

$$z(t') = \int_0^\infty L(t', s') g(s') ds' \tag{2.127}$$

where $g(s')$ is any integrable and continuous function on the real line. More explicitly, Eq. (2.123) indicates the system response $x(t)$ to an input forcing function $f(t)$ for the system described by Eq. (2.116) whose matrix weighting function is given as $\kappa(t, s)$. By the same token $z(t')$ is the response to the forcing function $g(s')$ for the system whose weighting function $L(t', s')$. It remains to develop the differential equations for the system of Eq. (2.127) which will be like the differential Eqs. (2.116). To proceed with this, we may rewrite Eq. (2.127) as

$$z(t') = \int_0^{t'} L(t', s') g(s') ds' \tag{2.128}$$

since $L(t', s') = 0$ for $t' < s'$. Differentiating Eq. (2.128)

$$\begin{aligned} \dot{z}(t') &= \int_0^{t'} \frac{\partial}{\partial t'} L(t', s') g(s') ds' + L(t', t') g(t') \\ &= \int_0^{t'} \frac{\partial}{\partial t'} L(t', s') g(s') ds' + g(t') \end{aligned} \tag{2.219}$$

since $L(t', t') = 1$, as can be seen from Eq. (2.126). We now evaluate $\frac{\partial}{\partial t'} L(t', s')$ for $s' \leq t'$.

In this case, $s \geq t$ and $L(t', s') = [M^{-1}(t)]' M'(s)$. Hence

$$\begin{aligned} \frac{\partial}{\partial t'} L(t', s') &= \frac{\partial}{\partial t} \left\{ [M^{-1}(t)]' \cdot M'(s) \right\} \frac{dt}{dt}, \\ &= - \left[\frac{d}{dt} \{M^{-1}(t)\}' \right] \cdot M'(s) \end{aligned} \tag{2.130}$$

since $\frac{dt}{dt'} = -1$. Now

$$\frac{d}{dt} \{M^{-1}(t)\}' = \left[\frac{d}{dt} \{M^{-1}(t)\} \right]' \tag{2.131}$$

It thus remains for us to evaluate $\frac{d}{dt} \{M^{-1}(t)\}$. Since

$$M(t)M^{-1}(t) = 1$$

$$\dot{M}(t)M^{-1}(t) + M(t)\frac{d}{dt}\{M^{-1}(t)\} = 0 \tag{2.132}$$

and hence

$$\frac{d}{dt}\{M^{-1}(t)\} = -M^{-1}(t)\dot{M}(t)M^{-1}(t) \tag{2.133}$$

But $\dot{M}(t) = A(t)M(t)$, so that

$$M^{-1}(t)\dot{M}(t)M^{-1}(t) = M^{-1}(t)A(t) \tag{2.134}$$

Hence

$$\frac{d}{dt}\{M^{-1}(t)\} = -M^{-1}(t)A(t) \tag{2.135}$$

Thus we may conclude that

$$\frac{\partial}{\partial t'}L(t', s') = A'(t)\left[M^{-1}(t)\right]'M'(s) = A'L(t', s') \tag{2.136}$$

Then $\dot{z}(t')$ becomes

$$\dot{z}(t') = \int_0^{t'} A'(t)L(t', s')g(s')\,ds' + g(t') \tag{2.137}$$

If we let $B(t') = A'(t)$, we can rewrite $\dot{z}(t')$ as

$$\dot{z}(t') = B(t')z(t') + g(t') \tag{2.138}$$

It is clear that the weighting function matrix of this last equation is $L(t', s')$. This system of Eq. (2.138) is what may be referred to as the adjoint system of Eq. (2.116). Actually it is more common to refer to a slightly different form of Eq. (2.138) as the adjoint system. Suppose we define a function $y(t)$ by the relation

$$y(t) = z(t') \tag{2.139}$$

Then we have that $\dot{y}(t) = \dot{z}(t')\dfrac{dt'}{dt} = -\dot{z}(t')$. Then the equation for $\dot{z}(t')$ is transformed into

$$\dot{y}(t) = -A'(t)y(t) + h(t) \tag{2.140}$$

It is actually this system which is referred to as the adjoint of Eq. (2.116). Thus if we need to evaluate $\kappa(t, s)$ as a function of s for a given fixed value of t (we have now reverted to the original definitions for t and s in this expression for $\kappa(t, s)$, we need only evaluate

$L(s', t')$ for the corresponding t'. In particular if we wish to evaluate $\kappa(T, s)$ we may evaluate $L'(s', 0)$, then

$$\kappa(T, s) = L'(s', 0) \tag{2.141}$$

This, then, leads us to the following procedure for evaluating $\kappa(T, t)$ by the use of the adjoint method:

(1) Simulate the system

$$\dot{z}(t') = B(t')z(t') + g(t)$$

(2) Set all initial conditions to zero.

(3) Start the computer; at time $t = 0$ apply the impulse (or step function) to the first variable $g_1(t)$ and record.

(4) Repeat the process, applying at time $t = 0$ the impulse (or step function) to each successive variable $g_k(t)$ and record.

(5) The n solutions thus obtained form the rows of $\kappa(T, t)$ provided it is assumed that the computer has been running backward in time, starting at time T.

The principal problem encountered in this procedure is to simulate the adjoint system

$$\dot{z}(t) = B(t')z(t') + g(t')$$

when the computer is set to simulate the system

$$\dot{x}(t) = A(t)x(t) + f(t)$$

That is, however, easily done. The matrix $B(t')$ and the matrix $A(t)$ are related by the identity $B(t') = A'(t)$. Hence the simulation of the adjoint system can be derived from the simulation of the original system in the following two steps:

(1) The function generators for the components of $A(t)$ (or, to say the same thing another way, the elements of the matrix $A(t)$) are disconnected and reconnected adjointly. That is, where the generator for the component $a_{jk}(t)$ has been multiplied by $\xi_k(t)$, it is now multiplied by $\xi_j(t)$.

(2) The function generators for the components of $A(t)$ are set to start at time T, and run in the reverse direction to the way they run in the original system of Eq. (2.116).

For further discussions of the adjoint technique, including a detailed development of a number of excellent examples, see reference [63]. The adjoint technique, originally developed by Laning and Battin [2], has been adopted with considerable saving in time by many research and development groups around the country.

2.14 Concluding Remarks

This chapter has presented a number of techniques in the general area of system synthesis when the inputs are random processes. When synthesis techniques are specifically dealt with, the result of the synthesis procedure is the weighting function or transfer function of the optimum system. This weighting function still has to be developed as a feedback system for any given control problem. Part I has discussed this question and presented necessary techniques for multipole systems. The weighting functions which result when dealing with nonstationary random processes are discussed in Section 2.12; the necessary techniques for realizing the feedback system are taken up in Chapter 1.

There are many other questions which arise when attempting to apply these techniques to actual problems [55, 61, 64, 65, 66, 67, 68]. For example, in the synthesis of a particular missile fire control system [64], it was observed that the application of Wiener synthesis techniques called for control surface deflections whose rms values were ∞! A technique for dealing with this from a practical point of view, surprisingly enough without degrading the error performance too seriously, was presented in the course of the development of the necessary feedback control system [64]. This is only one illustration of the type of practical considerations which inevitably arise. Others will be found in the references listed.

Finally, there is the question of other error criteria. For instance, in some systems it is more desirable to keep the error within specified limits rather than minimize a mean-square error. Techniques for synthesizing from this point of view are presented in Bergen's report [69] as well as in some of the references listed in Zadeh's survey paper [56].

REFERENCES

1. Weiner, N., "The Interpolation, Extrapolation and Smoothing of Stationary Time Series," John Wiley and Sons, Inc., New York, 1949.
2. Lanning, J. H., Jr., and R. H. Battin, "Random Processes in Automatic Control," McGraw-Hill Book Co., Inc., New York, 1956.
3. Davenport, W. L., Jr., and W. L. Root, "Random Signals and Noise," McGraw-Hill Book Co., Inc., New York, 1958.
4. Bendat, J. S., "Principles and Applications of Random Noise Theory," John Wiley and Sons, Inc., New York, 1958.
5. Hsieh, H. C. and C. T. Leondes, "On the Optimum Synthesis of Multipole Control Systems in the Wiener Sense," IRE Trans. on Automatic Control, Vol. AC-4, No. 2, pp. 16-29, November, 1959.
6. Hsieh, H. C. and C. T. Leondes, "Techniques for the Optimum Synthesis of Multipole Control Systems with Random Processes as Inputs," IRE Trans. on Automatic Control, Vol. AC-4, No. 3, pp. 212-231, December, 1959.

7. Amara, R. C., "The Linear Least Square Synthesis of Multivariable Control Systems," AIEE Transactions, Part II, (Applications and Industry), Vol. 78, pp. 115-119, 1959.
8. Wiener, N., and P. Masani, "The Prediction Theory of Multivariate Stochastic Processes, Part I," Acta Math., Vol. 98, pp. 111-150, 1957.
9. Wiener, N., and P. Masani, "The Prediction Theory of Multivariate Stochastic Processes, Part II," Acta Math., Vol. 99, pp. 93-137, 1958.
10. Youla, D. C., "On the Factorization of Rational Matrices," IRE Trans. on Information Theory, Vol. IT-7, No. 3, pp. 172-189, July, 1961.
11. Davis, M. C., "On Factoring the Spectral Matrix," Preprints, 1963 Joint Automatic Control Conference, pp. 459-467.
12. Cramer, H., "On the Theory of Stationary Processes," Ann. Math., Vol. 41, Ser. 2, 1940.
13. Maytyash, I., and Ya. Shilkhanek, "A Generator of Random Processes from Their Given Spectral Density Matrices," Automation and Remote Control (English Translation), Vol. 21, No. 1, pp. 18-22, August, 1960.
14. Freeman, H., "A Synthesis Method for Multipole Control Systems," AIEE Transactions, Part II, Vol. 76, pp. 28-31, March, 1957.
15. Horowitz, I. M., "Synthesis of Feedback Systems," Academic Press, New York and London, 1963, Chapter 10.
16. Newton, G. C., Jr., L. A. Gould, and J. F. Kaiser, "Analytical Design of Linear Feedback Controls," John Wiley and Sons, Inc., New York, 1957.
17. Blackman, R. and J. Tukey, "The Measurement of Power Spectra," Dover Publications, Inc., New York, 1958.
18. Bennett, W. R., "Electrical Noise," McGraw-Hill Book Company, Inc., New York, 1960.
19. Muchmore, Robert B., "Aircraft Scintillation Spectra," IRE Transactions on Antennas and Propagation, March, 1960, pp. 201-212 (see "Theoretical Scintillation Spectra" HAC TM-271, March 1, 1952).
20. Freeman, J. J., "Principles of Noise," John Wiley and Sons, Inc., New York, 1958.
21. Hammon, Robert L., "An Application of Random Process Theory to Gyro Drift Analysis," IRE Transactions on Aeronautical and Navigational Electronics, September, 1960.
22. Hammon, R. L., "Effects on Inertial Guidance Systems of Random Error Sources," IRE Transactions on Aeronautical and Navigational Electronics, December, 1962.
23. Dushman, A., "On Gyro Drift Models and Their Evaluation," IRE Transactions on Aeronautical and Navigational Electronics, December, 1962.
24. Newton, George C., Jr., "Inertial Guidance Limitations Imposed

by Fluctuation Phenomena in Gyroscopes," IRE Proceedings, April, 1960.
25. Weaver, C. S., "Thresholds and Tracking Ranges in Phase-Locked Loops," IRE Transactions on Space Electronics and Telemetry, Vol. SET-7, No. 3, September, 1961.
26. Livingston, M. L., "The Effect of Antenna Characteristics on Antenna Noise Temperature and System SNR," IRE Transactions on Space Electronics and Telemetry, Vol. SET-7, No. 3, September, 1961.
27. Develet, J. A., Jr., "Fundamental Accuracy Limitations in a Two-Way Coherent Doppler Measurement System," IRE Transactions on Space Electronics and Telemetry, Vol. SET-7, No. 3, September, 1961.
28. Dworetsky, L. H. and A. Edwards, "Principles of Doppler-Inertial Guidance," ARS Journal, Vol. 29, No. 12, pp. 967-72, December, 1959.
29. Dunn, H. H. and D. D. Howard, "The Effects of Automatic Gain Control Performance on the Tracking Accuracy of Monopulse Radar Systems," Proc. of the IRE, Vol. 47, No. 3, March, 1959.
30. Dunn, J. H., D. D. Howard, and A. M. King, "Phenomena of Scintillation Noise in Radar-Tracking Systems," Proc. of the IRE, Vo. 47, No. 5, May, 1959.
31. Delano, R. H., "Angular Scintillation of Radar Targets," HAC TM-233, dated 24 April 1950 (Hughes Aircraft Co.).
32. Muchmore, R. B., "Review of Scintillation Measurements," HAC TM-272, dated December 1952 (Hughes Aircraft Co.).
33. Delano, Richard H., "A Theory of Target Glint or Angular Scintillation in Radar Tracking," Proc. of the IRE, pp. 1778-84, December, 1953.
34. Muchmore, R. B., "Theoretical Scintillation Spectra," HAC TM-271, dated 1 March 1952 (Hughes Aircraft Co.).
35. Delano, R. H., Irwin Pfeffer, "The Effect of AGC on Radar Tracking Noise," Proc. of the IRE, pp. 801-810, June, 1956.
36. Favreau, R. R., H. Low and I. Pfeffer, "Evaluation of Complex Statistical Functions by an Analog Computer," Hughes Aircraft Co., presented at Project Typhoon Symposium III on Simulation and Computing Techniques, October, 1953, University of Pennsylvania.
37. "Angular Scintillation of Radar Targets with Monopulse and Interferometer Target Seekers," HAC TM-257, dated November, 1950 (Hughes Aircraft Co.).
38. Povejsil, D., R. Raven, and P. J. Waterman, "Airborne Radar," Van Nostrand Company, Inc., 1958.
39. Develet, Jean A., "Thermal-Noise Errors in Simultaneous-Lobing and Conical-Scan Angle-Tracking Systems," IRE Transactions on Space Electronics and Telemetry, Vol. SET-7, No. 2, June, 1961.

40. Meade, J. E., A. E. Hastings, and H. L. Gerwin, "Noise In Tracking Radars," Naval Research Lab Report 3759, November 15, 1950.
41. Chittenden, R. W., R. J. Massa and J. F. Frazer, "Evaluation of Satellite Tracking System Performance In the Presence of Noise and Interference," Proceedings of the Sxith Conference on Radio Interference Reduction, ASTIA No. AD 244.264.
42. "Time Series Analysis" Edited by M. Rosenblatt, John Wiley and Sons, Inc., 1962.
43. Stecca, A. J. and N. V. O'Neal, "Target Noise Simulator-Closed-Loop Tracking," NRL Report 4770, July 27, 1956.
44. Howard, D. D. and B. L. Lewis, "Tracking Radar External Range Noise Measurements and Analysis," NRL Report 4602, August 31, 1955.
45. Lewis, B. L., A. J. Stecca, and D. D. Howard, "The Effect of An Automatic Gain Control on the Tracking Performance of a Monopulse Radar," NRL Report 4796, July 31, 1956.
46. Stecca, A. J., N. V. O'Neal, and J. J. Freeman, "A Target Simulator," NRL Report 4694, February 9, 1956.
47. Leshnover, S., "Prediction of Anisoelastic and Vibropendulous Effects on Inertial Navigation System Performance in Linear Random Vibration Environments," Proceedings of the National Specialists Meeting on Guidance of Aerospace Vehicles, May, 1960.
48. Stewart, R. M., "Some Effects of Vibration and Rotation on the Drift of Gyroscopic Instruments," ARS Journal, January, 1959.
49. Blasingame, B. P., "Optimum Parameters for Automatic Airborne Navigation," D.Sc. Thesis, M.I.T., 1950 (formerly classified secret, now declassified).
50. Press, H. and J. C. Houbolt, "Some Applications of Generalized Harmonic Analysis to Gust Loads on Airplanes," Jour. Aerospace Sciences, Vol. 22, pp. 17-26, 1955.
51. Marks, W. and W. Pierson, "The Power Spectrum Analysis of Ocean Wave Records," Trans. American Geophysical Union, Vol. 33, pp. 834-844, 1952.
52. Goodman, N. R., "On the Joint Estimation of the Spectra, Cospectrum, and Quadrature Spectrum of a Two-Dimensional Stationary Gaussian Process," Scientific Paper No. 10, Engineering Statistics Laboratory, New York University, March, 1957 (Also available as ASTIA Document No. AD 134 919).
53. Balakrishnan, A. V., "A Note on the Sampling Principle for Continuous Signals," IRE Trans. PGIT, June, 1957.
54. Rosenblatt, M., "Random Processes," Oxford University Press, New York, 1962.
55. Stewart, E. C. and G. L. Smith, "The Synthesis of Optimum Homing Missile Guidance Systems with Statistical Inputs," NASA Memo 2-13-59A, April, 1959.

56. Zadeh, L., "Progress in Information Theory in the USA, 1957-1960," Part 5: Prediction and Filtering, Trans. PGIT, July, 1961.
57. Kalman, R. E. and R. S. Bucy, "New Results in Linear Filtering and Predition Theory," Transactions of the ASME, Journal of Basic Engineering, March, 1961.
58. Kalman, R. E., "A New Approach to Linear Filtering and Prediction Problems," Trans. ASME, Journal of Bais Engineering March, 1960, pp. 35-45.
59. Hsieh, H. C. and C. T. Leondes, "On the Optimum Synthesis of Sampled Data Multipole Filters with Random and Nonrandom Inputs," IRE Trans. on Automatic Control, Vol. AC-5, No. 3, pp. 193-208, August, 1960.
60. Shinbrot, Marvin, "Optimization of Time Varying Linear Systems with Nonstationary Inputs," Trans. ASME, Vol. 80, 1958, pp. 457-462.
61. Broniwitz, L., "A New Approach to the Design of Optimal Homing Missile Guidance Systems," Raytheon Report BM-2056, August 22, 1961.
62. Courant, R. and D. Hilbert, "Methods of Mathematical Physics," Vol. I, Interscience Publishers, New York, 1953.
63. Fifer, S., "Analogue Computation," Vol. IV., McGraw-Hill Book Company, Inc., New York, 1961.
64. Stewart, E. C., "Aplication of Statistical Theory to Beam Rider Guidance in the Presence of Noise. I—Wiener Filter Theory," NACA, RM A55E11, 1955.
65. Stewart, E. C. "Application of Statistical Theory to Beam Rider Guidance in the Presence of Noise. II—Modified Wiener Filter Theory," NACA, TN 4278, 1958.
66. IEEE Transactions on Aerospace and Navigational Electronics, Vol. ANE-10, No. 1, March, 1963.
67. Battin, R. H., "A Statistical Optimizing Navigation Proceedure for Space Flight," ARS Journal, November, 1962.
68. McLean, J. D., S. F. Schmidt, and L. A. McGee, "Optimal Filtering and Linear Prediction Applied to a Space Navigation System for the Circumlunar Mission," NASA TN D-1208, March, 1962.
69. Bergen, A. R., "A Non Mean-Square Error Criterion for the Synthesis of Optimum Sample Data Filters," Technical Report T-2/133, Electronics Research Labs., Columbia University, 1956 (Also available as ASTIA Document No. AD 110 180).

3

Functional Analysis and its Applications to Mean-Square Error Problems

H. C. HSIEH*
DECEASED, FORMERLY ASSISTANT PROFESSOR
ELECTRICAL ENGINEERING
NORTHWESTERN UNIVERSITY, EVANSTON, ILLINOIS

R. A. NESBIT
BECKMAN CORPORATION
SANTA MONICA, CALIFORNIA

PART I Some Basic Concepts of Functional Analysis

3.1 Introduction

There exists a fairly extensive mathematical theory of functions, and it is natural to inquire whether this theory can lead to the solution of control problems. The history of control technology could scarcely be told without reference to the "advances" made when a particular mathematical theory is found to have practical application. The realm of practical problems is increased by the availability of computing machines, and the problem of using them effectively cannot be completely ignored.

The discussion below is intended to introduce the basic ideas of function analysis and to indicate how they may be used to "solve" certain control problems. No attempt will be made to give a basic course in functional analysis such as may be found in "Functional Analysis" by Kolmogorov and Fomin [10].

There are many parallels between vector analysis and functional analysis. Many physical problems can be formulated without the use of vector analysis; in the same way, the formality of

*Previously with the Department of Engineering, University of California, Los Angeles.

functional analysis can be avoided. Those who deal with the vector equations of physical systems can readily appreciate the value of studying vector analysis, and if one is interested in problems involving functions, a study of functional analysis is certainly useful.

This discussion represents an attempt to highlight the subject of functional analysis and show its application to certain problems. The steepest-descent method of solving certain mean-square problems, and in particular, the approach due to A. V. Balakrishnan, will be discussed.

3.2 Types of Spaces

The starting point for many basic mathematical discussions is the concept of a set. The elements of an abstract set can be of a most general sort, but for most purposes the elements can be thought of as points, numbers, or functions. There are certain statements or theorems dealing with the properties of sets which can be proved without making too many assumptions about the exact nature of these elements. By establishing theorems with a minimum of assumptions, the theorems can be applied in many special cases or realizations. In most discussions, there is a set which is taken to be the universe or set of elements used to compose all the sets under consideration. This set is also referred to as the *space*. By making additional assumptions or definitions, various types of spaces are possible.

A *metric space* consists of a set X and a distance function ρ defined for each pair of elements in the set. This distance function must have the following properties for each x and y contained in X (written $x, y \in X$)

$$\rho(x, y) > 0 \qquad x \neq y \qquad \text{positive}$$

$$\rho(x, x) = 0$$

$$\rho(x, y) = \rho(y, x) \qquad \text{symmetry}$$

$$\rho(x, y) + \rho(y, z) \geq \rho(x, z) \quad \text{triangle inequality}$$

One of many examples of a metric space is the set of all continuous functions $x(t)$ on the closed interval $a \leq t \leq b$ with the distance function

$$\rho(x, y) = \left[\int_a^b (x - y)^2 dt\right]^{1/2}.$$

Other examples can be found in Kolmogorov and Fomin [10].

Questions of convergence of a sequence can be treated in a metric space. A metric space is said to be complete if every Cauchy sequence [13] converges to an element of the space.

A *linear space* is a set R of elements $x, y, z, \ldots$ for which the following operations are defined:

1. Addition. For each pair of elements x, y there is a unique element

$$z = x + y$$

such that

a) $x + y = y + x$
b) $x + (y + z) = (x + y) + z$
c) there is an element $0 \epsilon R$
$x + 0 = x$ for all $x \epsilon R$
d) for each $x \epsilon R$ there is an element $-x \epsilon R$
$x + (-x) = 0$

2. Scalar multiplication. For $x \epsilon R$ there is an element $\alpha x \epsilon R$.
a) $\alpha(\beta x) = (\alpha\beta)x$
b) $1 \cdot x = x$

3. Relation between addition and scalar multiplication.
a) $(\alpha + \beta)x = \alpha x + \beta x$
b) $\alpha(x + y) = \alpha x + \alpha y$

A *normed linear space* is a linear space R with a nonnegative number $\|x\|$ associated with each $x \epsilon R$. This number is called the norm of x and must have the properties

1) $\|x\| = 0$ if and only if $x = 0$
2) $\|\alpha x\| = |\alpha| \, \|x\|$
3) $\|x + y\| \leq \|x\| + \|y\|$

By using $\rho(x, y) = \|x - y\|$ it is seen that a normed linear space is a metric space. Thus the theorems of convergence which apply to any metric space apply in particular to the normed linear space.

A complete, normed linear space is called a *Banach space* or B-space. One example of a normed linear space is the space of all continuous functions $x(t)$ defined for $a \leq t \leq b$ with addition and scalar multiplication defined as usual and with the norm

$$\|x(t)\| = \left[\int_a^b x^2(t)\,dt\right]^{1/2}$$

An operator A establishes the rule by which elements x of one Banach space R are associated with the elements y of a second Banach space R'. This is indicated by

$$y = Ax$$

One normed linear space which has many useful properties is *Hilbert space*. Hilbert space is a finite or infinite dimensional set H with the following properties:

1. H is a linear space
2. An inner product is defined for each pair of elements, $f, g \epsilon H$.

The following properties define an inner product. There is a number, $\langle f, g \rangle$, associated with each pair and

$$\langle f, g \rangle = \langle g, f \rangle$$

$$\langle af, g \rangle = a \langle f, g \rangle$$

$$\langle f, ag \rangle = a \langle f, g \rangle$$

$$\langle f_1 + f_2, g \rangle = \langle f_1, g \rangle + \langle f_2, g \rangle$$

$$\langle f, f \rangle > 0 \quad \text{if } f \neq 0$$

The quantity $\| f \| = \sqrt{\langle f, f \rangle}$ is taken as the norm, and thus H is a normed linear space.*

3. The space H is complete in the metric $\rho(f, g) = \| f - g \|$.

When the space H contains an everywhere dense [13] denumerable subset, this property of a space is described by saying that the space is separable.

It can be shown that all Hilbert spaces of the same cardinal dimension are isomorphic [13]. This space is the natural one in which to study functions, and is also the natural extension of Euclidean space to infinite dimension. Thus one realization of Hilbert space is the space of all countable, ordered sequences $x = (x_1, x_2, x_3, \ldots, x_n, \ldots)$ with the property that $\sum_{i=1}^{\infty} x_i^2 < \infty$, and with the following definitions of addition, multiplication by a scalar, and inner product:

$$x + y = (x_1 + y_1, x_2 + y_2, \ldots, x_n + y_n, \ldots)$$

$$ax = (ax_1, ax_2, \ldots, ax_n \ldots)$$

$$\langle x, y \rangle = \sum_{i=1}^{\infty} x_i y_i$$

This space is called ℓ_2.

Another realization of Hilbert space also often used is the space of square integrable functions L_2. This is the space of functions $\int_R |f(t)|^2 \, d\mu(t) < \infty$ with the usual definitions of addition and multiplication, and with an inner product defined as

$$\langle f, g \rangle = \int_R f(t) \, g(t) \, d\mu(t).$$

*A suggested exercise: show that $\sqrt{\langle f, f \rangle}$ has the three properties required of the norm.

The expansion of functions in Fourier series is one familiar operation which is justifiable because of the isomorphism between ℓ_2 and L_2.

3.3 Fundamental Inequalities

In addition to the triangle inequality

$$\|f + g\| \leq \|f\| + \|g\|$$

mentioned above, another fundamental inequality can be shown using only the properties of Hilbert space.

The Schwarz inequality is used repeatedly in analysis. It is satisifed in every Hilbert space and has the form

$$|\langle f, g\rangle|^2 \leq \langle f, f\rangle\langle g, g\rangle$$

or

$$|(f, g)| \leq \|f\| \ \|g\|$$

It is important to know when the equality sign holds since this gives the answer to an optimization problem.

Problem: Given $f, g \epsilon H$, show that $\|f + \lambda g\|$ is minimized when

$$\lambda = -\frac{(f, g)}{(g, g)}.$$

3.4 Fourier Series

To become more familiar with the notation, consider the usual problem of expanding a square integrable function $f(t)\ a \leq t \leq b$ in a Fourier series. In this case $\langle f, g\rangle = \int_a^b f(t)\, g(t) dt$, and we have a set of functions $[\emptyset_1\ \emptyset_2 \ldots \emptyset_N(t)]$ satisfying the relations

$$\langle \emptyset_i, \emptyset_j\rangle = \delta_{ij} = \begin{cases} 0 & i \neq j \\ 1 & i = j \end{cases}$$

and are therefore orthonormal. The coefficients a_i of the finite Fourier series $\sum_{i=1}^{N} a_i \emptyset_i = a_i \emptyset_i$ (by this equation we mean that repeated indices in an expression such as $a_i \emptyset_i$ correspond to a summation of i from 1 to N) are to be chosen such that $\|f - \sum_{i=1}^{N} a_i \emptyset_i\|$ is minimized. The identity $\|f - a_i\emptyset_i\|^2 = \|f - c_i\emptyset_i + c_i\emptyset_i - a_i\emptyset_i\|^2$ is independent of the c_i and is expanded as follows:

$$\begin{aligned}\|f - a_i\emptyset_i\|^2 &= \|f - c_i\emptyset_i + c_i\emptyset_i - a_i\emptyset_i\|^2 \\ &= \|f - c_i\emptyset_i\|^2 + \langle f - c_i\emptyset_i, c_i\emptyset_i - a_i\emptyset_i\rangle \\ &\quad + \langle c_i\emptyset_i - a_i\emptyset_i, f - c_i\emptyset_i\rangle \\ &\quad + \|(c_i - a_i)\emptyset_i\|^2\end{aligned}$$

At this point, the coefficients c_i are chosen to simplify the computation. Since the identity is true for all c_i, the result is not affected by this choice.

Let

$$c_i = \langle f, \emptyset_i\rangle$$

Then

$$\begin{aligned}\langle f - c_i\emptyset_i, b\emptyset_j\rangle &= b\left[\langle f, \emptyset_j\rangle - c_i\langle\emptyset_i, \emptyset_j\rangle\right] \\ &= b\left[\langle f, \emptyset_j\rangle - c_j = 0\right]\end{aligned}$$

and

$$\begin{aligned}b\emptyset_j, f - c_i\emptyset_i &= b\left[\langle\emptyset_j, f\rangle - c_i\langle\emptyset_j, \emptyset_i\rangle\right] \\ &= 0\end{aligned}$$

Thus

$$\|f - a_i\emptyset_i\|^2 = \|f - c_i\|^2 + \sum_{i=1}^{N} |(a_i - c_i)|^2$$

The c_i are already fixed and the right-hand side of the equation, considered as a function of the a_i, clearly has a minimum for $a_i = c_i$.

The quantities $a_i = (f, \emptyset_i)$ are called the Fourier coefficients. The function $g = \sum_{i=1}^{N} a_i\emptyset_i$ is called the projection of the function f on the linear subspace L. A linear subspace can be generated by any subset of H, $\{f_1, f_2, \ldots, f_m\}$. Elements of the subspace consist of all functions of the form

$$x = \sum_{i=1}^{m} b_i f_i$$

Problem: In the finite dimensional analog of the above argument an n-dimensional vector F is to be approximated by an r-dimensional $(r \leq n)$ vector G such that the magnitude of the error,

$\|F - G\|$ is minimized. How should the components of G be selected?

3.5 Projection Operator

To summarize the result given above without reference to a particular basis in the linear subspace, a projection operator P is defined. Let L be a linear subspace of H. Then any element of H can be reduced to two components, one in L and the other orthogonal to every element of L: $f = g + h$, $f \epsilon H$, $g \epsilon L$, and $\langle h, x \rangle = 0$ for all $x \epsilon L$. The projection operator P will be defined as a linear operator with $Pg = g$ and $Ph = 0$. (Note that $P(Pg) = P^2 g = Pg$, so that $P^2 = P$ is sometimes taken as the definition of a projection operator.) The projection argument above can be carried out with the projection operator as follows:

Find the function $g \epsilon L$ which minimizes $\|f - g\|^2$ with $f \epsilon H$

$$\|f - g\|^2 = \|f - Pf + Pf - g\|^2$$

$$= \|f - Pf\|^2 + \langle f - Pf, Pf \rangle - \langle f - Pf, g \rangle$$

$$\langle Pf, f - Pf \rangle - \langle g, f - Pf \rangle + \|Pf - g\|^2$$

Since $Pf, g \epsilon L$, and $f - Pf \perp L$,

$$\|f - g\|^2 = \|f - Pf\|^2 + \|Pf - g\|^2 .$$

This is minimized for

$$g = Pf$$

3.6 Application of Projection Theory to Mean Square Minimization

The expected value of two random variables has the properties of an inner product. The projection theory above can be used for random processes by taking

$$\langle f, g \rangle = E(f g) .$$

Random processes with finite variance defined on a linear space are often reasonable models for the statistical description of certain control processes and of certain communication processes. In these problems, the projection theory yields the defining equations for optimum filters in the mean-square sense. These equations must still be solved, and the Hilbert space analysis is also useful for the solution of the integral equations.

Consider the following control system:

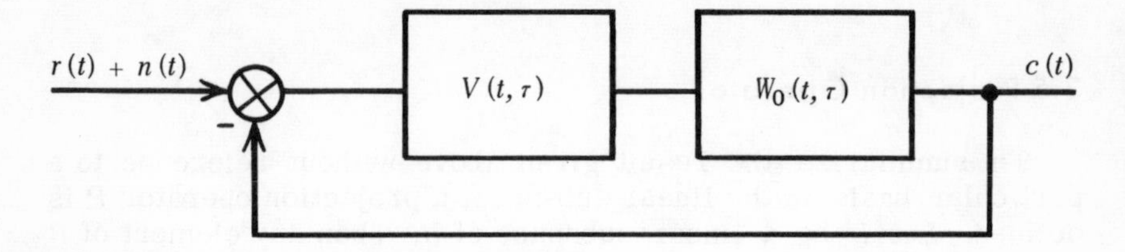

It is desired to minimize

$$\|c - r\|^2 = E\{(c - r)(c - r)\} = E\{ee\} = E|e|^2$$

for all $t: a \leq t \leq b$ (i.e., $t \in T$).

The problem is of the form

x (t) = r (t) + n (t) W (t) c (t)

and if the optimum W is known, the solution for V in terms of W and W_0 may be obtained for the linear system shown.*

The associated filtering problem is to estimate the signal $r(t)$ from the signal plus noise $x(t)$ by the best linear operation on $x(t)$.

$$c(t_0) = \int_{-a}^{t_0} W(t_0, \sigma)\, x(\sigma_n)\, d\sigma$$

$$= \lim_{\substack{N \to \infty \\ \delta_n \to 0}} \sum_{n=1}^{N} a_n\, x(\sigma_n) \qquad a \leq \sigma \leq t_0$$

For this problem, consider the closed linear subspace $\mathfrak{L}$ generated by $\{x(\sigma): a \leq \sigma \leq t_0\}$. It is required to find $c \in \mathfrak{L}$ so that

$$\|r - c\|^2$$

is minimized. From the general arguments above, we know that $c = Pr$ and that $r - c$ must be orthogonal to every element of $\mathfrak{L}$. This means that

$$E\left\{\left[r(t_0) - \int_{a}^{t_0} W(t_0, \sigma)\, x(\sigma)\, d\sigma\right] x(\tau)\right\} = 0$$

for each $\tau: a \leq \tau \leq t_0$. With the usual definition

*Not always true. $V(t,\tau)$ may be unstable. See Newton, Gould and Kaiser's book.

$$E[r(t_0)x(\tau)] = R_{rx}(t_0, \tau)$$

the equation above becomes

$$R_{rx}(t_0, \tau) = \int_a^{t_0} W(t_0, \sigma) R_{xx}(\sigma, \tau) d\sigma$$

This is the general Wiener-Hopf equation, and the solution of this equation for $W(t_0, \sigma)$ is one of the major problems of linear filter design. Another important problem is the determination of the statistics of the various processes. As the above equation indicates, only second-order correlation functions are required for the design of the optimum linear filter.

A polynomial filter may also be considered. Consider the quadratic filter for which

$$c(t_0) = \int_a^{t_0} W(t_0, \sigma) x(\sigma) d\sigma + \int_a^{t_0} \int_a^{t_0} K(t_0, \sigma, \tau) x(\sigma) x(\tau) d\tau d\sigma$$

The key problem in applying the projection theory is the selection of the appropriate linear manifold [13]. For the above problem the linear manifold $\mathfrak{L}$ generated by $\{x(\sigma), x(\sigma)x(\tau)$ for $a \leq \sigma \leq t_0$, $a \leq \tau \leq t_0\}$ is suitable.

In this case the equations

$$R_{rx}(t_0, t_1) = \int_a^{t_0} W(t_0, \sigma) R_{xx}(\sigma, t_1) d\sigma + \int_a^{t_0} \int_a^{t_0} K(t_0, \sigma, \tau) R_{xxx}(\sigma, \tau, t_1) d\tau d\sigma$$

$$R_{rxx}(t_0, t_1, t_2) = \int_a^{t_0} W(t_0, \sigma) R_{xxx}(\sigma, t_1, t_2) d\sigma + \int_a^{t_0} \int_a^{t_0} K(t_0, \sigma, \tau) R_{xxxx}(\sigma, \tau, t_1, t_2) d\tau d\sigma$$

must be solved simultaneously for K and W. Not only are the second order statistics required, but the third and fourth order statistics as well.

Another type of functional optimization problem occurs when the optimum is not in a linear subspace but in a bounded convex subspace [13]. The projection theory is not sufficient in this case.

As another example, suppose we are to determine the gain k of a transfer function with a known step response $s(t)$ by applying a step, and measuring the output of the system. The measurement $x(t)$ is corrupted by additive noise so that $x(t) = ks(t) + n(t)$. If only linear operations on the data are considered,

$$\hat{k} = \int_a^b x(t)h(t)\,dt \qquad \int_0^T |h|^2\,dt < \infty$$

If the noise has zero mean, then

$$E(\hat{k}) = k\int_a^b s(t)h(t)\,dt,$$

and it is reasonable to require that $\int_a^b s(t)h(t)\,dt = 1$ so that the $E(\hat{k})$ is the desired value k. The additional requirement on $h(t)$ is that the variance of the estimate $E\left[(k-\hat{k})^2\right]$ is minimized.

$$\begin{aligned} E(k-\hat{k})^2 &= E\left[\left(\int_a^b n(t)h(t)\,dt\right)^2\right] \\ &= \int_a^b\int_a^b h(s)R(s,t)h(t)\,ds\,dt \end{aligned}$$

By considering the real Hilbert space L_2 on the interval $a \leq t \leq b$ with the inner product

$$(f,g) = \int_a^b f(t)g(t)\,dt$$

one may formulate the above problem as follows:

Minimize (Rh, h) under the constraint $(h, s) = 1$. (R is a covariance function and thus a nonnegative operator.)

The properties of Hilbert space are seen to be useful for formulating the functional problems when minimizing mean-square error. Also, the solution of these problems by steepest-descent techniques can be formulated using the theorems of functional analysis, as will be discussed below.

The foregoing introduction is intended to explain some of the concepts of functional analysis and their relation to control problems. The following development uses these concepts and others not here explained. The interested reader may use references [1, 9, 10, 11, 12] for further exposition.

PART II Mean-Square Error Problems

3.7 Nonnegative Operators in Hilbert Space and Minimization of the Quadratic Functional

Let H denote a real-valued vector Hilbert space. A linear, bounded operator R mapping H into H is said to be symmetric or self-adjoint if, for every two elements x and y in H,

$$\langle R\mathrm{x}, \mathrm{y} \rangle = \langle \mathrm{x}, R\mathrm{y} \rangle \tag{3.1}$$

The operator R is called a nonnegative definite operator if, in addition to the symmetric property,

$$\langle R\mathrm{x}, \mathrm{x} \rangle \geqq 0 \tag{3.2}$$

Furthermore, it is called a positive definite operator if

$$\langle R\mathrm{x}, \mathrm{x} \rangle > 0 \qquad x \neq 0 \tag{3.3}$$

and $\langle R\mathrm{x}, x \rangle = 0$ if and only if $x = 0$ (except for a set of measure zero) [13].

In what follows, we shall restrict our discussions to the square integrable vector function space over a finite interval (O, T) or the $L_2(T)$ space.

For most of the problems considered, the linear operator R is not only nonnegative definite but also compact [14].* This operator is then characterized by a set of denumerable eigenvalues, all nonnegative. Arranging them in nonincreasing order, the nonzero eigenvalues will converge to zero if they are infinite in number. Let $\{\lambda_i\}$ be the set of nonincreasing eigenvalues and $\{\emptyset_i\}$ be the set of corresponding eigenfunctions. Then, from the Hilbert and Schmidt Theorem [1], we have, for any symmetric operator R, the expression

$$R\mathrm{x} = \sum_{i=1}^{\infty} \lambda_i \quad \mathrm{x}, \emptyset_i \quad \emptyset_i \tag{3.4}$$

for each x in L_2. Here $\{\lambda_i\}$ are the set of nonzero eigenvalues. The development of Eq. (3.4) is valid in the sense of mean convergence [10] for all transformations of the form

$$R\mathrm{x} = \int_T \mathrm{R}(s, t)\, \mathrm{x}(t)\, dt$$

where T is a parameter set. If, now, the integral operator R is nonnegative definite and with continuous kernel, then from Mercer's theorem [1], we can expand the integral kernel

*An operator A is compact if it transforms every bounded set into a compact set.

$$R(s,t) = \sum_{i=1}^{\infty} \lambda_i \phi_i(s)\phi_i'(t) \tag{3.5}$$

where the prime is used to denote the transpose of a matrix.

Here the mean convergence of this series is uniform with respect to both variables s and t.

Abstractly, all the mean-square error problems can be treated as problems of minimizing the quadratic functional of the form

$$Q(\mathrm{x}) = \langle R\mathrm{x}, \mathrm{x}\rangle - 2\langle \mathrm{x}, \mathrm{g}\rangle \tag{3.6}$$

Here R is, in general, a nonnegative definite operator in L_2, and g is a given element in L_2. The minimization is made with respect to x in L_2. Suppose now that there is an element in L_2 such that

$$R\mathrm{h} = \mathrm{g} \tag{3.7}$$

Then

$$Q(\mathrm{x}) = \langle R\mathrm{x}, \mathrm{x}\rangle - 2\langle \mathrm{x}, R\mathrm{h}\rangle$$

After adding and subtracting a term $R\mathrm{h}, \mathrm{h}$ and using the symmetric property of R, we have

$$\begin{aligned} Q(\mathrm{x}) &= \langle R\mathrm{x}, \mathrm{x}\rangle - \langle R\mathrm{x}, \mathrm{h}\rangle - \langle R\mathrm{h}, \mathrm{x}\rangle + \langle R\mathrm{h}, \mathrm{h}\rangle - \langle R\mathrm{h}, \mathrm{h}\rangle \\ &= \langle R(\mathrm{x}-\mathrm{h}), \mathrm{x}-\mathrm{h}\rangle - \langle R\mathrm{h}, \mathrm{h}\rangle \end{aligned} \tag{3.8}$$

Assume now that R is actually positive definite. Specifically, this means that there is no nonzero element in L_2 such that

$$R\mathrm{x} = 0 \tag{3.9}$$

and thus zero is not an eigenvalue of R. It is evident, then, that Q will be minimized if we choose

$$\hat{\mathrm{x}} = \mathrm{h}$$

Thus

$$Q_{\min}(\mathrm{x}) = -\langle \mathrm{g}, \mathrm{h}\rangle \tag{3.10}$$

and Eq. (3.7) becomes

$$R\hat{\mathrm{x}} = \mathrm{g} \tag{3.11}$$

This is the Wiener-Hopf integral equation for this minimization problem. If R is compact, then, as is well known, the necessary and sufficient condition for Eq. (3.11) to have a solution in L_2 is [1, 10]

$$\sum_{i=1}^{\infty} \frac{\langle g, \phi_i \rangle^2}{\lambda_i^2} < \infty \tag{3.12}$$

When the kernel of the integral operator R is stationary—i.e., is a function of the difference of its two arguments only—and its Fourier transform is a rational function, the general solution for Eq. (3.11) contains δ functions and their various orders of derivatives at the end points of the interval T.

The case where Eq. (3.9) is satisfied for $x \neq 0$ is usually referred to as the singular case and will be treated first [2, 5]. Let us now denote the null-space [10] of R by H_0, i.e.,

$$Rx = 0 \qquad x \in H_0$$

Let g_0 be the projection of g on this null-space. Then

$$g = g_0 + g_1 \qquad g_0 \in H_0 \tag{3.13}$$

To minimize Q, we can choose

$$x = K\, g_0 \tag{3.14}$$

Since $\langle g, g_0 \rangle = \langle g_0, g_0 \rangle$, we actually have

$$Q(x) = -2K \langle g_0, g_0 \rangle \tag{3.15}$$

It is evident that by choosing K as large as possible

$$Q_{\min}(x) = -\infty \tag{3.16}$$

Let us now consider the case where there is no solution in L_2 for Eq. (3.7). We shall show that there exists a sequence of elements $\{h_n\}$ in L_2 such that

$$\| Rh_n - g \| \to 0 \tag{3.17}$$

and it will be enough to give

$$Q_{\mathrm{Inf}}(x) = \lim_{n \to \infty} Q(h_n) \tag{3.18}$$

To see this we have only to define

$$h_n = \sum_{i=1}^{n} \frac{\langle g, \phi_i \rangle}{\lambda_i} \phi_i \tag{3.19}$$

It is evident that Eq. (3.17) is satisfied.

Now we have, after adding and subtracting the quantity $2\langle R\mathbf{h}_n, \mathbf{x}\rangle + \langle R\mathbf{h}_n, \mathbf{h}_n\rangle$,

$$\begin{aligned} Q(\mathbf{x}) &= \langle R\mathbf{x}, \mathbf{x}\rangle - 2\langle \mathbf{x}, \mathbf{g}\rangle + 2\langle R\mathbf{h}_n, \mathbf{x}\rangle + \langle R\mathbf{h}_n, \mathbf{h}_n\rangle - 2\langle R\mathbf{h}_n, \mathbf{x}\rangle \\ &\quad - \langle R\mathbf{h}_n, \mathbf{h}_n\rangle \\ &= \langle R(\mathbf{x}-\mathbf{h}_n), \mathbf{x}-\mathbf{h}_n\rangle - \langle R\mathbf{h}_n, \mathbf{h}_n\rangle + 2\langle \mathbf{x}, R\mathbf{h}_n - \mathbf{g}\rangle \end{aligned}$$

Thus

$$\begin{aligned} Q(\mathbf{x}) - Q(\mathbf{h}_n) &= \langle R(\mathbf{x}-\mathbf{h}_n), \mathbf{x}-\mathbf{h}_n\rangle + 2\langle R\mathbf{h}_n - \mathbf{g}, \mathbf{x}-\mathbf{h}_n\rangle \\ &\geqq 2\langle R\mathbf{h}_n - \mathbf{g}, \mathbf{x}-\mathbf{h}_n\rangle = 2\langle R\mathbf{h}_n - \mathbf{g}, \mathbf{x}\rangle \end{aligned}$$

Now

$$\langle R\mathbf{h}_n - \mathbf{g}, \mathbf{x}\rangle \leqq \|R\mathbf{h}_n - \mathbf{g}\| \, \|\mathbf{x}\|.$$

Therefore,

$$\langle R\mathbf{h}_n - \mathbf{g}, \mathbf{x}\rangle \to 0 \qquad n \to \infty.$$

Hence we have

$$Q(\mathbf{x}) \geqq \lim_{n\to\infty} Q(\mathbf{h}_n)$$

Also it follows that

$$Q_{\mathrm{Inf}}(\mathbf{x}) = -\lim_{n\to\infty} \langle R\mathbf{h}_n, \mathbf{h}_n\rangle \doteq -\sum_{i=1}^{\infty} \frac{\langle \mathbf{g}, \phi_i\rangle^2}{\lambda_i} \tag{3.20}$$

The solution of the minimization of quadratic functional as introduced above obviously depends on a knowledge of the eigenvalues and eigenfunctions of the operator R. The determination of these quantities is itself difficult. Thus this approach is not too constructive in solving the synthesis problem. In the next section, we shall introduce a successive approximation solution based on the method of steepest descent in Hilbert space. This method will be shown to be the most general approach to the solution of all mean-square error problems.

3.8 The Method of Steepest Descent in Hilbert Space

The use of the steepest descent method for solving the minimization problem of a quadratic functional was first introduced by

Kantarovich in Russian [3]. Not until recently has the application of this method for solving engineering problems appeared in the literature [4, 5, 6, 7, 8]. In particular, Balakrishnan has done a great deal of original work emphasizing and elaborating this method in coping with a large class of engineering problems [5, 6]. The essential idea of this method is contained in the following. In Hilbert space, a quadratic function $Q(x)$ is considered. In seeking its minimum we shall take x_0 as an arbitrary initial approximation. We then try to find the "gradient" at the point x_0, i.e., to select an element z such that $\frac{d}{d\epsilon} Q(x_0 + \epsilon z)$ is maximized for $\epsilon = 0$. Let z_0 be such an element. Since $Q(x_0 + \epsilon z_0)$ is a polynomial of second degree in ϵ, it will attain a minimum for certain ϵ_0. Then the element $x_1 = x_0 + \epsilon_0 z_0$ will be adopted as the next approximation and the process can be repeated as many times as required for accuracy.

Now we consider the quadratic functional

$$Q(x) = \langle Rx, x \rangle - 2\langle x, g \rangle \tag{3.21}$$

Here R is a positive definite operator [10]. Let z (nonzero) be an arbitrary element in H. Then, for any real parameter ϵ, we have

$$Q(x + \epsilon z) = Q(x) + 2\epsilon \langle Rx - g, z \rangle + \epsilon^2 \langle Rz, z \rangle \tag{3.22}$$

Let us assume that x_0 is the initial approximation. In seeking the gradient, we have to maximize the expression

$$\frac{d}{d\epsilon}\Big[Q(x_0 + \epsilon z_0)\Big]_{\epsilon=0} = 2\langle Rx_0 - g, z_0 \rangle \tag{3.23}$$

By using Schwarz's inequality, Eq. (3.23) will be maximized, for $\|z_0\| = 1$, if

$$z_0 = Rx_0 - g \tag{3.24}$$

With this choice of z_0, Eq. (3.22) will attain its minimum with respect to ϵ if

$$\epsilon_0 = -\frac{\|z_0\|^2}{\langle Rz_0, z_0 \rangle} \tag{3.25}$$

Thus, the next approximation will be taken as

$$x_1 = x_0 + \epsilon_0 (Rx_0 - g)$$

and

$$Q(x_1) = Q(x_0) - \frac{\|z_0\|^4}{\langle Rz_0, z_0 \rangle}$$

It is evident that at the n-th step, we have

$$x_n = x_{n-1} - \epsilon_{n-1} z_{n-1} \tag{3.26}$$

where

$$\epsilon_{n-1} = \frac{\| z_{n-1} \|^2}{\langle R z_{n-1}, z_{n-1} \rangle}$$

$$z_{n-1} = R x_{n-1} - g$$

and

$$Q(x_n) = Q(x_{n-1}) - \frac{\| z_{n-1} \|^4}{\langle R z_{n-1}, z_{n-1} \rangle} \tag{3.27}$$

We shall now show that $Q(x_n)$ actually converges to the true infimum [5, 13].

$$Q_{\text{Inf}}(x) = \lim_{n \to \infty} Q(x_n) \tag{3.28}$$

It is evident from Eq. (3.27) that

$$Q(x_n) = Q(x_0) - \sum_{i=0}^{n-1} \frac{\| z_i \|^4}{\langle R z_i, z_i \rangle} \tag{3.29}$$

Now if the infinite series

$$\sum_{n=0}^{\infty} \frac{\| z_n \|^4}{\langle R z_n, z_n \rangle} = +\infty$$

then $Q(x_n)$ will converge monotonically to minus infinity, and certainly the implication is that the infimum has been reached. On the other hand, we have

$$\sum_{n=0}^{\infty} \frac{\| z_n \|^4}{\langle R z_n, z_n \rangle} < +\infty$$

Since

$$\langle R z_n, z_n \rangle \leqq \| R \| \, \| z_n \|^2$$

where

$$\| R \| = \sup_{\|x\| = 1} \langle R x, x \rangle$$

then we have

$$\frac{\|z_n\|^4}{\langle Rz_n, z_n\rangle} \geqq \frac{1}{\|R\|}\|z_n\|^2$$

This certainly implies that

$$\sum_{n=1}^{\infty} \|z_n\|^2 < +\infty \tag{3.30}$$

Thus we have

$$\|z_n\| = \|Rx_n - g\| \to 0 \tag{3.31}$$

However, it can be shown that Eq. (3.31) is enough to imply [5, 8]

$$Q_{\text{Inf}}(x) = \lim_{n\to\infty} Q(x_n)$$

It should be noticed that the sequence $\{x_n\}$ does not necessarily converge strongly.

3.9 Nonlinear Filtering of Random Processes

It has already been pointed out that all mean-square error problems can be viewed as the minimization of a quadratic functional. By using the method of steepest descent for solving these problems, no assumption is made about the stationarity of the processes. In this section, we shall show that even for nonlinear filtering problems, this same procedure is also applicable [5].

Let ξ be a random variable. We shall try to get the best mean-square estimation of this random variable after observing a random process $x(t)$ over a period $0 \le t \le T$. By best estimation, we mean a nonlinear approximation of the form

$$C + \sum_{n=1}^{N} \int_0^T \cdots \int_0^T W_n(t_1, t_2, \ldots t_n)x(t_1)x(t_2)\ldots x(t_n)\,dt_1\ldots dt_n \tag{3.32}$$

where C is a constant. Here $W_n(t_1, t_2, \ldots t_n)$ is symmetric in its variables and is required to be in $L_2(T^n)$. Let us now define an N-dimensional process vector $\mathbf{x}(t)$ by

$$\mathbf{x}(t) = \begin{bmatrix} x(t_1) \\ x(t_1)x(t_2) \\ \cdot \\ \cdot \\ \cdot \\ \cdot \\ \cdot \\ x(t_1)x(t_2)\ldots x(t_n) \end{bmatrix} \tag{3.33}$$

and an N-dimensional weighting function vector $\mathbf{K}(t)$ by

$$\mathbf{K}(t) = \begin{bmatrix} W_1(t_1) \\ W_2(t_1, t_2) \\ \cdot \\ \cdot \\ \cdot \\ \cdot \\ W_N(t_1, t_2, \ldots t_N) \end{bmatrix} \tag{3.34}$$

where $t = (t_1, t_2, \ldots, t_N)$ is in the product space T^N. Then Eq. (3.32) can be written as

$$C + \int_{T^N} \mathbf{K}'(t)\mathbf{x}(t)\,d|t| \tag{3.35}$$

where the prime is used to denote the transpose of a matrix.

The problem is now to choose C and $\mathbf{K}(t)$ such that

$$E\left\{\xi - \left(C + \int_{T^N} \mathbf{K}'(t)\mathbf{x}(t)\,d|t|\right)\right\} = 0 \tag{3.36}$$

and

$$I(\mathbf{K}) = E\left\{\xi - \left(C + \int_{T^N} \mathbf{K}'(t)\mathbf{x}(t)\,d|t|\right)\right\}^2 \tag{3.37}$$

is minimized. Let us define two new quantities such that

$$\tilde{\xi} = \xi - E\{\xi\} \tag{3.38}$$

$$\tilde{\mathbf{x}} = \mathbf{x} - E\{\mathbf{x}\} \tag{3.39}$$

If we replace ξ and $\mathbf{x}$ by $\tilde{\xi}$ and $\tilde{\mathbf{x}}$ respectively in Eq. (3.36), we have

$$C = 0$$

since $E\left\{\tilde{\xi}\right\} = 0$ and $E\{\tilde{\mathbf{x}}\} = 0$. Then Eq. (3.37) becomes

$$I(\mathbf{K}) = E\left\{\tilde{\xi} - \int_{T^N} \mathbf{K}'(t)\tilde{\mathbf{x}}(t)\,d|t|\right\}^2 \tag{3.40}$$

Thus it is convenient to consider these two new quantities instead.

Now the right-hand side of Eq. (3.40) can be expanded

$$I(\mathbf{K}) = E\left\{\tilde{\xi}^2\right\} - 2\int_{T^N} \mathbf{K}'(t)E\left\{\tilde{\xi}\,\tilde{\mathbf{x}}(t)\right\} d|t|$$
$$+ \int_{T^N}\int_{T^N} \mathbf{K}'(s)E\left\{\tilde{\mathbf{x}}(s)\tilde{\mathbf{x}}'(t)\right\}\mathbf{K}(t)\,d|s|\,d|t|$$

This obviously can be written in the operator form as

$$I(\mathbf{K}) = \langle R\mathbf{K}, \mathbf{K}\rangle - 2\langle \mathbf{g}, \mathbf{K}\rangle + E\left\{\tilde{\xi}^2\right\} \tag{3.41}$$

where

$$\mathbf{R}(s,t) = E\left\{\tilde{\mathbf{x}}(s)\tilde{\mathbf{x}}'(t)\right\} \qquad N \times N \tag{3.42}$$
$$\mathbf{g}(t) = E\left\{\tilde{\xi}\,\tilde{\mathbf{x}}(t)\right\}$$

and R is a compact and nonnegative operator in N-dimensional $L_2(T^N)$ space. It is evident that the quantity to be minimized is again a quadratic functional.

$$Q(\mathbf{K}) = \langle R\mathbf{K}, \mathbf{K}\rangle - 2\langle \mathbf{g}, \mathbf{K}\rangle \tag{3.43}$$

The Wiener-Hopf equation for this problem is

$$R\mathbf{K} = \mathbf{g}$$

This equation can also be derived by using the Projection Theorem. The possiblity that there may not be a solution for $\mathbf{K}$ in $L_2^N(T^N)$ still exists. However, by using the same argument as given in Sections 3.1 and 3.2, we can always obtain a sequence $\{\mathbf{K}_n\}$ such that

$$\|R\mathbf{K}_n - \mathbf{g}\| \to 0$$

and

$$Q_{\mathrm{Inf}}(\mathbf{K}) = \lim_{n=\infty} Q(\mathbf{K}_n)$$

Hence the method of steepest descent can certainly be used to generate an approximation sequence. The condition for strong convergence of this sequence can also be expressed in terms of the eigenvalues of R as given in Eq. (3.12). It should be noticed that in nonlinear estimation, higher moments of the input process are required.

3.10 Illustrative Example

Let us now consider the problem of detection of signals in noise. The available input is assumed to be of the form

$$I(t) = S(t) + N(t) \tag{3.44}$$

Here $S(t)$ is a known function of time and $N(t)$ is a stationary random noise with covariance $R_N(\tau)$. It is desired to specify a linear filter to operate on this input over a finite time interval T such that the output signal-to-noise ratio is maximized at some chosen time $t = t_1$. This is a matched-filter problem.

For this simple problem, the analytic solution for the optimal filter can be obtained [9]. Our purpose, then, is to make a comparison between the exact solution and the approximation sequence obtained through the steepest descent method.

Let us denote the output components due to signal $S(t)$ and noise $N(t)$ by $S_0(t)$ and $N_0(t)$ respectively. Then

$$S_0(t_1) = \int_0^T W(\tau)S(t_1 - \tau)\,d\tau \tag{3.45a}$$

$$\triangleq \langle W, S_1 \rangle \tag{3.45b}$$

$$N_0(t_1) = \langle W, N_1 \rangle$$

and

$$E\{N_0^2(t_1)\} = \int_0^T \int_0^T W(u)W(\tau)R_n(u - \tau)\,du\,d\tau \tag{3.46}$$

$$\triangleq \langle RW, W \rangle$$

The problem now is to maximize the ratio

$$\rho = \frac{S_0^2(t_1)}{E\{N_0^2(t_1)\}} \tag{3.47}$$

It is clear that the maximization of Eq. (3.47) is equivalent to the minimization of

$$Q(W) = E\{N_0^2(t)\} - \lambda S_0(t_1) \tag{3.48}$$

subject to the constraint on $S_0(t_1)$. Here x is a Lagrangian multiplier. Equation (3.48) can be conveniently written as

$$Q(W) = \langle RW, W \rangle - 2\left[\frac{\lambda}{2}S_1, W\right] \tag{3.49}$$

Now, from the previously developed theory, we obtain the Weiner-Hopf equation as

$$R\hat{W} = \frac{\lambda}{2} S_1 \qquad 0 \leqq \tau \leqq T \tag{3.50}$$

As far as the signal-to-noise ratio ρ is concerned, the specific number for λ is immaterial. Hence Eqs. (3.49) and (3.50) can be normalized as

$$Q(W) = \langle RW, W \rangle - 2\langle S_1, W \rangle \tag{3.51}$$

and

$$R\hat{W} = S_1 \tag{3.52}$$

For this example, we shall assume that the covariance of noise process is given by

$$R_N(\tau) = e^{-|\tau|} \tag{3.53}$$

The signal portion of the input is given by

$$S(t) = \cos^2 2\pi t \tag{3.54}$$

Let us choose $T = 1$ and $t_1 = T$. Then the integral equation to be solved is

$$\int_0^1 e^{-|t-\tau|} W(\tau) d\tau = \sin^2 2\pi t \tag{3.55}$$

In general, the solution of this integral equation will contain δ-functions at both ends. However, with this particular choice of $S(t)$, no δ-function need be included. Hence the solution is in L_2. It can be shown that the optimal weighting function of the filter is

$$\hat{W}(t) = \frac{1}{2} \sin^2 2\pi t - 4\pi^2 \cos 4\pi t \tag{3.56}$$

The minimum Q is then given by

$$Q_{\min} = -\langle S_1, W \rangle$$

$$= -10.0572$$

Now let us apply the method of steepest descent to this problem. It can be shown that, with the noise covariance as given by

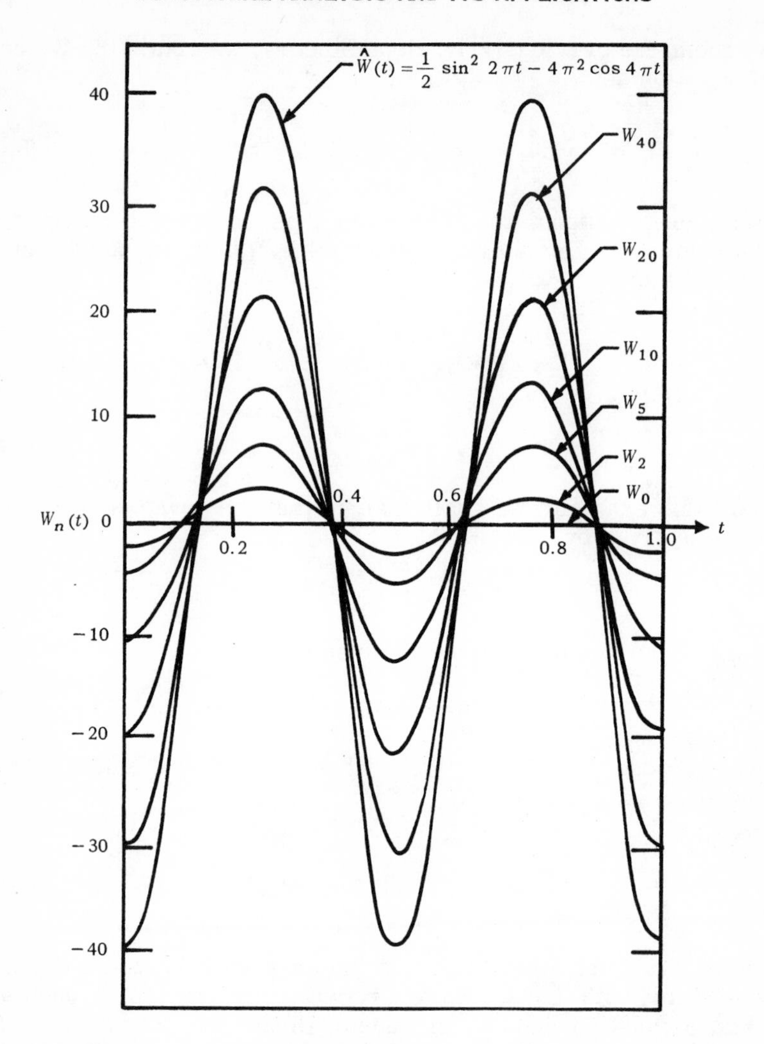

FIG. 3-1. Sequence of approximate solutions for the example by using the steepest descent method.

Eq. (3.53), the operator R considered here is positive definite [9]. We shall then use repeatedly Eq. (3.26). The sequence of approximate solutions is plotted in Fig. 3.1. Some computational results are listed in Table 1. The initial approximation is arbitrarily chosen to be $W_0(t) = 0$. It is interesting to see that after the first iteration, the desired shape of the solution is completely shown. If we define the approximation error as

$$E_n = 1 - \frac{Q(W_n)}{Q_{\min}} \tag{3.57}$$

then we see that

$$E_{40} = 5.36\%$$

$$E_{70} = 1.16\%$$

It should be noted that there are other successive approximation methods which can also be used to solve these problems. However, the steepest descent method is the most fundamental and the simplest one.

Table 3.1. Some Computational Results for the Example by Using the Steepest Descent Method

$$Q_{\min} = -10.0572$$

Number of Iterations n	$Q(W_n)$	$ER_n = 1 - \dfrac{Q(W_n)}{Q_{\min}}$
1	-0.742040	0.926218
2	-1.427547	0.858057
3	-2.061064	0.795066
4	-2.646916	0.736814
5	-3.18820	0.682932
6	-3.690287	0.633070
7	-4.154406	0.586922
8	-4.584077	0.544200
9	-4.981896	0.504644
10	-5.350292	0.468014
15	-6.822408	0.321639
20	-7.826654	0.221786
25	-8.512110	0.153630
30	-8.980070	0.107100
35	-9.299579	0.075331
40	-9.517746	0.053639
50	-9.768479	0.028708
60	-9.885469	0.017075
70	-9.940115	0.011642
80	-9.965699	0.009098
90	-9.977733	0.007902

REFERENCES

1. F. Riesz and B. Sz-Nagy, "Functional Analysis," Frederick Ungar Publishing Co., New York, 1955.
2. A. V. Balakrishnan, "Estimation and Detection Theory for Multiple Stochastic Processes," Journal of Mathematical Analysis and Applications, December, 1960.

3. L. V. Kantarovich, "Functional Analysis and Applied Mathematics," Usp. Mat. Nauk, Vol. 3, 1948.
4. E. Parzen, "A New Approach to the Synthesis of Optimal Smoothing and Prediction Systems," Technical Report No. 34, Applied Mathematics and Statistics Laboratories, Stanford University, July, 1960.
5. A. V. Balakrishnan, "A General Theory of Nonlinear Estimation Problems in Control Systems," presented at the Symposium on Mathematical Problems in Control Systems, Washington, D. C., November, 1961.
6. A. V. Balakrishnan, "An Operator Theoretic Formulation of a Class of Control Problems and a Steepest Descent Method of Solution," J. Soc. Indust. Appl. Math. Ser. A: On Control, Vol. 1, No. 2, 1963, pp. 109-127.
7. H. C. Hsieh, "Synthesis of Optimum Multivariable Control Systems by the Method of Steepest Descent," IEEE Trans. on Application and Industry, Vol. 82, No. 66, 1963, pp. 125-130.
8. H. C. Hsieh, "Synthesis of Adaptive Control Systems by the Function Space Methods," Ph. D. Dissertation, Department of Engineering, University of California, Los Angeles, June, 1963.
9. W. B. Davenport, Jr., and W. L. Root, "An Introduction to the Theory of Random Signals and Noise," McGraw-Hill Book Company, Inc., New York, 1958.
10. A. N. Kolmogorov and S. V. Komin, "Functional Analysis," Vol. 1 and Vol. 2, Graylock Press, Albany, New York, 1961.
11. P. R. Halmos, "Introduction to Hilbert Space," Chelsea Publishing Company, New York, 1957.
12. J. L. Doob, "Stochastic Processes," John Wiley and Sons, Inc., New York, 1960.
13. G. James and R. James, Editors, "Mathematics Dictionary," van Nostrand, Princeton, New Jersey, 1959.
14. M. A. Krasnosel'skiy," Topological Methods in the Theory of Nonlinear Integral Equations," The MacMillan Company, New York, 1964.

4

Shaping Filters for Stochastic Processes

E. B. STEAR
MANAGER, CONTROL AND COMMUNICATION LABORATORY
LEAR SIEGLER INC., RESEARCH LABORATORIES

The ability to represent, in some appropriate statistical sense, a given stochastic process in terms of a related stochastic process whith a simpler probabilisitc structure, has played and continues to play an important role in the development of the general theory of stochastic processes and in the application of this theory to scientific and engineering problems. An outstanding example of such a representation is the spectral representation theorem for stationary stochastic processes. This theorem gives a harmonic (i.e., Fourier) decomposition of the sample functions of continuous in the mean, stationary stochastic processes, which has the property that the various (Fourier) components of the decomposition are uncorrelated. If the components of the decomposition are themselves considered to be a related stochastic process, then it is clear that this harmonic decomposition is indeed a representation of the type mentioned above. The great utility of this particular representation is that it allows one to apply the powerful methods of harmonic analysis to the study of the effects of operations (filtering, etc.) on stochastic processes. That the components of the harmonic decomposition are uncorrelated is of crucial importance in this regard.

Another useful representation of the type mentioned above is representation of a given stochastic process in terms of a linear "filtering" operation made on a related "white noise" process. Stochastic processes with representations of this form are frequently called processes of moving averages (especially by mathematicians). The utility of this representation also depends on the fact that the components of the "white noise" process, i.e., its values at different times, are uncorrelated. The operator which performs the linear "filtering" operation for this representation is often called a *shaping filter,* and here it will be referred to as such. The term *shaping filter* is motivated by the view of the linear "filtering" operation as shaping the spectrum; i.e., the covariance of the components of the harmonic decomposition, of the "white noise" process for the case of stationary stochastic processes. The problem of characterizing the shaping filter for given stochastic processes is naturally referred to as the shaping-filter problem. It is to this problem that this chapter is devoted.

The shaping-filter problem has not been solved for the general case where the given stochastic process has an arbitrary, continuous, covariance function. However, by suitably restricting the class of admissible covariance functions, certain fairly definitive results have been obtained. These results are presented in this chapter. As will be seen later, the spectral representation theorem plays a key role in the results for stationary stochastic processes.

4.1 Motivation

From a control systems engineering point of view, the motivation for research on the shaping-filter problem arises primarily from three sources: the desire to simulate stochastic processes with prescribed covariances; the desire to solve linear, least-square filtering and prediction problems; and the desire to identify linear systems through the use of "white noise" processes as system inputs.

In the simulation of stochastic processes with prescribed covariances, advantage is taken of the availability of noise sources whose outputs are reasonable engineering approximations to "white noise" processes. These sources are applied to the inputs of the simulations (on computers) of appropriate shaping filters to obtain simulations of stochastic processes with the given covariances. Given the desired covariances, this obviously requires the solution of the shaping-filter problem.

In the solution of linear, least-square filtering and prediction problems by traditional methods, the problem reduces to that of solving the Wiener-Hopf integral equation

$$\bar{\Gamma}_{XY}(t_2, t_1) = \int_T W(t_1, \tau)\, \Gamma_{XX}(\tau, t_2)\, d\tau$$

where the covariances $\bar{\Gamma}_{XY}(t_2, t_1)$ and $\Gamma_{XX}(\tau, t_2)$ are given, and the weighting function $W(t_1, \tau)$ of the least-square filter or predictor is sought. If the observed stochastic process $\{X(t), t \in T\}$ is a "white noise" process, then $\Gamma_{XX}(\tau, t_2) = \delta(\tau - t_2)$, and the solution of the Wiener-Hopf equation is immediate. The result is $W(t_1, t_2) = \bar{\Gamma}_{XY}(t_2, t_1)$. This observation leads to the notion of representing the observed stochastic process in terms of a related "white noise" process by a shaping filter, treating the related "white noise" process as if it were the observed stochastic process and solving the "related" Wiener-Hopf equation whose solution is now immediate, and finally, operating on the resulting solution of this "related" problem to obtain $W(t_1, \tau)$. This is, in essence, the Bode-Shannon method for solving such least-square problems. That this method requires the solution of the shaping filter problem is apparent. A particular form for the solution of the shaping-filter problem is also required in the newer method of solution of linear, least-square filtering and prediction problems given by Kalman and Bucy. In both methods, the lack of correlation among the components of the "white noise" processes is extensively exploited.

Finally, the predominance of the role of the shaping-filter problem in identifying linear systems (through the application of "white noise" processes to their inputs and the determination of the covariances of the resulting outputs) is obvious.

4.2 Historical Sketch of the Shaping-Filter Problem

The first significant result relative to the shaping-filter problem was apparently obtained by Wold [1] (1938) for discrete parameter stationary* processes and is part of his fundamental decomposition theorem. Kolmogorov [2, 3, 4] (1939; 1941; 1941) then put Wold's decomposition theorem in an analytic setting and obtained some new theorems for discrete parameter stationary processes, parts of which again pertain to the shaping-filter problem. Independently, Wiener [5] (1942) obtained Kolmogorov's results for discrete parameter stationary processes with absolutely continuous spectral distribution functions, and generalized the results to include continuous-parameter stationary processes with absolutely continuous spectral distribution functions. Wiener thus obtained, as part of his results, the first solution of the shaping-filter problem for continuous-parameter stationary processes. Then Hanner [6] (1949) and Karhunen [7] (1950) obtained, by different methods, the continuous-parameter analog of the Wold decomposition theorem, parts of which again pertain to the shaping-filter problem. Bode and Shannon [8] (1950), in their simplified heuristic derivation of Wiener's results on linear, least-square prediction and filtering theory, stressed the solution of the shaping-filter problem as an important step in their method. The next significant result relative to the shaping-filter problem was obtained by

*"Stationary" should always be interpreted as "wide-sense stationary." The term "processes" is used in this chapter rather than the more cumbersome phrase "stochastic processes."

Darlington [9] (1959) for nonstationary processes, and is a generalization of the method of factorization of rational power spectra used for stationary processes. An earlier, less general result was given by Dolph and Woodbury [10] (1952). Two months later, Batkov [11] (1959) published a paper presenting three methods for solving the shaping-filter problem, including an *algebraic* method using various partial derivatives of the covariance function, for a certain class of continuous-parameter nonstationary processes. As will be pointed out later, Batkov's algebraic procedure only works for a rather specialized subclass of the class claimed. Leonov [12] (1960) has presented a rather nice mathematical solution to the shaping-filter problem for continuous parameter processes (both stationary and nonstationary) in terms of expansions in orthogonal functions.

Finally, Kalman [13] (1961) has recently given a nice representation theorem for Gaussian Markov processes which is of considerable practical utility for both stationary and nonstationary processes.

4.3 Mathematical Preliminaries and Formulation of the Problem

In this chapter, a stochastic process will be defined as a family of random variables $\{X(t), t \epsilon T\}$ where t is the parameter of the family, and T is the set over which the parameter ranges. In practice, t is usually "time," and T is some "time interval." When the set T is obvious from the discussion, the shorter notation $\{X(t)\}$ will be used. Without the brackets, $X(t)$ will be used to denote a sample function of the stochastic process $\{X(t), t \epsilon T\}$. It is assumed that all processes occurring in this chapter have a zero mean unless stated otherwise (i.e., $EX(t) = 0$ for all $t \epsilon T$, etc.), and the covariance function $\Gamma_{XY}(t_2, t_1)$ is defined as $E[X(t_2)\overline{Y}(t_1)]$ where $\overline{Y}(t_1)$ is the complex conjugate of $Y(t_1)$. It is further assumed that all processes except the "white noise" processes are continuous in the mean. That is, $E|X(t) - X(s)|^2 \to 0$ as $|t - s| \to 0$ for all $t, s \epsilon T$. This implies that $\Gamma_{XX}(t_2, t_1)$ is continuous on TXT. All limits of random variables, such as those occurring in the definitions of derivatives and integrals of stochastic processes, are taken to be limits in the mean, and the equality $X(t) = Z(t)$, where $X(t)$ and $Z(t)$ are random variables, is to be interpreted as meaning that $E|X(t) - Z(t)|^2 = 0$ unless stated otherwise. For a detailed discussion of this calculus in the mean, the reader is referred to Loeve [14] (pp. 464–490). When $\Gamma_{XX}(t_2, t_1)$ is only a function of $t_2 - t_1$ for all $t_2, t_1 \epsilon T$, then the process $\{X(t), t \epsilon T\}$ is said to be wide-sense stationary. Otherwise, it is said to be nonstationary. The adjective "wide-sense" will be omitted from here on for convenience. A process $\{Z(t), t \epsilon T\}$ will be called a process with orthogonal increments whenever $E[Z(t_4) - Z(t_3)]\overline{[Z(t_2) - Z(t_1)]} = 0$ for all $t_1, t_2, t_3,$

$t_4 \epsilon T$ such that $t_4 > t_3 \geq t_2 > t_1$. Processes with orthogonal increments have the property that their "derivatives" can formally be considered to be "white noise" processes. If $E|Z(t_2) - Z(t_1)|^2$ depends only on $t_2 - t_1$ for all $t_1, t_2 \epsilon T$, $\{Z(t)\}$ is said to have stationary increments, and its "derivative" can be considered a stationary "white noise" process.

With these preliminaries, the shaping-filter problem can be mathematically formulated as follows. *Given a real valued, continuous in the mean process* $\{Y(t), t \epsilon T\}$, *show that* $\{Y(t)\}$ *can be represented in the form*

$$Y(t) = \int_T W(t, \tau) U(\tau) d\tau$$

where the impulse response of the shaping filter $W(t, \tau)$ *is continuous and* $\{U(t), t \epsilon T\}$ *is a stationary "white noise" process, and give a constructive method for finding* $W(t, \tau)$ *(or its corresponding differential equation if one exists).* In some cases, such as that corresponding to the existence of a differential equation for the shaping filter, the addition of a linear combination $\sum_{i=1}^{n} q_i(t) Y_i$ of real-valued random variables Y_i (corresponding to random initial conditions) to the above may be required. Here the $q_i(t)$ are assumed to be continuous.

If $W^{-1}(t, \tau)$ is defined to be the inverse filter corresponding to $W(t, \tau)$, then it can be shown formally by direct evaluation that $E|Y(t) - \int_T W(t, \tau) U(\tau) d\tau|^2 = 0$ if

$$U(t) = \int_T W^{-1}(t, \tau) Y(\tau) d\tau$$

and $W(t, \tau)$ satisfies the integral equation

$$\Gamma_{YY}(t_2, t_1) = \int_T W(t_2, \tau) W(t_1, \tau) d\tau \tag{4.I}$$

This, in essence, reduces the shaping filter problem to that of solving the integral equation (4.I). However, since $W^{-1}(t, \tau)$ does not normally exist as an ordinary function, and since the integral $\int_T W^{-1}(t, \tau) Y(\tau) d\tau$ does not converge in the mean even with an appropriate interpretation of $W^{-1}(t, \tau)$, this procedure for verifying the representation given above is not mathematically rigorous. On the other hand, the procedure given can be rigorized by treating

$W^{-1}(t, \tau)$ as a generalized function [15] and treating $\int_T W^{-1}(t, \tau) Y(\tau) d\tau$ as a generalized stochastic process [16], where $W(t, \tau)$ is still taken to be the solution of the integral equation (4.I).

In view of the above comments, *the shaping-filter problem can be reduced to that of solving the integral equation (4.I).* If the additional initializing random variables Y_i are required, then Eq. (4.I) is replaced by

$$\begin{aligned} \Gamma_{YY}(t_2, t_1) &= \int_T W(t_2, \tau) W(t_1, \tau) d\tau \\ &+ \sum_{i=1}^{n} \sum_{j=1}^{n} \Gamma_{ij} q_i(t_2) q_j(t_1) \end{aligned} \tag{4.II}$$

where $\Gamma_{ij} = E\, Y_i Y_j$. This is the formulation to be used from here on. It should be noted that the shaping-filter problem, as formulated here, only requires knowledge of the covariance $\Gamma_{YY}(t_2, t_1)$ because the representation is required to hold only in the mean.

There are corresponding problems and results for the discrete-parameter case (i.e., T is a denumerable set), but they are completely analogous to those presented here for the continuous-parameter case and will not be presented.

4.4 The Classical Results for Scalar Stationary Processes

Since no useful purpose would be served insofar as this discussion is concerned, no attempt will be made to present the results for scalar stationary processes in terms of individual contributions as cited in the historical sketch. Rather, an overall summary will be given, the details of which can be found in the books by Doob [17] (pp. 527-559, 569-590) and Grenander and Rosenblatt [18] (pp. 65-82). Only the continuous-parameter case is considered, and naturally, $T = (-\infty, \infty)$.

If the stochastic process $\{Y(t)\}$* is stationary and continuous in the mean, then it has the spectral representation

$$Y(t) = \int_{-\infty}^{\infty} e^{i2\pi\lambda t}\, dZ(\lambda) \tag{4.1}$$

where the process $\{Z(\lambda)\}$ has orthogonal increments and $E|dZ(\lambda)|^2 = dF_Y(\lambda)$. $F_Y(\lambda)$ is called the spectral distribution function of $\{Y(t)\}$ and

*$\{Y(t)\}$ is assumed to be real-valued.

$$\Gamma_{YY}(\tau) = E\left[Y(t+\tau)\overline{Y(t)}\right] = \int_{-\infty}^{\infty} e^{i2\pi\lambda\tau}\, dF_Y(\lambda) \tag{4.2}$$

Furthermore, $F_Y(\lambda)$ is nondecreasing, and since

$$\int_{-\infty}^{\infty} dF_Y(\lambda) = \Gamma_{YY}(0) < \infty \tag{4.3}$$

$F_Y(\lambda)$ is also of bounded variation. Hence, $F_Y(\lambda)$ can be decomposed into the sum of three nondecreasing functions

$$F_Y(\lambda) = F_{Y_1}(\lambda) + F_{Y_2}(\lambda) + F_{Y_3}(\lambda) \tag{4.4}$$

where $F_{Y_1}(\lambda)$ is the jump function part of $F_Y(\lambda)$, $F_{Y_2}(\lambda)$ is the absolutely continuous part of $F_Y(\lambda)$, and $F_{Y_3}(\lambda)$ is the continuous singular part of $F_Y(\lambda)$. This decomposition of $F_Y(\lambda)$ corresponds to a decomposition of $\{Y(t)\}$ into three mutually orthogonal processes $\{Y_1(t)\}$, $\{Y_2(t)\}$, and $\{Y_3(t)\}$ with spectral distribution functions $F_{Y_1}(\lambda)$, $F_{Y_2}(\lambda)$, and $F_{Y_3}(\lambda)$ respectively.

If $\{Y(t)\}$ is applied to the input of a stable linear system whose frequency response function $G(\lambda)$* satisfies the condition

$$\int_{-\infty}^{\infty} |G(\lambda)|^2\, dF_Y(\lambda) < \infty \tag{4.5}$$

then the system output $\{X(t)\}$ will be a continuous in the mean, stationary process whose spectral distribution function, $F_X(\lambda)$, is given by

$$F_X(\lambda) = \int_{-\infty}^{\lambda} |G(\lambda)|^2\, dF_Y(\lambda) \tag{4.6}$$

It should be noted that $G(\lambda)$ is not, in general, required to be in L_2 (i.e., it is not required that $\int_{-\infty}^{\infty} |G(\lambda)|^2\, d\lambda < \infty$). From Eq. (4.6), it follows that $F_X(\lambda)$ is absolutely continuous if $F_Y(\lambda)$ is. If $F_Y(\lambda)$ is absolutely continuous and if $|f_Y(\lambda)|^2 = F'(\lambda)$, then Eq. (4.1) can be replaced by

$$Y(t) = \int_{-\infty}^{\infty} e^{i2\pi\lambda t} f_Y(\lambda)\, d\widetilde{Z}(\lambda) \tag{4.7}$$

where $\{\widetilde{Z}(\lambda)\}$ has orthogonal increments and $E|d\widetilde{Z}(\lambda)|^2 = d\lambda$.

*$G(\lambda)$ is Doob's gain function $C(\lambda)$.

On the other hand, suppose $\{Y(t)\}$ is generated from a process $\{V(t)\}$ according to the equation

$$Y(t) = \int_{-\infty}^{\infty} W(\tau)\, dV(t-\tau) \tag{4.8}$$

where $\{V(t)\}$ has orthogonal increments with $E|dV(t)|^2 = dt$ and $\int_{-\infty}^{\infty} |W(\tau)|^2\, d\tau < \infty$. Then

$$\Gamma_{YY}(t_1, t_2) = \int_{-\infty}^{\infty} W(t_1 - \theta)\, W(t_2 - \theta)\, d\theta \tag{4.9}$$

From Eq. (4.9), it follows that $\{Y(t)\}$ is stationary and continuous in the mean. Furthermore, it is easily shown, by taking the Fourier transform of both sides of Eq. (4.9), that

$$F_Y(\lambda) = \int_{-\infty}^{\lambda} |G(\lambda)|^2\, d\lambda \leq \int_{-\infty}^{\infty} |G(\lambda)|^2\, d\lambda = \int_{-\infty}^{\infty} |W(\tau)|^2\, d\tau \tag{4.10}$$

where $G(\lambda)$ is the Fourier transform of $W(\tau)$*, and hence $F_Y(\lambda)$ is absolutely continuous and $F'_Y(\lambda) = |G(\lambda)|^2$. Formally considering the increments of $\{V(t)\}$ to be given by

$$V(t_2) - V(t_1) = \int_{t_1}^{t_2} U(t)\, dt \tag{4.11}$$

where $\{U(t)\}$ is a "white noise" process, Eq. (4.8) represents the response of a linear system with weighting function $W(\tau)$ to a "white noise" input process, and Eqs. (4.9) and (4.10) represent well-known results usually obtained by engineers in a less rigorous way.**

As a consequence of the above results, a simple necessary and sufficient condition for the existence of a solution to the shaping-filter problem for continuous in the mean, stationary processes can be stated, providing physical realizability of the shaping filter is not required: If $F_Y(\lambda)$ is the spectral distribution function corresponding to the covariance function $\Gamma_{YY}(\tau)$ of the process $\{Y(t)\}$,

*The Plancherel (or L_2) theory of the Fourier transform is appropriate here.

**The usual engineering procedure can also be made rigorous at the expense of introducing generalized linear functionals and processes.

then $\{Y(t)\}$ can be represented in the form given in Eq. (4.8) if and only if $F_Y(\lambda)$ is absolutely continuous. Moreover, any stable linear system whose frequency response function $G(\lambda)$ satisfies the equality $|G(\lambda)|^2 = F'_Y(\lambda)$, almost everywhere (a.e.), can be used as the shaping filter for such a process $\{Y(t)\}$. More generally, even if $F_Y(\lambda)$ is not absolutely continuous, the above still applies to the absolutely continuous part of $F_Y(\lambda)$; i.e., to $F_{Y_2}(\lambda)$ in the decomposition given above. If the requirement of physical realizability is not waived, then the above condition must be strengthened somewhat. Many years ago, Paley and Wiener [19] (p. 16, Theorem XII) showed that if $\int_{-\infty}^{\infty} |G(\lambda)|^2\, d\lambda < \infty$,* where $G(\lambda)$ is the frequency response function of a stable linear system, then the system is physically realizable if and only if

$$\int_{-\infty}^{\infty} \frac{\left|\log|G(\lambda)|^2\right|}{1+\lambda^2}\, d\lambda < \infty \tag{4.12}$$

In view of the required equality $|G(\lambda)|^2 = F'_Y(\lambda)$, a.e., physical realizability of the linear system (the shaping filter) requires the additional condition

$$\int_{-\infty}^{\infty} \frac{|\log F'_Y(\lambda)|}{1+\lambda^2}\, d\lambda < \infty \tag{4.13}$$

That is, for physical realizability of the shaping filter, $F_Y(\lambda)$ must be absolutely continuous and must satisfy Eq. (4.13). Since the above conditions depended only on the magnitude of $G(\lambda)$, it is clear that they do not uniquely determine the shaping filter. A desirable** way of rendering the shaping filter essentially unique (to determine the weighting function uniquely except on a set of Lebesque-measure zero) is to require it to be a minimum phase filter; i.e., to require that $G(\lambda) \neq 0$ for $\operatorname{Im}\lambda < 0$. Such a $G(\lambda)$ is given by the (loss-phase) integral

$$G(\lambda) = \exp\left[-\frac{1}{2\pi i}\int_{-\infty}^{\infty} \frac{(1+\lambda\omega)\log F'_Y(\omega)}{(\lambda-\omega)(1+\omega^2)}\right] d\omega \tag{4.14}$$

*The assumed continuity of $\Gamma_{YY}(\tau)$ guarantees that $\int_{-\infty}^{\infty} |G(\lambda)|^2\, d\lambda < \infty$.

**The inverse filter corresponding to a physically reliable, minimum-phase filter is physically realizable and stable. This is important for applications to linear, least-square filtering and prediction theory.

An important special case occurs when $F'_Y(\lambda)$ is a rational function of λ^2. In this case, $G(\lambda)$ turns out to be a rational function whose poles and zeros are confined to the region $\text{Im}\,\lambda > 0$, and the shaping filter can be characterized by a constant-coefficient ordinary differential equation whose coefficients are the coefficients of the polynomials of $G(\lambda)$. As is well known, use of the loss-phase integral in this case to construct $G(\lambda)$ can be avoided by "simple" factorization of the polynomials of $F'_Y(\lambda)$ and the exclusive association of all poles and zeros of $F'_Y(\lambda)$ in the region $\text{Im}\,\lambda > 0$ with $G(\lambda)$. This is just the procedure discussed in all introductory engineering texts on stochastic processes.

Except for some brief comments which appear at appropriate places throughout the remainder of this chapter, this concludes the discussion of stationary processes.* Clearly, for continuous in the mean, scalar, stationary processes and $T = (-\infty, \infty)$, the shaping-filter problem had been resolved in rather definitive terms prior to 1950.

4.5 Generalization of "Rational Spectrum Factorization" for Nonstationary Processes

For a certain restricted class of nonstationary processes, it is possible in principle to generalize the operation of factorization of a rational power spectrum for handling nonstationary processes. The essential results here, due primarily to Darlington [9], will be described below beginning with certain background information.

If $W(t, \tau)$ denotes the weighting function of a linear filter, then the covariance function $\Gamma_{YY}(t_1, t_2)$ of the output $\{Y(t)\}$ of the filter when the input is a "white noise" process is given, when it exists, by the expression

$$\Gamma_{YY}(t_1, t_2) = \int_{-\infty}^{\infty} W(t_1, \tau)\, W(t_2, \tau)\, d\tau \tag{4.15}$$

Note that since the lower limit of the integral is $-\infty$, it has tacitly been assumed that the "white noise" input has been applied to the filter continuously throughout the infinite past. If the filter is physically realizable, then $W(t, \tau) = 0$ for $\tau > t$ and the upper limit of the integral in Eq. (4.15) can be replaced by $\min(t_1, t_2)$. Letting $W^a(t, \tau)$ denote the weighting function of the adjoint filter, then $W^a(t, \tau) = W(\tau, t)$, and $\Gamma_{YY}(t_1, t_2)$ can be expressed in the equivalent form

$$\Gamma_{YY}(t_1, t_2) = \int_{-\infty}^{\infty} W(t_1, \tau)\, W^a(\tau, t_2)\, d\tau \tag{4.16}$$

*The shaping-filter problem is apparently still unresolved for stationary processes which are not continuous in the mean.

Since Eq. (4.16) expresses $\Gamma_{YY}(t_1, t_2)$ as the convolution of two weighting functions, $\Gamma_{YY}(t_1, t_2)$ can be interpreted as the weighting function of the nonphysically realizable (self-adjoint) system composed of the original filter in cascade with its corresponding adjoint filter.

When the filter is completely characterized by a finite-order linear differential equation, then its response V is related to its excitation E by an expression of the form

$$B(p, t)\, V(t) = H(t)\, A(p, t)\, E(t) \tag{4.17}$$

where $B(p, t)$ and $A(p, t)$ are polynomials in p with time-varying coefficients, where $p = \dfrac{d}{dt}$; i.e.,

$$\begin{aligned} B(p, t) &= p^n + b_{n-1}(t)\, p^{n-1} + \cdots + b_0(t) \\ A(p, t) &= p^m + a_{m-1}(t)\, p^{m-1} + \cdots + a_0(t) \end{aligned} \tag{4.18}$$

and $H(t)$ is a time-varying scale factor. Any set of n linearly independent solutions, say $U_i(t)$, $i = 1, \ldots, n$, of

$$B(p, t)\, V(t) = 0 \tag{4.19}$$

form a set of basis functions (bf's) for $B(p, t)$ and for the filter. Similarly, any set of m linearly independent solutions of

$$A(p, t)\, E(t) = 0 \tag{4.20}$$

form a set of basis functions for $A(p, t)$ and are called the zero-response functions (zrf's) of the filter. If the filter is also time-invariant (stationary), then the bf's and zrf's* are exponentials $e^{S_\sigma t}$, where the S_σ are the familiar poles and zeros of the filter transfer function. The bf's and zrf's of nonstationary systems play analogous, equally important roles even though they cannot be represented by simple coefficients like S_σ.

When two filters are cascaded, where both are completely characterized by a finite-order differential equation, then the overall filter is completely characterized by a finite-order differential equation corresponding to the "product" of the differential equations of the two given filters. In terms of operators, the "product" may be represented by**

$$B_1 V_1 = H_1 A_1 E, \quad B_2 V = H_2 A_2 V_1, \quad BV = HAE \tag{4.21}$$

where $BV = HAE$ is the differential equation of the overall filter. The operators B, A, and H can be determined from B_1, B_2, A_1, A_2, H_1, and H_2 be means of derivative and algebraic operations.*** This corresponds formally to the convolution of the weighting

*or linear combinations of them.

**Suppressing the arguments for convenience in notation.

***For details on this see Chapter 1 of this book.

functions of the two given filters. Similarly, corresponding formally to the sum of weighting functions, there is a suitably defined "sum" of their corresponding differential equations represented by

$$B_1 V_1 = H_1 A_1 E \ , \ B_2 V_2 = H_2 A_2 E$$
$$V = V_1 + V_2 \ , \ BV = HAE \tag{4.22}$$

The operators B and A and the scale factor H can also be determined from $B_1, B_2, A_1, A_2, H_1,$ and H_2 by means of derivative and algebraic operations.* Further, the bf's of B are those of B_1 plus those of B_2, but the bf's of A (the zrf's of the "sum" filter) are not related to those of A_1 and A_2 in any simple way.

Corresponding to the filter characterized by Eq. (4.17) is its related adjoint filter, which is completely characterized by the adjoint differential equation

$$B^a(p, t)\, V(t) = \pm H(t)\, A^a(p, t)\, E(t) \tag{4.23}$$

corresponding to Eq. (4.17), the operators $B^a(p, t)$ and $A^a(p, t)$ being easily determined from $B(p, t)$ and $A(p, t)$. When the filter is physically realizable, the weighting function corresponding to Eq. (4.17) can be expressed in the form [9]

$$W(t, \tau) = \begin{cases} \sum_{i=1}^{n} \dfrac{U_i(t)}{U_i(\tau)} J_i(\tau) \ , & t > \tau \\ 0 \ , & t < \tau \end{cases} \tag{4.24}$$

and that corresponding to Eq. (4.23)—i.e., that of the nonphysically realizable adjoint filter—in the form

$$W^a(t, \tau) = \begin{cases} \sum_{i=1} \dfrac{U_i(\tau)}{U_i(t)} J_i(t) \ , & t < \tau \\ 0 \ , & t > \tau \end{cases} \tag{4.25}$$

The "product" of Eq. (4.17) and (4.23) corresponds to the convolution of $W(t, \tau)$ and $W^a(t, \tau)$, as in Eq. (4.16), and is written as

$$B(p, t)\, V(t) = \pm H^2(t)\, A(p, t)\, E(t) \tag{4.26}$$

From the discussion following Eq. (4.16), it is clear that the weighting function of the filter characterized by Eq. (4.26) is $\Gamma_{YY}(t_1, t_2)$, which, from Eqs. (4.16), (4.24), and (4.25), can be written in the form

$$\Gamma_{YY}(t_1, t_2) = \begin{cases} \sum_{i=1}^{n} \dfrac{U_i(t_1)}{U_i(t_2)} Q_i(t_2) \ , & t_1 > t_2 \\ \sum_{i=1}^{n} \dfrac{U_i(t_2)}{U_i(t_1)} Q_1(t_1) \ , & t_1 < t_2 \end{cases} \tag{4.27}$$

*For details on this see Chapter 1 of this book.

The symmetry of $\Gamma_{YY}(t_1, t_2)$ expresses the fact that Eq. (4.26) is a self-adjoint equation.

With this background, the shaping-filter problem as encountered in the Bode-Shannon model can be resolved if it is assumed that the "signal" $S(t)$ and "noise" $N(t)$ are generated from uncorrelated "white noise" sources by means of physically realizable filters characterized by finite-order, linear differential equations. If the bf's and zrf's of the filters are known, then the weighting functions of the filters $W_S(t, \tau)$ and $W_N(t, \tau)$ are easily determined, and $\Gamma_{SS}(t_1, t_2)$ and $\Gamma_{NN}(t_1, t_2)$ can be found from Eq. (4.16). Also, corresponding differential equations and their adjoints are easily determined as described above (even if the bf's and zrf's of the generating filters are unknown). If $F = S + N$, then $\Gamma_{FF}(t_1, t_2) = \Gamma_{SS}(t_1, t_2) + \Gamma_{NN}(t_1, t_2)$, and a differential equation of the form of Eq. (4.26) whose corresponding weighting function is $\Gamma_{FF}(t_1, t_2)$ can be found from $\Gamma_{FF}(t_1, t_2)$ itself (or by "summing" the differential equations corresponding to $\Gamma_{SS}(t_1, t_2)$ and $\Gamma_{NN}(t_1, t_2)$ if $\Gamma_{FF}(t_1, t_2)$ is unknown). In this way, we determine

$$\begin{aligned} B_S(p, t)\, V(t) &= \pm H_S^2(t)\, A_S(p, t)\, E(t) \\ B_N(p, t)\, V(t) &= \pm H_N^2(t)\, A_N(p, t)\, E(t) \\ B_F(p, t)\, V(t) &= \pm H_F^2(t)\, A_F(p, t)\, E(t) \end{aligned} \tag{4.28}$$

The bf's of $B_F(p, t)$ are those of $B_S(p, t)$ and $B_N(p, t)$ and are even in number, half of them the bf's of the systems used to generate $S(t)$ and $N(t)$, and the other half the bf's of the corresponding non-physically realizable adjoint filters. On the other hand, the bf's of $A_F(p, t)$, again even in number, are not simply related to the bf's of $A_S(p, t)$ and $A_N(p, t)$ and must be found as the solutions of

$$A_F(p, t)\, E(t) = 0 \tag{4.29}$$

This corresponds to the calculation of the zeros of the rational signal-plus-noise spectral density function in the stationary case, in which the spectral densities of S and N are added [corresponding to forming the sum of the differential equations for $\Gamma_{SS}(t_1, t_2)$ and $\Gamma_{NN}(t_1, t_2)$] to get the spectral density of F. The addition retains the poles but the zeros must be calculated as the zeros of a polynomial [corresponding to finding the solutions of Eq. (4.29)].

Now, the shaping-filter problem, as considered here, is that of finding a weighting function $W_F(t, \tau)$ such that the filters corresponding to both the function and its inverse are physically realizable and behave suitably (i.e., are stable) as $\tau \to -\infty$ for all t and such that

$$\Gamma_{FF}(t_1, t_2) = \int_{-\infty}^{\infty} W_F(t_1, \tau)\, W_F^a(\tau, t_2)\, d\tau \tag{4.30}$$

To do this, one finds the bf's of $B_F(p, t)$ and $A_F(p, t)$ from the "known" bf's of $B_S(p, t)$ and $B_N(p, t)$, and by solving Eq. (4.29). The problem then is to assign half of them to $W_F(t, \tau)$ and the remaining half to $W_F^a(t, \tau)$ so that the requirements demanded of $W_F(t, \tau)$ as stated above are met, if possible. It can be shown that it is possible provided the coefficients of the differential equations characterizing the filters used to generate S and N are regular at $t = \infty$, are periodic, or are of moderate variation. In these cases, the bf's either become exponentials as $t \to \pm\infty$, are exponentials multiplied by periodic coefficients, or are dominated by exponentials as $t \to \pm\infty$; and those bf's associated with exponentials $e^{S_\sigma t}$, where $\operatorname{Re} S_\sigma < 0$, are assigned to $W_F(t, \tau)$ just as in the stationary case. The $W_F(t, \tau)$ thereby obtained will then have the required properties.

Before proceeding to the next section, it should be noted that in this section it was assumed that $T = (-\infty, \infty)$, and that $\Gamma_{FF}(t_1, t_2)$ was known to be the sum of two processes generated from uncorrelated "white noise" sources by *physically realizable* finite-order linear filters. Further, no terms due to initial conditions are present in $\Gamma_{SS}(t_1, t_2)$ or $\Gamma_{NN}(t_1, t_2)$ because of the assumption of stability of the S and N shaping filters and the choice of interval T. Although it appears that the method might be more widely applicable than shown so far, its exact generality is apparently unknown at present, and in view of Kalman's work [13] may not be worth determining.

4.6 A Further Generalized Result

The results of the preceding section can be generalized further and in a slightly different way. The result here is due to Batkov [11] and is presented primarily for the sake of completeness. If it is assumed that a given process $\{Y(t)\}$ has a covariance function of the form*

$$\Gamma_{YY}(t_1, t_2) = \begin{cases} \displaystyle\sum_{i=1}^{n} q_i(t_1)\, p_i(t_2) \; ; & t_1 > t_2 \\ \displaystyle\sum_{i=1}^{n} q_i(t_2)\, p_i(t_1) \; ; & t_2 > t_1 \end{cases} \tag{4.31}$$

and if the $q_i(t)$ have n continuous derivatives, then the $q_i(t)$ can be considered as basis functions for the differential operator $B(p, t)$ in Eq. (4.19). Further, the $b_i(t)$ in Eq. (4.18) can be easily determined algebraically from the $q_i(t)$ by a well-known procedure [20].

*Equations (4.27) and (4.24) have been rewritten in different form here for convenience.

(Theorem 6.2). The problem remaining is to find the $a_i(t)$ in the differential operator $A(p, t)$. If it is assumed that $\Gamma_{YY}(t_1, t_2)$ is such that it is possible to solve for the $a_i(t)$ from Eq. (4.31), then

$$\begin{aligned} B(p, t_1)\Gamma_{YY}(t_1, t_2) &= 0 \quad ; \quad t_1 > t_2 \\ B(p, t_1)\Gamma_{YY}(t_1, t_2) &= A(p, t_1)\, W(t_2, t_1) \;;\quad t_2 > t_1 \end{aligned} \tag{4.32}$$

where $W(t, \tau)$ is

$$W(t, \tau) = A^a(p, \tau)\, G(t, \tau) \tag{4.33}$$

and

$$B(p, t)\, G(t, \tau) = \delta(t - \tau) \tag{4.34}$$

Using these results, Eq. (4.32) can be rewritten as

$$B(p, t_1)\Gamma(t_1, t_2) = A(p, t_1)\, A^a(p, t_1)\, G(t_2, t_1) \;;\; t_2 > t_1 \tag{4.35}$$

Since $G(t_2, t_1)$ can also be found algebraically once the $q_i(t)$ are known, the only unknown in Eq. (4.35) is the product operator $A(p, t_1)\, A^a(p, t)$ which can be determined from Eq. (4.35). The remaining step is to decompose it into its adjoint factors. As Batkov points out, this is very difficult unless $A(p, t)$ is a scalar, in which case it is simple. Of course, in this case the solution has been known for a long time [10]. It will be observed that the problem of factoring $A(p, t)\, A^a(p, t)$ is just the problem of factoring $A_F(p, t)$ in Darlington's work.

Because of this difficulty, Batkov [11] developed another method for determining the $a_i(t)$* which involves the use of the discontinuities of the partial derivatives of $\Gamma_{YY}(t_1, t_2)$ at $t_1 = t_2$. He arrives at his equation (41) which, he insists, introduces the $a_i(t)$ recursively and algebraically as μ varies beginning with $a_m(t)$ and ending with $a_0(t)$. However, if the steps he outlines are carried out in detail, it is found that every other one of the recursive equations is strictly dependent on those before it, and the recursive method fails. Batkov did not discover this because he did not carry out the details and only considered simple examples. All attempts to modify the procedure have produced sets of (Ricatti-like) nonlinear differential equations in the $a_i(t)$ instead of algebraic equations as desired.

4.7 Some Fundamental Results for Nonstationary Processes

As is well known, *physically realizable* shaping filters do not, in general, exist for processes with arbitrary covariance functions.

*Our $a_i(t)$ are Batkov's $b_i(t)$ and vice versa.

For example, as pointed out in Section 4.4 for the stationary case, physical realizability of the shaping filter requires that the spectral distribution function be absolutely continuous and satisfy the Paley-Wiener criterion given in Eq. (4.13). Apparently, no simple criterion analogous to that of Paley and Wiener has been developed for the general nonstationary case. This is not too surprising, considering the difficulty of the problem. In this section, the question of the existence of *physically realizable* shaping filters is discussed for the class of separable covariance functions, and for this class it is seen that, providing one remarkably simple requirement is met, a *physically realizable shaping filter does indeed exist*. In addition, the question of uniqueness of the shaping filter is also discussed. These results are discussed further in a recent report [21]. The restriction to the class of separable covariance functions certainly seems reasonable in view of the fact that in this case the resulting shaping filter is usually rather easily realized physically. This is, of course, of importance in engineering applications.

It is rather interesting and enlightening to examine the treatment of this question for the nonstationary case as given in the two preceding sections. Darlington was apparently well aware of the physical realizability problem and did provide answers for two rather restrictive cases. They were restrictive in the sense that he *assumed physical realizability* of the underlying signal- and noise-shaping filters, and either periodicity or regularity at ∞ of the corresponding differential equations. His answers clearly leave much to be desired. Batkov simply avoided the problem by implicitly making the assumption that the covariance function was of the required form. Leonov, whose work is discussed below, was not concerned about physical realizability of the shaping filter (and neither discussed nor obtained it) because it was not required for his application. Kalman, whose work is also discussed below, does obtain a physically realizable shaping filter.

Neglecting initial conditions for the moment, establishing the existence of a physically realizable shaping filter amounts to establishing the existence of a solution of the nonlinear Volterra integral equation of the first kind

$$\Gamma(t_1, t_2) = \int_0^{t_2} d\tau W(t_1, \tau) W(t_2, \tau) \quad , \quad t_1 \geq t_2 \geq 0 \tag{4.36}$$

When the covariance function is separable; i.e.,

$$\Gamma(t_1, t_2) = \sum_{i=1}^{n} q_i(t_1) p_i(t_2) \quad ; \quad t_1 \geq t_2 \geq 0 \tag{4.37}$$

it is reasonable, in view of Eqs. (4.24) and (4.27), to consider solutions of the form

$$W(t, \tau) = \begin{cases} \sum_{i=1}^{n} q_i(t) \beta_i(\tau) \ ; & t \geq \tau \geq 0 \\ 0 \ ; & t < \tau \end{cases} \tag{4.38}$$

In this case, the integral equation becomes

$$\sum_{i=1}^{n} q_i(t_1) p_i(t_2) = \sum_{i=1}^{n} q_i(t_1) \sum_{j=1}^{n} q_j(t_2) \int_0^{t_2} d\tau \, \beta_i(\tau) \beta_j(\tau) \tag{4.39}$$

and, upon making use of the linear independence of the $q_i(t_1)$ and adding initial condition terms, there results

$$p_i(t) = \sum_{j=1}^{n} q_j(t) \left[\int_0^t d\tau \, \beta_i(\tau) \beta_j(\tau) + \Gamma_{ij} \right] ; \quad t \geq 0 \, , \ i = 1 \, ; \ \ldots . , n \tag{4.40}$$

Thus, for the case of separable covariance functions, the problem has been reduced to establishing the existence of a solution of the simultaneous nonlinear Volterra integral equations of the first kind given in Eq. (4.40). This certainly represents a reduction over Eq. (4.36) since Eq. (4.36) actually represents an infinite set of simultaneous integral equations, one for each value of t_1.

The usual procedure in the study of Volterra integral equations of the first kind is first to convert the integral equation into an integral equation of the second kind, and then apply the standard techniques known for Volterra integral equations of the second kind. For linear equations, this conversion is easily carried out [22]. That such a procedure can also be carried out for the equations in (4.40) is perhaps not obvious, but nevertheless it can be accomplished as follows.

Upon differentiating the equations in (4.40), there results

$$p_i^{(1)}(t) = \sum_{j=1}^{n} q_j^{(1)}(t) \left[\int_0^t d\tau \beta_i(\tau) \beta_j(\tau) + \Gamma_{ij} \right] + \beta_i(t) \sum_{j=1}^{n} q_j(t) \beta_j(t) \tag{4.41}$$

Now, examination of Eq. (4.41) shows that the multiplier of $\beta_i(t)$ is the same for all i; namely, $\sum_{i=1}^{n} q_j(t) \beta_j(t)$. Let $k(t) = \sum_{j=1}^{n} q_j(t) \beta_j(t)$. The only problem is that $k(t)$ is unknown. If $k(t)$ were known and nonzero for all $t > 0$, then the desired conversion to integral equations of the second kind would be complete upon division by $k(t)$. While at first glance it may appear that since $k(t)$ involves the unknown $\beta_i(t)$, there is no hope of being able to determine it; it can, nevertheless, be determined. Solving Eq. (4.41) for $\beta_i(t)$ it is found that

$$\beta_i(t) = k^{-1}(t)\left\{p_i^{(1)}(t) - \sum_{j=1}^{n} q_j^{(1)}(t)\left[\int_0^t d\tau\beta_i(\tau)\beta_j(\tau) + \Gamma_{ij}\right]\right\} \tag{4.42}$$

Substituting back in Eq. (4.41), there results

$$\begin{aligned} p_i^{(1)}(t) - \sum_{j=1}^{n} q_j^{(1)}(t)\left[\int_0^t d\tau\beta_i(\tau)\beta_j(\tau) + \Gamma_{ij}\right] \\ = k^{-2}(t)\left\{p_i^{(1)}(t) - \sum_{j=1}^{n} q_j^{(1)}(t)\left[\int_0^t d\tau\beta_i(\tau)\beta_j(\tau) + \Gamma_{ij}\right]\right\} \times \\ \left\{\sum_{j=1}^{n} q_j(t)\left(p_j^{(1)}(t) - \sum_{k=1}^{n} q_k^{(1)}(t)\left[\int_0^t d\tau\beta_j(\tau)\beta_k(\tau) + \Gamma_{jk}\right]\right)\right\} \end{aligned} \tag{4.43}$$

which, upon cancelling and rearranging terms, yields

$$\begin{aligned} k^2(t) = \sum_{j=1}^{n} q_j(t)\, p_j^{(1)}(t) - \sum_{k=1}^{n} q_k^{(1)}(t) \sum_{j=1}^{n} q_j(t) \times \\ \left[\int_0^t d\tau\beta_j(\tau)\beta_k(\tau) + \Gamma_{jk}\right] \end{aligned} \tag{4.44}$$

But the second term in Eq. (4.44) is just $p_k(t)$; hence

$$k(t) = \pm\sqrt{\sum_{j=1}^{n} q_j(t)\, p_j^{(1)}(t) - q_j^{(1)}(t)\, p_j(t)} = \pm\sqrt{J_{0,1}(t)} \tag{4.45}$$

Thus $k(t)$ *can be determined* from $\Gamma(t_1, t_2)$.

It is quite important to note that if the $\beta_i(t)$ are to be real, then Eq. (4.42) together with Eq. (4.45) requires that $J_{0,1}(t) \geq 0$ for all $t \geq 0$. This completes conversion of the integral equations of the first kind given in Eq. (4.40) to integral equations of the second kind as given in Eq. (4.42).

On course, there still remains the problem of what to do in case $k(t) \equiv 0$. In this case, the equations given in Eq. (4.41) reduce to a set of integral equations of the first kind of the form given in Eq. (4.40), with $q_i(t)$ and $p_i(t)$ replaced by $q_i^{(1)}(t)$ and $p_i^{(1)}(t)$ respectively. Hence, the logical thing to do is to reapply the conversion procedure used in the previous paragraph on Eq. (4.40). When this is done, one obtains

$$\beta_i(t) = k_1^{-1}(t)\left\{p_i^{(2)}(t) - \sum_{j=1}^{n} q_j^{(2)}(t)\left[\int_0^t d\tau\beta_i(\tau)\beta_j(\tau) + \Gamma_{ij}\right]\right\} \tag{4.46}$$

where

$$k_1(t) = \pm \sqrt{\sum_{i=1}^{n} q_i^{(1)}(t)\, p_i^{(2)}(t) - q_i^{(2)}(t)\, p_i^{(1)}(t)} = \pm\sqrt{J_{1,2}(t)} \qquad (4.47)$$

For the same reason as before, it is required that $J_{1,2}(t) \geq 0$ for all $t \geq 0$. Naturally, if $k_1(t) \equiv 0$, then one reapplies the procedure to the new equations, etc.

When $k(t) = 0$ (or $k_1(t) = 0$, etc.) for some values of t but not identically, then one is dealing with the more complicated type of integral equation Picard called an equation of the third kind. Such cases have been studied for linear equations by Lalesco [23].

The problem now has been reduced to establishing the existence of a solution of the integral equations of the second kind given in Eq. (4.42). This can be done by making use of some results due to T. Sato [24], who treats the existence of solutions of nonlinear Volterra integral equations by means of a fixed-point theorem.

The basic idea underlying fixed-point theorems can be nicely demonstrated by the following simple example. Let C be the set $\{x : 0 \leq x \leq 1\}$ and let $\sigma(x)$ be a continuous, single-valued transformation of C into itself (i.e., $\sigma(x)$ is a continuous, single-valued function defined on [0, 1], for which $\sigma(x) \epsilon [0,1]$ for all $x \epsilon [0,1]$). Then there exists an $x_0 \epsilon C$ such that $x_0 = \sigma(x_0)$. For transformation $\sigma(x)$, x_0 is called a fixed point. The truth of this result is obvious from Fig. 4-1. It is also obvious that x_0 may be either 1 or 0 and that it is not necessarily unique (there are five fixed points in Fig. 4-1).

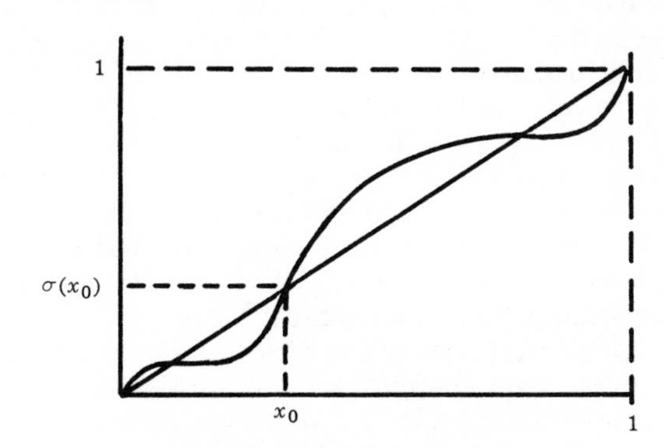

FIG. 4-1. Illustration of fixed point theorem.

The generalization of this simple result to more general sets C in more general underlying topological spaces has led to the development of rather powerful (fixed-point) theorems for establishing the existence of solutions (fixed points) of functional equations in general and integral equations in particular. For integral equations, C becomes a class of functions and σ is an integral operator, e.g., the right-hand side of Eq. (4.42), and asserting the existence of a fixed point for σ is clearly equivalent to asserting the existence of a solution for the corresponding integral equation. One of the most general fixed-point theorems and the one apparently used by Sato was proven by Schauder [25] (p. 260), and can be stated as follows:

Schauder's Theorem—Let C be a nonempty, compact, convex set from a locally convex space X, and let σ be a continuous, single-valued transformation of C into C. Then there exists an $x_0 \epsilon C$ such that $\sigma(x_0) = x_0$.

In applying Schauder's Theorem, the essential problem is, of course, to find an appropriate class C for the problem at hand.

It is interesting to note that the requirements of compactness and convexity of C stated in the theorem could have been anticipated on the basis of the simple example given above.

By a straightforward application of Sato's results to the set of integral equations given in Eq. (4.42), the following important result is easily deduced.

Suppose that $\Gamma_{YY}(t_1, t_2)$ *is of the form given by Eq. (4.37), that* $q_i^{(1)}(t)$ *and* $p_i^{(1)}(t)$ *exist and are continuous on* $[0,T]$, *that* $J_{0,1}(t) > 0$ *on* $[0,T]$, *and that there exists a nonnegative-definite matrix* Γ_{ij} *such that* $p_i(0) - \sum_{j=1}^{n} \Gamma_{ij}\, q_j(0) = 0$ *for all* i. *Then a physically realizable shaping filter exists and* $W(t, \tau)$ *is of the form given in Eq. (4.38), where the* $\beta_i(t)$ *are continuous on* $[0,T]$.

The only possible difficulty with this result is that it may be impossible to extend the solution to $[0, T]$. It can be shown that such an impossibility would imply that the $\beta_i(t)$ become unbounded, and hence discontinuous on $[0, T]$. *In the statement of this result and in those to follow, it has been assumed that it is possible to extend the solution to* $[0, T]$. In any case, the result can be shown to hold on a subinterval $[0, T']$ where $0 < T' \leq T$. Cases where the $\beta_i(t)$ are unbounded are not of great importance in engineering applications; furthermore, they involve computational problems.

When $J_{0,1}(t) = 0$ on $[0, T]$ but $J_{1,2}(t) > 0$ on $[0, T]$, then the following modified form of the above result holds.

Suppose that $\Gamma_{YY}(t_1, t_2)$ *is of the form given by Eq. (4.37), that* $q_i^{(2)}(t)$ *and* $p_i^{(2)}(t)$ *exist and are continuous on* $[0,T]$, *that* $J_{1,2}(t) > 0$ *on* $[0, T]$, *and that there exists a nonnegative-definite matrix* Γ_{ij} *such that* $p_i(0) - \sum_{i=1}^{n} \Gamma_{ij} q_j(0) = 0$, *and* $p_i^{(1)}(0) - \sum_{i=1}^{n} \Gamma_{ij} q_j^{(1)}(0) = 0$, *for all* i. *Then a physically realizable shaping filter exists on* $[0, T]$ *whose weighting function is of the form given in Eq. (4.38), the* $\beta_i(t)$ *are continuous on* $[0, T]$, *and* $W(t, t) \equiv 0$ *for* $t \epsilon [0,T]$.

Further modification of the first result, when $J_{0,1}(t) = J_{1,2}(t) \equiv 0$ on $[0, T]$ but $J_{2,3}(t) > 0$ on $[0, T]$, is obvious. The case where $J_{0,1}(t) = 0$ for some $t \epsilon [0, T]$, but not identically, is not discussed, but satisfactory results could possibly be obtained by following up Lalesco's work [23].

Finally, suppose that in addition to satisfying the hypotheses for the above results, the $q_i(t)$ and $p_i(t)$ have n continuous derivatives on $[0, T]$ and the Wronskian of the $q_i(t)$ doesn't vanish on $[0, T]$. Then by successive differentiation of Eqs. (4.42) it follows that the $\beta_i(t)$ have $n-1$ continuous derivatives on $[0, T]$. It then follows that the shaping filter can be characterized by a differential equation of the form given in Eq. (4.17) where the $a_i(t)$ and $b_j(t)$ are continuous on $[0, T]$.

Having resolved the question of the existence of a *physically realizable shaping filter*, the question of the uniqueness of the shaping filter naturally arises. Examining Eq. (4.40), it is clear that there is no unique solution because, if $W(t, \tau)$ is a solution, then $-W(t, \tau)$ is also a solution. Note that if $W(t, \tau)$ is the solution associated with the plus sign in Eq. (4.45), then $-W(t, \tau)$ is the solution associated with the minus sign. However, the question still remains as to whether the solution is unique, say, to within a multiplicative factor of absolute value 1. The answer again is no, because there may be more than one nonnegative-definite matrix Γ_{ij} which meets the requirements for obtaining the results, and which leads to different solutions. As an example, consider the covariance function

$$\Gamma(t_1, t_2) = 4/3e^{-|t_1 - t_2|} - 5/12e^{-2|t_1 - t_2|}; \; t_1, t_2 \geq 0$$

The two matrices

$$\Gamma'_{ij} = \begin{bmatrix} 2 & -2/3 \\ -2/3 & 1/4 \end{bmatrix}, \quad \Gamma''_{ij} = \begin{bmatrix} 8 & -20/3 \\ -20/3 & 25/4 \end{bmatrix}, \quad J_{0,1}(t) \equiv 1$$

meet the requirements of the first result. Furthermore, as direct substitution in Eq. (4.42) shows, $\beta_1(t) = 2e^t$ and $\beta_2(t) = -e^{2t}$ are solutions on $[0, \infty)$ for Γ'_{ij}, while $\beta_1(t) = -4e^t$ and $\beta_2(t) = 5e^{2t}$ are solutions on $[0, \infty)$ for Γ''_{ij}. Hence, $W_1(t, \tau) = 2e^{-(t-\tau)} - e^{-2(t-\tau)}$ and $W_2(t, \tau) = -4e^{-(t-\tau)} + 5e^{-2(t-\tau)}$ are, respectively, the weighting functions of the physically realizable shaping filters for these matrices. Taking the Laplace transforms of $W_1(t-\tau)$ and $W_2(t-\tau)$, there results $G_1(S) = \dfrac{S+3}{(S+1)(S+2)}$ and $G_2(S) = \dfrac{S-3}{(S+1)(S+2)}$. It is interesting to note that the transfer function of the system associated with Γ'_{ij} has its zero in the left half-plane while that associated with Γ''_{ij} has its zero in the right half-plane. In the light of this example, the question now arises as to whether the solution is unique if, say, the plus sign in Eq. (4.45) is used and a matrix Γ_{ij} which meets the hypotheses for the results is specified. The answer this time is yes, as the following result, which can be established by successive approximations, shows.

If the hypotheses of the results stated in italics above are satisfied, then a physically realizable shaping filter exists on $[0, T]$, *and, if the sign in Eq. (4.45) is chosen and the matrix* Γ_{ij} *specified, the shaping filter is unique.*

This concludes the discussion of existence and uniqueness of *physically realizable* shaping filters for stochastic processes with separable covariances. It is clear that Eq. (4.42) can be used for purposes of calculation.

4.8 Results Based on Series Expansions

When a process $\{X(t)\}$ can be represented in terms of an infinite series with uncorrelated coefficients, the shaping filter problem can be solved in a fairly straightforward manner, provided the requirement of physical realizability is waived. The main result here is due to Leonov [12].

In this case, the shaping-filter problem can be reformulated as follows. Given a "white noise" process $\{Y(t)\}$* where $-\infty < t < \infty$ [i.e., $T_Y = (-\infty, \infty)$], and a nonstationary process $\{X(t)\}$ where $0 < t < T$ [i.e., $T_X = (0,T)$], it is required to show that the random function $X(t)$ (a sample function of $\{X(t)\}$) can, under certain conditions, be represented in the form

$$X(t) = A_X Y(t) \tag{4.48}$$

where the (linear) operator A_X is defined if the function $X(t)$ is given.** The corresponding inverse problem is that of representing $Y(t)$ in the form

$$Y(t) = A_X^{-1} X(t) \tag{4.49}$$

where A_X^{-1} is the operator inverse to A_X. It can be shown that this can be done by explicitly constructing a suitable A_X and a suitable A_X^{-1} as follows:

As is well known [26], a random function $Z(t)$, $T_Z = (a,b)$, can be represented as a series (canonical expansion)

$$Z(t) = \sum_{i=1}^{\infty} B_i z_i(t)^{***} \tag{4.50}$$

where the B_i are random variables which satisfy the conditions

$$E B_i B_j = \delta_{ij} D_j \tag{4.51}$$

*It is always assumed that $E\,Y(t) = 0$ for all "white noise" processes considered here.

**Pugachev [26, 27] calls Eq. (4.48) the integral canonical representation of $x(t)$.

***The $z_i(t)$ are not necessarily orthogonal, and $-\infty \le a < b \le \infty$.

and the $z_i(t)$ are some regular (nonrandom) functions. So that the series in Eq. (4.50) converge in the mean to $Z(t)$, it is necessary and sufficient that the series

$$\Gamma_{ZZ}(t_1, t_2) = \sum_{i=1}^{\infty} D_i \, z_i(t_1) \, z_i(t_2) \tag{4.52}$$

converge to $\Gamma_{ZZ}(t_1, t_2)$ pointwise. The definition of convergence in the mean is, of course, meaningful only for random functions with finite variances.

Now, to solve the problem, it is necessary to represent $Y(t)$ by a series of the form of Eq. (4.50). However, since $Y(t)$ does not have a finite variance, convergence in the mean cannot be used, and a new concept of convergence must be introduced. Leonov introduces the concept of weak convergence in the mean.* A sequence of random functions $U_n(t)$ is said to converge weakly in the mean to the random function $U(t)$ if the integral

$$a_n(T) = \int_0^T R(t)\, U_n(t)\, dt \tag{4.53}$$

converges in the mean as $n \to \infty$ for any "sufficiently smooth" random function $R(t)$, i.e., for any $R(t)$ which has finite variance, is continuous in the mean, has the necessary number of continuous stochastic derivatives, and whose covariance function $\Gamma_{RR}(t_1, t_2)$ satisfies the inequality $\int_{-\infty}^{\infty} \Gamma_{RR}^2(t, t)\, dt < \infty$. With this definition of convergence, it can be shown that $Y(t)$ can be represented in the form

$$Y(t) = \sum_{i=1}^{\infty} C_i y_i(t) \tag{4.54}$$

where $E\, C_i C_j = \delta_{ij}$ and the $y_i(t)$ are any complete (in L_2) set of orthonormal functions over $(-\infty, \infty)$, and where the series in Eq. (4.54) converges weakly in the mean to the "white noise" random function $Y(t)$.

It is now fairly easy to solve the basic problem. The C_i in Eq. (4.54) are defined

$$C_i = \frac{V_i}{\sqrt{D_i}} \tag{4.55}$$

*This is clearly analogous to the ordinary concept of weak convergence in Hilbert space [28].

where the random variables V_i are the coefficients in the series expansion of $X(t)$

$$X(t) = \sum_{i=1}^{\infty} V_i X_i(t) \tag{4.56}$$

and $D_i = E V_i^2$. The linear operator A_X is then defined as

$$A_X Y(t) = \int_{-\infty}^{\infty} W_X(t, \tau) Y(\tau) d\tau \tag{4.57}$$

where

$$W_X(t, \tau) = \sum_{i=1}^{\infty} \sqrt{D_i} X_i(t) y_i(\tau) \tag{4.58}$$

Then, from Eqs. (4.57) and (4.58),

$$A_X Y(t) = \int_{-\infty}^{\infty} W_X(t, \tau) Y(\tau) d\tau = \sum_{i=1}^{\infty} V_i X_i(t) = X(t) \tag{4.59}$$

where the integral in Eq. (4.59) is taken in the mean.

The functions $W_X(t, \tau)$ and $Y(\tau)$ in Eq. (4.59) can be defined in infinitely many ways by using any other representation of $X(t)$ in the form Eq. (4.56), as is shown to be possible by Pugachev [26]. However, Leonov shows that if $Y(t)$ is so chosen that Eq. (4.59) holds, then there is one and only one $W_X(t, \tau)$; i.e., $W_X(t, \tau)$ is unique.

Finally, the inverse problem is easily solved. Let A_X^{-1} be defined as

$$A_X^{-1} X(t) = \int_{-\infty}^{\infty} W_X^{-1}(t, \tau) X(\tau) d\tau \tag{4.60}$$

where

$$W_X^{-1}(t, \tau) = \sum_{i=1}^{\infty} y_i(t) a_i(\tau) / \sqrt{D_i} \tag{4.61}$$

and the $a_i(\tau)$ are chosen so that

$$\int_0^T a_i(\tau) X_j(\tau) d\tau = \delta_{ij} \tag{4.62}$$

As before, the $y_i(t)$ are any complete (in L_2) set of orthonormal functions over $(-\infty, \infty)$. Then from Eqs. (4.53), (4.60), (4.61), and (4.62) it follows that

$$Y(t) = \int_0^T W_X^{-1}(t,\tau)\, X(\tau)\, d\tau = \sum_{i=1}^{\infty} \frac{V_i}{\sqrt{D_i}}\, y_i(t) \qquad (4.63)$$

The series in Eq. (4.63) converges weakly in the mean to the "white noise" random function $Y(t)$ as noted above.

This completes the discussion of the solution to the shaping-filter problem and the corresponding inverse shaping-filter problem by series expansion. It is clear that the weighting function for the shaping filter can be written down immediately in series form once the $X_i(t)$ and D_i for the canonical expansion of $X(t)$ are known. In this book [26], Pugachev presents several techniques for finding the first n terms of expansions of the form of Eq. (4.56) rather simply, which avoid the need for determining the eigenvalues and eigenfunctions of an integral equation as required in the well-known Karhunen-Loeve Representation Theorem. However, it should be noted that Leonov's solution is *always obtained in the form of an infinite series* and, further, there is *no guarantee of physical realizability of the shaping filter or its inverse.*

4.9 A Representation for Vector Processes

All stochastic processes considered in this chapter up to the present section have been scalar processes; i.e., the random variable $X(t)$ has been a scalar-valued variable for all $t \epsilon T$. In contrast, this section considers the shaping filter problem for vector processes. Specifically, a very useful representation is developed for a class of vector processes, and this representation is then used to derive Kalman and Bucy's results [29] for linear, least-square filtering and prediction problems.* The particular representation developed here is equivalent to that developed by Kalman [13] for Gaussian-Markov processes whenever the process being represented is Gaussian, but both the development and the form of the result differ from that given by Kalman. The reason for presenting this development as opposed to Kalman's is that it is simpler and more consistent with the previous sections of this chapter.

As motivation for the development of the representation, consider the case where a real vector process $\{\mathbf{X}(t),\ t \epsilon [0, \infty)\}$ is generated from a "white noise" process $\{\mathbf{W}(t),\ t \epsilon [0, \infty)\}$ according to the first-order vector differential equation

$$\dot{\mathbf{X}}(t) = \mathbf{F}(t)\,\mathbf{X}(t) + \mathbf{W}(t) \qquad (4.64)$$

where the matrix $F(t)$ is continuous in t. Then

*For a statement of Kalman and Bucy's results, see Section 2.12 of Chapter 2 of this book.

$$\mathbf{X}(t) = \phi(t)\,\mathbf{X}(0) + \phi(t)\int_0^t \phi^{-1}(s)\,\mathbf{W}(s)\,ds \tag{4.65}$$

where the (fundamental) matrix $\phi(t)$ is the solution of the matrix differential equation

$$\dot{\phi}(t) = F(t)\,\phi(t)\;;\quad \phi(0) = I \tag{4.66}$$

If the covariance matrix $\Gamma_{xx}(t,\tau)$ of $\{\mathbf{X}(t)\}$ is defined as

$$\Gamma_{xx}(t,\tau) = E\left[\mathbf{X}(t)\,\mathbf{X}^T(\tau)\right] \tag{4.67}$$

where $\mathbf{X}^T(\tau)$ is the transpose of $\mathbf{X}(\tau)$, then it follows from Eq. (4.65) that*

$$\Gamma_{xx}(t,\tau) = \phi(t)\left[\Gamma_{xx}(0,0)\,\phi^T(\tau) + \int_0^\tau \phi^{-1}(s)Q(s)\;\phi^{-1}(s)^{\;T} ds\;\phi^T(\tau)\right];\quad t \geq \tau \geq 0 \tag{4.68}$$

where

$$\Gamma_{ww}(t,\tau) = E\left[\mathbf{W}(t)\,\mathbf{W}^T(\tau)\right] = Q(t)\,\delta(t-\tau) \tag{4.69}$$

Here, $Q(t)$ is assumed to be a continuously differentiable, symmetric, nonnegative-definite matrix. Equation (4.68) can be written in the simpler form

$$\Gamma_{xx}(t,\tau) = \phi(t)\,\psi(\tau)\;;\quad t \geq \tau \geq 0 \tag{4.70}$$

where $\psi(\tau)$ is the term in square brackets in Eq. (4.68). These results lead to the conjecture that if $\Gamma_{xx}(t,\tau)$ is of the form given in Eq. (4.70), where $\phi(t)$ and $\psi(\tau)$ both have continuous derivatives and $\phi(t)$ is nonsingular, than $\{\mathbf{X}(t), t\epsilon[0,\infty)\}$ can be represented in the form given in Eq. (4.64) where $\{\mathbf{W}(t), t\epsilon[0,\infty)\}$ is a vector "white noise" process.** The validity of this conjecture is established formally in the following paragraph.

Given a process whose covariance function has the properties stated above, consider the process $\{\mathbf{W}(t), t\epsilon[0,\infty)\}$ defined by the relationship

$$\mathbf{W}(t) = \dot{\mathbf{X}}(t) - F(t)\,\mathbf{X}(t) \tag{4.71}$$

*Assuming that $E\left[\mathbf{X}(0)\,\mathbf{W}^T(t)\right] = 0$ for all t.

**It is assumed here that $\phi(0) = I$. If not, then it is redefined as $\phi(t)\phi^{-1}(0)$, and $\psi(\tau)$ is redefined as $\phi(0)\psi(\tau)$; then the assumption is satisfied.

where $F(t)\,X(t)$ is the linear, least-squares estimate of $\dot{X}(t)$ given $X(t)$. Strictly speaking, the limit used to define $\dot{X}(t)$ in Eq. (4.71) does not normally exist as a limit in the mean, and, hence, it is not rigorously correct to speak of a least-squares estimate of $\dot{X}(t)$ because $\dot{X}(t)$ does not have a finite variance. This is analogous to the difficulty noted in Section 4.3, and it can be overcome by the use of generalized processes as noted in that section. Alternatively and preferably, the limit in Eq. (4.71) can be interpreted as a weak limit in the mean (see Section 4.8), with the estimate of $\dot{X}(t)$ being interpreted accordingly. Assuming that the latter course is taken, the following formal derivation can be rigorized in a straightforward manner. Proceeding formally, the problem now is to show that $F(t)$ can be determined uniquely from $\Gamma_{xx}(t,\tau)$ and that $\{W(t), t\epsilon[0,\infty)\}$ is a vector "white noise" process (i.e., the right-hand side of Eq. (4.71) converges weakly to a vector "white noise" process). Since $F(t)\,X(t)$ is the linear, least-square estimate of $\dot{X}(t)$, it satisfies the Wiener-Hopf equation which, for this case, becomes

$$E\left[\dot{X}(t)\,X^T(t)\right] - F(t)\,E\left[X(t)\,X^T(t)\right] = 0 \tag{4.72}$$

With Eq. (4.70), Eq. (4.72) becomes

$$\dot{\phi}(t)\,\psi(t) - F(t)\,\phi(t)\,\psi(t) = 0 \tag{4.73}$$

If it is assumed that $\Gamma_{xx}(t,t)$ is nonsingular, then $\psi(t)$ is nonsingular and Eq. (4.73) can be solved for $F(t)$ with the result

$$F(t) = \dot{\phi}(t)\,\phi^{-1}(t) \tag{4.74}$$

Even if $\Gamma_{xx}(t,t)$ is singular for some $t > 0$, $F(t)$ as given in Eq. (4.74) is still a solution of Eq. (4.73) as direct substitution shows. The reason for this is that when $\Gamma_{xx}(t,t)$ is singular, Eq. (4.73) will have many solutions, one of which will be that given by Eq. (4.74).* Equation (4.74) is consistent with Eq. (4.66) and is determined uniquely by $\phi(t)$. To show that $\{W(t)\}$ is a "white noise" process, consider $E[W(t)\,X^T(\tau)]$ for $t > \tau$. Making use of Eqs. (4.71) and (4.74), it follows that

$$\begin{aligned} E\left[W(t)X^T(\tau)\right] &= E\left[\dot{X}(t)X^T(\tau)\right] - F(t)E\left[X(t)X^T(\tau)\right] \\ &= \dot{\phi}(t)\psi(\tau) - \dot{\phi}(t)\phi^{-1}(t)\phi(t)\psi(\tau) = 0; \quad t > \tau \end{aligned} \tag{4.75}$$

*Kalman [13] takes $F(t) = \dot{\phi}(t)\,\psi(t)[\phi(t)\,\psi(t)]^{\#}$ where $[\phi(t)\,\psi(t)]^{\#}$ is the (Penrose) pseudo inverse of $[\phi(t)\,\psi(t)]$. While this gives a solution involving a pseudo inverse with the minimum norm (norm = sum of squares of elements) there is no good physical reason to use it here rather than the solution given in Eq. (4.74).

Differentiating Eq. (4.75) with respect to τ yields

$$E\left[\mathbf{W}(t)\,\dot{\mathbf{X}}^T(\tau)\right] = 0; \; t > \tau \tag{4.76}$$

Hence, making use of Eqs. (4.71), (4.75), and (4.76)

$$E\left[\mathbf{W}(t)\mathbf{W}^T(\tau)\right] = E\left[\mathbf{W}(t)\dot{X}(\tau)\right] - E\left[\mathbf{W}(t)\mathbf{W}^T(\tau)\right]F^T(\tau) = 0, \quad t > \tau \tag{4.77}$$

Finally, the symmetry of $E\left[\mathbf{W}(t)\mathbf{W}^T(\tau)\right]$ implies that

$$E\left[\mathbf{W}(t)\mathbf{W}^T(\tau)\right] = 0, \quad t \neq \tau \tag{4.78}$$

Equation (4.78) shows that $\{\mathbf{W}(t)\}$ is a "white noise" process, and hence that $\Gamma_{ww}(t,\tau)$ is of the form given in Eq. (4.69). To find $Q(t)$, it is observed from Eq. (4.68) that the given $\psi(t)$ will be obtained if

$$Q(t) = \phi(t)\frac{d}{dt}\left[\psi(t)\left(\phi^T(t)\right)^{-1}\right]\phi^T(t) \tag{4.79}$$

as can be proven by direct substitution of Eq. (4.79) in Eq. (4.68). [Note that since $\phi(0) = I$, $\Gamma_{xx}(0,0) = \psi(0)$.]

It is probably worth noting, at this point, the two major differences between the above derivation and Kalman's [13]. The first difference is that Kalman assumes that $\{X(t)\}$ is a Gaussian-Markov process, whereas it is assumed above that the covariance matrix of $\{\mathbf{X}(t)\}$ is of the form given in Eq. (4.70). The second difference is that Kalman defines $\{\mathbf{W}(t)\}$ by the equation

$$\mathbf{W}(t) = \dot{\mathbf{X}}(t) - E\left[\dot{\mathbf{X}}(t) \mid \mathbf{X}(t)\right] \tag{4.71$'$}$$

instead of by Eq. (4.71). The Gaussian assumption implies that the conditional expectation in Eq. (4.71′) is a linear function of $\mathbf{X}(t)$ [i.e., $F(t)\mathbf{X}(t)$] and is a least-squares estimate of $\dot{\mathbf{X}}(t)$. And the Markovian assumption is used to derive Eq. (4.75) instead of using the assumption that $\Gamma_{xx}(t,\tau)$ is of the form given in Eq. (4.70). Finally, it is also worth noting that Kalman's result implies that $\Gamma_{xx}(t,\tau)$ is of the form given in Eq. (4.70).

This concludes the discussion of the shaping-filter problem for vector processes. In the remainder of this section, the results will be applied to the linear, least-square filtering and prediction problem as discussed by Kalman and Bucy [29].

One formulation of the linear, least-square filtering and prediction problem is as follows. Consider a stochastic process $\{\mathbf{X}(t)\}$ generated from a "white noise" process $\{\mathbf{W}(t)\}$ by the shaping filter

$$\dot{\mathbf{X}}(t) = F(t)\mathbf{X}(t) + \mathbf{W}(t); \quad t \geq 0 \tag{4.80}$$

as discussed above. And suppose that a process $\{\mathbf{Z}(t)\}$ is observed where $\mathbf{Z}(t)$ is related to $\mathbf{X}(t)$ by the expression

$$\mathbf{Z}(t) = H(t)\mathbf{X}(t) + \mathbf{V}(t) = \mathbf{y}(t) + \mathbf{V}(t)\,; \quad t \geq 0 \tag{4.81}$$

where $H(t)$ is continuous and $\{\mathbf{V}(t)\}$ is also a "white noise" process. Let

$$\Gamma_{vv}(t, \tau) = E\left[\mathbf{V}(t)\mathbf{V}^T(\tau)\right] = R(t)\delta(t - \tau)\,; \quad t, \tau \geq 0 \tag{4.82}$$

and assume that $R(t)$ is positive definite. Also assume that

$$E\left[\mathbf{W}(t)\mathbf{V}^T(\tau)\right] = 0\,; \quad t, \tau \geq 0 \tag{4.83}$$

The problem is to find the linear, least-square estimate, $\hat{\mathbf{X}}(t_1 \mid t)$, of $\mathbf{X}(t_1)$ given the observed values of $\mathbf{Z}(\tau)$ in the interval $t \geq \tau \geq 0$ where $t_1 \geq t$. That is, find $\hat{\mathbf{X}}(t_1 \mid t)$ where

$$\mathbf{X}(t_1 \mid t) = \int_0^t A(t_1, \tau)\mathbf{Z}(\tau)\,d\tau \tag{4.84}$$

and where the matrix $A(t_1, \tau)$ is continuously differentiable in both arguments. Now the necessary and sufficient condition for $\mathbf{X}(t_1 \mid t)$ to be a least-square estimate of $\mathbf{X}(t_1)$ is that $A(t_1, \tau)$ satisfy the Wiener-Hopf integral equation*

$$E\left[\mathbf{X}(t_1)\mathbf{Z}^T(\tau)\right] - \int_0^t A(t_1, s)E\left[\mathbf{Z}(s)\mathbf{Z}^T(\tau)\right]ds = 0 \tag{4.85}$$

To derive Kalman and Bucy's results (as stated in Section 2.12 of this book) proceed as follows.

Assume for the moment that $t_1 = t$; i.e., that the problem is one of filtering. Differentiating Eq. (4.85) with respect to t, interchanging the order of the operations $\partial/\partial t$ and $E[\quad]$, and making use of Eq. (4.80) yields

$$\frac{\partial}{\partial t}E\left[\mathbf{X}(t)\mathbf{Z}^T(\tau)\right] = F(t)E\left[\mathbf{X}(t)\mathbf{Z}^T(\tau)\right] + E\left[\mathbf{W}(t)\mathbf{Z}^T(\tau)\right]; \; t > \tau \geq 0 \tag{4.86}$$

and

$$\frac{\partial}{\partial t}\int_0^t A(t, s)E\left[\mathbf{Z}(s)\mathbf{Z}^T(\tau)\right]ds = \int_0^t \frac{\partial A(t, s)}{\partial t}E\left[\mathbf{Z}(s)\mathbf{Z}^T(\tau)\right]ds + A(t,t)E\left[\mathbf{Z}(t)\mathbf{Z}^T(\tau)\right]; \quad t > \tau \geq 0 \tag{4.87}$$

*For the statement and derivation of the Wiener-Hopf integral equation for vector processes, see Section 3.3 of this book.

However, by Eqs. (4.80) through (4.83), $\mathbf{W}(t)$ is uncorrelated with $\mathbf{V}(\tau)$ and $\mathbf{X}(\tau)$ for $t > \tau$, and $\mathbf{X}(t)$ and $\mathbf{V}(\tau)$ are uncorrelated for all t, $\tau \geq 0$. Hence.

$$E\left[\mathbf{W}(t)\,\mathbf{Z}^T(\tau)\right] = 0\,;\; t > \tau \tag{4.88}$$

and

$$\begin{aligned} E\left[\mathbf{Z}(t)\,\mathbf{Z}^T(\tau)\right] &= E\left[\left(\mathbf{y}(t) + \mathbf{V}(t)\right)\left(\mathbf{y}(\tau) + \mathbf{V}(\tau)\right)^T\right] \\ &= E\left[\mathbf{y}(t)\,\mathbf{y}^T(\tau)\right] \\ &= H(t)E\left[\mathbf{X}(t)\,\mathbf{Z}^T(\tau)\right] - E\left[\mathbf{y}(t)\,\mathbf{V}^T(\tau)\right] \\ &= H(t)E\left[\mathbf{X}(t)\,\mathbf{Z}^T(\tau)\right];\; t \geq \tau \end{aligned} \tag{4.89}$$

Combining Eqs. (4.86) and (4.87), and making use of Eqs. (4.85), (4.88), and (4.89) yields

$$\int_0^t \left[F(t)A(t,s) - \frac{\partial A(t,s)}{\partial t} - A(t,t)H(t)A(t,s)\right] E\left[\mathbf{Z}(s)\,\mathbf{Z}^T(\tau)\right] = 0\,; \quad t > \tau \geq 0 \tag{4.90}$$

Now, Eq. (4.90) is clearly satisfied if $A(t,\tau)$ is a solution of the equation

$$F(t)A(t,\tau) - \frac{\partial}{\partial t}A(t,\tau) - A(t,t)H(t)A(t,\tau) = B(t,\tau) = 0\,; \quad t \geq \tau \geq 0 \tag{4.91}$$

If, as assumed, $R(\tau)$ is positive definite for $t \geq \tau \geq 0$, then Eq. (4.91) is also a necessary condition for $A(t,\tau)$. To see this, note that because of Eq. (4.90), $A(t,\tau) + B(t,\tau)$ is also a solution of the Weiner-Hopf equation. This implies that

$$\hat{\mathbf{X}}(t\,|\,t) + \int_0^t B(t,\tau)\,\mathbf{Z}(\tau)\,d\tau \tag{4.92}$$

is also an optimal estimater for $\mathbf{X}(t)$. Now, it can be easily shown by means of the Wiener-Hopf equation, that the covariance matrix of the difference of two optimal estimates vanishes for $t = \tau$. Hence

$$\int_0^t\int_0^t B(t,s)E\left[\mathbf{Z}(s)\,\mathbf{Z}^T(\sigma)\right]B^T(t,\sigma)\,ds\,d\sigma = 0 \tag{4.93}$$

Further, by Eqs. (4.80) through (4.84), it follows that

$$E\left[\mathbf{Z}(s)\mathbf{Z}^T(\sigma)\right] = R(s)\,\delta(s-\sigma) + E\left[\mathbf{y}(s)\mathbf{y}^T(\sigma)\right] \tag{4.94}$$

Substituting Eq. (4.94) in Eq. (4.93), we find that the contribution of the second term is nonnegative, while that of the first term is positive [because of the positive-definiteness of $R(t)$] unless Eq. (4.91) is satisfied. Hence $B(t, \tau) = 0$.

Differentiating Eq. (4.84) with respect to t gives

$$\frac{d\,\hat{\mathbf{X}}(t \mid t)}{dt} = \int_0^t \frac{\partial A(t,\tau)}{\partial t}\mathbf{Z}(\tau)\,d\tau + A(t,t)\mathbf{Z}(t) \tag{4.95}$$

Letting $K(t) = A(t,t)$ and making use of Eqs. (4.84) and (4.91), we can rewrite Eq. (4.95) in the form

$$\frac{d\,\hat{\mathbf{X}}(t \mid t)}{dt} = F(t)\,\hat{\mathbf{X}}(t \mid t) + K(t)\;\mathbf{Z}(t) - H(t)\,\hat{\mathbf{X}}(t \mid t) \tag{4.96}$$

This is the differential equation for the optimal filter. The only step remaining to complete the solution of the least-square filtering problem is to find a way to compute $K(t)$. This can be done as follows.

Let $\tilde{\mathbf{X}}(t)$ be defined by the relationship

$$\tilde{\mathbf{X}}(t \mid t) = \mathbf{X}(t) - \hat{\mathbf{X}}(t \mid t) \tag{4.97}$$

Then, from Eqs. (4.80) and (4.96), it follows that the differential equation for the estimation error $\tilde{\mathbf{X}}(t \mid t)$ is

$$\frac{d\,\tilde{\mathbf{X}}(t \mid t)}{dt} = \left[F(t) - K(t)H(t)\right]\tilde{\mathbf{X}}(t \mid t) + \mathbf{W}(t) - K(t)\mathbf{V}(t) \tag{4.98}$$

Furthermore, as is easily shown by means of the Wiener-Hopf equation, Eq. (4.85) implies that

$$E\left[\tilde{\mathbf{X}}(t \mid t)\mathbf{Z}^T(\tau)\right] = 0\,; \quad t > \tau \geq 0 \tag{4.99}$$

which implies

$$E\left[\tilde{\mathbf{X}}(t \mid t)\hat{\mathbf{X}}^T(t)\right] = 0 \tag{4.100}$$

From Eq. (4.99) and the fact that $\mathbf{Z}(t) = \mathbf{y}(t) + \mathbf{V}(t)$, it follows that

$$E\left[\mathbf{X}(t)\mathbf{y}^T(\tau)\right] - \int_0^t A(t,s)E\left[\mathbf{y}(s)\mathbf{y}^T(\tau)\right]ds = A(t,\tau)R(\tau)\,; \quad t > \tau \geq 0 \tag{4.101}$$

Now, both sides of Eq. (4.101) are continuous in τ at t for all t, hence limits may be taken on both sides of the equation. Noting that $E\left[y(s)y^T(\tau)\right] = E\left[Z(s)y^T(\tau)\right]$ and making use of Eq. (4.100), taking the limit yields

$$
\begin{aligned}
K(t)R(t) &= A(t,t)R(t) = E\left[\widetilde{X}(t \mid t)y^T(t)\right] \\
&= E\left[\widetilde{X}(t)X^T(t)\right]H^T(t) = E\left[\widetilde{X}(t \mid t)\widetilde{X}^T(t \mid t)\right]H^T(t) \\
&= P(t)H^T(t)
\end{aligned}
\tag{4.102}
$$

Since $R(t)$ is positive definite, it is nonsingular and

$$
K(t) = P(t)H^T(t)R^{-1}(t) \tag{4.103}
$$

This is an explicit expression for $K(t)$ in terms of $P(t)$.

To find an equation for computing $P(t)$, consider

$$
\frac{dP(t)}{dt} = E\left[\left(\frac{d\widetilde{X}(t \mid t)}{dt}\right)\widetilde{X}^T(t \mid t)\right] + E\left[\widetilde{X}(t \mid t)\left(\frac{d}{dt}\widetilde{X}^T(t \mid t)\right)\right] \tag{4.104}
$$

The right-hand side of Eq. (4.104) can be evaluated by means of Eq. (4.98). Consider

$$
\begin{aligned}
E\left[\left(\frac{d\widetilde{X}(t \mid t)}{dt}\right)\widetilde{X}^T(t \mid t)\right] &= [F(t) - K(t)H(t)]E\left[\widetilde{X}^T(t \mid t)\widetilde{X}^T(t \mid t)\right] \\
&\quad + E\left[\left(W(t) - K(t)V(t)\right)\widetilde{X}^T(t \mid t)\right]
\end{aligned}
\tag{4.105}
$$

The solution $\widetilde{X}(t \mid t)$ of Eq. (4.98) can be expressed in the form

$$
\begin{aligned}
\widetilde{X}(t \mid t) &= \Omega(t)\Omega^{-1}(0)\widetilde{X}(0 \mid 0) \\
&\quad + \Omega(t)\int_0^t \Omega^{-1}(\tau)[W(\tau) - K(t)V(\tau)]\,d\tau
\end{aligned}
\tag{4.106}
$$

where $\Omega(t)$ is any nonsingular solution of the matrix differential equation

$$
\dot{\Omega}(t) = [F(t) - K(t)H(t)]\Omega(t) \tag{4.107}
$$

Substituting Eq. (4.106) in Eq. (4.105) and using Eq. (4.103) gives

$$
\begin{aligned}
E\left[\left(\frac{d}{dt}\widetilde{X}(t \mid t)\right)\widetilde{X}^T(t \mid t)\right] &= F(t)P(t) - P(t)H^T(t)R^{-1}(t)H(t)P(t) \\
&\quad + \frac{1}{2}Q(t) + \frac{1}{2}P(t)H^T(t)R^{-1}(t)H(t)P(t)
\end{aligned}
\tag{4.108}
$$

where the factor of $\frac{1}{2}$ comes from the symmetry of the δ function and the fact that it occurs at the upper limit of the integrals. The terms in Eq. (4.108) due to initial conditions vanish because

$$\tilde{\mathbf{X}}(0 \mid 0) = \mathbf{X}(0) - \hat{\mathbf{X}}(0 \mid 0) = \mathbf{X}(0) \tag{4.109}$$

and because $\mathbf{X}(0)$ is uncorrelated with $\mathbf{W}(t)$ and $\mathbf{V}(t)$ for all t. The vanishing of $\hat{\mathbf{X}}(0 \mid 0)$ follows immediately from Eq. (4.84). The other term in Eq. (4.104) is simply the transpose of that evaluated in Eq. (4.108). Thus, making use of the symmetry of the matrices $P(t)$, $Q(t)$, and $R(t)$, Eq. (4.104) becomes

$$\frac{dP(t)}{dt} = F(t)P(t) + P(t)F^T(t) - P(t)H^T(t)R^{-1}(t)H(t)P(t) + Q(t) \tag{4.110}$$

Equation (4.110) is a nonlinear (Ricatti) differential equation for computing $P(t)$ and it is called the variance equation. The initial condition for $P(t)$ follows directly from Eqs. (4.102) and (4.109), and is

$$P(0) = E\left[\mathbf{X}(0)\mathbf{X}^T(0)\right] \tag{4.111}$$

This completes the derivation of the results for the filtering problem. It is seen that the optimal filter is completely determined by the solution of the variance equation and is independent of the observed process $\{\mathbf{Z}(t)\}$. This is not surprising because the variance equation was derived from and is equivalent to the Wiener-Hopf equation, and the optimal filter as specified by the Wiener-Hopf equation (i.e., $A(t, \tau)$) does not depend on $\{\mathbf{Z}(t)\}$. The optimal filter is given by Eq. (4.96) where $K(t)$ is computed by Eqs. (4.103), (4.110), and (4.111), and the initial condition for the optimal filter is $\hat{\mathbf{X}}(0 \mid 0) = 0$ as noted above.

When $t_1 \geq t$, one is dealing with a prediction problem, and it can be easily shown (see Kalman [13]) that

$$\hat{\mathbf{X}}(t_1 \mid t) = \phi(t_1)\phi^{-1}(t)\hat{\mathbf{X}}(t \mid t), \quad t_1 \geq t \tag{4.112}$$

where $\phi(t)$ comes from $\Gamma_{xx}(t, \tau)$ as given in Eq. (4.70).

The questions of existence, uniqueness, and stability of the solutions of Eq. (4.110) will not be discussed here. The interested reader is referred to Kalman [13] or to Kalman and Bucy [29].

REFERENCES

1. Wold, H., *A Study in the Analysis of Stationary Time Series*, Uppsala, 1938.

2. Kolmogorov, A., "Sur l'Interpolation et Extrapolation des Suites Stationnaires," *C. R. Acad. Sci.*, Paris 208, pp. 2043-2045, 1939.
3. Kolmogorov, A., "Stationary Sequences in Hilbert Space," (Russian), *Bull. Math. Univ.* Moscow 2, no. 6, 40 pp., 1941.
4. Kolmogorov, A., "Interpolation und Extrapolation von Stationaren Zufalligen Folgen," *Bull. Acad. Sci. U. R. S. S.* Ser. Math. 5, 3-14, 1941.
5. Wiener, N., *Extrapolation, Interpolation, and Smoothing of Stationary Time Series,* John Wiley and Sons, Inc., New York, 1949.
6. Hanner, O., "Deterministic and Nondeterministic Stationary Random Processes," *Ark. Mat. 1,* pp. 161-177, 1949.
7. Karhunen, K., "Uber die Struktur Stationarer Zufalliger Funktionen," *Ark. Mat. 1,* pp. 141-160, 1960.
8. Bode, H. and C. Shannon, "A Simplified Derivation of Linear Least-Square Smoothing and Prediction Theory," *Proc. IRE* 38, pp. 417-426, April, 1950.
9. Darlington, S., "Nonstationary Smoothing and Prediction Using Network Theory Concepts," *Transactions IRE,* I T-5, May, 1959, pp. 1-11.
10. Dolph, C. and M. Woodbury, "On the Relation Between Green's Functions and Covariances of Certain Stochastic Processes and Its Application to Unbiased Linear Prediction," *Trans. Amer. Math. Soc.*, 72, May, 1952, pp. 519-550.
11. Batkov, A., "Generalization of the Shaping-Filter Method to Include Nonstationary Random Processes," *Automation and Remote Control,* Vol. 20, No. 8, Aug., 1959, pp. 1049-1062.
12. Leonov, Y., "The Problem of Shaping Filters and Optimal Linear Systems," *Automation and Remote Control,* Vol. 21, No. 6, June, 1960, pp. 467-471.
13. Kalman, R., "New Methods and Results in Linear Prediction and Filtering Theory," *RIAS Technical Report 61-1.* Also an Appendix to ASD-TR-61-27.
14. Loeve, M., *Probability Theory,* D. Van Nostrand, Princeton, N. J., 1960.
15. Gel'fand, I. and G. Shilov, *Theory of Distributions* (book, in Russina), Vol. 4, Fizmatiz, Moscow, 1960.
16. Gel'fand, I., "Generalized Stochastic Processes," Doklady Akad. Nauk USSSR, *100,* (1955), pp. 853-856.
17. Doob, J., *Stochastic Processes,* John Wiley and Sons, Inc., New York, 1953.
18. Grenander, U. and M. Rosenblatt, *Statistical Analysis of Stationary Time Series,* John Wiley and Sons, Inc., New York, 1957.
19. Paley, R. E. A. and N. Wiener, *Fourier Transforms,* An American Math. Society Colloquium Publication, 1932.
20. Coddington, E. and N. Levinson, *Theory of Ordinary Differential Equations,* McGraw-Hill Book Company, Inc., New York, 1955.

21. Stear, E., "Synthesis of Shaping Filters for Nonstationary Stochastic Processes and Their Uses." *AFOSR Report 61-50*, August, 1961.
22. Tricomi, F., *Integral Equations*, Intersicence Publishers, Inc., New York, 1957.
23. Lalesco, T., *Introduction a la Theorie des Equations Integrales*, Gauthier-Villars, Paris, 1922.
24. Sato, T., "Sur l'Equation Integrale Nonlinear de Volterra," *Compositio Mathematica*, 11, 1953, pp. 271-290.
25. Berge, C., *Espaces Topologiques*, Dunod, Paris, 1959.
26. Pugachev, V., *Theory of Random Functions and Their Application to Automatic Control Problems* (Russian), Gostekhizdat, 1957.
27. Pugachev, V., "Integral Canonical Representation of Random Functions and Their Application in Deriving Optimal Linear Systems," *Automation and Remote Control*, Vol. 18, No. 11, November, 1957, pp. 1017-1031.
28. Natanson, I., *Theory of Functions of a Real Variable*, Frederick Ungar Publishing Co., New York, 1955.
29. Kalman, R. E., and Bucy, R. S., "New Results in Linear Filtering and Prediction Theory," *J. Basic Engr.* (ASME Trans.), *83D*, March, 1961, pp. 95-108.

5

Lyapunov's Direct Method in the Analysis of Nonlinear Control Systems

DALE D. DONALSON
HEAD, FUNCTIONAL DESIGN SECTION
HUGHES AIRCRAFT COMPANY, CULVER CITY, CALIFORNIA

5.1 Introduction

5.1.1 Historical Background

The "direct method" of Lyapunov is a general method which may be used to determine the stability of sets of ordinary differential equations. It is applicable to autonomous or nonautonomous and to linear or nonlinear ordinary differential equations. This technique is often referred to as Lyapunov's "second method." These names are used interchangeably in the literature and sometimes cause confusion to the beginning student of this material. The technique was first introduced in Russia just prior to 1900 by the mathematician A. M. Lyapunov. A French translation of his work appeared in 1907. In his work Lyapunov described all stability analysis techniques for ordinary differential equations as belonging to essentially one of two methods. The first method was referred to as the indirect method since it required the actual solutions to the set of differential equations under investigation. The second method was referred to as the direct method since it dealt with stability criteria that could be applied directly to the set of ordinary differential equations without requiring solutions of the equations. This, then, explains the historical origin of the two names for this technique. It will be referred to as Lyapunov's direct method in the material presented in this chapter because this name appears to be more descriptive. The spelling of Lyapunov has undergone several different mutations in its translation into English. Most of these variations are easily recognizable as having the same Russian ancestor and no one of them appears to be preferred to the others. Examples of common spellings are: Lyapunov, Liapunov, and Ljapunov.

The direct method of Lyapunov was relatively dormant until the first part of the 1940-1950 decade. At this time, the Russians began to recognize its value in the analysis of nonlinear automatic control systems. In 1947, the Princeton University Press republished the early French translation of Lyapunov's work. Publications in English began to appear around 1955. These articles represent both original contributions written in English and translations from Russian and German articles.

5.1.2 Basic Concept

Before we can discuss the stability of either autonomous or nonautonomous nonlinear systems, we should define the meaning of stability in greater detail than is required for linear systems. A linear free dynamic system will have only one equilibrium point. If it is stable, all solution trajectories will approach this equilibrium point as time goes to infinity. If it is unstable, all solution trajectories not initially at the equilibrium point will go to infinity as times goes to infinity. A special case occurs if one pair of roots of the characteristic equation is purely imaginary. In this case, if the system is not initially at the origin, it will exhibit sustained oscillations for all time. The latter situation is seldom encountered in engineering systems since the roots of the characteristic equation are not likely to be purely imaginary, even if there has been an attempt to cause them to be. The stability of the system is independent of the magnitude of the initial conditions; moreover, the response of a stable system to any continuously acting bounded input will be bounded. In fact, the latter may be used as a definition of stability for linear systems. Stability for nonlinear systems is stratafied into a number of degrees or levels. These definitions are given in Section 5.2 where they are treated with greater mathematical rigor than is necessary for this introduction to Lyapunov's direct method.

To present a very simple analogy to Lyapunov's direct method, the system shown in Fig. 5-1 is considered.

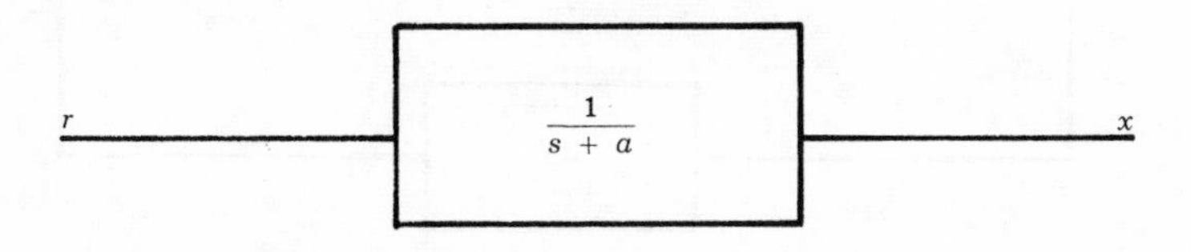

FIG. 5-1. Simple first order system.

The output is related to the input by

$$\dot{x} + ax = r .$$

If the input, r, is set equal to zero, then one can consider the response of this system to various initial conditions. For this case the system equation may be written as

$$\dot{x} = -ax .$$

From very elementary differential equation theory it is evident that the system will be stable for a equal to any real, positive, finite constant. This is not new to any control system engineer; however, consideration of the physical reasoning associated with this equation has a clear analogy with the direct method of Lyapunov. For $0 < a < +\infty$, note that regardless of the value assumed by x the rate of change of x is such that its absolute value $|x|$ is always decreasing except when $x = 0$. Then, $\dot{x} = 0$ also. Therefore, $x = 0$ is an equilibrium point for the system and the system is stable since, for any finite initial condition for x (either positive or negative), x_0, the solution of the equation $x(t)$, will approach zero as time approaches infinity. Furthermore, it can easily be reasoned that if $-\infty < a < 0$ then $|x|$ is always increasing for $x \neq 0$ and therefore for these values of a the system is unstable.

It is not possible to perform such an analysis directly in terms of the system variable x for second and higher order systems. However, the various theorems based on Lyapunov's direct method do employ the Lyapunov function, V, to which a very similar type of reasoning is applied. As one would expect, there are extra considerations which must receive attention to retain mathematical rigor and assure the validity of the resulting conclusions. A collection of theorems dealing with the "direct method" are presented in conjunction with an example of its application to a simple linear system. The system to be considered is shown in Fig. 5-2.

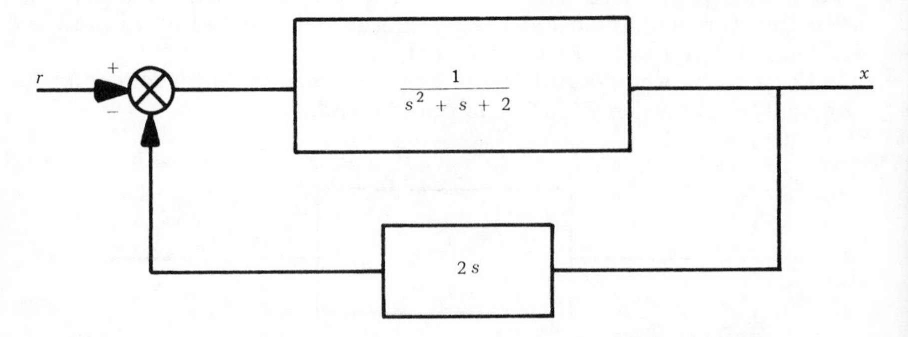

FIG. 5-2. Simple second order system.

The input-output relation for this system may be written as

$$\ddot{x} + 3\dot{x} + 2x = r.$$

The input r will be set equal to zero, that is

$$r = 0,$$

then the equilibrium point is $x = 0$. The analysis will be concerned with the response of the unforced system for various initial conditions on x. The differential equation which represents the system is then

$$\ddot{x} + 3\dot{x} + 2x = 0,$$

or

$$\ddot{x} = -3\dot{x} - 2x.$$

The theorems of Lyapunov's direct method are usually stated in a form applicable to a set of first-order differential equations rather than a single higher order equation. Higher order equations are easily transformed into a set of first-order differential equations. For the case in question, the following set of variables is defined:

$$x = x_1;$$

$$\dot{x} = \dot{x}_1 = x_2.$$

The unforced second-order system may then be represented by the following set of first-order differential equations:

$$\begin{aligned}\dot{x}_1 &= x_2\\ \dot{x}_2 &= -3x_2 - 2x_1.\end{aligned} \tag{5.1-1}$$

In general, variables x_1 and x_2 are called "state" variables. In the case where they represent the output of the system and its various derivatives with respect to time, they may also be referred to as "phase" variables. The multidimensional space which has the state variables or phase variables along its rectangular coordinate axes is referred to as the state space or the phase space. In the particular case being considered, it is the phase plane which is familiar to anyone who has done phase-plane plots for simple second-order nonlinear systems.

The system equations are now in a form to which Lyapunov's direct method may be applied. The statement of the applicable theorem is given next.

Theorem: For a set of first-order differential equations of the form

$$\dot{x}_i = X_i(x_1, \ldots x_n), \; i = 1, 2, \ldots n$$

and such that:

(1) $\dot{x}_i = X_i(0, 0, \ldots, 0) = 0 \quad i = 1, 2, \ldots n$

(2) The functions X_i are continuous with respect to all variables x_i in the entire state space.

Then, if there exists a real-valued scalar function $V(x_1, x_2, \ldots, x_n)$ with the properties:

(a) $V(x_1, x_2, \ldots, x_n)$ is continuous and has continuous first partial derivatives;

(b) $V(x_1, x_2, \ldots, x_n) > 0$ except when $x_i = 0$ for $i = 1, \ldots, n$;(i.e., V is positve-definite)

(c) $V(0, 0, \ldots, 0) = 0$;

(d) $V(x_1, x_2, \ldots, x_n) \to \infty \quad \text{for} \quad \left(\sum_{i=1}^{n} x_i^2\right)^{1/2} \to \infty$;

(e) $\dot{V} = \dfrac{dV}{dt} = \sum_{i=1}^{n} \left(\dfrac{\partial V}{\partial x_i} \cdot \dfrac{dx_i}{dt}\right) < -\epsilon \sum_{i=1}^{n} x_i^2$ (i.e., $\dot{V}$ is negative-definite).

the system is asymptotically stable in the large.

Asymptotic stability in the large assures that, for any real finite initial conditions on the system, the output of the system will approach the equilibrium state of $x = 0, \dot{x} = 0$ as $t \to \infty$. For linear systems, this matches the usual definition of stability.

The set of first-order ordinary differential equations for the system under consideration Eq. (5.1-1) satisfies both conditions (1) and (2) of the theorem and is in the appropriate form. The problem, then, is to find a scalar function $V(x_1, x_2)$ to satisfy conditions (a) through (e) of the theorem. A number of techniques for generating the Lyapunov function V exist and they are discussed in Section 5.3. For this introductory example no attention will be given to the method by which the function V is generated. To establish stability, it is enough merely to determine whether a function exists that satisfies the conditions of the theorem. The problem of how one synthesizes this function is, however, usually the most difficult part of the analysis.

For the system in question, let

$$V = 1/2\left(5x_1^2 + 2x_1x_2 + x_2^2\right) \tag{5.1-2}$$

This function satisfies conditions (a), (c), and (d) of the theorem as one can easily verify by inspection. $V(x_1, x_2)$, as given by Eq. (5.1-2), is a quadratic form. To determine whether V satisfies condition (b) of the theorem, it is necessary to determine whether the quadratic form is positive-definite. For a general quadratic function of two variables,

$$Q(x_1, x_2) = a x_1^2 + 2 b x_1 x_2 + c x_2^2 , \tag{5.1-3}$$

a necessary and sufficient condition for it to be positive-definite is that

$$\text{(a)} \quad a > 0 , \tag{5.1-4}$$

and

$$\text{(b)} \quad ac - b^2 > 0 . \tag{5.1-5}$$

For the quadratic form in equation (5.1-2)

$$a = 2.5 ,$$
$$b = 0.5 ,$$

and

$$c = 0.5 .$$

Therefore it is clear that the function given by Eq. (5.1-2) also satisfies condition (b) of the theorem. If the derivative of this function with respect to time also satisfies condition (e), then $V(x_1, x_2)$ is a Lyapunov function for the system in question, and thereby establishes a sufficient condition by which the system may be declared asymptotically stable in the large.

In order to determine whether condition (e) is satisfied, Eq. (5.1-2) is differentiated with respect to time, and Eq. (5.1-1) is used to provide the time derivatives of the state variables. Therefore

$$\begin{aligned} \dot{V}(x_1, x_2) &= 5 x_1 \dot{x}_1 + x_1 \dot{x}_2 + x_2 \dot{x}_1 + x_2 \dot{x}_2 \\ &= 5 x_1 x_2 + x_1 (-3 x_2 - 2 x_1) + x_2^2 + x_2 (-3 x_2 - 2 x_1) \\ &= -2 \left(x_1^2 + x_2^2 \right) \end{aligned} \tag{5.1-6}$$

Observing the last line of Eq. (5.1-6), it is clear that it is also a quadratic form, and is furthermore negative-definite. Hence condition (e) of the theorem is satisfied. The system has thus been proved to be asymptotically stable in the large by Lyapunov's direct method.

Some observations are in order with regard to the intuitive or heuristic meaning of this analysis. A function of the state variables was employed which was always positive except when the variables were all identically equal to zero; then the function V was also equal to zero. The derivative of the function V with respect to time (with the control system equations substituted for the time derivatives of the state variables) was found to be always negative except when the state variables were both identically equal to zero; then $\dot{V}$ was also zero. Hence, it is clear that for any non-zero initial values of the state variables, $V > 0$ and decreases as time increases since $\dot{V} < 0$. Hence, a function V, which may encompass a large number of state variables, is used in Lyapunov's direct method. If this function V is of the appropriate form, one can apply reasoning very similar to the reasoning associated with the simple first-order system given earlier. The reader should be warned, however, that there are certain additional requirements in the theorem stated above, and hence that it is dangerous to attempt analysis on the basis of these intuitive ideas. They may serve, however, as an intuitive aid in understanding the technique. The rigorous proof of the theorem used above parallels the reasoning of this paragraph; however, the above discussion is by no means intended as a proof.

The Lyapunov function V may also be considered analogous to the potential energy of the system under consideration. In this analogy, the positive V and negative $\dot{V}$ correspond to a system that dissipates energy. In this way, the response to any initial condition will cause the system to dissipate energy until the energy is zero, at which time the state variables will all be zero. The latter condition is the equilibrium condition for a system that satisfies conditions (1) and (2) in the above theorem. Some references have indicated that the potential energy analogy is sometimes useful in selecting functions that may serve as Lyapunov functions. In view of the availability of more analytical methods of generating Lyapunov functions, the potential energy analogy appears to have very limited value for this purpose.

The above theorem also lends itself to a geometric interpretation. The discussion to follow will be based on a system with only two state variables, since this permits demonstrating geometric relationships on a plane surface. The ideas may easily be extended to n-dimensional space. Consider a Lyapunov function that satisfies conditions (a), (b), (c), and (d) of the theorem. If $V(x_1, x_2)$ is set equal to a constant K, then the resulting equation describes a closed bounded curve in the x_1, x_2 state plane. If $V(x_1, x_2)$ is a quadratic form, then the curves are ellipses. For increasing values of the constant K, that is, for $0 < K_1 < K_2 < K_3 < K_4$, the curves for $V(x_1, x_2) = K_i$ would appear as shown in Fig. 5-3.

The time derivative of $V(x_1, x_2)$ is

$$\frac{dV}{dt} = \frac{\partial V}{\partial x_1}\frac{dx_1}{dt} + \frac{\partial V}{\partial x_2}\frac{dx_2}{dt}$$

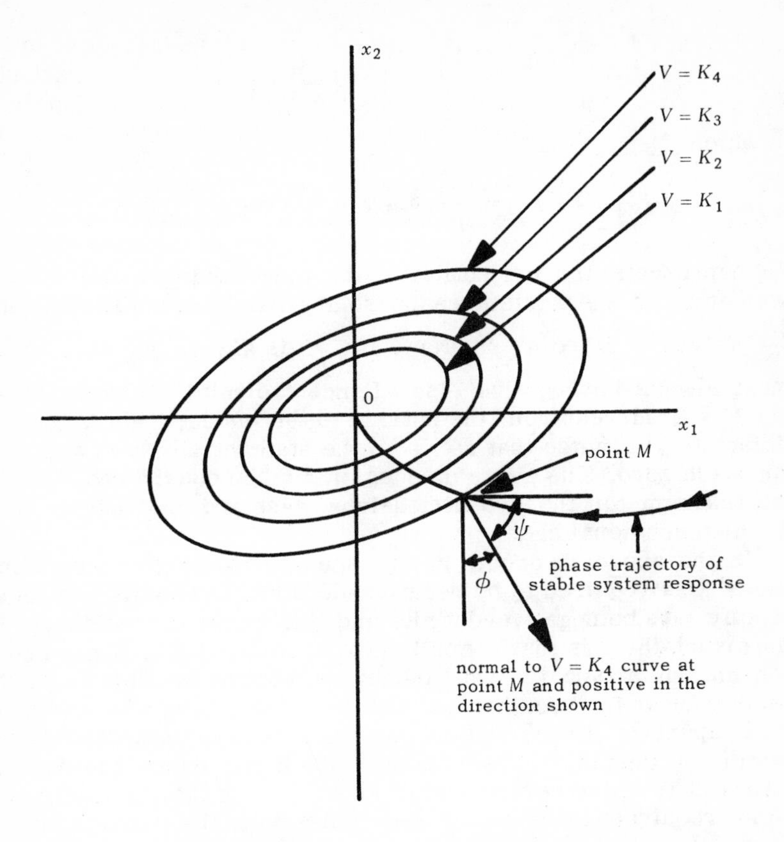

FIG. 5-3. Geometry associated with Lyapunov criteria.

Considering now the point M, where a particular phase trajectory crosses the curve $V = K_4$, it is possible to write the partial derivatives of V as

$$\frac{\partial V}{\partial x_1} = \left[\left(\frac{\partial V}{\partial x_1}\right)^2 + \left(\frac{\partial V}{\partial x_2}\right)^2\right]^{1/2} \cos \Psi$$

and

$$\frac{\partial V}{\partial x_2} = \left[\left(\frac{\partial V}{\partial x_1}\right)^2 + \left(\frac{\partial V}{\partial x_2}\right)^2\right]^{1/2} \cos \phi$$

The geometric relations are shown in Fig. 5-3. Using these relations $\frac{dV}{dt}$ may be written as

$$\frac{dV}{dt} = \left[\left(\frac{\partial V}{\partial x_1}\right)^2 + \left(\frac{\partial V}{\partial x_2}\right)^2\right]^{1/2} V_n$$

in which

$$V_n = \frac{dx_1}{dt} \cos \Psi + \frac{dx_2}{dt} \cos \phi .$$

V_n represents the projection of the tangential velocity along the trajectory of the system response at point M on the normal of the $V = K_4$ curve at point M. Now, if $\frac{dV}{dt}$ is always negative, then V_n must always be negative also. Hence the point M always crosses the $V = K$ curves from the outside to the inside. This is then sufficient to guarantee that as $t \to \infty$, the state variables x_1 and x_2 will approach zero. The above discussion may be considered a proof of the theorem for the two-dimensional case and may be extended to the n-dimensional case.

In the above theorem, V was specified to be positive-definite, and $\dot{V}$ was required to be negative-definite. One could just as well specify V to be negative-definite and $\dot{V}$ to be positive-definite. The important thing is that V and $\dot{V}$ have opposite signs. It has become conventional to specify V as positive in most of the literature. This convention is followed here.

Lyapunov's direct method may also be used to establish that a system is unstable. The theorem which may be used to do this is identical to the theorem stated above except that in condition (e), $\dot{V}$ is required to be positive-definite (i.e., the same sign as V). These theorems as well as a number of others are formally stated in Section 5.2.

It should be pointed out that although the stability criteria supplied by Lyapunov's direct method are sufficient to establish stability, they are not necessary. Therefore, failure to discover a function which will satisfy the conditions for either stability or instability for a given system yields no information about the system stability or instability. Furthermore, a Lyapunov function which establishes stability or instability for a system is not unique. In fact, it can be shown that if one Lyapunov function can be discovered which establishes the stability or instability of a system, then there exists an infinite number of functions capable of being used to establish the stability or instability of the system.

5.1.3 Organization of the Material

This introductory section has given a brief historical background to Lyapunov's direct method. It has also presented the basic concept of this method of stability analysis in the form of a single theorem. The theorem was applied to a simple second-order linear system as an example. The presentation was intentionally formulated in a rather tutorial manner and a number of intuitive analogies were made. The geometric interpretation for the case of two state variables provides the essence of a proof for the theorem.

In the remainder of the material to be presented, no proofs will be given. In all cases, references will be cited and the reader will be referred to the references for the proofs. For the sake of simplicity in this introduction, the theorem presented above is more restrictive than is necessary. Hence, it has not been assigned a theorem number as is done in the following material. Its less restrictive counterpart is given in Section 5.2.

Section 5.2 presents the notation, definitions, and theorems which may be useful in the stability analysis of automatic control systems by Lyapunov's direct method. The notation and definitions are required to facilitate understanding and application of the theorems. They are most concentrated at the beginning of the section; however, as new notation or definitions are required, they will be included throughout the text. Each theorem is given a number for easy reference and a descriptive title that will help the user identify the theorem that may be applicable to a given problem. Credit to the originator is given through reference numbers rather than by incorporating his name into the theorem title. Theorem 5-1 deals with the conditions required for existence, uniqueness, and continuity for the solution of a set of differential equations. Theorem 5-2 deals with the stability of a linear system. Theorems 5-3 through 5-8 deal with the stability of nonlinear systems in an arbitrarily small region about an equilibrium point. It is often advantageous to apply these theorems first to determine whether the equilibrium point for the system in question is stable or unstable. The results thus obtained are helpful in directing the analysis of the system in an expanded region of (and hopefully, the entire) state space. The next set of theorems deals with variations and extensions of the technique referred to as Lyapunov's direct method. At the end of the section, some algebraic theorems are included which are useful in the application of the stability theorems.

Section 5.3 contains a collection of the more successful methods for applying the stability theorems and generating Lyapunov functions. Examples of the application of some of the techniques are given. Included in this section are Lure's canonic transformations, Aizerman's method, Krasovskiy's method, Pliss' method, and the Variable Gradient method. An application of one of the theorems relating to eventual stability for an adaptive control system problem is also given. The section concludes with a technique for establishing an upper bound on the transient response of a system by Lyapunov's direct method.

Section 5.4 gives a brief summary of the material presented, and concludes with some remarks about the present status of this method of stability analysis for nonlinear control systems.

5.2 Theorems, Definitions, and Notation

In the discussion to follow, it will be convenient to use matrix notation to describe systems of differential equations that represent the control systems being analyzed. A system of first-order linear differential equations of the form

$$\begin{aligned}
\dot{x}_1 &= a_{11}x_1 + a_{12}x_2 + \cdots + a_{1n}x_n \\
\dot{x}_2 &= a_{21}x_1 + a_{22}x_2 + \cdots + a_{2n}x_n \\
&\cdots\cdots\cdots\cdots \\
\dot{x}_n &= a_{n1}x_1 + a_{n2}x_2 + \cdots + a_{nn}x_n
\end{aligned} \qquad (5.2\text{-}1)$$

may be represented as the following matrix equation

$$\dot{\mathbf{x}} = \mathbf{A}\mathbf{x} \qquad (5.2\text{-}2)$$

in which

$$\dot{\mathbf{x}} = \begin{bmatrix} \dot{x}_1 \\ \dot{x}_2 \\ - \\ - \\ - \\ \dot{x}_n \end{bmatrix}, \quad \mathbf{x} = \begin{bmatrix} x_1 \\ x_2 \\ - \\ - \\ - \\ x_n \end{bmatrix}$$

and

$$\mathbf{A} = \begin{bmatrix} a_{11} a_{12} \cdots a_{1n} \\ a_{21} a_{22} \cdots\cdots \\ \cdots\cdots\cdots \\ a_{n1} \cdots\cdots a_{nn} \end{bmatrix}$$

The usual matrix algebra operations are implied.

Definition 5-1: *Linear Systems.* Systems which may be written in the form of Eq. (5.2-2) are referred to as linear systems.

If, in addition to the linear terms in Eq. (5.2-2) above, there are nonlinear terms also, the system may be represented by

$$\dot{\mathbf{x}} = \mathbf{A}\mathbf{x} + \mathbf{f}(\mathbf{x}) \qquad (5.2\text{-}3)$$

in which

$$\mathbf{f}(\mathbf{x}) = \begin{bmatrix} f_1(\mathbf{x}) \\ f_2(\mathbf{x}) \\ - \\ - \\ - \\ f_n(\mathbf{x}) \end{bmatrix} \tag{5.2-4}$$

where the $f_i(\mathbf{x})$ are some nonlinear functions of the variables $x_1, x_2, \ldots, x_n$.

Definition 5-2: Nonlinear Systems. Systems which may be written in the form of Eq. (5.2-3) are referred to as nonlinear systems.

Definition 5-3: State Variables. The variables $x_1, x_2, \ldots, x_n$ are referred to as state variables since at any given instant of time they represent the condition of the system.

Definition 5-4: State Space. The n-dimensional Euclidean space with axes labeled $x_1, x_2, \ldots, x_n$ is referred to as the state space. Each point in this space represents a unique and particular set of values that the state variables may assume. That is, each point represents a particular state of the system.

If the state of the system is given for $t = 0$, then for $t > 0$ the state variables describe a trajectory in the state space. The form of this trajectory is completely determined by the initial state of the system and the form of Eq. (5.2-2) or Eq. (5.2-3). Clearly, Eq. (5.2-2) is merely a special case of Eq. (5.2-3).

Definition 5-5: Stationary Systems. Systems described by Eq. (5.2-2) or Eq. (5.2-3), in which the elements of the matrix $\mathbf{A}$ and the coefficients in the function $\mathbf{f}$ do not change with time, are referred to as stationary systems.

Definition 5-6: Nonstationary Systems. Systems described by Eq. (5.2-2) or Eq. (5.2-3), in which either the elements of the matrix $\mathbf{A}$ or the coefficients in the function $\mathbf{f}$ or both vary with time, are referred to as nonstationary systems.

Definition 5-7: Free or Unforced Systems. Systems that may be described by Eq. (5.2-2) or Eq. (5.2-3), and which are either stationary or nonstationary, are referred to as free or unforced systems.

Definition 5-8: Forced System. A system represented by

$$\dot{\mathbf{x}} = \mathbf{A}\mathbf{x} + \mathbf{f}(\mathbf{x}) + \mathbf{R}(t) \tag{5.2-5}$$

in which

$$\mathbf{R}(t) = \begin{bmatrix} r_1(t) \\ r_2(t) \\ - \\ - \\ - \\ r_n(t) \end{bmatrix}$$

is referred to as a forced system.

Definition 5-9: Autonomous Systems. A system both free and stationary is referred to as an autonoumous system.

Definition 5-10: $\mathbf{P}_\theta(\mathbf{x})$, *Vector Power Series in* $\mathbf{x}$. The vector $\mathbf{P}_\theta(\mathbf{x})$ has as elements either zero or functions which involve powers or products of the state variables $x_1, x_2, \ldots, x_n$ of order equal to or greater than θ.

Definition 5-11: Norms $\|\mathbf{x}\|$ *and* $|\mathbf{x}|$. The forms $\|\mathbf{x}\|$ and $|\mathbf{x}|$ are referred to as norms. There are two types of norms, the first of which is referred to as the Euclidian norm, and is given by

$$\|\mathbf{x}\| = \left(\sum_{i=1}^{n} x_i^2\right)^{1/2} \tag{5.2-6}$$

The section type of norm is referred to as the non-Euclidean norm and is given by

$$|\mathbf{x}| = \sum_{i=1}^{n} |x_i| \tag{5.2-7}$$

Definition 5-12: $\phi(t;\mathbf{x}_0,t_0)$ *Solution or Trajectory of State Variables* [11]. If there exists a unique vector function $\phi(t;\mathbf{x}_0,t_0)$ differentiable in t, such that for any fixed $\mathbf{x}_0$, t_0

$$\text{(a)} \quad \phi(t_0;\mathbf{x}_0,t_0) = \mathbf{x}_0 \tag{5.2-8}$$

$$\text{(b)} \quad \frac{d}{dt}\phi(t;\mathbf{x}_0,t_0) = \mathbf{A}\phi(t;\mathbf{x}_0,t_0) + \mathbf{f}\big(\phi(t;\mathbf{x}_0,t_0)\big) \tag{5.2-9}$$

in some interval $|t - t_0| \leqq a(t_0)$, then the function ϕ is said to be a solution of Eq. (5.2-3). The solution to Eq. (5.2-2) is a special case of this definition.

Theorem 5-1: Existence, Uniqueness, and Continuity (Ref. 6, Theorems 2.3 and 7.1, Chapter 1; also Ref. 11)

Let $\mathbf{A}\mathbf{x} + \mathbf{f}(\mathbf{x})$ be continuous in x, t, and satisfy the Lipschitz condition in some region about any $\mathbf{x}_0$, t_0:

$$R(x_0, t_0) = \begin{cases} \|x - x_0\| \leqq b(x_0) \\ |t - t_0| \leqq c(t_0) \end{cases} \qquad b, c > 0, \tag{5.2-10}$$

that is, the following condition is satisfied for (x, t), (y, t) in $R(x_0, t_0)$:

$$\|(Ax + f(x)) - (Ay + f(y))\| \leqq k\|x - y\| \tag{5.2-11}$$

in which k is a positive constant that depends only on b and c. Then

(a) There exists a unique solution $\phi(t; x_0, t_0)$ of Eq. (5.2-3) starting at x_0, t_0 for all $|t - t_0| \leqq a(t_0)$. The constant $a(t_0)$ may be estimated as

$$a(t_0) \geqq \min\{c(t_0), b(x_0)/M(x_0, t_0)\} \tag{5.2-12}$$

where $M(x_0, t_0)$ is the maximum assumed by the continuous function $\|Ax + f(x)\|$ in the closed and bounded set $R(x_0, t_0)$.

(b) In some small neighborhood of x_0, t_0 the solution is a continuous function of its arguments.

The Lipschitz condition implies that $Ax + f(x)$ must be continuous in x, but not necessarily in t. This local Lipschitz condition implies that a solution for Eq. (5.2-3) exists only in the region near x_0, t_0. If the above Lipschitz condition holds for $Ax + f(x)$ everywhere, then the system is described as being Lipschitz-in-the-large. If a system is Lipschitz-in-the-large, the possibility of a solution or state trajectory going to infinity in a finite time is excluded (i.e., no finite escape time). The terminology "Lipschitz-in-the-large" is equivalent to "Global Lipschitz," which is also common in the literature.

Definition 5-13: Continuous-Time Dynamic System [11]. If the trajectory ϕ is such that

(a) $\phi(t_0; x_0, t_0) = x_0$ for all x_0, t_0, (5.2-13)
(b) $\phi(t_2; \phi(t_1; x_0, t_0), t_1) = \phi(t_2; x_0, t_0)$ for all x_0, t_0, t_1, t_2 (existence and uniqueness requirement),
(c) ϕ is continuous with respect to all arguments, and
(d) ϕ is defined for all x_0, t_0, t (no finite escape time).

then the system is referred to as a continuous time dynamic system.

Definition 5-14: Equilibrium State. A state x_e of a free dynamic system [Eq. (5.2-3)] is an equilibrium state if

$$Ax_e + f(x_e) = 0 \quad \text{for all } t \tag{5.2-15}$$

or equivalently, if

$$\phi(t; x_e, 0) = x_e \quad \text{for all } t. \tag{5.2-16}$$

The following definitions are concerned with the various types of stability and follow those given in reference 11.

Definition 5-15: Stable [11]. An equilibrium state x_e of a free dynamic system is said to be stable if for every real number $\epsilon > 0$ there exists a real number $\delta(\epsilon, t_0) > 0$, such that

$$\| x_0 - x_e \| \leqq \delta \tag{5.2-17}$$

implies

$$\| \phi(t; x_0, t_0) - x_e \| \leqq \epsilon \text{ for all } t \geqq t_0. \tag{5.2-18}$$

Definition 5-16: Asymptotic Stability [11]. An equilibrium state x_e of a free dynamic system is said to be asymptotically stable if

(a) it is stable,

(b) every motion starting sufficiently near x_e converges to x_e as $t \to \infty$. That is, there is some real constant $r(t_0) > 0$, and for every real number $\mu > 0$ there corresponds a real number $T(\mu, x_0, t_0)$ such that $\| x_0 - x_e \| \leqq r(t_0)$ implies $\| \phi(t; x_0, t_0) - x_e \| \leqq \mu$ for all $t \geqq t_0 + T$.

It should be noted that asymptotic stability is a local concept since the size of the region $r(t_0)$ is not specified and may be arbitrarily small.

Definition 5-17: Equiasymptotic Stability [11]. An equilibrium state x_e of a free dynamic system is equiasymptotically stable if

(a) it is stable and

(b) every motion starting sufficiently near x_e converges to x_e as $t \to \infty$ uniformly in x_0. That is, for a precise mathematical statement, T in Definition 5-16 is of the form $T(\mu, r(t_0), t_0)$.

Definition 5-18: Asymptotic Stability in the Large [11]. An equilibrium state x_e of a free dynamic system is said to be asymptotically stable in the large if

(a) it is stable and

(b) every motion converges to x_e as $t \to \infty$.

Definition 5-19: Equiasymptotic Stability in the Large [11]. An equilibrium state x_e of a free dynamic system is said to be equiasymptotically stable in the large if

(a) it is stable and,

(b) every motion converges to x_e uniformly in x_0 for $\| x_0 \| \leqq r$ where r is fixed but arbitrarily large as $t \to \infty$.

Definition 5-20: Bounded Trajectory [11]. A trajectory is said to be bounded for every x_0, t_0 if there is some constant $B(x_0, t_0)$ such that $\| \phi(t; x_0, t_0) \| \leqq B$ for all $t \geqq t_0$.

Definition 5-21: Equibounded Trajectory [11]. A trajectory is said to be equibounded if, in Definition 5-20, $B(\mathbf{x}_0, t_0) \leqq B(r, t_0)$ for all $\|\mathbf{x}_0\| \leqq r$.

It should be noted that bounded trajectories are implied by asymptotic stability in the large, and equibounded trajectories are implied by equiasymptotic stability in the large.

Definition 5-22: Uniform Stability [11]. If, in the above definitions, δ does not depend on t_0, then uniform stability is obtained.

Definition 5-23: Uniform Asymptotic Stability [11]. If equiasymptotic stability is such that δ, r, and T are independent of t_0, then uniform asymptotic stability is said to exist.

Definition 5-24: Uniform Asymptotic Stability in the Large [11]. An equilibrium state $\mathbf{x}_e$ of a free dynamic system is uniformly asymptotically stable in the large if

(a) it is uniformly stable,

(b) it is uniformly bounded; that is, given any $r > 0$, there is some $B(r)$ such that $\|\mathbf{x}_0 - \mathbf{x}_e\| \leqq r$ implies $\|\boldsymbol{\phi}(t; \mathbf{x}_0, t_0) - \mathbf{x}_e\| \leqq B$ for all $t \geqq t_0$, and

(c) every motion converges to $\mathbf{x}_e$ as $t \to \infty$ uniformly in t_0 and $\|\mathbf{x}_0\| \leqq r$ when r is fixed but arbitrarily large; that is, given any $r > 0$ and $\mu > 0$, there exists some $T(\mu, r)$ such that $\|\mathbf{x}_0 - \mathbf{x}_e\| \leqq r$ implies $\|\boldsymbol{\phi}(t; \mathbf{x}_0, t_0) - \mathbf{x}_e\| \leqq \mu$ for all $t \geqq t_0 + T$.

With the above definitions established, the various theorems regarding stability may be set forth. The first group of theorems are summarized in reference 19 and treated in greater detail in references 6 and 3.

Theorem 5-2: Stability of Linear Autonomous systems. Every trajectory for a system described by

$$\dot{\mathbf{x}} = \mathbf{A}\mathbf{x} \tag{5.2-19}$$

is uniformly asymptotically stable in the large with $\mathbf{x}_e = \mathbf{0}$ if, and only if, all the characteristic roots of $\mathbf{A}$ have negative real parts. The symbol $\mathbf{0}$ is used to represent the n element vector in which each element is 0. Note that $\mathbf{A}$ is a matrix with constant elements for this type of system.

Theorem 5-3: Stability of Nonlinear Autonomous Systems. For a system of the form

$$\dot{\mathbf{x}} = \mathbf{A}\mathbf{x} + \mathbf{f}(\mathbf{x}) \tag{5.2-20}$$

if $\mathbf{x}(0) = \mathbf{c}$ and

(a) every solution of $\mathbf{x} = \mathbf{A}\mathbf{x}$ is uniformly asymptotically stable in the large,

(b) $\mathbf{f}(\mathbf{x})$ is continuous in some region about $\mathbf{x} = \mathbf{0}$, and

(c) $\dfrac{|\mathbf{f}(\mathbf{x})|}{|\mathbf{x}|} \to 0$ as $|\mathbf{x}| \to 0$, then $x_e = 0$ is an asymptotically stable solution of Eq. (5.2-20) provided $|x(0)| = |c|$ is sufficiently small.

Reference 6, pp. 314-317, and ref. 3, pp. 79-80 give the proof of the above theorem.

Theorem 5-4: Conditions for $x_e = 0$ *Unstable.* For a system described by Eq. (5.2-20) in which conditions (b) and (c) of Theorem 5-3 are satisfied, if the matrix Λ possesses at least one characteristic root with positive real part, then the point $x_e = 0$ is unstable.

The proof of Theorem 5-4 is given in ref. 6, pp. 317-318 and ref. 3, pp. 88-89.

Theorem 5-5: Conditional Stability of Nonlinear Autonomous Systems. For a system described by Eq. (5.2-20), if

(a) k of the characteristic roots of Λ have negative real parts (where $k \leq n$),

(b) $\dfrac{|f(x)|}{|x|} \to 0$ as $|x| \to 0$,

(c) $|f(x_1) - f(x_2) \leq c_1 |x_1 - x_2|$ for $|x_1|$ and $|x_2| \leq c_2$ where $c_1 \to 0$ as $c_2 \to 0$.

then there is a k-parameter family of solutions of Eq. (5.2-20) that are asymptotically stable with $x_e = 0$.

The implication of this theorem is that although solutions which start anywhere in a region of the state space about the point $x_e = 0$ may not be stable, there is a portion of this region about the origin in which trajectories may initiate and be asymptotically stable.

Theorem 5-5 is proved in ref. 6, pp. 329-333 and ref. 3, pp. 90-91.

Theorem 5-6: Stability of Nonlinear Autonomous Systems—A Special Case. For a system described by equations of the form

$$\dot{x} = \Lambda x + f(x, y), \quad x(0) = c \tag{5.2-21a}$$

$$\dot{y} = g(x, y), \quad y(0) = d \tag{5.2-21b}$$

the point $x_e = 0$, $y_e = 0$ is stable, provided that

(a) every solution of $\dot{x} = \Lambda x$ is asymptotically stable in the large,

(b) $f(x, y)$ and $g(x, y)$ are continuous in the region of the state space about $x = 0$,

(c) $\dfrac{|f(x, y)|}{|x|} \to 0$ as $|x| \to 0$,

(d) $|x(0)| = |c|$ is sufficiently small,

(e) $\dfrac{|g(x, y)|}{|x|} \to 0$ as $|x| \to 0$; $g(x, 0) = P_2(x)$.

Moreover, $x_e = 0$ is an asymptotically stable solution of Eq. (5.2-21a).

A heuristic argument is given as proof of Theorem 5-6 in ref. 19, pp. 39-40.

Theorem 5-7: Stability of Nonlinear Nonautonomous Systems. The solution of the system of equations

$$\dot{x} = Ax + f(x) + R(t), \quad x(0) = c \tag{5.2-22}$$

in which

(a) every solution of $\dot{x} = Ax$ is asymptotically stable in the large,
(b) $f(x)$ is continuous in the region $x_e = 0$ of the state space,
(c) $\dfrac{|f(x)|}{|x|} \to 0$ as $|x| \to 0$, and
(d) $|R(t)| < M$

is bounded if $|c|$ and M are sufficiently small.

Both a heuristic argument for and a formal proof of Theorem 5-7 are given in ref. 19, pp. 40-43.

Theorem 5-8: Stability of Nonlinear Nonautonomous Systems—A Special Case. For a system described by equations of the form

$$\dot{x} = Ax + f(x, y) + M(t), \quad x(0) = c \tag{5.2-23a}$$

$$\dot{y} = g(x, y), \quad y(0) = d \tag{5.2-23b}$$

in which

(a) every solution of $\dot{x} = Ax$ is asymptotically stable in the large,
(b) $f(x, y)$ and $g(x, y)$ are continuous in the region of the state space about the origin,
(c) $\dfrac{|f(x, y)|}{|x|} \to 0$ as $|x| \to 0$, and
(d) $\dfrac{|g(x, y)|}{|x|} \to 0$ as $|x| \to 0$; $g(x, 0) = P_2(x)$,

the solution are bounded provided that $|x|$ and $\int_0^t |x(\tau)|\, d\tau$ are bounded.

The substantiating arguments for Theorem 5-8 are given in ref. 19 pp. 43-44.

Theorems 5-3 through 5-8 deal only with stability or instability in an arbitrarily small region about the origin of the state space. They give no information as to how large the region may be. Moreover, they are of limited value with regard to establishing good or optimum values of system parameters in the design of a control system, since for the most part they ignore the effect of the parameters on the nonlinear portion of the system of differential equations. They are useful, however, in determining the stability of the equilibrium point of a system. This information may be useful in determining which of the theorems pertaining of Lyapunov's direct method should be applied. Since the above theorems are relatively simple to apply, they may significantly reduce the number of blind alleys encountered in attempting to apply Lyapunov's direct method to the stability analysis.

The next set of theorems includes both the basic theorems of Lyapunov's direct method and most of the extensions of this method that may be of interest to the control system engineer. There are a number of different forms for these theorems, and some additional theorems and interpretations in the literature. For the reader who is interested in extending his familiarity with the literature on this subject, an auxiliary bibliography is included. One fact must be carefully noted about the theorems based on the Lyapunov direct method. This fact is that they establish sufficient conditions but not necessary ones, and hence, in general, failure to satisfy the conditions of these theorems does not imply that the system of equations being analyzed is necessarily unstable.

Theorem 5-9: Uniform Asymptotic Stability in the Large for Free Nonlinear Nonautonomous Systems [11]. For the continuous-time, free dynamic system

$$\dot{\mathbf{x}} = \mathbf{A}\mathbf{x} + \mathbf{f}(\mathbf{x}) \tag{5.2-24}$$

in which both $\mathbf{A}$ and $\mathbf{f}$ may be time varying and where

$$\mathbf{A}\mathbf{0} + \mathbf{f}(\mathbf{0}) = \mathbf{0} \quad \text{for all} \quad t \tag{5.2-25}$$

($\mathbf{0}$ is the n-element zero vector), suppose there exists a scalar function $V(\mathbf{x}, t)$ with continuous first partial derivatives with respect to $\mathbf{x}$ and t such that $V(\mathbf{0}, t) = 0$ and

(a) $V(\mathbf{x}, t)$ is positive-definite; that is, there exists a continuous, nondecreasing scalar function $\alpha(\|\mathbf{x}\|)$ such that $\alpha(0) = 0$, and, for all t and all $\mathbf{x} \neq \mathbf{0}$

$$0 < \alpha\left(\|\mathbf{x}\|\right) \leqq V(\mathbf{x}, t) \tag{5.2-26}$$

(b) there exists a continuous scalar function $\gamma(\|\mathbf{x}\|)$ such that $\gamma(0) = 0$ and the derivative of V with respect to time $\dot{V}$ along the motion starting at t, $\mathbf{x}$ satisfies, for all t and all $\mathbf{x} \neq \mathbf{0}$

$$\begin{aligned} \dot{V}(\mathbf{x}, t) &\equiv \frac{dV(\mathbf{x}, t)}{dt} \quad \text{along motion starting at} \quad t, \mathbf{x} \\ &= \left.\frac{dV\left[\boldsymbol{\phi}(\tau; \mathbf{x}_0, t), \tau\right]}{d\tau}\right|_{\tau = t} \\ &= \lim_{h \to 0} \frac{V \ \mathbf{x} + h \ \mathbf{A}\mathbf{x} + \mathbf{f}(\mathbf{x}) \ , t = h \ - V(\mathbf{x}, t)}{h} \\ &= \frac{\partial V}{\partial t} + (\nabla V)'[\mathbf{A}\mathbf{x} + \mathbf{f}(\mathbf{x})] \leqq -\gamma\left(\|\mathbf{x}\|\right) < 0 \end{aligned} \tag{5.2-27}$$

(c) there exists a continuous, nondecreasing scalar function $\beta \|\mathbf{x}\|$ such that $\beta(0) = 0$ and, for all t,

$$V(\mathbf{x}, t) \leqq \beta(\|\mathbf{x}\|) \tag{5.2-28}$$

(d) $(d)\ \alpha(\|\mathbf{x}\|) \to \infty$ with $\|\mathbf{x}\| \to \infty$ (5.2-29)

then the equilibrium state $\mathbf{x}_e = 0$ is uniformly asymptotically stable in the large. $V(\mathbf{x}, t)$ is called a Lyapunov function for the system of differential equations given by Eq. (5.2-24).

Definition 5-25: ∇V—*Gradient of a Scalar Function.* For a scalar function $V(x_1, \ldots x_n, t)$, the gradient, ∇V, is defined as the column vector

$$\nabla V = \begin{bmatrix} \dfrac{\partial V}{\partial x_1} \\ \dfrac{\partial V}{\partial x_2} \\ \cdot \\ \cdot \\ \cdot \\ \dfrac{\partial V}{\partial x_n} \end{bmatrix} \tag{5.2-30}$$

Several combinations of the conditions of Theorem 5-9 yield sufficient conditions for various weaker types of stability. These are stated in the following theorems.

Theorem 5-10: Uniform Asymptotic Stability [11]. Conditions (a), (b), and (c) of Theorem 5-9 are sufficient for uniform asymptotic stability.

Theorem 5-11: Equiasymptotic Stability in the Large [11]. Conditions (a), (b), and (d) of Theorem 5-9 are sufficient for equiasymptotic stability in the large.

Theorem 5-12: Equiasymptotic Stability [11]. Conditions (a) and (b) of Theorem 5-9 are sufficient for equiasymptotic stability.

Theorem 5-13: Uniform Stability [11]. Conditions (a) and (c) of Theorem 5-9 and

$$\dot{V}(\mathbf{x}, t) \leqq 0 \text{ for all } \mathbf{x}, t \tag{5.2-31}$$

are sufficient for uniform stability.

Theorem 5-14: Stability [11]. Condition (a) of Theorem 5-9 and $\dot{V}(\mathbf{x}, t) \leqq 0$ for all $\mathbf{x}, t$ are sufficient for stability.

Theorem 5-15: No Finite Escape Time [11]. Conditions (a) and (d) of Theorem 5-9 and

$$\dot{V}(\mathbf{x}, t) \leqq cV(\mathbf{x}, t) \text{ for all } \mathbf{x}, t \text{ and } c \text{ a positive constant} \tag{5.2-32}$$

are sufficient for no finite escape time.

Theorem 5-16: Equiasymptotic Stability in the Large for Continuous-Time Nonlinear Autonomous Dynamic Systems [11]. For a continuous-time autonomous dynamic system

$$\dot{\mathbf{x}} = \mathbf{A}\mathbf{x} + \mathbf{f}(\mathbf{x}), \ \mathbf{A}\mathbf{0} + \mathbf{f}(\mathbf{0}) = \mathbf{0} \tag{5.2-33}$$

equiasymptotic stability in the large is assured by the existence of a scalar function $V(\mathbf{x})$ which has continuous first partial derivatives with respect to $\mathbf{x}$, such that

(a) $V(\mathbf{0}) = 0$
(b) $V(\mathbf{x}) > 0$ for all $\mathbf{x} \neq \mathbf{0}$
(c) $\dot{V}(\mathbf{x}) < 0$ for all $\mathbf{x} \neq \mathbf{0}$
(d) $V(\mathbf{x}) \to \infty$ with $\|\mathbf{x}\| \to \infty$

Condition (c) may be replaced by the conditions

(c-1) $\dot{V}(\mathbf{x}) \leqq 0$ for all $\mathbf{x}$ (i.e., $\dot{V}(\mathbf{x})$ negative semidefinite)
(c-2) $\dot{V}[\boldsymbol{\phi}(t; \mathbf{x}_0, t_0)]$ does not vanish identically in $t \geqq t_0$ for any t_0 and any $\mathbf{x}_0 \neq \mathbf{0}$.

Proofs for Theorems 5-9 through 5-16 are given in ref. 11, although in some cases different notation is used.

Theorem 5-17: Instability in the Large for Continuous-Time Nonlinear Autonomous Dynamic Systems [8, 28]. For a continuous-time autonomous dynamic system of the form shown in Eq. (5.2-33), if there exists a scalar function $V(\mathbf{x})$ which is continuous and has continuous first partial derivatives with respect to $\mathbf{x}$ such that

(a) $V(\mathbf{0}) = 0$
(b) $V(\mathbf{x}) \geqq 0$ for all $\mathbf{x} \neq \mathbf{0}$
(c) $\dot{V}(\mathbf{x}) > 0$ for all $\mathbf{x} \neq \mathbf{0}$
(d) $\dot{V}(\mathbf{x}) \to \infty$ with $\|\mathbf{x}\| \to \infty$

then the system is unstable in the large.

The statement of this theorem is somewhat different in refs. 8 and 28 where it is proved.

Theorem 5-18: Existence of a Lyapunov Function [11]. Let the function $\mathbf{A}\mathbf{x} + \mathbf{f}(\mathbf{x})$ of Eq. (5.2-3) be Lipschitzian. Assume further that $\mathbf{A}\mathbf{0} + \mathbf{f}(\mathbf{0}) = \mathbf{0}$ and that the equilibrium state $\mathbf{x}_e = \mathbf{0}$ is uniformly asymptotically stable in the large. Then there exists a Lyapunov function $V(\mathbf{x}, t)$ which is infinitely differentiable with respect to $\mathbf{x}$, t, and satisfies all of the conditions of Theorem 5-9.

Theorem 5-18 is formulated and proved in ref. 20. It implies that the existence of a Lyapunov function as required by the conditions of Theorem 5-9 is both necessary and sufficient for uniform asymptotic stability in the large.

Theorem 5-19: Linear Systems with Forcing Functions [11]. Consider a continuous-time, linear dynamic system

$$\dot{\mathbf{x}} = \mathbf{A}\mathbf{x} + \mathbf{D}\mathbf{R}(t) \tag{5.2-35}$$

in which $\mathbf{R}(t)$ represents the input or forcing function and the matrices $\mathbf{A}$ and $\mathbf{D}$ may have elements that vary with time, subject to the restrictions

(a) $\|\mathbf{A}\mathbf{x}\| \leqq c_1 < \infty$ for all $\|\mathbf{x}\| = 1$, all t, and*

(b) $0 < c_2 \leqq \|\mathbf{D}\mathbf{x}\| \leqq c_3 < \infty$ for all $\|\mathbf{x}\| = 1$, all t.

For this case, the general solution to Eq. (5.2-35) may be written

$$\phi(t\,; \mathbf{x}_0 t_0) = \mathbf{\Phi}(t, t_0)\mathbf{x}_0 + \int_{t_0}^{t} \mathbf{\Phi}(t, \tau)\mathbf{D}(\tau)\mathbf{R}(\tau)\,d\tau^{**}$$

The following propositions concerning this system are then equivalent:

(1) Any uniformly bounded excitation

$$\|\mathbf{R}(t)\| \leqq c_4 < \infty, \quad (t \geqq t_0)$$

gives rise to a uniformly bounded response for all $t \geqq t_0$

$$\|\mathbf{x}(t)\| = \left\|\mathbf{\Phi}(t, t_0)\mathbf{x}_0 + \int_{t_0}^{t} \mathbf{\Phi}(t, \tau)\mathbf{D}(\tau)\mathbf{R}(\tau)\,d\tau\right\|$$

$$\leqq c_5\left(c_4, \|\mathbf{x}_0\|\right) < \infty$$

(2) For all $t \geqq t_0$, $\displaystyle\int_{t_0}^{t} \|\mathbf{\Phi}(t, \tau)\|\, d\tau \leqq c_6 < \infty$

(3) The equilibrium state $\mathbf{x}_e = \mathbf{0}$ of the free system is uniformly asymptotically stable.

(4) There exist positive constants c_7 and c_8 such that whenever $t \geqq t_0$

$$\|\mathbf{\Phi}(t, t_0)\| \leqq c_7\, e^{-c_8(t - t_0)}$$

(5) Given any symmetric positive-definite matrix $\mathbf{Q}(t)$ continuous in t and such that for all $t \geqq t_0$

$$[\mathbf{Q}(t) - c_9\mathbf{I}],$$

$$[c_{10}\mathbf{I} - \mathbf{Q}(t)],$$

and

$$(c_{10} - c_9)\mathbf{I}$$

*All c_i are fixed finite positive constants.

**$\mathbf{\Phi}(t, t_0)$ is referred to as the "transition matrix" of the unforced system. Its elements may be considered the system impulse responses under appropriate excitations and observations [11].

are positive-definite, then the scalar function defined by

$$V(\mathrm{x}, t) = \int_t^{\infty} \left\{[\Phi(\tau, t)\mathrm{x}]'[\mathrm{Q}(\tau)][\Phi(\tau, t)\mathrm{x}]\right\} d\tau = \mathrm{x}'[\mathrm{P}(t)]\mathrm{x} \quad (5.2\text{-}35)$$

exists and is a Lyapunov function for the free system satisfying the conditions of Theorem 5-9, with its derivative along the free motion starting at x, t

$$\dot{V}(\mathrm{x}, t) = -\mathrm{x}'[\mathrm{Q}(t)]\mathrm{x} \quad (5.2\text{-}36)$$

Theorem 5-20: Asymptotic Stability of a Continuous-Time, Linear, Autonomous Dynamic System [11]. The equilibrium state $\mathrm{x}_e = 0$ of a continuous-time linear autonomous dynamic system

$$\dot{\mathrm{x}} = \mathrm{A}\mathrm{x} \quad (5.2\text{-}37)$$

is asymptotically stable (a) if and (b) only if given any symmetric, positive-definite matrix Q there exists a symmetric, positive-definite matrix P which is the unique solution of the set of $n(n+1)/2$ linear equations

$$\mathrm{A}'\mathrm{P} + \mathrm{P}\mathrm{A} = -\mathrm{Q} \quad (5.2\text{-}38)$$

moreover $\|\mathrm{x}\|^2{}_P$ is a Lyapunov function for the system.

Theorem 5-21: Extension of Theorem 5-20 [11]. The real parts of the characteristic roots of a constant matrix A (see Theorem 5-20) are less than σ if and only if given any symmetric positive-definite matrix Q there exists a symmetric, positive-definite matrix P which is the unique solution of the set of $n(n+1)/2$ linear equations

$$-2\sigma\mathrm{P} + \mathrm{A}'\mathrm{P} + \mathrm{P}\mathrm{A} = -\mathrm{Q} \quad (5.2\text{-}39)$$

Proofs for Theorems 5-19 through 5-21 are given in ref. 11 along with references as to their origin and some discussion of their implications and applicability.

Theorem 5-22: Regions of Asymptotic Stability for Nonlinear Autonomous Systems [13]. Let Ω be a bounded closed region about $\mathrm{x}_e = 0$ in the state space for a system of differential equations of the form

$$\dot{\mathrm{x}} = \mathrm{A}\mathrm{x} + \mathrm{f}(\mathrm{x}), \quad \mathrm{A}0 + \mathrm{f}(0) = 0 \quad (5.2\text{-}40)$$

where $\mathrm{A}\mathrm{x} + \mathrm{f}(\mathrm{x})$ satisfies the conditions for a continuous-time *autonomous* dynamic system. Further, let Ω have the property that every solution $\phi(t; \mathrm{x}_0, t_0)$ that starts in Ω remains for all future time in Ω. If there exists a scalar function $V(\mathrm{x})$ which is continuous and has continuous first partial derivatives in Ω and such that, in Ω

(a) $V(0) = 0$
(b) $V(\mathbf{x}) > 0$ for all $\mathbf{x} \neq 0$
(c) $\dot{V}(\mathbf{x}) < 0$ for all $\mathbf{x} \neq 0$

Condition (c) may be replaced by the conditions
(c-1) $\dot{V}(\mathbf{x}) \leqq 0$ for all $\mathbf{x}$ (i.e., $\dot{V}(\mathbf{x})$ negative-semidefinite)
(c-2) $\dot{V}[\boldsymbol{\phi}(t; \mathbf{x}_0, t_0)]$ does not vanish identically in $t \geqq t_0$ for any t_0 and any $\mathbf{x}_0 \neq 0$.

Then, every solution starting in Ω is asymptotically stable, and the point $\mathbf{x}_e = 0$ is the equilibrium point.

The proof of Theorem 5-22 is given in ref. 13 pp. 9-10. It should be noted that Theorem 1 on page 9 of ref. 13 and the asociated proof are more general in statement than Theorem 5-22.

Theorem 5-23: Method of Determining a Region of Asymptotic Stability for Nonlinear Autonomous Systems [13]. Let Ω denote the closed region defined by $V(\mathbf{x}) \leqq L$ where $V(\mathbf{x})$ is a scalar function which is continuous and has continuous partial first derivatives in Ω. For a system of the form of Eq. (5.2-40), if, in the region Ω, $V(\mathbf{x})$ is such that conditions (a), (b), and (c) of Theorem 5-22 are satisfied and the region Ω is bounded, then every solution of Eq. (5.2-40) which originates in or enters Ω is asymptotically stable, and the point $\mathbf{x}_e = 0$ is the equilibrium point.

In Theorem 5-23, a sufficient condition for the region Ω defined by $V(\mathbf{x}) \leqq L$ to be bounded for all L is the $V(\mathbf{x}) \to \infty$ as $\|\mathbf{x}\| \to \infty$. If

$$\lim_{\|\mathbf{x}\| \to \infty} \inf V(\mathbf{x}) = L_0$$

then Ω is bounded for all $L < L_0$.

Reference 13, pp. 10-14a, gives a discussion and some examples of the use of Theorem 5-23.

Theorem 5-24: Boundedness of Solutions by Lyapunov's Direct Method (Lagrange Stability Theorem) [13]. For a system represented by Eq. (5.2-40), let Ω be a bounded neighborhood of the origin and let Ω^c be its complement (that is, Ω^c is the set of all points outside Ω). Assume that $W(\mathbf{x})$ is a scalar function with continuous first partial derivatives in Ω^c and satisfying the conditions:
(a) $W(\mathbf{x}) > 0$ for all $\mathbf{x}$ in Ω^c
(b) $\dot{W}(\mathbf{x}) \leqq 0$ for all $\mathbf{x}$ in Ω^c
(c) $W(\mathbf{x}) \to \infty$ as $\|\mathbf{x}\| \to \infty$

then each solution of Eq. (5.2-40) is bounded for all $t \geqq 0$.

Reference 13, pp. 16-19, gives a proof of Theorem 5-24 along with a discussion of its application to two simple examples.

Theorem 5-25: Regions of Eventual Asymptotic Stability for Free Nonlinear Nonautonomous Systems [14]. Consider the continuous-time, free dynamic system

$$\dot{\mathbf{x}} = \mathbf{A}\mathbf{x} + \mathbf{f}(\mathbf{x}) \tag{5.2-41}$$

in which both $\mathbf{A}$ and $\mathbf{f}$ may be time-varying and where

$$\mathbf{A0} + \mathbf{f}(\mathbf{0}) = \mathbf{0} \text{ for all } t \qquad (5.2\text{-}42)$$

and $\mathbf{f}(\mathbf{x})$ has continuous first partial derivatives. The solution satisfying $\boldsymbol{\phi}(t_0; \mathbf{x}_0, t_0) = \mathbf{x}_0$ is denoted by $\boldsymbol{\phi}(t; \mathbf{x}_0, t_0)$.

Let Ω be a bounded closed region containing the origin. Further, let Ω_0 be a subregion of Ω with the property that solutions which start in Ω_0 at a time $t_0 \geqq T_0$ remain thereafter in Ω. Then, if there exists a scalar function $V(\mathbf{x}, t)$ such that:

(a) $V(\mathbf{x}, t) \to U(\mathbf{x})$ as $t \to \infty$ uniformly for $\mathbf{x}$ in Ω

(b) $\dot{V}(\mathbf{x}, t) \to -W(\mathbf{x})$ as $t \to \infty$ uniformly for $\mathbf{x}$ in Ω

(c) $U(\mathbf{x})$ and $W(\mathbf{x})$ are positive-definite for $\mathbf{x}$ in Ω.

Then there exists a $T_0 > 0$ with the property that $\boldsymbol{\phi}(t; \mathbf{x}_0, t_0) \to 0$ as $t \to \infty$ for all $\mathbf{x}_0$ in Ω_0 and all $t_0 \geqq T_0$.

Theorem 5-26: Method of Determining a Region of Eventual Asymptotic Stability for Free Nonlinear Nonautonomous Systems [14]. Consider a continuous-time, free dynamic system of the type described in Theorem 5-25.

Let Ω denote the closed bounded region defined by $U(\mathbf{x}) \leqq L$ $(L > 0)$ and assume that conditions (a), (b), and (c) of Theorem 5-25 are satisfied. Now, for any $\delta > 0$, let Ω_δ be the region defined by $U(\mathbf{x}) \leqq L - \delta$. Then there is a $T_\delta > 0$ such that

$$\boldsymbol{\phi}(t; \mathbf{x}_0 t_0) \to 0 \text{ as } t \to \infty$$

for all $\mathbf{x}_0$ in Ω_δ and all $t_0 \geqq T_\delta$.

It should be noted, in Theorem 5-26, that a sufficient condition for the region Ω defined by $U(\mathbf{x}) \leqq L$ to be bounded for all L is that $U(\mathbf{x}) \to \infty$ as $\|x\| \to \infty$. Moreover, if

$$\lim_{\|\mathbf{x}\| \to \infty} \inf U(\mathbf{x}) = L_0$$

then Ω is bounded for all $L < L_0$.

Theorems 5-25 and 5-26 are due to Dr. J. P. LaSalle and are presented in reference 14.

I. G. Malkin has proved a very useful theorem with regard to the form which the Lyapunov function may be caused to assume if the form of Eq. (5.2-40) satisfies certain restrictions. This theorem, given in ref. 5, page 112, is stated next.

Theorem 5-27: Conditions for Quadratic Form for the Lyapunov Function [5]. If a system can be represented by a set of differential equations of the form

$$\dot{\mathbf{x}} = \mathbf{A}\mathbf{x} + \mathbf{P}(t, \mathbf{x}) \qquad (5.2\text{-}43)$$

in which $\mathbf{A}$ is an $n \times n$ matrix whose coefficients are continuous bounded functions of t, and $\mathbf{P}$ is a vector function of t and $\mathbf{x}$ whose components are power series of $x_1, \ldots, x_n$, convergent for all $\|x\| < k$ for some $k > 0$ (uniformly with respect to $t \geqq 0$), and whose coefficients are continuous bounded functions of t. Then, if there exists a positive definite function $V(x)$ such that $\dot{V}(x)$ is negative-definite, there also exists a $V(x)$, a quadratic form in the state variables $x_1, \ldots x_n$ with the same properties. Moreover, this quadratic $V(x)$ satisfies the conditions

$$V(x) \geqq a^2 \Sigma x_i^2, \; \dot{V}(x) \leqq -b^2 \Sigma x_i^2 \tag{5.2-44}$$

for some $a, b > 0$.

This concludes the statement of theorems relating to extensions and variations of the direct method of Lyapunov. There are a number of theorems and corollaries which deal with symmetric matrices and quadratic forms which will be useful in selecting and manipulating Lyapunov functions. Some of these are repeated below for the reader's convenience. Most of them were obtained from ref. 4, but they also appear in numerous other books on modern algebra.

Definition 5-26: Matrix Notation of a General Quadratic Form. A quadratic form Q in n state variables may be written as a matrix equation of the form

$$Q(x) = x'Bx \tag{5.2-45}$$

in which x is the column vector of the n state variables [see Eq. (5.2-2)] and x' is the row vector which is the transpose of x. B is an $n \times n$ matrix which may have either time-varying or constant elements. It need not be symmetric; if it is not, there exists another symmetric matrix which will produce the identical quadratic form Q. Therefore, in this application, the matrix B in quadratic forms will always be symmetric.

Theorem 5-28: Reduced Form for a Quadratic Function [p. 271, ref. 4]. Any quadratic form Q over the field of real numbers can be reduced by nonsingular linear transformations of the variables to a form

$$Q(\xi) = z_1^2 + \ldots + z_p^2 - z_{p+1}^2 - \ldots - z_r^2 \tag{5.2-46}$$

Theorem 5-29: Invariant Property of a Quadratic Function [p. 272, ref. 4]. The number p of positive squares which appears in the reduced form is an invariant of the given form in the sense that p depends only on the form and not on the method used to reduce it.

Definition 5-27: Positive Definite Matrix B. If the matrix B is of such form that the quadratic function $Q(x)$, Eq. (5.2-45), is positive-definite (that is, $Q(x) > 0$ for all $x \neq 0$, all t) then the matrix B is said to be a positive-definite matrix.

Theorem 5-30: Positive-Definite Quadratic Function—Criterion I [p. 273, ref. 4]. A real quadratic form is positive-definite if, and only if, all the squares in the reduced form (5.2-46) are positive.

Theorem 5-31: Positive-Definite Symmetric Matrix [p. 273, ref. 4]. A real symmetric matrix B is positive-definite if, and only if, there exists a real nonsingular matrix P such that $B = PP'$.

Theorem 5-32: Diagonal Form for a Quadratic Function [p. 277, ref. 4]. Any real quadratic form in n variables assumes a diagonal form relative to a suitable normal orthogonal basis.

Corollary 1 (to Theorem 5-32): Any real homogeneous quadratic function of n variables can be reduced to the diagonal form by an orthogonal point-transformation.

Corollary 2 (to Theorem 5-32): For any real symmetric matrix B, there is a real orthogonal matrix P such that $PBP' = PBP^{-1}$ is diagonal.

Corollary 3 (to Theorem 5-32): Every nonsingular real matrix B can be expressed as a product $B = SR$, where S is a symmetric positive-definite matrix and R is orthogonal.

Theorem 5-33: Characteristic Roots or Eigenvalues for a Matrix [Lemma, p. 311, ref. 2]. The characteristic roots, or eigenvalues, of a matrix B are scalars λ such that $|\lambda I - B| = 0$.

Definition 5-28: The Characteristic Function of B *is given by* $|\lambda I - B|$.

Definition 5-29: The Characteristic Equation of B *is given by* $|\lambda I - B| = 0$.

Theorem 5-34: Roots of Characteristic Equation [p. 313, ref. 4]. The characteristic roots, or eigenvalues, of a matrix B are the roots of the characteristic equation of B.

Theorem 5-35: Reduction of Quadratic Function to Diagonal Form [p. 314, ref. 4]. Any real quadratic form $x'Bx$ may be reduced by an orthogonal transformation to a diagonal form $\lambda_1 z_1^2 + \ldots + \lambda_n z_n^2$, in which the coefficients λ_i are the roots of the characteristic equation

$$|\lambda I - B| = (\lambda - \lambda_1)(\lambda - \lambda_2) \ldots (\lambda - \lambda_n) = 0 \text{ of } B.$$

Corollary (to Theorem 5-35): All characteristic roots of a real symmetric matrix are real.

Theorem 5-36: Positive Definite Quadratic Function—Criterion II. Any real quadratic form $x'Bx$ is positive-definite if the eigenvalues of the real symmetric matrix B are positive.

Proof: Follows directly from Definition 5-26 and Theorem 5-35.

Theorem 5-37: Positive Definite Quadratic Function—Criterion III. Any real quadratic form $x'Bx$ is positive-definite if the determinants of all the principal minors of B are positive.

Theorem 5-38: Positive Definite Quadratic Function—Criteria IV. Any real quadratic form $x'Bx$ is positive-definite if there exist positive-definite matrices C and D such that $b_{ij} = c_{ij}d_{ij}$ for all i, j, all t.

The theorems presented in this section do not represent all the available theorems relating to Lyapunov's direct method. They do, however, include most of the theorems of interest to the control system engineer. These same theorems as well as others appear in a variety of different forms in the literature. In the above presentation, an attempt has been made to state the theorems in as homogeneous a manner as possible. A few additional theorems will be given in Section 5.3. The latter are more closely related to the application of Lyapunov's direct method, or deal with the generation of Lyapunov functions. Hence they are in a more appropriate context there.

5.3 Application of Lyapunov's Direct Method and the Generation of Lyapunov Functions

5.3.1 *Introduction*

The major difficulty in applying the theorems of Lyapunov's direct method lies in the selection or generation of a scalar function $V(x)$ which satisfies the conditions of the theorems. In many of the English publications, the ability to determine a Lyapunov function is treated as an art which depends on the skill, experience, and very often the luck of the investigator. At one time, this was actually the situation. However, a number of fairly proceduralized and analytical techniques are now available for generating Lyapunov functions. Some of the more generally applicable techniques are given in this section. In some cases examples of their application are given. The techniques to be treated are the Canonic Transformations [8, 16, 17, 18, 22], Aizerman's method [1, 8], Krasovskiy's method [12, 8], Pliss' method [21, 8], and the Variable Gradient method [23, 24]. The last-named technique appears to handle all cases to which the methods of Ingwerson [9, 10, 24] and Szego [24, 25, 26] are applicable, and so these two are not treated. The techniques will be discussed in the order mentioned above.

5.3.2 *The Canonic Transformations* [8, 16, 17, 18, 22]

A transformation which changes the form of a system of differential equations into a set of canonic differential equations is called a canonic transformation. Lyapunov functions have been developed which apply to systems that can be transformed into one of two canonic forms. These canonic forms are referred to as the "first canonic form" and the "second canonic form." It is possible to use certain standard Lyapunov functions to establish secondary stability criteria which may be applied to systems represented by one of the canonic forms. The secondary stability criteria are easier to apply and do not require the generation of a Lyapunov function for each system investigated. The details of these forms are given below. In the presentation of Lure [17] and Letov [16], in Russian, reference is made to control systems of "direct control" and "indirect control." There seems to be no equivalent classification for control systems in the English literature; and hence this terminology will not be used here.

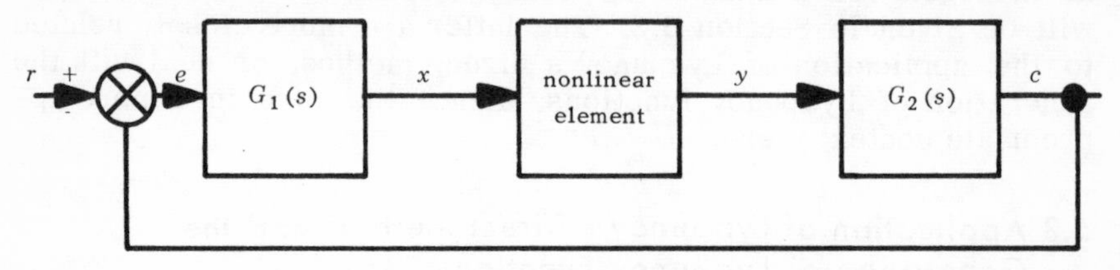

FIG. 5-4. Block diagram of a closed-loop system with a single nonlinear element.

The systems to which this method may be applied must be reducible to the form shown in Fig. 5-4. However, the fact that a system is reducible to this form does not guarantee that it may be analyzed by this technique. In the discussion which is to follow, it will be assumed that the input to the system is defined by

$$r(t) = 0 \text{ for all } t > 0 \tag{5.3-1}$$

The block diagram for the system may then be placed in the form shown in Fig. 5-5.

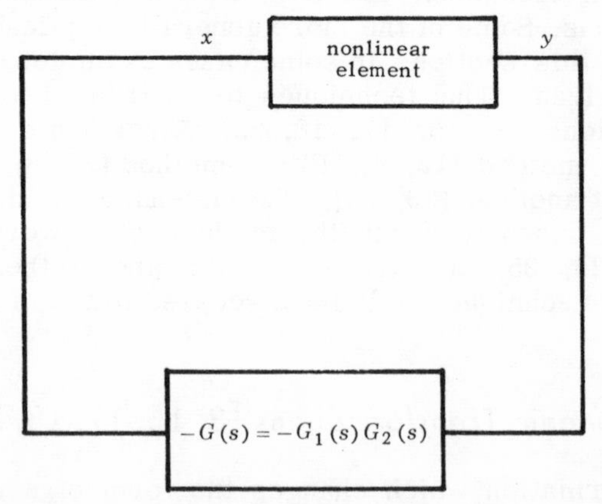

FIG. 5-5. Simplified block diagram of a closed-loop system with a single nonlinear element.

The input-output characteristics of the nonlinear gain element will be restricted to those which can be described by a continuous function

$$y = f(x); \; f(0) = 0 \tag{5.3-2}$$

where x is the input to the element, y is the output, and $f(x)$ is single-valued and analytical in a sufficiently small region about the point $x = 0$. It should be noted that this element has neither integration, nor differentiation, nor any form of energy storage associated with it.

The First Canonic Transformation [8, 8a, 18]

The discussion given below uses the block diagram of Fig. 5-5 as the basic system configuration. If $G(s) = G_1(s)\,G_2(s)$ does not have any multiple poles, and the number of poles exceeds the number of zeros, then it is possible to represent the system by a set of differential equations referred to as the first canonic form. For the general case in which $G(s)$ has n poles, the first canonic form of system differential equations is

$$\frac{dz_i}{dt} = \lambda_i z_i + f(x) \qquad i = 1, 2, \ldots n \tag{5.3-3a}$$

and

$$x = \sum_{i=1}^{n} a_i z_i \tag{5.3-3b}$$

in which the λ_i's are the poles of $G(s)$ and the a_i's are the negative of the residue of $G(s)$ as λ_i. The z_i are referred to as the canonic variables and x is still the single input to the nonlinear element. An additional equation is obtained by differentiating Eq. (5.3-3b) with respect to time, and then substitution of Eq. (5.3-3a):

$$\frac{dx}{dt} = \sum_{i=1}^{n} \beta_i z_i - rf(x) \tag{5.3-3c}$$

where

$$\beta_i = a_i \lambda_i \qquad i = 1, 2, \ldots n \tag{5.3-4}$$

and

$$r = -\sum_{i=1}^{n} a_i \tag{5.3-5}$$

Of the set of equations that represents the first canonic transformation, Eq. (5.3-3a) is referred to as the principal part, and Eqs.

(5.3-3b) and (5.3-3c) are called the complementary part. A block diagram representation of the system as described by the first canonic transformation is shown in Fig. 5-6. From this block diagram it may be seen that the first canonic transformation merely represents a partial fraction expansion of the linear portion of the system, $G(s)$

$$G(s) = -\sum_{i=1}^{n} \frac{a_i}{s - \lambda_i} \tag{5.3-6}$$

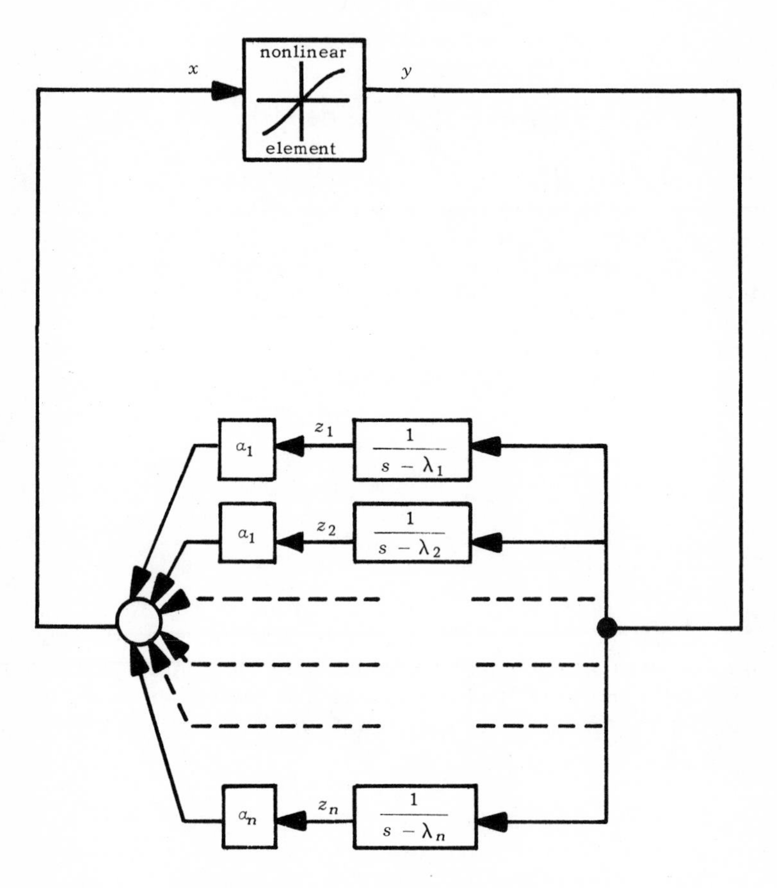

FIG. 5-6. Block diagram representation of canonic transformation.

Secondary Stability Criteria

For systems that may be transformed into the first canonic form, it is possible to derive secondary stability criteria by applying the theorems of Lyapunov's direct method. Unfortunately these secondary stability criteria are excessively restrictive and hence reject many systems which are actually stable. They have the advantage, however, that they are easier to use and do not require the generation of a Lyapunov function for each different application.

To establish the secondary stability criteria, Lure [18] employed a Lyapunov function of the form

$$V = -\sum_{i=1}^{n} \sum_{j=i}^{n} \frac{a_i a_j z_i z_j}{\lambda_i + \lambda_j} + \int_0^x f(x)\,dx \tag{5.3-7}$$

This function is positive-semidefinite for the conditions:

(a) $\int_0^x f(x)\,dx = 0,\ f(0) = 0;$ (5.3-8)

(b) the constants a_i are real for corresponding real λ_i's and are in complex conjugate pairs for corresponding complex conjugate pairs of λ_i's.

(c) $\text{Re}\,\lambda_i < 0$ for $i = 1, \ldots n$.

The time derivative of Eq. (5.3-7), with Eqs. (5.3-3a), (5.3-3b), and (5.3-3c) substituted as required, is

$$\begin{aligned} \frac{dV}{dt} &= -rf(x)^2 - \left(\sum_{i=1}^{n} a_i z_i\right)^2 \\ &\quad + f(x) \sum_{i=1}^{n} z_i \left(\beta_i - 2a_i \sum_{j=1}^{n} \frac{a_j}{\lambda_i + \lambda_j}\right) \end{aligned} \tag{5.3-9}$$

Equation (5.3-9) is negative-semidefinite for the conditions:

(d) $2a_i \sum_{j=1}^{n} \frac{a_j}{\lambda_i + \lambda_j} = \beta_i \qquad i = 1, 2, \ldots n$ (5.3-10)

(e) $r = -\sum_{i=1}^{n} a_i \geq 0$ (5.3-11)

It is also possible to make Eq. (5.3-9) negative-definite by adding to Eq. (5.3-7) a term of the form

$$\begin{aligned} \phi &= A_1 z_1^2 + A_2 z_2^2 + \ldots + A_s z_s^2 + C_1 z_{s+1} z_{s+2} \\ &\quad + C_3 z_{s+3} z_{s+4} + \ldots C_{s-n-1} z_{n-1} z_n \end{aligned} \tag{5.3-11}$$

in which the constants A and C are infinitesimally small positive numbers. The constants A_i are associated with the real canonic variables $z_i (i = 1, 2, \ldots s)$ and the constants C_i are associated with the complex variables $z_i (i = s, s+1, \ldots n)$.

In the above line of reasoning, Lyapunov's direct method was used to derive a secondary set of stability criteria. This may be stated as a theorem referred to as Lure's Theorem.

Theorem 5-39: Secondary Stability Criterion I (First Canonic Forms). A system which may be transformed into the first canonic form [Eqs. (5.3-3)], and for which conditions (a) through (e) above are satisfied, is asymptotically stable in the large. Reduced ranges of stability for the variable x can be established by means of this theorem as long as the range includes the equilibrium point $x = 0$ and condition (a) [Eq. (5.3-8)] is satisfied over the range.

It should be noted that in applying Theorem 5-39, the problem has been changed from requiring the generation of a Lyapunov function to that of solving the set of equations represented by Eq. (5.3-9) for acceptable set of $a_i (i = 1, \ldots, n)$. In general, this criterion is excessively restrictive. A number of investigators have used different Lyapunov functions in place of Eq. (5.3-7) to derive additional secondary stability criteria analogous to conditions (a) through (e) above [8, 16, 17, 22]. A summary and tabulation of these stability criteria is presented in refs. 8 and 22, where the first canonic transformation is treated in considerable detail. Although these various forms increase the scope of applicability of this technique, it is still quite restrictive by virtue of its basic formulation, and may reject many systems which are actually stable. It is relatively easy to understand why this is the case. In all the variations of this approach, the restriction on the nonlinear element [condition (a)] is one of the following two forms

$$\int_0^x f(x)\,dx \geq 0, \quad f(0) = 0$$

or

$$x\,f(x) \geq 0, \quad f(0) = 0$$

The second of these is the more restrictive and only limits $f(x)$ to the first and third quadrants (unshaded portion) of the nonlinear element input-output plane shown in Fig. 5-7. Three typical gain curves are shown in the figure. A linear gain element may be considered a special case of the nonlinear gain element. In the case of linear element $y = kx$, the restriction merely requires that $0 < k < \infty$. Therefore these secondary stability criteria would reject all linear systems which were not stable for all positive values of open-loop gain. Because of the looseness of the restrictions on $f(x)$, the analysis cannot descriminate between linear and nonlinear gain functions as long as they remain within the unshaded region of Fig. 5-7. Consequently, this secondary stability criterion will reject all systems for which the root-locus of $G(s)$ enters the right half of the s-plane. Because of this, it is always wise to plot a root-locus for $G(s)$ before attempting to apply the secondary stability criterion. A summary of reasons for which a system of the form of Fig. 5-5 may be rejected by this technique are summarized as follows:

(a) Some of the poles of $G(s)$ are in the right half of the s-plane.
(b) Some of the zeros of $G(s)$ are in the right half of the s-plane.
(c) The root-locus of $G(s)$ is not confined to the left half of the s-plane.
(d) $G(s)$ has poles at the origin of the s-plane.
(e) $G(s)$ has multiple poles.
(f) The difference between the number of poles and zeros of $G(s)$ is equal or greater than 2 (i.e., $n - m \geq 2$).
(g) The constant r (Eq. 5.3-5) is nonpositive.

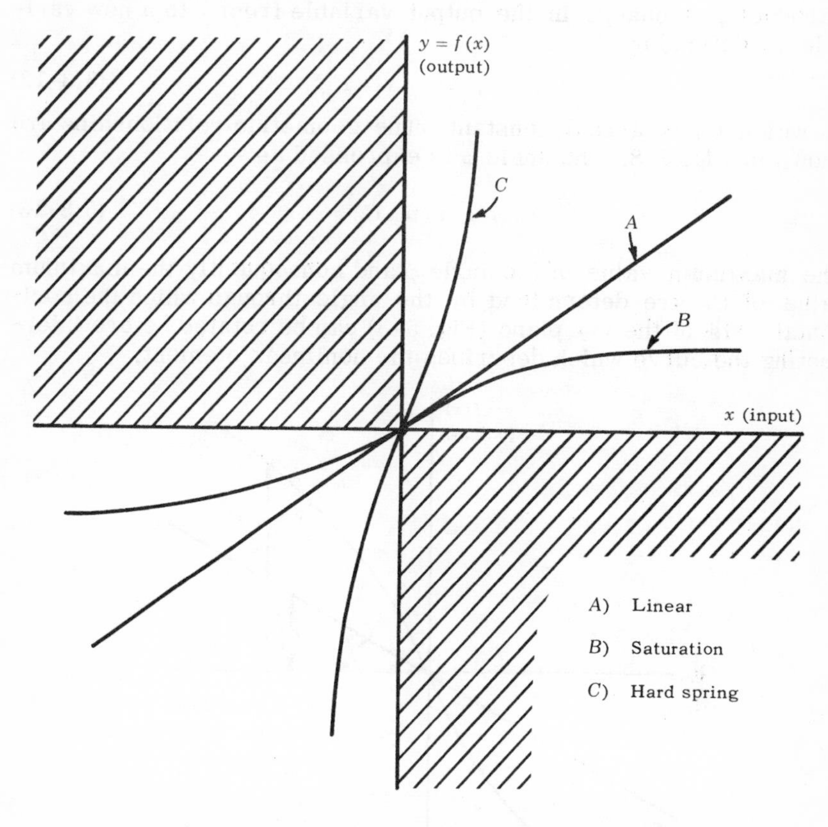

FIG. 5-7. Restrictions on the nonlinear gain element.

In view of the summary above, a large number of systems of interest would be rejected by the secondary stability criterion based on the first canonic form.

The range of applicability of the technique may be increased by restricting the gain characteristics of the nonlinear element to only a portion of the first and third quadrants of the input-output plane of the nonlinear element. This is done by applying pole- and zero-shifting techniques.

The Pole-Shifting Technique

It is possible to place a lower limit on the gain of the nonlinear element in order to expand the scope of applicability of the secondary stability criterion so that the criterion will no longer reject stable systems whose gain does not fall below this lower limit. This is accomplished by rotating the input axis of the input-output characteristic plane for the nonlinear element in the counterclockwise direction through an angle ϕ. The result is equivalent to introducing a change in the output variable from y to a new variable y', defined by

$$y' = f'(x) = y - C_p x \tag{5.3-12}$$

in which C_p is a real constant. The geometric relationships are shown in Fig. 5-8. The angle ϕ is expressed as

$$\phi = \arctan C_p \tag{5.3-13}$$

The maximum value of the angle ϕ and consequently the maximum value of C_p are determined by the angle through which the horizontal axis of the x-y plane (Fig. 5-8) can be rotated before intersecting the curve which describes the nonlinear element.

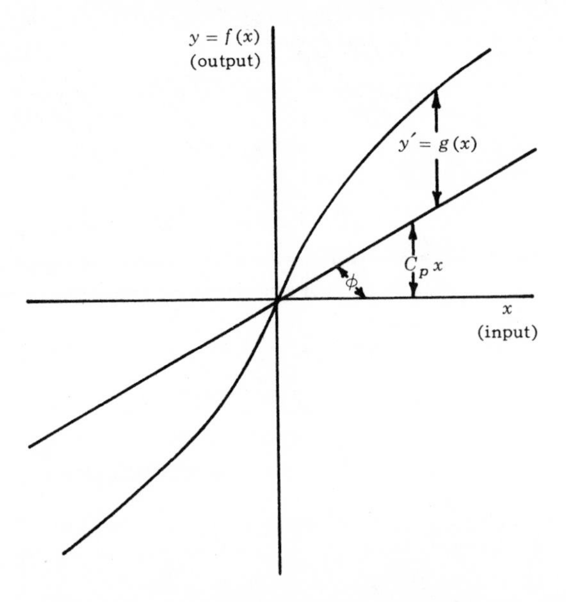

FIG. 5-8. Geometry of the pole-shifting technique.

The new variable y' is used for y in the secondary stability criterion and the function $f'(x)$ must satisfy the criterion previously established for $f(x)$. When the variable y' is introduced into the block diagram of Fig. 5-5, then the dynamic elements are represented by $G'(s)$, which is related to $G(s)$ by the expression

$$G'(s) = \frac{G(s)}{1 + C_p G(s)} \tag{5.3-14}$$

Equation (5.3-14) shows the facility with which pole-shifting may be interpreted on a root-locus plot to alleviate some of the difficulties associated with the first canonic transformation. It also shows why this technique is referred to as pole-shifting.

The Zero-Shifting Technique

A limit may also be placed on the maximum gain of the nonlinear element so that stable systems whose gain does not exceed this maximum value will not be rejected by the secondary stability criterion. In this case, the y (output) axis of the input-output characteristic plane for the nonlinear element is rotated in the clockwise direction through an angle θ. The geometric relationships are shown in Fig. 5-9. The result of this axis rotation is equivalent to introducing a change of the input variable from x to x', where $x' = x - C_z y$, and C_z is a real positive constant related to the angle θ by

$$\theta = \arctan C_z \tag{5.3-15}$$

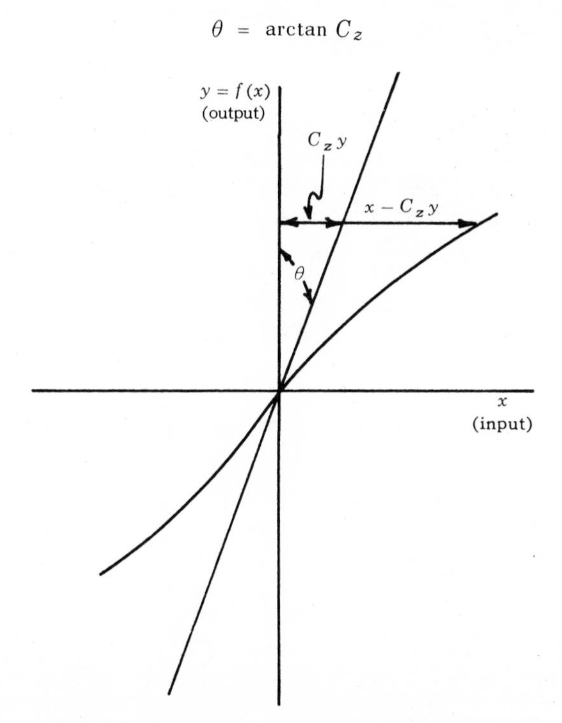

FIG. 5-9. Geometry of the zero-shifting technique.

The maximum value of the angle θ and consequently the maximum value of C_z are determined by the angle through which the vertical axis of the x-y plane (Fig. 5-9) can be rotated without intersecting the curve which represents the nonlinear element. When the variable x' is introduced into the block diagram of Fig. 5-5, then the dynamic elements are represented by $G''(s)$ which is related to $G(s)$ by the expression

$$G''(s) = G(s) + C_z = \frac{N(s) + C_z D(s)}{D(s)} \tag{5.3-16}$$

in which $N(s)$ and $D(s)$ represent the numerator and denominator polynomials of $G(s)$. Equation (5.3-16) shows that $G''(s)$ will have the same number of zeros as poles. Consequently, the system cannot be transformed into the first canonic form after the application of the zero-shifting technique. It is possible, however, to modify the first canonic form to accommodate this situation. Secondary stability criteria are then determined for the modified first canonic form. This is treated in considerable detail in references 8 and 22. Because of its rather limited range of applicability, it is not treated in detail here.

The Second Canonic Transformation

The second canonic form has not received as much attention in the literature as the first. It appears to offer little additional advantage over the first canonic form when the latter is augmented by the pole-shifting and zero-shifting techniques. Nevertheless, it is directly applicable to systems of the form of Fig. 5-5, in which $G(s)$ contains either multiple poles or poles in the right half of the s-plane, or both. For this reason it is given brief treatment here.

For a system that may be represented in the block diagram form of Fig. 5-5, the second canonic form for the system differential equations is

$$\frac{dz_i}{dt} = \omega_i z_i + x \qquad i = 1, 2, \ldots m, \tag{5.3-17}$$

$$\frac{dx}{dt} = \sum_{i=1}^{m} \gamma_i z_i + \delta x - f(x), \tag{5.3-18}$$

in which:

(a) the ω_i are the zeros of $G(s)$

(b) $\delta = \sum_{i=1}^{n} \lambda_i - \sum_{i=1}^{m} \omega_i$ (5.3-19)

(c) the λ_i are the poles of $G(s)$

$$\text{(d)}\quad \gamma_i = \frac{-\prod_{j=1}^{n}(\omega_i - \lambda_j)}{\prod_{\substack{j=1\\ j\neq 1}}^{m}(\omega_i - \omega_j)} \qquad i = 1, 2, \ldots m. \tag{5.3-20}$$

If $G(s)$ satisfies certain conditions, then the system represented by Fig. 5-5 may be represented by the second canonic form. These conditions are stated concisely in the following theorem.

Theorem 5-40: Transformation into the Second Canonic Form. A system of the form shown in Fig. 5-5 can be described by the second canonic form of differential equations if, and only if, the following conditions hold:

a. all the zeros, ω_i, of $G(s)$ are simple;
b. the number of poles n of $G(s)$ is greater by one than the number m of its zeros, i.e., if $n = m + 1$.

Simplfied stability criteria based on particular forms of a Lyapunov function have been obtained for the second canonic form of system differential equations (see refs 16, pp. 192-195, and 8, pp. 86-90). This information is not presented here since it appears that most systems which satisfy the conditions for the second canonic transformation may be treated just as easily by the first cononic transformation augmented with the pole-shifting and zero-shifting techniques [8].

5.3.3 Aizerman's Method

One of the simplest functions $V(x)$ that may serve as a Lyapunov function for a nonlinear autonomous system is the generalized quadratic form. For a specific set of system equations, it is desirable to have a procedure for selecting the constant coefficients such that the quadratic form will satisfy the criteria of one of the stability theorems in Sec. 5-2 and thereby establish the stability or instability of the system in question. One method for doing this is to:

(a) approximate the nonlinear elements by linear elements; $V(x)$

(b) determine the coefficients in the quadratic form so that it will serve as a Lyapunov function for the linearized system;

(c) apply the $V(x)$ determined in (b) to the system with the nonlinearities, and use the restrictions which apply to $\frac{dV}{dt}$ to establish limits about the linear approximation which the nonlinear elements may not exceed.

This technique is referred to as Aizerman's method [8], and is demonstrated in the following example.

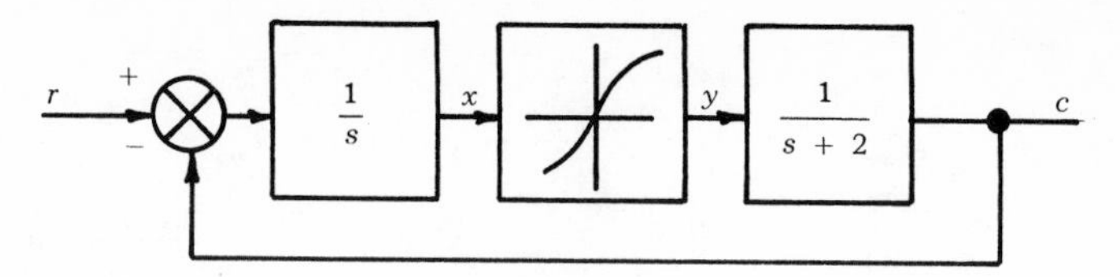

FIG. 5-10. Block diagram of the nonlinear system of example 5-1.

Example 5-1: The stability of the system shown in Fig. 5-10 is to be investigated. If the input is set equal to zero, that is,

$$r(t) = 0 \text{ for } t > 0$$

then the system may be represented by

$$\begin{aligned} \ddot{x} + 2\dot{x} + y &= 0 \\ y &= f(x) \end{aligned} \tag{5.3-21}$$

By introducing a change in variables of the form

$$\begin{aligned} x_1 &= x \\ x_2 &= \dot{x} \end{aligned} \tag{5.3-22}$$

it is possible to represent the system as a set of first-order differential equations in the state variables x_1 and x_2. They are

$$\begin{aligned} \dot{x}_1 &= x_2 \\ \dot{x}_2 &= -2x_2 - f(x_1) \end{aligned} \tag{5.3-23}$$

If the nonlinear element has the input-output characteristic of Fig. 5-11, then it may be approximated by a straight line of the form

$$y = f(x_1) \approx 2x_1 \tag{5.3-24}$$

Then

$$\begin{aligned} \dot{x}_1 &= x_2 \\ \dot{x}_2 &= -2x_2 - 2x_1 \end{aligned} \tag{5.3-25}$$

For two state variables, the generalized quadratic form which is to serve as a Lyapunov function is

$$V(x_1, x_2) = b_{11}x_1^2 + 2b_{12}x_1x_2 + b_{22}x_2^2 \tag{5.3-26}$$

Differentiating Eq. (5.3-26) with respect to time and substituting Eq. (5.3-25) yields

$$\frac{dV}{dt} = (-4b_{12})x_1^2 + (2b_{11} - 4b_{12} - 4b_{22})x_1x_2 + (2b_{12} - 4b_{22})x_2^2 \tag{5.3-27}$$

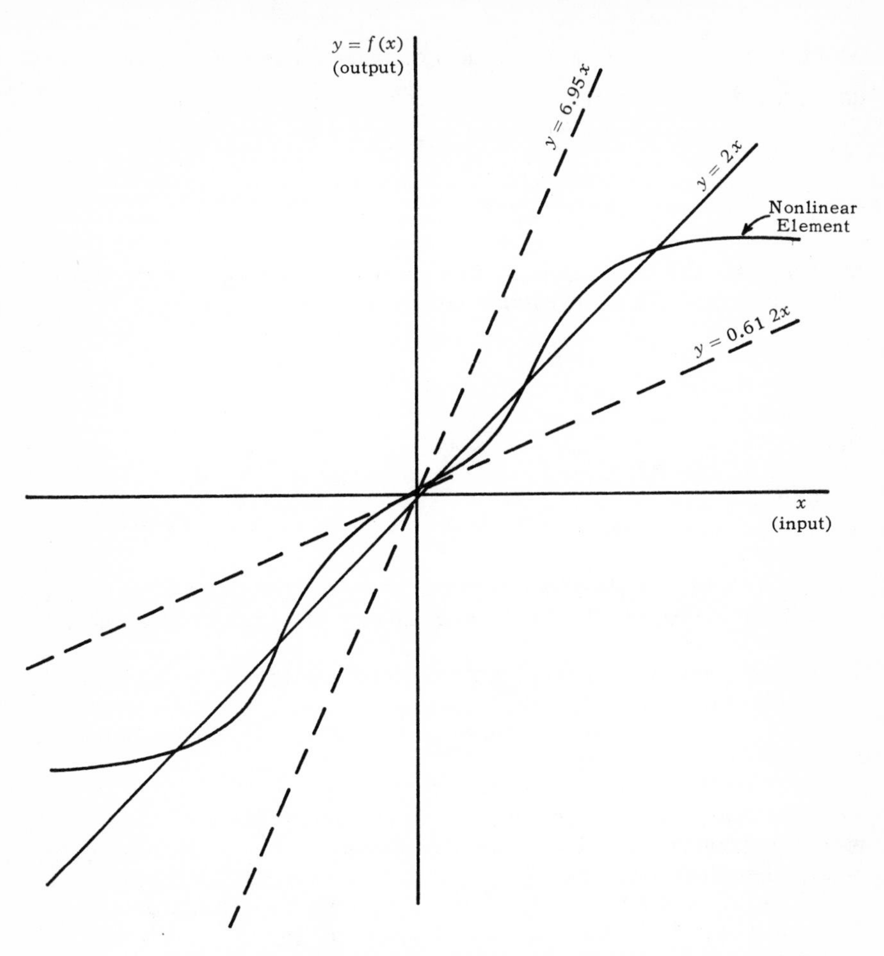

FIG. 5-11. Input-output characteristics of the nonlinear element of example 5-1.

The time derivative $\frac{dV}{dt}$ will satisfy the conditions of Theorem 5-37 if it is constrained to be

$$\frac{dV}{dt} = -x_1^2 - x_2^2 \tag{5.3-28}$$

For this case, b_{11}, b_{12}, and b_{22} may be determined from a set of equations obtained by equating coefficients of corresponding terms in Eqs. (5.3-27) and (5.3-28). The result yields

$$V(x_1, x_2) = \frac{5}{4}x_1^2 + \frac{1}{2}x_1x_2 + \frac{3}{8}x_2^2 \tag{5.3-29}$$

which is positive-definite. Using Eq. (5.3-29) and evaluating $\frac{dV}{dt}$ by using Eq. (5.3-23) yields

$$\frac{dV}{dt} = -\left(\frac{1}{2}\frac{f(x_1)}{x_1}\right)x_1^2 - \left(\frac{3}{4}\frac{f(x_1)}{x_1} - \frac{3}{2}\right)x_1 x_2 - x_2^2 \tag{5.3-30}$$

In this case, the conditions of Theorem 5-37, which assure asymptotic stability in the large, are satisfied if

$$\frac{f(x_1)}{x_1} > 0 \text{ or } x_1 f(x_1) > 0 \tag{5.3-31}$$

and

$$0.612 < \frac{f(x_1)}{x_1} < 6.95 \tag{5.3-31a}$$

These limits of permissible deviation from the straight-line approximation for the nonlinearity are shown in Fig. 5-11 as dashed lines.

The advantages of Aizerman's method are [8]:

(a) Its simplicity.

(b) Its applicability to systems with more than one nonlinear element.

(c) Its utility in placing bounds on the nonlinear element of a slightly nonlinear system in order to justify the use of a linear approximation for stability considerations.

Its disadvantages are [8]:

(a) It is applicable only if the input-output characteristics of the nonlinear elements do not deviate too far from a linear approximation.

(b) If the system contains differentiation (zeros in the transfer function of the linear part of the system), the restrictions on the nonlinear element characteristics in terms of $y, \frac{dy}{dt}$, etc., become rather complicated.

It should be pointed out that a nonlinear system is not necessarily stable in the large even if its linearized model ($y = kx$) is stable for all values of the equivalent linear gain k.

In applying Aizerman's method, the restrictions on the input-output function of the nonlinear elements will usually be least severe if

a. The linear approximation $y = kx$ is selected so that the line represented by this approximation bisects the angle formed by the lines representing the upper and lower bounds of gain for the nonlinear function, $f(x)$.

b. The time derivative of $V(\mathbf{x})$ is constrained to

$$\frac{dV}{dt} = -\sum_{i=1}^{n} x_i^2$$

5.3.4 Application of Krasovskiy's Theorem [12]

It is possible to derive forms of secondary stability criteria other than those associated with the canonic transformations. One such method is known as Krasovskiy's Theorem and is applicable to autonomous systems which may be expressed in the form

$$\dot{\mathbf{x}} = \mathbf{A}\mathbf{x} + \mathbf{f}(\mathbf{x}) = \mathbf{X} = \begin{bmatrix} X_1(x_1, \ldots x_n) \\ X_2(x_1, \ldots x_n) \\ \cdot \\ \cdot \\ \cdot \\ X_n(x_1, \ldots x_n) \end{bmatrix} \tag{5.3-32}$$

in which the X_i are required to be continuous and differentiable functions of the state variables in the entire state space, and the equilibrium point is assumed to be at the origin of the state space.

Theorem 5-41: [12]: *Secondary Stability Criterion II (Krasovskiy's Theorem).* A sufficient condition to establish asymptotic stability in-the-large for a system represented by Eq. (5.3-32) is that there exist a positive-definite symmetric matrix $\mathbf{B}$ such that the symmetric matrix

$$[\mathbf{BJ} + (\mathbf{BJ})'] \tag{5.3-33}$$

has characteristic roots (eigenvalues), $\lambda_i(x_1, \ldots, x_n)$, which satisfy the inequality

$$\lambda_i < -\delta, \quad i = 1, 2, \ldots, n \tag{5.3-34}$$

where δ is a real positive constant. The matrix $\mathbf{J}$ is the Jacobian matrix of the system function $\mathbf{X}$ and is expressed as

$$\mathbf{J} = \begin{bmatrix} \dfrac{\partial X_1}{\partial x_1} & \cdots & \dfrac{\partial X_1}{\partial x_n} \\ \cdots & \cdots & \cdots \\ \cdots & \cdots & \cdots \\ \dfrac{\partial X_n}{\partial x_1} & \cdots & \dfrac{\partial X_n}{\partial x_n} \end{bmatrix} \tag{5.3-35}$$

The practical problem of determining the existence of a matrix B which satisfies the conditions of this theorem is approached by forming a positive definite function of the system functions X_i (note that these are not the state variables). This function may be written as

$$V = \mathbf{X}'\mathbf{B}\mathbf{X} \tag{5.3-36}$$

The time derivative of this function is

$$\frac{dV}{dt} = \mathbf{X}'[\mathbf{BJ} + (\mathbf{BJ})']\mathbf{X} = \mathbf{X}'\mathbf{C}\mathbf{X} \tag{5.3-37}$$

in which C is a symmetric matrix whose elements, c_{ij}, are functions of the state variables x_i. It may be shown that if Eq. (5.3-37) is negative-definite, then the system will be asymptotically stable in-the-large. Therefore the criterion for selecting the elements of B is that they must cause Eq. (5.3-37) to be negative-definite for all real values of the state variables $x_1, \ldots, x_n$.

The procedure described above may be considered an application of the criterion of Lyapunov's direct method to the time derivatives of the state variables rather than applying the criterion directly to the state variables. This theorem is proved in ref. 12. The application of this technique is demonstrated by using the same system as in Example 5-1.

Example 5-2 [8]: After performing the transformation of variables indicated by Eq. (5.3-22), the system is represented by

$$\begin{aligned} \dot{x}_1 &= X_1(x_1,x_2) = x_2 \\ \dot{x}_2 &= X_2(x_1,x_2) = -2x_2 - f(x_1)\,. \end{aligned} \tag{5.3-37}$$

For general coefficients in the matrix B, the matrix C with the Jacobian matrix of Eqs. (5.3-37) substituted will be

$$\mathbf{C} = [\mathbf{BJ} + (\mathbf{BJ})'] = \begin{bmatrix} \left(-2b_{12}\dfrac{df(x_1)}{dx_1}\right) & \left(b_{11} - 2b_{12} - b_{22}\dfrac{df(x_1)}{dx_1}\right) \\ \left(b_{11} - 2b_{12} - b_{22}\dfrac{df(x_1)}{dx_1}\right) & \left(2b_{12} - 4b_{22}\right) \end{bmatrix} \tag{5.3-38}$$

If Eq. (5.3-38) is constrained to be

$$\mathbf{C} = \begin{bmatrix} -\left(\dfrac{1}{2}\dfrac{df(x_1)}{dx_1}\right) & -\left(\dfrac{3}{8}\dfrac{df(x_1)}{dx_1} - \dfrac{3}{4}\right) \\ -\left(\dfrac{3}{8}\dfrac{df(x_1)}{dx_1} - \dfrac{3}{4}\right) & -1 \end{bmatrix} \tag{5.3-39}$$

then

$$B = \begin{bmatrix} \frac{5}{4} & \frac{1}{4} \\ \frac{1}{4} & \frac{3}{8} \end{bmatrix} \tag{5.3-40}$$

This is a positive-definite matrix, which may be verified by application of Theorem 5-37, and therefore the function V as given by Eq. (5.3-36) is positive-definite. The matrix C, and consequently $\frac{dV}{dt}$, are negative-definite for $f(x)$ constrained by

$$0.573 < \frac{df(x)}{dx} < 6.98 \tag{5.3-41}$$

The constraint on $f(x)$ defined by Eq. (5.3-41) provides a sufficient, though not necessary, condition for the stability of the system investigated.

Krosovskiy's Theorem has approximately the same advantages and disadvantages of Aizerman's method. It should be noted, however, that in cases where it is not possible to decide on the stability of a system with one of these methods, the other may still yield results.

5.3.5 *Pliss' Method* [21]

Another method which initially employs a linear approximation to a nonlinear system in order to determine a Lyapunov function from which secondary stability may be derived is refered to as Pliss' method. This method may yield results when applied to systems that can be represented by a set of differential equations of the form

$$\dot{x}_j = \sum_{i=1}^{n} a_{ji}x_i + h_j f(z) \qquad j = 1, 2, \ldots n \tag{5.3-42}$$

where

$$z = \sum_{j=1}^{n} k_j x_j \tag{5.3-43}$$

and in which the a_{ij}, h_j, and k_j are constants. The nonlinear function, $f(z)$, is constrained to satisfy the conditions:

$$\begin{aligned} &\text{(a)} \quad f(0) = 0\,; \\ &\text{(b)} \quad c_1 z^2 < z f(z) < c_2 z^2 \end{aligned} \tag{5.3-44}$$

The linear approximation to Eq. (5.3-42) to be used is

$$\dot{x}_j = \sum_{i=1}^{n} a_{ji}x_i + h_j cz \qquad j = 1, 2, \ldots n \tag{5.3-45}$$

A Lyapunov function is selected to be of the form

$$V = \frac{1}{2} \sum_{i=1}^{n} \sum_{j=1}^{n} b_{ij}x_i x_j + \frac{1}{2} z^m \tag{5.3-46}$$

On the basis of the Lyapunov function, it is possible to derive a set of secondary stability criteria which may be stated in the form of the following theorem.

Theorem 5-42: Secondary Stability Criterion III. A system which may be represented by Eqs. (5.3-42) and (5.3-43) is asymptotically stable in-the-large if:

a. for all $c = c_1 + \epsilon$ and $c = c_2 - \epsilon$, where ϵ is an arbitrary small real positive constant, the linear approximation to the system, Eq. (5.3-45), is asymptotically stable;

b. there exist real numbers, β and $m_{ij} = m_{ji} (i, j = 1, 2, \ldots n)$, producing either a positive-definite or negative-definite quadratic form which may be represented by

$$Q(x_1, \ldots x_n) = \sum_{i=1}^{n} \sum_{j=1}^{n} r_{ij}x_i x_j \tag{5.3-47}$$

where coefficients r_{ij} are calculated from the equation

$$\begin{aligned} r_{ij} = {} & \sum_{k=1}^{n} m_{ik}b_{kj} + \sum_{k=1}^{n} m_{kj}b_{kj} + c\left(a_j \sum_{k=1}^{n} m_{ki}h_k + a_i \sum_{k=1}^{n} m_{kj}h_k\right) \\ & + c\beta\left(a_j \sum_{k=1}^{n} a_k b_{ki} + a_i \sum_{k=1}^{n} a_k b_{kj}\right) + c^2 a_i a_j \sum_{k=1}^{n} a_k h_k ; \\ & \qquad c_1 < c < c_2. \end{aligned} \tag{5.3-48}$$

It should be noted that it is not possible to establish asymptotic stability in-the-large for the nonlinear system simply because the linear approximation to the system is stable for all values of c in the interval $c_1 < c < c_2$. In terms of application to practical problems this technique suffers the disadvantage that it requires rather complicated algebraic maniputations.

5.3.6 The Variable Gradient Method of Generating Lyapunov Functions [23, 24]

The variable gradient method of generating Lyapunov functions appears to be the most flexible of the available techniques for applying the theorems of Lyapunov's direct method. As was pointed out in the Introduction to this section, this method appears to handle all cases to which Ingwerson's method [9, 10, 24] and Szego's method [24, 25, 26] are applicable, as well as other cases. Therefore, the latter two methods are not presented here.

In the development of this technique it is assumed that the physical system under consideration is *autonomous* and is represented by

$$\dot{\mathbf{x}} = \mathbf{A}\mathbf{x} + \mathbf{f}(\mathbf{x}) = \mathbf{X}(\mathbf{x}) \tag{5.3-49}$$

where

$$\mathbf{A}\mathbf{0} + \mathbf{f}(\mathbf{0}) = \mathbf{X}(\mathbf{0}) = \mathbf{0}\,. \tag{5.3-50}$$

This is the same representation as Eq. (5.2-3), p. 16; however, it will be more convenient to combine both the linear and nonlinear terms in one vector $\mathbf{X}(\mathbf{x})$ for the algebraic manipulations to follow.

Referring to Theorem 5-18, the Lipschitz* condition implies continuity of $\mathbf{X}(\mathbf{x})$ in the state space of $\mathbf{x}$. Hence all physical systems that are asymptotically stable in the large, and whose nonlinearities satisfy the Lipschitz condition, will then satisfy the conditions of Theorem 5-18. That theorem could be reworded to say that if a physical system with a continuous nonlinearity whose derivative exists and is bounded everywhere is asymptotically stable in-the-large, then there exists an infinitely differentiable $V(\mathbf{x})$ capable of proving this type of stability by Lyapunov's direct method.

The theorems of Lyapunov's direct method require that $V(\mathbf{x})$ be continuous with continuous first partial derivatives. If the scalar $V(\mathbf{x})$ has first partial derivatives with respect to $\mathbf{x}$, this is equivalent to saying that the gradient of $V(\mathbf{x})$ exists. This ∇V (gradient of V) is a unique n-dimensional vector with n components ∇V_i in the x_i direction. Thus, if a physical system with continuous nonlinearities is asymptotically stable in the large, at least one ∇V exists which can be determined from a $V(\mathbf{x})$ capable of providing such stability.

Instead of assuming a knowledge of V, from which ∇V may be determined, this technique assumes that ∇V is known. It is shown in standard texts on vector calculus (Lass, 15, pp. 297-301) that for a scalar function V to be obtained uniquely from a line integral of a vector function, ∇V, the following $(n-1)n/2$ equations must be satisfied

$$\frac{\partial \nabla V_i}{\partial x_j} = \frac{\partial \nabla V_j}{\partial x_i} \qquad i, j = 1, 2, \ldots n \tag{5.3-51}$$

*$\mathbf{X}(\mathbf{x})$ satisfies the Lipschitz condition in a region R if the following condition is satisfied

$$\| \mathbf{X}(\gamma) - \mathbf{X}(\delta) \| \leq K \| \gamma - \delta \|$$

Equation (5.3-51) is a necessary and sufficient set of conditions that the scalar function V be independent of the path of the line integration. In the three-dimensional case, the above set of equations is identical to those obtained from setting the curl of a vector equal to zero. This form of Stokes theorem is familiar to electrical engineers from field theory. Equation (5.3-51) is thus an n-dimensional representation of Stokes theorem, and will be referred to hereafter as curl equations.

A ∇V determined from a $V(\mathbf{x})$ capable of proving asymptotic stability in the large necessarily meets the conditions of Eq. (5.3-51). This is shown as follows. Theorem 5-18 guarantees that

$$\frac{\partial^2 V(\mathbf{x})}{\partial x_i \, \partial x_j} \quad \text{and} \quad \frac{\partial^2 V(\mathbf{x})}{\partial x_j \, \partial x_i} \tag{5.3-52}$$

exist and are continuous, as V is infinitely differentiable. A theorem from advanced calculus (Taylor, 27, p. 220) states that if the expressions of Eq. (5.3-52) are continuous in the whole region, then in the whole region,

$$\frac{\partial^2 V(\mathbf{x})}{\partial x_i \, \partial x_j} = \frac{\partial^2 V(\mathbf{x})}{\partial x_j \, \partial x_i} \tag{5.3-53}$$

This is simply a restatement of Eq. (5.3-51). Hence a knowledge of either $V(\mathbf{x})$ or ∇V uniquely defines the other. The conclusion from the above is stated as a theorem.

Theorem 5-43: Existence of the Gradient ∇V of a Lyapunov Function. If the system described by Eq. (5.3-49) and Eq. (5.3-50) is Lipschitzian, and if the equilibrium state, $\mathbf{x}_e = \mathbf{0}$, is asymptotically stable in the large, then a ∇V exists, from which $V(\mathbf{x})$ may be obtained by line integration, and the $V(\mathbf{x})$ so obtained is capable of establishing asymptotic stability in the large.

This is the rather powerful existence theorem. If a given autonomous system has nonlinearities that can be represented by continuous functions, and if that system is asymptotically stable in the large, then a gradient capable of establishing this stability exists.

Since the knowledge of either V or ∇V uniquely determines the other, then the Lyapunov theorems may be restated in terms of the gradient function. Theorem 5-16 is the applicable theorem for the autonomous case. It becomes

Theorem 5-44: Asymptotic Stability-In the Large for Autonomous Systems Based on the Gradient of a Lyapunov Function ∇V. If for Eq. (5.3-49), which satisfies Eq. (5.3-86), there exists a real vector function ∇V with elements ∇V_i such that

1. $\dfrac{\partial \nabla V_i}{\partial x_j} = \dfrac{\partial \nabla V_j}{\partial x_i}$

2. $\nabla V' \mathbf{X}(\mathbf{x}) \leq 0$, but not identically zero on a solution of Eq. (5.3-49), other than the origin, and such that the scalar function $V(\mathbf{x})$ formed by a line integration of ∇V is continuous with continuous first partials, and
3. $V(\mathbf{x}) > 0$ for $\mathbf{x} \neq 0$
4. $V(\mathbf{x}) \to \infty$ as $\|\mathbf{x}\| \to \infty$

then Eq. (5.3-49) is asymptotically stable in the large.

This theorem is new only in the sense that it is an extension or a generalization of an existing theorem. However, in this restatement of Theorem 5-16, the role of the gradient function is emphasized.

If condition 4 above is not satisfied, or if condition 2 is not satisfied in the whole space, it is impossible to conclude asymptotic stability in the large, and Theorem 5-22 or 5-23 may be used to prove stability in a smaller region. Theorem 5-23 will be of greater practical utility.

A comparison of Theorems 5-16 and 5-44 clearly indicates a shift in emphasis. The problem of determining a V function which satisfies Lyapunov's theorem is transformed into the problem of finding a ∇V such that the n-dimensional curl of this gradient is equal to zero, or, in other words, Eq. (5.3-51) is satisfied. Further, the V and dV/dt determined from ∇V must be sufficient to prove stability according to either theorem, since the theorems are equivalent. On the surface, it may appear that the problem is actually being made more difficult; but the reverse is true. The existence of the auxiliary curl equations permits a solution of the stability problem, starting with ∇V.

As the term "variable gradient" implies, the task of implementing Theorem 5-44 is accomplished by the assumption of a vector, ∇V, with n undetermined components. To make this vector general enough to embrace all possible solutions, each of the n undetermined components of the gradient is further assumed to be made up of n elements of the form $a_{ij}x_i$. The a are assumed to be general functions of $\mathbf{x}$, or polynomials with an unspecified number of terms such that

$$\nabla V = \begin{Bmatrix} a_{11}x_1 + a_{12}x_2 + \cdots a_{1n}x_n \\ a_{21}x_1 + a_{22}x_2 + \cdots \\ \vdots \\ \cdot \\ a_{n1}x_1 + \qquad \cdots a_{nn}x_n \end{Bmatrix} = \begin{Bmatrix} \nabla V_1 \\ \nabla V_2 \\ \vdots \\ \cdot \\ \nabla V_n \end{Bmatrix} \tag{5.3-54}$$

The x's are assumed to be made up of a constant portion and a portion which is a function of the state variables, and is represented by

$$a_{ij} = a_{ijk} + a_{ijv}(x) \tag{5.3-54}$$

Hence

$$\nabla V = \begin{Bmatrix} [a_{11k} + a_{11v}(x)]x_1 + [a_{12k} + a_{12v}(x)]x_2 + \dots [a_{1nk} + a_{1nv}(x)]x_n \\ [a_{21k} + a_{21v}(x)]x_1 + \dots \\ \cdot \\ \cdot \\ \cdot \\ [a_{n1k} + a_{n1v}(x)]x_1 + \dots \qquad [a_{nnk} + a_{nnv}(x)]x_n \end{Bmatrix} \tag{5.3-55}$$

Several interesting facts are apparent from an examination of the ith element of the gradient

$$\nabla V_i = [a_{i1k} + a_{i1v}(x)]x_1 + \dots [a_{iik} + a_{iiv}(x)]x_i + \dots [a_{ink} + a_{inv}(x)]x_n$$

The solution of a given problem may require that ∇V_i contain terms with more than one state variable as factors. It is evident that such terms may be found from terms such as $a_{ij}(x)x_i$. Therefore, the $a_{iiv}(x)$ may be restricted to be $a_{iiv}(x_i)$.

V is to be determined as a line integral of ∇V, as shown by

$$\begin{aligned} V = \int_{\text{line}} \nabla V' dx &= \int_0^{x_1} \nabla V_1(y_1, 0 \dots 0)\, dy_1 \\ &+ \int_0^{x_2} \nabla V_2(x_1, y_2, 0 \dots 0)\, dy_2 + \dots \\ &+ \int_0^{x_n} \nabla V_n(x_1, x_2, \dots x_{n-1}, y_n)\, dy_n \end{aligned} \tag{5.3-56}$$

Note that the a_{ii} coefficients give rise to terms such as

$$\frac{a_{iik} x_i^2}{2} \quad \text{and} \quad \int_0^{x_i} a_{iiv}(y_i)\, y_i\, dy_i$$

Here it has been assumed that $a_{iiv}(x)$ has been set equal to $a_{iiv}(x_i)$, as mentioned above. For V to be positive-definite in the neighborhood of the origin, a_{iik} must be always positive. For V to represent a closed surface in the whole space, or for V to be always positive, $a_{iiv}(x_i)$ must be an even function of x_i and > 0 for large x_i. Also, if $a_{iik} = 0$, $a_{iiv}(x_i)$ must be even and greater than zero for all x_i.

The assertions made above in regard to the a_{ii} relate to the requirements that must be met by the resulting V function if

Theorem 5-44 is to apply. This line of thinking is pursued further in the following paragraphs.

Since the a_{ijv} are allowed to be functions of the state variables, it is expected that V may well contain higher order terms in the state variables. Since this is the case, the question of the positive-definiteness of the resulting V becomes important.

The term positive-definiteness is usually used in reference to quadratic forms, although the concept does have meaning for a form of arbitrary order. Geometric means of insuring that a scalar function equated, like $V(\mathbf{x})$, to a constant represents a closed surface, are discussed in the appendix of ref. 24. The geometric method used requires that one of the state variables in V be raised to the second order, and no higher. This is accomplished by forcing *one* a_{ii} to be equal to a constant, and by forcing the remaining a_{ijv} not to be functions of x_i. Although this choice is actually unnecessarily restrictive, it does satisfy the geometric considerations which insure that $V(\mathbf{x}) = \text{constant}$ are closed surfaces.

In problems involving automatic control systems, the x_n term frequently appears linearly in the n first-order equations that describe the motion of the system. For this reason, the assumptions of the previous paragraph are applied to the x_n variable. Specifically, a_{nn} is set equal to 2. This seemingly arbitrary choice of a_{nn} in the gradient is equivalent to the assumption of an arbitrary constant, or scale factor, in V. The choice of $a_{nn} = 2$ insures that V will contain a term in x_n^2.

In view of the above discussion, ∇V is now

$$\nabla V = \begin{pmatrix} [a_{11k} + a_{11v}(x_1)]x_1 + [a_{12k} + a_{12v}(x_1, x_2, \ldots x_{n-1})]x_2 \\ \quad + \cdots [a_{1nk} + a_{1nv}(x_1, x_2, \ldots x_{n-1})]x_n \\ [a_{21k} + a_{21v}(x_1, x_2, \ldots x_{n-1})]x_1 + [a_{22k} + a_{22v}(x_2)]x_2 + \cdots \\ \vdots \\ [a_{n1k} + a_{n1v}(x_1, x_2, \ldots x_{n-1})]x_1 + \cdots 2x_n \end{pmatrix} \tag{5.3-57}$$

Through an examination of the requirements on V, the most general gradient of Eq. (5.3-55) has been somewhat simplified in form to that of Eq. (5.3-57). Without loss in generality, the a_{ii} have been constrained to be functions of x_i alone. With slight loss of generality, one of the a_{ii}, here a_{nn}, has been set equal to an arbitrary constant, and the a_{ijv} have been constrained to be $a_{ijv}(x_1, x_2, \ldots x_{n-1})$. This has been accomplished in view of the future requirements of V. Further knowledge of the unknown coefficients in ∇V is obtainable from an examination of the generalized curl equations, Eq. (5.3-51).

Consider the expanded form of Eq. (5.3-51)

$$\frac{\partial \nabla V_i}{\partial x_j} = \frac{\partial a_{i1v}(x_1, x_2, \ldots x_{n-1})x_1}{\partial x_j} + \ldots$$

$$\frac{\partial a_{ijk} x_j}{\partial x_j} + \frac{\partial a_{ijv}(x_1, x_2, \ldots x_{n-1})x_j}{\partial x_j} + \ldots \quad (5.3\text{-}58)$$

$$\frac{\partial a_{inv}(x_1, x_2, \ldots x_{n-1})x_n}{\partial x_j}$$

and

$$\frac{\partial \nabla V_j}{\partial x_i} = \frac{\partial a_{j1v}(x_1, x_2, \ldots x_{n-1})x_1}{\partial x_i} + \ldots$$

$$\frac{\partial a_{jik} x_i}{\partial x_i} + \frac{\partial a_{jiv}(x_1, x_2, \ldots x_{n-1})x_i}{\partial x_i} + \ldots$$

$$\frac{\partial a_{jnv}(x_1, x_2, \ldots x_{n-1})x_n}{\partial x_i}$$

Here $\frac{\partial a_{jik} x_i}{\partial x_i}$ and $\frac{\partial a_{ijk} x_j}{\partial x_j}$ result in constant terms. If the constant terms on either side of the equal sign are equated, it is seen that

$$a_{ijk} = a_{jik}$$

Thus, further knowledge of the variable gradient is provided, this time from the curl equations. A knowledge of the necessary values of the remaining unknowns in ∇V can be acquired from joint consideration of the generalized curl equations and dV/dt.

We determine dV/dt from the variable gradient by means of the relation

$$\frac{dV}{dt} = \nabla V' \dot{\mathbf{x}} = \nabla V' \mathbf{X} \quad (5.3\text{-}59)$$

To satisfy either Theorem 5-16 or 5-44, dV/dt must necessarily be constrained to be at least negative-semidefinite. In general, an attempt is made to make dV/dt negative-semidefinite in as simple a way as possible. This is accomplished if

$$\frac{dV}{dt} = -Kx_i^2 \quad (K > 0) \quad (5.3\text{-}60)$$

where K is initially assumed to be a constant. If dV/dt is constrained as in Eq. (5.3-60), the remaining terms in dV/dt must be forced to cancel. This is accomplished by grouping terms of similar state variables and choosing the a_{ij}'s to force cancellation. The a_{ij}'s are assumed to be constants, unless cancellation or the generalized curl equations require a more complicated form.

Grouping of terms is guided by the restrictions on the a_{ij}'s stated above. For example, if in a third-order system dV/dt contains the terms $a_{11}x_1x_2$, $a_{12}x_2^2$ and $-x_1x_2^3$, the indefinite term $-x_1x_2^3$ could not be grouped with $a_{11}x_1x_2$, since a_{11} can only be a function of x_1. However, if $-x_1x_2^3$ were grouped with $a_{12}x_2^2$, it could be eliminated by letting $a_{12} = x_1x_2$.

The choice of the a_{ij}'s to force cancellation is not arbitrary, since the generalized curl equations must be satisfied. In fact, if one coefficient is chosen through the necessity to eliminate undesirable terms in dV/dt, information concerning the required value of one or more of the unknown coefficients is often supplied directly from the generalized curl equations. Thus dV/dt is constrained to be at least negative-semidefinite in conjunction with and subject to the requirements of the generalized curl equations, Eq. (5.3-51).

If it proves to be impossible to constrain dV/dt as in Eq. (5.3-60), it is necessary to attempt to constrain dV/dt to be negative-semidefinite in terms of two state variables, then three, etc., until the final attempt is made to force dV/dt to be negative-definite. If still no solution is available, it may be necessary to revert to the more general gradient function of Eq. (5.3-55), or an attempt at a proof of instability may be in order.

In summary, we offer the following outline for the form application of the variable gradient method:

1. Assume a gradient of the form of Eq. (5.3-57).
2. From the variable gradient, form dV/dt, as $\dfrac{dV}{dt} = \nabla V' \dot{\mathbf{x}}$ [Eq. (5.3-59)].
3. In conjunction with and subject to the requirements of the generalized curl equations, Eq. (5.3-51), constrain dV/dt to be at least negative-semidefinite.
4. From the known gradient, determine V and the region of closedness of V.
5. Invoke the necessary theorem to establish stability.

This procedure is illustrated by means of an example.

Example 5-3: Assume the system is given by the block diagram of Fig. 5-12 such that the equations of motion written in state variable form become, with $x_1 = x$

$$\dot{x}_1 = x_2$$

$$\dot{x}_2 = -x_2 - x_1^3$$

Step 1

$$\nabla V = \begin{pmatrix} a_{11}x_1 + a_{12}x_2 \\ \\ a_{21}x_1 + 2x_2 \end{pmatrix} = \begin{pmatrix} \nabla V_1 \\ \\ \nabla V_2 \end{pmatrix}$$

Step 2

$$\frac{dV}{dt} = \nabla V' \dot{\mathbf{x}}$$

$$= x_1 x_2 \left(a_{11} - a_{21} - 2x_1^2\right) + x_2^2(a_{12} - 2) - a_{21}x_1^4$$

Step 3

If the system is stable, there is a large or even infinite, number of V functions, with a corresponding number of dV/dt. In fact, it is the existence of this large number of suitable Lyapunov functions as opposed to the one unique solution of the initial nonlinear differential equation that gives the Lyapunov method its advantage over classical methods in the determination of stability.

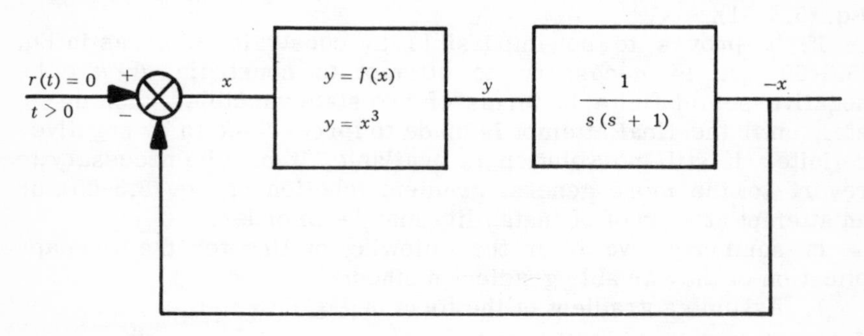

FIG. 5-12. Block diagram of the control system of example 5-3.

Here there are a large number of ways in which dV/dt might be constrained in order to prove stability. However, in order to be able to conclude anything about stability, dV/dt must be at least negative-semidefinite. For the system in question, this can be accomplished by setting the coefficient of x_1x_2 equal zero, and by assuring that x_2^2 and x_1^4 have zero or negative coefficients. The latter can be accomplished if a_{12} is any positive number from 0 to 2, and if a_{21} is any positive number whatever. Hence, a_{12} is assumed to be a constant between 0 and 2; and since it is constant, $a_{21} = a_{12}$. With the coefficient of x_1x_2 set equal to zero, dV/dt becomes

$$\frac{dV}{dt} = -x_2^2(2 - a_{12}) - a_{12}x_1^4$$

The requirement that the coefficient of $x_1 x_2$ be zero is satisfied if

$$a_{11} = a_{12} + 2x_1^2$$

Therefore, with these substitutions, ∇V becomes

$$\nabla V = \begin{Bmatrix} a_{12}x_1 + 2x_1^3 + a_{12}x_2 \\ a_{12}x_1 + 2x_2 \end{Bmatrix}, \quad 0 \le a_{12} \le 2$$

Step 4

V is determined from Eq. (5.3-56) to be the line integral

$$V = \int_0^x \nabla V' dx = \int_0^{x_1} a_{12}y_1 + 2y_1^3 \, dy_1 + \int_0^{x_2} a_{12}x_1 + 2y_2 \, dy_2$$

$$V = \frac{x_1^4}{2} + \frac{a_{12}x_1^2}{2} + a_{12}x_1x_2 + x_2^2, \qquad 0 \le a_{12} \le 2$$

Step 5

Here V is positive-definite and $\lim V \to \infty$ as the $\|x\| \to \infty$, so that V represents a closed surface in the whole space. Since dV/dt is also at least negative-semidefinite in the whole space, by either Theorem 5-16 or 5-44, the system of Fig. 5-12 is asymptotically stable in the large.

Although this illustrative example is quite simple, the technique has relatively wide applicability. Reference 24 contains a number of more complicated examples, including the analysis of a system which has a limit cycle. The technique appears to have its greatest applicability to autonomous systems, and the presentation here is restricted to such systems. A discussion of extending the variable gradient method to nonautonomous systems is given in ref. 24, however, the results are not as clear-cut or as general as for the autonomous case. The applicability to autonomous systems may be summarized as follows [24]:

1. As concerns nonlinearities, the method is applicable to single-valued, continuous nonlinearities where the nonlinearity is known as a polynomial, as a specific function of x, or as a curve determined from experimental results.

2. As concerns coordinate systems, the method is applicable independent of the particular state variable formulation used.

3. As concerns V functions, the method generates V functions to suit the problem at hand. V functions with higher order terms, integrals, and terms involving three state variables as the factors required for the particular situation are generated.

The question may be asked why this method of assuming a general gradient is better than a method assuming a general V. The answer is that if a V, sufficiently general, is selected as a starting point in the solution of a problem, then the number of terms resulting in dV/dt becomes prohibitively difficult to manipulate.

5.3.7 An Application of the Theorems Regarding Eventual Stability

As an example of the manner in which Theorem 5-26 may be applied, the system shown in Fig. 5-13 is considered. This block diagram represents a method of implementing an adaptive control system [7]. The input will be assumed to be a step function of magnitude R_0 applied at a time $t = 0$. The system is assumed to be at rest for $t < 0$. The set of differential equations which describes the operation of the system for $t > 0$ is

$$\dot{y} + a_0 y = R_0 \tag{5.3-61}$$

$$\dot{c} + a_0 c = R_0 \qquad (a_0 = g_0 + h_0) \tag{5.3-62}$$

$$\dot{u}_0 + a_0 u_0 = -y \tag{5.3-63}$$

and

$$\dot{a}_0 = -\ q_0 \epsilon + q_1 \dot{\epsilon} \quad q_0 u_0 + q_1 \dot{u}_0 \tag{5.3-64}$$

Equation (5.3-61) is a constant-coefficient differential equation with a Laplace-transformable input. Therefore, an explicit expression may be obtained for y. Once the explicit solution for y is available, it may be used in Eq. (5.3-63) to obtain an explicit solution for u_0. Assuming the initial conditions to be zero, the explicit solutions for y and u_0 and their derivatives with respect to time will be

$$y = \frac{R_0}{a_0}\left(1 - e^{-a_0 t}\right) \tag{5.3-65}$$

$$\dot{y} = R_0 e^{-a_0 t} \tag{5.3-66}$$

$$u_0 = -\frac{R_0}{a_0^{\,2}}\left[1 - (1 + a_0 t)e^{-a_0 t}\right] \tag{5.3-67}$$

and

$$\dot{u}_0 = -R_0 t e^{-a_0 t} \tag{5.3-68}$$

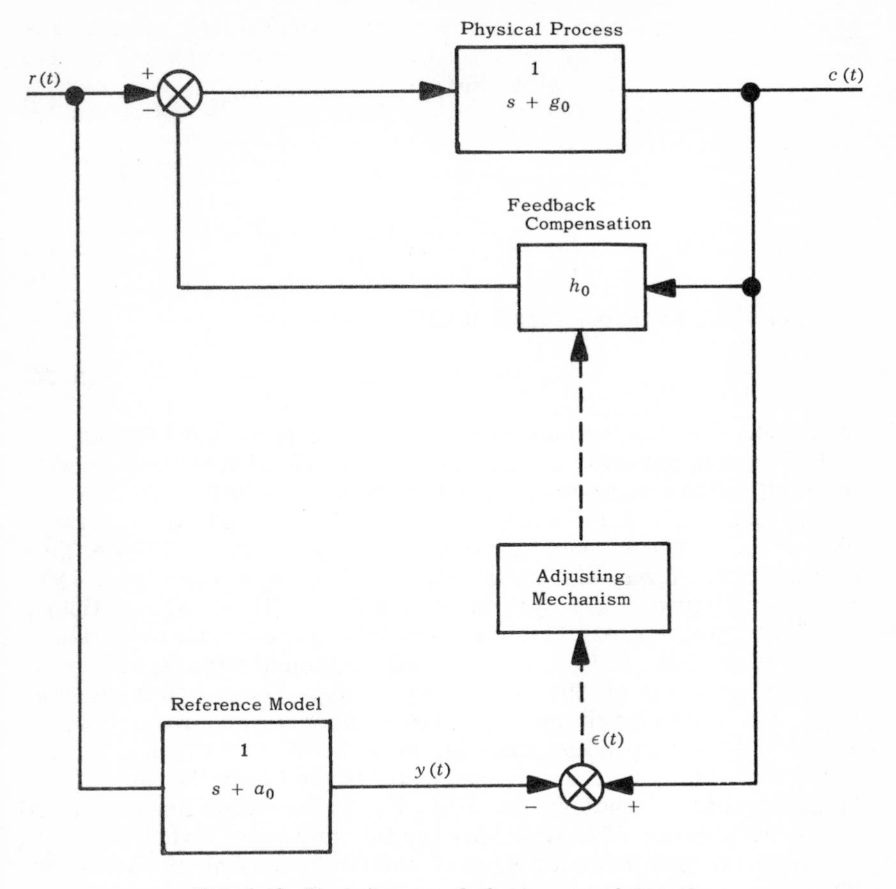

FIG. 5-13. Block diagram of adaptive control system.

To cast Eq. (5.3-62) and (5.3-64) in a form convenient for the application of Theorem 5-26, the following transformation of variables is introduced:

$$c = x_1 + y \tag{5.3-69}$$

and

$$a_0 = x_2 + a_0 \tag{5.3-70}$$

After substitution of Eqs. (5.3-69) and (5.3-70) in Eq. (5.3-62), and subsequent introduction of Eqs. (5.3-65) through (5.3-68), we have

$$\dot{x}_1 = -a_0 x_1 - \frac{R_0}{a_0}\left(1 - e^{-a_0 t}\right)x_2 - x_1 x_2 \tag{5.3-71}$$

and

$$\dot{x}_2 = M(q_0 - q_1 a_0)x_1 - \frac{Mq_1 R_0}{a_0}\left(1 - e^{-a_0 t}\right)x_2 - Mq_1 x_1 x_2 \qquad (5.3\text{-}72)$$

where

$$M = \frac{R_0}{a_0{}^2}\left\{q_0 - \left[q_0 + (q_0 - q_1 a_0)a_0 t\right]\right\}e^{-a_0 t} \qquad (5.3\text{-}73)$$

The potential Lyapunov function will be written as

$$V(\mathbf{x}, t) = \mathbf{x}'\mathbf{B}\mathbf{x} \qquad (5.3\text{-}74)$$

in which $\mathbf{x}$ is the column vector of the state variables x_1 and x_2, and B is a symmetric 2×2 matrix. The selection of the elements in $\mathbf{B}$ will be the main concern of the analysis to follow. In general, if the quadratic form given by Eq. (5.3-74) is positive-definite, and if its derivative with respect to time is negative-definite, then the quadratic form will satisfy the required conditions for a Lyapunov function and the system in question will be asymptotically stable in the large. This is certainly a desirable objective to strive for, since it is a very strong statement with regard to the stability of a set of differential equations. There are many systems for which such an objective cannot be achieved. Looking ahead, the system being analyzed is in the latter category. It may easily be shown that the system will be stable in the small for a large number of choices for q_1 and q_0 in the error functions, and for a wide range of magnitudes for the step input. What interests us, then is, how large a region of stability exists about the origin. Theorem 5-26 is useful in determining the size of the region of stability. The reason for stating that the system of equations cannot be shown to be stable in the large by the direct method of Lyapunov will become evident later in the analysis and will be discussed at that time.

Two criteria will be used to guide the selection of the elements in the Lyapunov matrix $\mathbf{B}$. The first criterion to be satisfied will be that $\mathbf{B}$ is to be positive-definite. The second is that, within the latitude permitted by the first criterion, the region about the origin in which the function $-W(\mathbf{x})$ = limit as $t \to \infty$ of $\dot{V}(\mathbf{x}, t)$ is negative, is to be as large as possible. When Theorem 5-26 is used, it is also desirable that the ellipse described by $U(\mathbf{x})$ = limit at $t \to \infty$ of $V(\mathbf{x}, t) = L$ (a constant) should not have high eccentricity. The elements of $\mathbf{B}$ will now be selected.

A sufficient set of conditions which the elements of $\mathbf{B}$ must satisfy for $\mathbf{B}$ to be positive-definite are

$$b_{11} > 0 \qquad (5.3\text{-}75)$$

and

$$b_{11}b_{22} - b_{12}^2 > 0 \tag{5.3-76}$$

for all t equal to or greater than T_0. Notice that this implies that the elements may be selected as functions of time if there is any advantage to be gained by such a selection. The derivative of Eq. (5.3-74) with respect to time may be written

$$\dot{V}(x) = 2x'B\dot{x} + x'\dot{B}x \tag{5.3-77}$$

If the right-hand side of Eqs. (5.3-70) and (5.3-72) are substituted in Eq. (5.3-77) and the indicated operations carried out, it is possible to write Eq. (5.3-77) as

$$\dot{V}(x) = c_{11}x_1^2 + 2c_{12}x_1x_2 + c_{22}x_2^2 + c_{112}x_1^2x_2 - c_{122}x_1x_2^2 \tag{5.3-78}$$

in which

$$c_{11} = -2b_{11}a_0 + 2b_{12}M(q_0 - q_1a_0) + \dot{b}_{11} \tag{5.3-79}$$

$$c_{12} = -b_{11}\frac{R_0}{a_0}\left(1 - e^{-a_0t}\right) - b_{12}a_0 - b_{12}\frac{Mq_1R_0}{a_0}\left(1 - e^{-a_0t}\right) + b_{22}M(q_0 - q_1a_0) + \dot{b}_{12} \tag{5.3-80}$$

$$c_{22} = -2b_{12}\frac{R_0}{a_0}\left(1 - e^{-a_0t}\right) - 2b_{22}\frac{Mq_1R_0}{a_0}\left(1 - e^{-a_0t}\right) + \dot{b}_{22} \tag{5.3-81}$$

$$c_{112} = -2(b_{11} + b_{12}Mq_1) \tag{5.3-82}$$

and

$$c_{122} = -2(b_{12} + b_{22}Mq_1) \tag{5.3-83}$$

Now, if

$$b_{12} = \frac{-b_{11}}{Mq_1} \tag{5.3-84}$$

then the coefficient of $x_1^2x_2$, c_{112}, in Eq. (5.3-78) vanishes. Next, in order to satisfy condition (5.3-76), b_{22} will be selected as

$$b_{22} = \frac{b_{11} + \gamma_1}{M^2q_1^2} \tag{5.3-85}$$

in which γ_1 is restricted to values greater than zero for all values of time. The coefficients represented by Eqs. (5.3-79) through (5.3-81) and (5.3-83) then become

$$c_{11} = -2\frac{q_0}{q_1} b_{11} + \dot{b}_{11} \quad (5.3\text{-}86)$$

$$c_{12} = -\frac{b_{11}a_0}{Mq_1} \frac{(b_{11} + \gamma_1)}{Mq_1^2}(q_0 - q_1 a_0) + \dot{b}_{12} \quad (5.3\text{-}87)$$

$$c_{22} = -2\frac{\gamma_1 R_0}{Mq_1 a_0}\left(1 - e^{-a_0 t}\right) + \dot{b}_{22} \quad (5.3\text{-}88)$$

and

$$c_{122} = -2\frac{\gamma_1}{Mq_1} \quad (5.3\text{-}89)$$

Next, γ_1 and b_{11} will be chosen respectively as

$$\gamma_1 = \gamma b_{11}^{*} Mq_1 \quad (5.3\text{-}90)$$

and

$$b_{11} = b_{11}^{*} Mq_1 \quad (5.3\text{-}91)$$

in which both γ and b_{11}^* are positive constants. Then, if q_1/q_0 is restricted to a range such that $M \geq 0$ for all $t \geq 0$, condition (5.3-72) is still satisfied. The elements of the matrix B and their derivatives may now be written as

$$b_{11} = b_{11}^{*}\frac{R_0 q_1}{a_0^2}\left\{q_0 - [q_0 + (q_0 - q_1 a_0)a_0 t]e^{-a_0 t}\right\} \quad (5.3\text{-}92)$$

$$\dot{b}_{11} = b_{11}^{*} R_0 q_1 [q_1 + (q_0 - q_1 a_0)t]e^{-a_0 t} \quad (5.3\text{-}93)$$

$$b_{12} = -b_{11}^{*} \quad (5.3\text{-}94)$$

$$\dot{b}_{12} = 0 \quad (5.3\text{-}95)$$

$$b_{22} = b_{11}^{*}\frac{(1 + \gamma)a_0^2}{R_0 q_1}\left\{q_0 - [q_0 + (q_0 - q_1 a_0)a_0 t]e^{-a_0 t}\right\}^{-1} \quad (5.3\text{-}96)$$

and

$$\dot{b}_{22} = b_{11}^{*}\frac{(1 + \gamma)a_0^4}{R_0 q_1}\left\{q_0 - [q_0 + (q_0 - q_1 a_0)a_0 t]e^{-a_0 t}\right\}^{-2} \times$$
$$\times [q_1 + (q_0 - q_1 a_0)t]e^{-a_0 t} \quad (5.3\text{-}97)$$

The coefficients in Eq. (5.3-78) are then

$$c_{11} = -\frac{b_{11}^* R_0}{a_0^2}\left\{2q_0^2 - \left[2q_0^2 + q_1^2 a_0^2 + a_0(2q_0 + q_1 a_0) \times (q_0 - q_1 a_0)t\right]e^{-a_0 t}\right\} \quad (5.3\text{-}98)$$

$$c_{12} = b_{11}^*\left[a_0 - \frac{(1+\gamma)}{q_1}(q_0 - q_1 a_0)\right] \quad (5.3\text{-}99)$$

$$c_{22} = b_{11}^*\left[2\gamma\frac{R_0}{a_0}\left(1 - e^{-a_0 t}\right)\right] - \frac{1 + \gamma a_0^4}{R_0 q_1}\left\{q_0 - [q_0 + (q_0 - q_1 a_0)a_0 t]e^{-a_0 t}\right\}^{-2} \times [q_1 + (q_0 - q_1 a_0)t]e^{-a_0 t} \quad (5.3\text{-}100)$$

and

$$c_{122} = -2b_{11}^*\gamma \quad (5.3\text{-}101)$$

The choice of b_{11}^* will not affect the analysis since it is a common factor throughout. Hence it will be set equal to unity.

At this point, it is possible to explain why $\dot{V}(\mathbf{x}, t)$ cannot be made negative-definite. First, observing the first three terms in Eq. (5.3-78), it is clear that they represent a quadratic form. The fourth term is zero by virtue of the manner in which the elements of the matrix $\mathbf{B}$ have been chosen. The last term is third-order in the state variables. If the quadratic form represented by the first three terms is negative-definite, then $\dot{V}(\mathbf{x}, t)$ will be negative in some region about the origin. This is true because, for small values of the state variables, the quadratic terms will dominate the third-order terms. Now, in view of Eq. (5.3-101), and recalling that γ has been defined as a positive constant, it is clear that the coefficient of the fifth term in Eq. (5.3-78) cannot be made to vanish. At some distance from the origin of the state space, the third-order term will dominate the quadratic terms. Furthermore, the third-order term can be caused to be either positive or negative by the choice of the direction of x_i in the state space. Therefore, there exists a region in the state space at some distance from the origin in which $\dot{V}(\mathbf{x}, t)$ is positive. On the basis of these considerations, the remainder of this analysis will be pointed toward the application of Theorem 5-26.

Rather than attempt to continue the analysis on the basis of general symbols for R_0, q_1, q_0, and a_0, a particular set of values for these system parameters will be selected, and our interest will be turned toward the selection of γ. The following set of numerical values will be used.

$$R_0 = 10 \tag{5.3-102}$$

$$a_0 = 0.1 \tag{5.3-103}$$

$$q_1 = 1 \tag{5.3-104}$$

and

$$q_0 = 1 \tag{5.3-105}$$

This choice is not arbitrary; however, the reasons for it will not be given here [7]. The coefficients of Eq. (5.3-74) then become

$$c_{11} = -10\left\{200 - [201 + 18.9t]e^{-0.1t}\right\} \tag{5.3-106}$$

$$c_{12} = (0.8 + 0.9\gamma) \tag{5.3-107}$$

$$c_{22} = -\Big[2\gamma + 10^2\left(1 - e^{-0.1t}\right) - (1 + \gamma) \times 10^{-5} \left\{1 - [1 + 0.09t]\, e^{-0.1t}\right\}^{-2} [1 + 0.9t]\, e^{-0.1t}\Big] \tag{5.3-108}$$

and

$$c_{122} = -2\gamma \tag{5.3-109}$$

The numerical choice for γ will be based on the condition of the system when $t \to \infty$. In this case, the coefficients in Eq. (5.3-78) become

$$c_{11} = -2 \times 10^3 \tag{5.3-110}$$

$$c_{12} = 0.8 + 0.9\gamma \tag{5.3-111}$$

$$c_{22} = -2\gamma \times 10^2 \tag{5.3-112}$$

and

$$c_{122} = -2\gamma \tag{5.3-113}$$

If γ is chosen as

$$\gamma = 10^5 \tag{5.3-114}$$

then, for t equal to infinity, the elements in the matrix B will be

$$b_{11} = 10^3 \tag{5.3-115}$$

$$b_{12} = -1 \tag{5.3-116}$$

and

$$b_{22} = 10^2 \tag{5.3-117}$$

If these coefficients are used in the right-hand side of Eq. (5.3-74), they will satisfy condition (a) of Theorem 5-25, and the resulting $U(\mathbf{x})$ will satisfy condition (c) of that theorem for any Ω. Moreover, in this case, the ellipse described by $U(\mathbf{x})$ = constant will not have high eccentricity.

Substitution of Eq. (5.3-114) in Eqs. (5.3-110) through (5.3-150), and these, in turn, substituted in Eq. (5.3-78) yield

$$\lim_{t \to \infty} \dot{V}(\mathbf{x}, t) = -W(\mathbf{x}) = -\Big[2 \times 10^3 x_1^2 - 1.8 \times 10^5 x_1 x_2 + 2 \times 10^7 x_2^2 + 2 \times 10^5 x_1 x_2^2\Big] \tag{5.3-118}$$

Here, $W(\mathbf{x})$ is the function defined in conditions (b) and (c) of Theorem 5-25, and will satisfy these conditions if the region Ω can be determined. The quadratic portion of Eq. (5.3-78) is negative-definite. Hence, for some region about the origin, $-W(\mathbf{x})$ will be negative. The locus of points which separates the state plane into the regions in which $-W(\mathbf{x})$ is positive and those in which it is negative may be obtained by setting $-W(\mathbf{x})$ equal to zero and plotting the locus which satisfies this equation. The numerical work is simplified if a transformation of variables to polar coordinates is introduced. This transformation is

$$x_1 = r \cos\theta \tag{5.3-119}$$

and

$$x_2 = r \sin\theta \tag{5.3-120}$$

Upon substitution of Eqs. (5.3-115) and (5.3-120) in Eq. (5.3-118) $-W(\mathbf{x})$ becomes $-W(r, \theta)$ and may be written

$$-W(r, \theta) = -2 \times 10^3 r^2 [\cos^2\theta - 0.9 \times 10^2 \sin\theta \cos\theta + 10^4 \sin^2\theta + 10^2 r \sin^2\theta \cos\theta] \tag{5.3-121}$$

Equating this to zero and solving for r in terms of θ yields

$$r = -\frac{\cos^2\theta - 0.9 \times 10^2 \sin\theta\cos\theta + 10^4 \sin^2\theta}{10^2 \sin^2\theta\cos\theta} \tag{5.3-122}$$

Negative values of r are meaningless and are to be taken as infinity. Therefore, the locus does not exist in the first and fourth quadrants of the state space. The locus which exists for the second and third quadrants is plotted in Fig. 5-14. The regions in which $-W(x)$ is positive and negative are labeled.

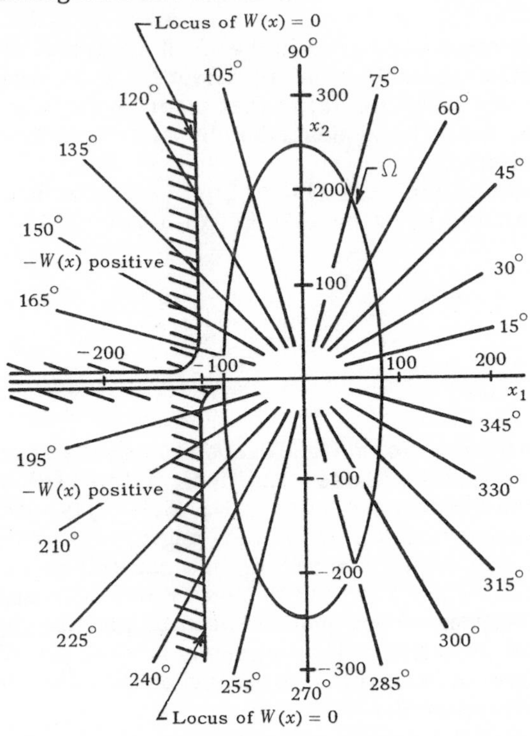

FIG. 5-14. State plane stability region for an adaptive control system.

The situation is now ripe for the application of Theorem 5-26. To establish the numerical choice appropriate for L, the change of variables given by Eqs. (5.3-119) and (5.3-120) is introduced into Eq. (5.3-74). Then the numerical values given by Eqs. (5.3-115) through (5.3-117) are substituted for the elements of $B, U(x)$, or $U(r,\theta)$ is equated to L, and finally an expression for r in terms of θ is obtained. The resulting expression is

$$r = \left(\frac{L}{10^3 \cos^2\theta - 2\sin\theta\cos\theta + 10^2 \sin^2\theta}\right)^{1/2} \tag{5.3-123}$$

This equation may be used to explore the state plane to determine the largest value L may assume, subject to the restriction that $-W(x)$ is negative within the region described by $U(\mathbf{x}) \leq L$. The resulting value is

$$L = 6.24 \times 10^6 \tag{5.3-124}$$

The region Ω, defined by $U(\mathbf{x}) \leq L$, is plotted in Fig. 5-14 and labeled. Within this region, conditions (a), (b), and (c) of Theorem 5-25 are satisfied, and hence the conditions of Theorem 5-26 are satisfied. Therefore regions Ω_δ may be defined by $U(\mathbf{x}) \leqq L - \delta$, with associated T_δ, such that any trajectory $x(t, t_0, \mathbf{x}_0)$ will approach zero as $t \to \infty$ for all $\mathbf{x}_0$ and all $t_0 \geqq T_\delta$.

Some comments are now in order with regard to the meaning of the above analysis. Initially, a relatively simple set of nonlinear differential equations which describe an adaptive control system were considered. The stability of the system was to be analyzed with a step-function input of magnitude R_0 at $t = 0$. A transformation of variables was introduced into the set of equations so that they represented the response of the system to perturbations about a desired trajectory. The resulting set of equations was both time-varying and nonlinear. The analysis was directed toward the application of Theorem 5-26 and specific numerical values were selected for the magnitude of the input and the system parameters. The analysis has been carried to the point where the conclusions of Theorem 5-26 may be applied. For values of time greater than the settling time of the reference model (approximately 30 seconds after the application of the step function) the value of δ may be made small, and hence the size of the region Ω_δ will be nearly as large as Ω. In this case, the resulting value of T_δ will be approximately 30 seconds. If the system has not suffered a perturbation from the desired trajectory up to this point in time, then Theorem 5-26 assures that any perturbation of the state variables that does not exceed the region Ω_δ, which is only slightly smaller than Ω, will result in an asymptotically stable return of the state variables to the origin. In terms of the operation of the adjusting mechanism of the original system, this implies that if the value of g is instantaneously changed by even as much as 200, the adjusting mechanism will adjust h_0 in a stable manner to compensate for the change. In view of the fact that the nominal value of a_0 is only 0.1, this is a strong statement with regard to the stability of the system. The statement is, of course, valid only for the condition of the system twenty or thirty seconds after the application of the step input. To consider perturbations occurring at smaller values of t (i.e., T_δ smaller) it is necessary for δ to be larger, and hence the region of assured asymptotic stability Ω_δ will be smaller. The shape of the region remains the same, however. There is no implication that trajectories that start outside this region will necessarily diverge, even if they start in the region where $-W(x)$ is positive. This is so because the Lyapunov criterion for stability is sufficient but not necessary.

On the basis of this analysis, it may be concluded that the control system under consideration with the prescribed input will adapt in an asymptotically stable manner to perturbations or variations in the system parameters with the possible exception of large perturbations occurring shortly after the step input is applied. Simulation of this adaptive control system on an analog computer yielded results closely matching the conclusions implied by the above analysis. That is, the system had excellent adaptive response to perturbations of the parameter g_0 with the exception of large perturbations that occurred shortly after the input was applied.

Generalization of the Procedure

The procedure used above to determine an appropriate value for L in Theorem 5-26 requires a graphical plot of the locus for $-W(\mathbf{x}) = 0$. If there are three or more state variables, such a procedure is impracticable. The following summarizes the general procedure for the application of Theorem 5-26 to systems which may be represented by Eq. (5.2-3), and which satisfy Eq. (5.2-43) when the Lyapunov function is restricted to quadratic forms. The application of this technique for Theorem 5-23 is discussed at the end of this section. The potential Lyapunov function may be written as

$$V(\mathbf{x}, t) = \mathbf{x}'\mathbf{B}\mathbf{x} \tag{5.3-125}$$

in which $\mathbf{B}$ is a symmetric matrix and $\mathbf{x}$ is the column vector of the state variables. The elements of $\mathbf{B}$ may be selected as functions of time. The derivative of the Lyapunov function may be written

$$\dot{V}(\mathbf{x}, t) = -\mathbf{x}'\mathbf{C}\mathbf{x} + \mathbf{x}'\mathbf{B}\mathbf{f}(\mathbf{x}) \tag{5.3-126}$$

in which C is a symmetric matrix defined by

$$\mathbf{C} = -\left[\mathbf{B}\mathbf{A} + (\mathbf{B}\mathbf{A})' + \dot{\mathbf{B}}\right] \tag{5.3-127}$$

The matrix $\mathbf{A}$ must be nonsingular in the present state of this stability analysis technique; otherwise, it is very difficult to select the elements in $\mathbf{B}$ so that $\mathbf{C}$ will be positive-definite.

The next step is the selection of the elements of $\mathbf{B}$. In general, the choice of the elements of $\mathbf{B}$ should satisfy four criteria. The first two criteria are that both $\mathbf{B}$ and $\mathbf{C}$ must be positive-definite. Subordinate of these requirements, but still important, are the second two requirements. The first of these is that the eigenvalues of the matrix B should not differ from each other by more than a factor of ten or twenty. The second is that none of the eigenvalues of C should be too small. Although these statements are rather general in nature, they will at least serve as guide lines in the selection of the elements in B. An aid to the manipulation of the eigenvalues of matrices B and C as a function of the elements of B can be obtained by arranging the characteristic equations of B and C so that their eigenvalues may be plotted as a root locus which has open-loop poles, zeros, and gain as functions of the elements of B. The details of this procedure are given elsewhere [7].

Once the elements of B have been selected, our interest then turns to the size of the region of asymptotic stability that may be established by Theorem 5-26. The size of the region is in turn dictated by the largest value of L which may be obtained subject to the restriction that $W(x) > 0$ within the region described by $U(x) \leqq L$. The value of L is most easily determined if a transformation of variables is introduced which permits writing $U(x)$ as

$$U(x) = x'Dx = z'Iz \,. \qquad (D = \lim_{t \to \infty} B) \tag{5.3-128}$$

In which I is the unity or identity matrix. It is convenient to determine this transformation of variables in two steps. The first step is to determine an orthogonal transformation which will permit $U(x)$ to be written as

$$U(x) = U(v) = v'\Lambda y \tag{5.3-129}$$

in which Λ is a diagonal matrix. The elements of Λ are the eigenvalues of the matrix D. The variables v are related to the x's by the relation

$$v = Px \tag{5.3-130}$$

in which P is the matrix of the orthogonal transformation. The elements of P may be determined by solving the set of linearly independent elemental equalities that are contained in the relations

$$PP' = I \tag{5.3-131}$$

and

$$PD = \Lambda P \tag{5.3-132}$$

Once P has been determined, it is a simple matter to define an additional transformation of variables that permits writing U as

$$U = x'Dx = v'\Lambda v = z'Iz \tag{5.3-133}$$

The relation between the variables in vectors v and z may be written as

$$z = \theta v \tag{5.3-134}$$

in which θ is a diagonal matrix. The elements of θ are easily obtained from the elemental equalities implied by

$$\theta^2 = \Lambda \tag{5.3-135}$$

The relation between x and z may be written as

$$x = P'\theta^{-1} z \tag{5.3-136}$$

The function $-W(x)$ then becomes

$$-W(x) = -W(z) = -z'\left[P'\theta^{-1}\right] E \left[P'\theta^{-1}\right] z + z\left[P'\theta^{-1}\right] D \left(f(x)\Big|_{x_i = g(z)} \right).$$

$$(\text{in which } E = \lim_{t \to \infty} C) \tag{5.3-137}$$

In terms of z variables, the boundary of the region described by $U \leq L$ will be a hypersphere in the $m + n$ variable space. The largest value of L which can be used will be the square of the Euclidian distance from the origin of the z coordinate system to the closest point which lies on one of the sheets of the surface defined by $W(z) = 0$. This point is most easily found by making a change in variables to the hyperspherical polar coordinates. The new variables will be the radius r, and $(m + n - 1)$ angular measures. When $W = 0$ is expressed in these new variables, it is possible to solve for r as a function of the $(n - 1)$ angular variables. This function may then be programmed on a high-speed digital computer and methods of descent or total space-sampling employed to find the minimum value of r. A word of caution is in order here. Each sheet of the surface defined by $W = 0$ will have a minimum point. Therefore, it is necessary to use the smallest of these minima in determining L. Let this minimum value of r be designated by r_0. In terms of the hyperspherical polar coordinates, the function U is expressed simply

$$U = r^2 \tag{5.3-138}$$

The appropriate value for L is then simply

$$L = r_0^2 \tag{5.3-139}$$

This value of L may then be used with $U(x)$ to plot two-dimensional sections through the region Ω in the application of Theorem 5-26.

Likewise, two-dimensional sections are easily plotted through the regions Ω_δ defined by $U(x) \leqq L - \delta$.

This technique may also be used for determining the region Ω for Theorem 5-23. In the latter case, the problem is simpler since, for the autonomous case, $\mathbf{B} = \mathbf{D}$ and $\mathbf{C} = \mathbf{E}$ for all t. Otherwise the procedure is identical.

5.3.8 *Establishing an Upper Boundary on Transient Response*

Once a Lyapunov function has been generated which establishes the asymptotic stability of the system, it is possible to establish an upper boundary on its transient response if $\dot{V}(x, t)$ is negative-definite. This is most easily understood by recalling the over-simplified analogy given in Section 5.1 between Lyapunov's direct method and the response of a first-order linear system. In the case of the first-order linear system, with the input set equal to zero ($r(t) = 0$), the system may be represented by

$$\dot{x} = -ax, \quad a = \text{a positive constant,} \tag{5.3-140}$$

or

$$a = -\frac{\dot{x}}{x} \tag{5.3-141}$$

For an initial condition of x_0, the transient response for $t > 0$ is

$$x(t) = x_0 e^{-at}, \quad t > 0 \tag{5.3-142}$$

In this case, x represents the measure of the deviation of the system output from the equilibrium point of $x = 0$. Equation (5.3-142) describes the time history of the approach of x to the equilibrium point. The positive constant a is referred to as the time constant of the system since it represents the time required for the output to become $\dfrac{x_0}{e}$.

If $V(\mathbf{x}, t)$ is considered as a measure of the distance from the equilibrium point to a given position on a solution trajectory, then by defining

$$\eta = \min\left\{-\frac{\dot{V}(\mathbf{x}, t)}{V(\mathbf{x}, t)}\right\} \tag{5.3-143}$$

over the interval of time of interest and in the region in which asymptotic stability has been established, the following inequality holds

$$V\big(\boldsymbol{\phi}(t; \mathbf{x}_0, t_0), t\big) \leqq V(\mathbf{x}_0, t_0)e^{-\eta(t-t_0)} \tag{5.3-144}$$

In the case of autonomous systems, if the transient response is not required for the entire state space for which asymptotic stability as been established, it is possible to evaluate Eq. (5.3-143) for smaller regions Ω defined by

$$V(\mathbf{x}) \leq L \tag{5.3-145}$$

where L is a positive constant.

This procedure is inapplicable if $\dot{V}$ is not negative-definite. This is easily seen since, if $\dot{V}$ is negative-semidefinite, η is zero, and the resulting conclusion from inequality (5.3-144) is of no value. It should also be noted that η will be affected by the initial choice of the Lyapunov function. For a given system, the value of η as estimated from two different Lyapunov functions may differ by orders of magnitude. In all cases, inequality (5.3-144) is valid; however, the apparent transient response may be much faster in one case than in the other. As a consequence, this technique is unacceptable for comparing the transient response of two systems. Nevertheless, it does provide a means of establishing an upper bound on the transient response.

5.4 Conclusions and Summary

The above sections have attempted a brief, though fairly complete, presentation of Lyapunov's direct (second) method. The presentation has been oriented toward automatic control systems. A number of the theorems presented in Section 5.2 have been restated in a form slightly different from the way they appeared in the original references. This was done in an attempt to standardize the notation and description of the physical system throughout the presentation. Theorems dealing with stability in an arbitrarily small region about an equilibrium position for a system were given. These theorems are applicable to a fairly large number of nonlinear systems. They are often of value as a preliminary step in the analysis of system stability, since the results of their application may help direct the choice of a theorem to prove stability in the large, stability in a defined region, or the instability of the system.

The definition of stability in nonlinear systems, which may be either autonomous or nonautonomous, is more complex than the definition of stability in linear systems. A number of subdivisions or levels of stability are defined in Section 5.2. Although these stratified definitions do have meaning, it is likely that most control engineers will be interested in asymptotic stability of some type and will feel no need for further specifiction.

The greatest difficulty in the application of Lyapunov's direct method is the generation or discovery of a function V which satisfies the conditions of one of the theorems. Section 5.3 presents a number of the available techniques which either directly or indirectly determine Lyapunov functions for various systems. None of these techniques is all-inclusive, and one may succeed where the others fail. The variable gradient method shows considerable potential in the application of Lyapunov's direct method to autonomous nonlinear systems; however, its application to nonautonomous systems is quite limited. In general, failure to determine the function V which serves to establish either stability or instability yields no information. The reason for this is that while the Lyapunov criteria provide sufficient conditions to establish stability, the conditions are not necessary for system stability. Another way of stating this is that the Lyapunov criteria are excessively restrictive. Hence, the failure to meet the conditions for stability does not necessarily mean instability, and the failure to meet the conditions for instability does not necessarily mean stability. It should also be noted that a Lyapunov function capable of proving a system stable or unstable is not unique. In fact, it can be shown that if one such function exists, there are an infinite number of other functions capable of accomplishing the same job.

A little reflection is also in order as to the type of information obtained by Lyapunov's direct method. If applicable, it does determine stability or instability. This is good information. However, the control engineer is often interested in information regarding tendencies to overshoot, and the settling time after introduction of a perturbation. It is also desirable to be able to use stability criteria as a guide in the selection of such system parameters as loop gain and coefficients in compensation elements. Although these capabilities are discussed in the literature [11] they are quite limited at the present time. For instance, $\frac{dV}{dt}$ may be used to estimate settling time for a system as is discussed in Section 5.3.8. Clearly, however, any estimate made in this manner is dependent both on the characteristics of the system in question and the form and parameters selected for V. This being the case, such an estimate may not do the system justice, since a different choice for V might imply a far shorter settling time. The technique does have utility, however, since it may be used to establish an upper bound on settling time. A similar problem arises in using the Lyapunov criteria to establish system parameters, since here again, the parameters and form of a given Lyapunov function affect the conclusions.

In general the Lyapunov criteria do not supply the versatile design and analysis tool for nonlinear systems that the Root-Locus technique does for linear systems. This is not surprising since the class of problems to which the former may be applied is far more general. It appears that the effectiveness in the use of the Lyapunov criteria would be significantly increased if a technique were developed to derive a Lyapunov function that not only satisfied the conditions of the theorems, but also "fitted tightly" (in some sense) the system in question. Such a technique would probably be so complicated that high-speed computing devices would be required for its implementation. It might, however, permit an investigator to establish the minimum upper bound on the settling time of a system.

REFERENCES

1. Aizerman, M. A., *Theory of the Automatic Control of Motors* (in Russian) GITTL, Moscow, 1952.
2. Beckenbach, E. P., *Modern Mathematics for the Engineer*, McGraw-Hill, N. Y., 1956.
3. Bellman, R., *Stability Theory of Differential Equations*, McGraw-Hill, N. Y., 1953.
4. Birkhoff, G., and MacLane, S., *A Survey of Modern Algebra*, Revised Edition, MacMillan Co., N. Y., 1953.
5. Cesari, L., "Asymptotic and Static Problems in Ordinary Differential Equations," *Ergebnisse der Mathematik Und Threr Grenzgebiete*, Neue Folge-Heft 16, Springer-Verlag, Berlin, 1959.
6. Coddington, E. A., and Levinson, N., *Theory of Ordinary Differential Equations*, McGraw-Hill, N. Y., 1955.
7. Donalson, D. D., "The Theory and Stability Analysis of a Model Referenced Parameter Tracking Technique for Adaptive Automatic Control System" Ph. D. Thesis, UCLA, May, 1961.
8. Gibson, J. E., et al., "Stability of Nonlinear Control Systems by the Second Method of Lyapunov," Purdue School of Electrical Engineering Report No. EE 61-5, Lafayette, Indiana, May, 1961.
9. Ingwerson, D. R., "A Modified Lyapunov Method for Nonlinear Stability Problems," Ph. D. Thesis, Stanford University, Nov., 1960.
10. Ingwerson, D. R., "A Modified Lyapunov Method for Nonlinear Stability Analysis," *IRE Trans. on Automatic Control*, Vol. AC-6, May, 1961.
11. Kalman, R. E. and Bertram, J. E., "Control System Analysis and Design via the Second Method of Lyapunov, I, Continuous-Time Systems," *Journal of Basic Engineering* (Series D, Trans. ASME), 82:371-393, June, 1960.
12. Krasovskiy, N. N., "Overall Stability of a Solution of a Nonlinear System of Differential Equations," *Prikladnaya Matematika i Mekhanika*, Vol. 18, pp. 735-737, 1954. (English translation, same title, STL-T-Ru-23, 60-5111-103, Literature Research Group, Technical Information Center, Space Technologies Laboratories, Los Angeles, Calif., August, 1960.)

13. LaSalle, J. P., "Some Extensions of Lyapunov's Second Method," RIAS Technical Report 60-5, AFOSR TN-60-22, Research Institute of Advanced Study, Baltimore, Md., 1960.
14. LaSalle, J. P., Rath, R. J., "A New Concept of Stability," Paper No. 415, Research Institute for Advanced Study, 7212 Bellona Ave., Baltimore 12, Maryland, August 29, 1962.
15. Lass, Harry, *Vector and Tensor Analysis*, McGraw-Hill Book Co., New York, 1950.
16. Letov, A. M., *Stability in Nonlinear Control Systems*, Princeton University Press, Princeton, New Jersey, 1961.
17. Lur'e, A. I., *Some Nonlinear Problems in the Theory of Automatic Control* (Translation from Russian), Her Majesty's Stationery Office, 1957.
18. Lur'e, A. I., Rozenvasser, E. N., "On Methods of Constructing Lyapunov Functions in the Theory of Nonlinear Control Systems," *Proc. of the International Federation of Automatic Control Congress*, Butterworth Scientific Publications, 1960.
19. Margolis, M., *On the Theory of Process Adaptive Control Systems, the Learning Model Approach*, Doctoral Dissertation, Department of Engineering, University of California, Lox Angeles, 1959.
20. Massera, J. L., "Contributions to Stability Theory," *Ann-Math*, 64:182-206, 1956.
21. Pliss, V. A., "Some Problems in the Theory of Stability of Motion in the Large, "Izdatel'stvo LGU, 1959.
22. Rekasius, Z. V., "Stability Analysis of Nonlinear Control Systems by the Second Method of Lyapunov," *IRE Trans. on Automatic Control*, Vol. AC-1, January, 1962.
23. Shultz, D. G., and Gibson, J. E., "The Variable Gradient Method for Generating Lyapunov Functions," AIEE Transaction Paper 62-81.
24. Shutlz, D. G., and Gibson, J. E., "The Variable Gradient Method of Generating Lyapunov Functions with Application to Automatic Control Systems," Purdue University, School of Electrical Engineering, Control Information Systems Laboratory, Report No. TR-EE 62-3, April, 1962.
25. Szego, G. P., "A Contribution to Lyapunov's Second Method: Nonlinear Autonomous Systems," *Journal of Basic Engineering, Trans. ASME (D) and Proceedings of the OSR RIAS International Symposium on Nonlinear Differential Equations and Nonlinear Mechanics.*
26. Szego, G. P., "A Contribution to Lyapunov's Second Method: Nonlinear Autonomous Systems," Paper No. 61-WA-192 presented at ASME Annual Winter Meeting, November, 1961.
27. Taylor, A. E., *Advanced Calculus*, Ginn and Company, Boston, 1955.
28. Zubov, V. I., *Mathematical Methods of Control System Analysis* (book: in Russian), Subpromizdat, 1959.

6

Review of Adaptive Control System Theories and Techniques

DALE D. DONALSON
HEAD, FUNCTIONAL DESIGN SECTION
HUGHES AIRCRAFT COMPANY, CULVER CITY, CALIFORNIA

F. H. KISHI
MEMBER OF TECHNICAL STAFF, TRW SPACE TECHNOLOGY LABORATORIES, REDONDO BEACH, CALIFORNIA
(FORMERLY WITH HUGHES AIRCRAFT COMPANY
CULVER CITY, CALIFORNIA)

6.1 Introduction

To effectively treat the subject of adaptive control systems, it is desirable to define this category of control systems. The term "adaptive" appeared in control literature around 1954, when Tsien [48] in his book described Ashby's [2] model of the human brain. About the same time, Benner and Drenick [5] presented a control design with "adaptive" characteristics. Since that time, the term "adaptive" has been attached to a wide variety of control systems. Numerous investigators have thus offered a variety of definitions for "adaptive control systems." Initially, the definitions tended to be limited to certain specific cases. More recently, the definitions have been generalized but not reduced in number, and no generally accepted definition exists at the present time.

Before specifying what is meant by adaptive control systems, let us define the expression "acceptably performing system."* This term describes the external manifestations of the system under investigation. The goal for a control system design should be clear to the designer as the first step in his design process.

*Zadeh [53] used "adaptivity" for "acceptably performing systems." His definition for adaptive system is here used to define acceptably performing systems.

The definitive delineation of "acceptably performing system" is an attempt to express quantitatively whether the designer has attained this goal.

We shall first explain some of the terminology. With reference to Fig. 6-1, let

$\mathcal{A}$ ≡ system including process and controller;

$v(t)$ ≡ vector function defined for the interval of operation $0 \le t \le T$, and composed of (some may be missing): (a) reference inputs, (b) known inputs to the process, (c) disturbances to the process, (d) measurable outputs;

γ ≡ parameter vector belonging to some set Γ which determines the portion of the set $v(t)$ (items (a), (b), and (c) above) to be impressed upon the system;

S_γ ≡ set of $v(t)$ generated for the particular γ;

$P(\gamma)$ ≡ performance criterion which can take on a range of values for a given γ;

W ≡ set to which it is desired to restrict $P(\gamma)$.

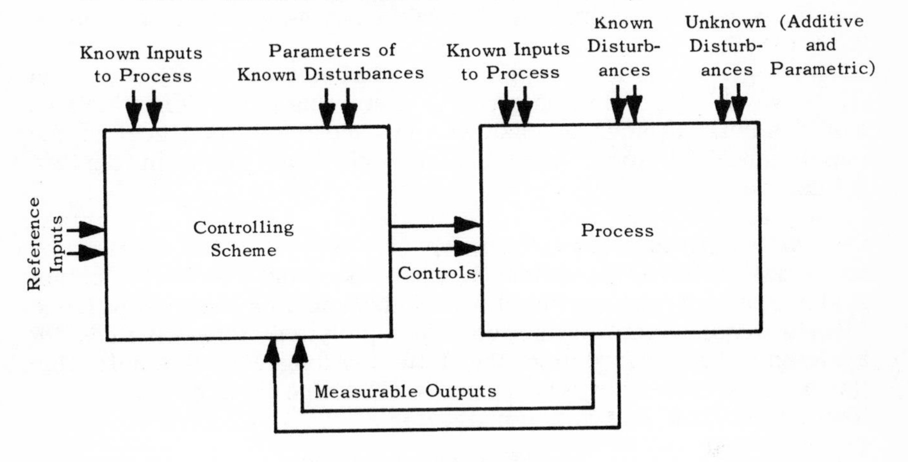

Fig. 6-1. Control system.

In terms of the above, we define a *criterion of acceptability* in the following manner. If $P(\gamma)$ is maintained in the set W, then the criterion of acceptability is satisfied. This notion leads to the definition:

Definition 6-1—A system $\mathcal{A}$ is an *acceptably performing system* (APS) with respect to S_γ and W if it satisfies the criterion of acceptability with every source in the family S_γ, γ belonging to Γ. To elaborate, we have as APS, if it can possibly be designed, a mechanism in $\mathcal{A}$ which can provide a control to maintain the performance criterion within the acceptable limits of W. This acceptable performance is to be maintained for the class of inputs represented by S_γ.

Even open-loop systems can be APS as long as the criterion of acceptability is maintained. A problem arises, however, when $P(\gamma)$ cannot be maintained in W. Here, it becomes necessary to consider more complex mechanisms within $\mathcal{A}$ to satisfy the criterion of acceptability. Therefore, one is led to many possible alternatives for the construction of the control mechanism, each with an attempt to satisfy the criterion of acceptability.

To illustrate the above notions, let us take the example of the pitch response of an aircraft control system. Given a command or reference input, signals are sent to the controlling surfaces which, through interaction with the airstream, create a torque on the aircraft. In this example, we designate the performance criterion $P(\gamma)$ as the percent overshoot to a step command, and γ the altitude. For a particular altitude γ, a set or ensemble of air-density variations is experienced, a set which corresponds to part of the $v(t)$ just described. For this set of air-density variations, there is a range of $P(\gamma)$, say from 0% to 30%. If this range of $P(\gamma)$ is within the allowable set as given by W, then we have an APS.

In a given application it may be difficult to specify Γ, W, and P. However, the definition gives us a starting point from which we can describe various mechanizations whose intended goal is the maintenance of some performance criterion within prescribed limits.

We are now prepared to define various forms of mechanization of $\mathcal{A}$. In this connection, we can describe what we mean by an adaptive control system.* Open-loop and feedback control systems are familiar control mechanizations. In contrast to these forms, we seek the distinctive characteristics of *adaptive control systems*. We shall define the latter. Along with this definition for *adaptive control systems*, we shall define the other two forms to point out their distinctive characteristics.

Definition 6-2—A system is an *open-loop system* if control action as a function of time is impressed upon the process on the basis of a priori knowledge of the process.

Definition 6-3—A system is a *feedback control system* if means is provided to monitor the variables depending upon the control action (state or controlled variables) to modify accordingly the subsequent control action in an attempt to obtain an acceptably performing system.

Definition 6-4—A system is an *adaptive control system* if a means is provided to monitor, in addition to the state variables, its

*Zadeh [53] used the term "adaptive" for the external manifestation while we choose to use adaptive for the internal mechanization.

performance and/or process (internal and/or external) characteristics to modify the control action accordingly in an attempt to make it an acceptably performing system.

Definition 6-3 is influenced by other defintions given in the past, perhaps most strongly by that given by Cooper and Gibson [11]. It is important, however, to establish a fairly general definition to encompass the many adaptive systems described in the literature, and still distinguish feedback control systems from the other two forms of mechanization. To elaborate, monitoring performance and/or process characteristics cause adaptive control systems to differ from feedback control systems. It should be noted that while adaptive control systems are feedback control systems, the converse is not necessarily true; therefore, it is expected that better performance can be achieved by adaptive control systems. And, it is for this reason that we study adaptive control systems. It is noted, however, that even an adaptive control system may not be an APS.

From the definition, it is observed that commonly described controllers whose modifying effects depend upon environmental measurements (e. g., air-data measurements) are classed as adaptive control systems.

Adaptive control systems take many forms. No attempt will be made in this section to survey all the different schemes devised in the past because several good survey articles are available [1, 11, 37, 44, 47]. However, three categories which appear to encompass a large proportion of adaptive control systems are (1) high-gain schemes, (2) model-referenced schemes, and (3) optimum-adaptive schemes. The order of the listing is that of increased complexity. It is expected that the range of performance varies with the different schemes; and complexity should be added only if improved performance is obtainable and mandatory. At present, the selection of a particular scheme appears to depend on loosely-defined qualitative judgment.

From the practical standpoint, the high-gain scheme, first proposed by the Minneapolis-Honeywell Company [7, 15, 46], has been widely discussed and tested. It has proved to be widely applicable. The gain in the feedback loop around the changing process is kept as high as possible to keep the input-output transfer close to unity. Because stability problems arise at high gain, the signal in the loop is monitored to check for oscillations. With this information, the loop gain is adjusted to keep the system on the verge of instability. A response close to that of a particular model is obtained regardless of the process parameters by placing the model in front of the feedback loop. A schematic diagram of the high-gain scheme is shown in Fig. 6-2. One of the objections to this appraoch is that the designer must have a large amount of a priori information about the process, i.e., he must know the general vicinity where the roots of the system go into the right half-plane. Of course, a frequency-insensitive unity gain can only be approached, implying that the output response will differ to some extent from the model response. Also, small oscillations are always present in the loop (this oscillation has been reported to be unobjectionable in some aerospace applications).

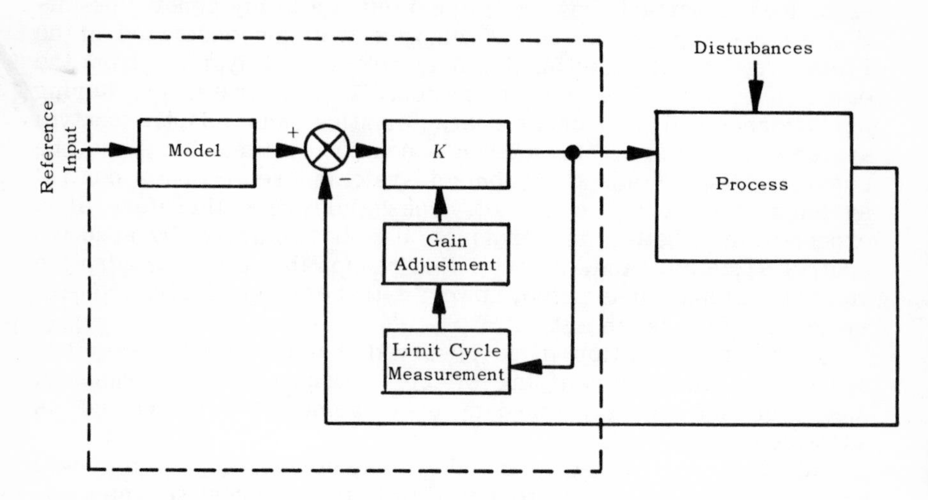

FIG. 6-2. High gain scheme.

The other two schemes will subsequently be described in more detail. The chapter is organized in the following way. This section gives introductory material including a definition for adaptive control systems. Section 6-2 describes model-referenced techniques, and Sec. 6-3 describes optimal adaptive schemes. Process and state estimations which are vital parts of optimal adaptive schemes are discussed separately in Sec. 6-4.

6.2 Model Reference Adaptive Techniques

One of the general categories into which the broad scope of adaptive control systems may be subdivided is referred to as the Model Reference Adaptive approach. A number of Model Reference techniques have been suggested as a means for implementing adaptive control systems. There are clear similarities between many of these concepts, yet, at the same time, most show significant differences. In most cases, the research or evaluation associated with each technique has been sufficient to demonstrate that the concept works. In some cases, a technique has been implemented in an actual experimental control system. Whitaker, Yarmon, and Kezer [49] report the results of implementing an adaptive control system in the pitch, yaw, and rudder coordination subsystems of an F-94A aircraft. Although this work represents the best practical demonstration to date, the data obtained from actual aircraft flights were quite limited due to a relatively short flight-test schedule and a number of equipment malfunctions. The flight tests were augmented by reasonably extensive analog computer simulation studies.

Although the empirical data in support of the model reference concepts are encouraging, they are not sufficient to establish the superiority of one technique over the rest. Therefore, at the present time, it does not appear possible to state a single unified theory for model reference adaptive control systems. It should be pointed out, however, that if the present trend continues, a few fairly unified approaches should soon be available. It does not appear to be either desirable or practical to attempt to present each of the noteworthy contributions in detail here. Instead, an attempt will be made to establish a central trend or framework, and then discuss the deviation of the various techniques from this framework. Some methods will be presented in greater detail than others; however, this should not necessarily be interpreted as the authors' measure of their significance.

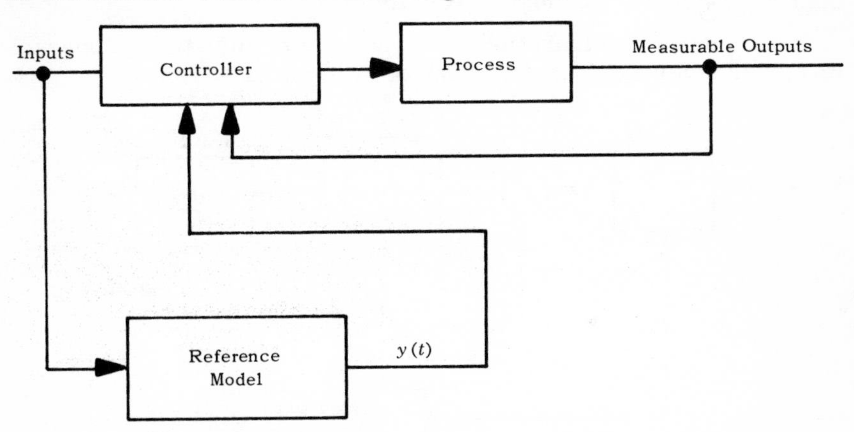

FIG. 6-3. Generalized model reference adaptive control system.

As is implied by the name, the model reference approach employs a reference model as part of the adaptive scheme. In most cases, the model is actually part of the hardware implementation; however, in some instances it enters in a more subtle manner. Figure 6-3 shows a basic block diagram which encompasses most of the model-referenced adaptive control system configurations in which the reference model is explicit in the mechanization. The reference model is an analog representation of the desired dynamic response of the overall control system. In terms of the definitions of Sec. 6-1, the $P(\gamma)$ for the reference model will be within the region of desired performance W. The objective then is to use the controller to compare the output of the process with that of the reference model and cause the controller to "adapt" or change so that the input-output transference of the controller-process combination closely matches that of the reference model. Whether the $P(y)$ for the actual process in also within W depends upon how well the adaptive mechanism does in achieving this objective.

It is at this point, in terms of detail, that the initial differences in the model reference adaptive techniques become evident. For the purpose of orderly presentation, it will be convenient to divide the model reference adaptive techniques into three major categories:

1. Parameter adjustment techniques;
2. Parameter perturbation techniques;
3. Control signal augmentation techniques.

Each of these is treated in some detail below. The exact categorization of techniques is somewhat arbitrary. Actually, in many instances, a particular technique may be placed in a different category depending on how one chooses to view the problem and its solution.

6.2.1 Parameter Adjustment Techniques

The general configuration for the parameter adjustment approach to model reference adaptive control systems is shown in Fig. 6-4.

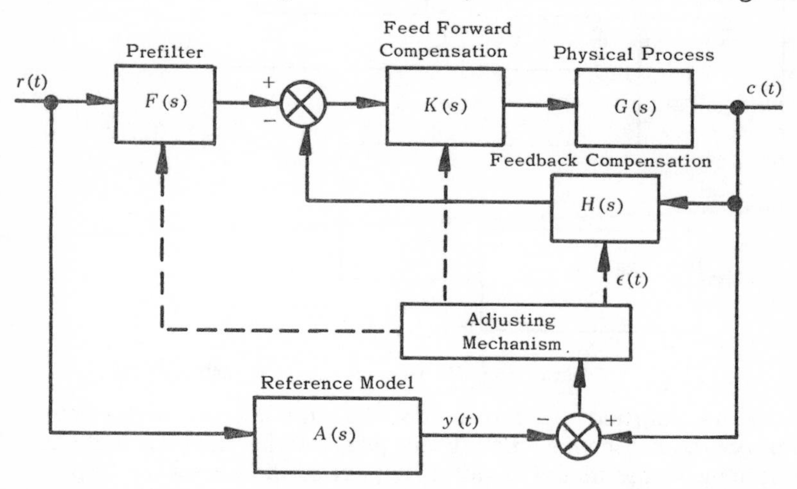

FIG. 6-4. Generalized model referenced adaptive control system.

The portion of the block diagram which involves the prefilter, feedforward compensation, physical process, and feedback compensation represents the general form for a closed-loop control system. The adaptive feature is achieved by adding the reference model and the adjusting mechanism. The reference model is an explicit analog or digital representation of the desired dynamic response of the overall control system. At the present state of development for this approach, the form selected for the reference model must be compatible with the basic form of the physical process and the available compensation. That is, given the form of the differential equation that describes the physical process, it must be possible to select physically realizable input-output characteristics for the compensation elements so that the form of the overall control system input-output relation is the same as that of the reference model. Furthermore, it must be possible to select the parameters in the compensation elements so that the differential equation which represents the overall control system is identical to that of the Reference Model for any values the parameters in the Physical Process may assume.

The input to the reference model is the same as the input to the control system. The output of the reference model and the output of the control system are applied to a summing device with opposite signs to form an error signal. This error signal then serves as the input to the adjusting mechanism. On the basis of this input, the adjusting mechanism then adjusts the parameters in the various compensation elements so that the error signal is zero. If the input r(t) is not zero, then for the error signal e(t) to be zero for a period of time, it is necessary for the differential equation which describes the overall control system to be identical with that of the reference model. Thus, if the adjusting mechanism operates as described, then the overall control system will adapt the parameters in the compensation elements when changes occur in the physical process so that the overall control system response remains the same. The parameter adjustment approach to the reference model concept as it has been described above appears to have its origin in the work done by Whitaker, Yarmon, and Kezer [49]. However, the concept has been extended and variations added by a number of other investigators. The key difference in each of these extensions is in the detailed operation of the adjusting mechanism. Three fairly general methods are presented below. These methods encompass the basic characteristics of most of the approaches that fall into this category.

Method 1—Based on the Gradient of an Error Function. Many of the techniques developed in the learning model approach to adaptive control systems [30, 31] have been applied to the model reference approach [12].

A simple second-order control system will serve as the basis for the development of this parameter adjustment method. The system considered is shown in Fig. 6-5, and is merely a specific example of the general adaptive system shown in Fig. 6-4. The input-output relationship for the control portion of the system (that is, between c and r) may be written

$$\ddot{c} + (g_1 + h_1)\dot{c} + (g_0 + h_0)c = r(t) \tag{6.2-1}$$

The dot notation represents differentiation with respect to time. The desired input-output relationship represented by the reference model may also be written in the form of a differential equation

$$\ddot{y} + a_1\dot{y} + a_0 y = r(t) \tag{6.2-2}$$

If the relations

$$h_1 + g_1 = a_1 \tag{6.2-3}$$

and

$$h_0 + g_0 = a_0 \tag{6.2-4}$$

are satisfied, then the control system will have the desired input-output relationship.

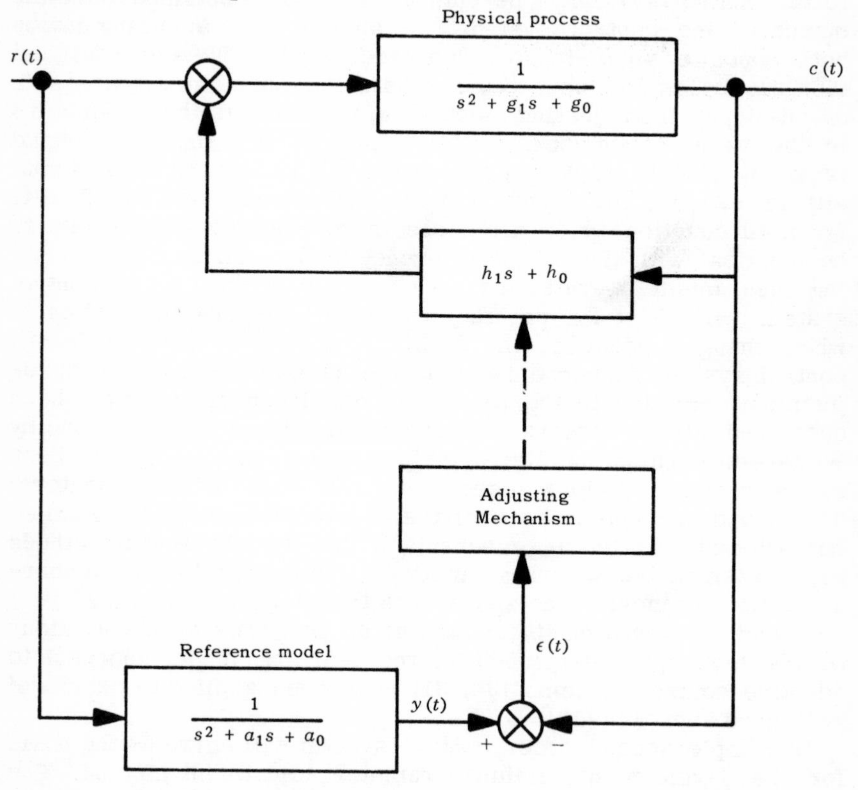

FIG. 6-5. Model referenced adaptive control system.

The adaptive portion of the system consists of the reference model, a summing device to form the error signal, and an adjusting mechanism. This loop provides rate information for the adjustment of parameters h_1 and h_0 so that they will "adapt" to changes in parameters g_1 and g_0. Here, one might also think in terms of the integral of the rate information over the period of time the system has been in operation. This integral then determines the instantaneous values for h_1 and h_0. The reference model may be considered an analog implementation of the differential equation to yield the desired input-output relation. This reference model is an explicit mechanization in terms of the adaptive control system hardware. The adjusting mechanism is that portion of the system which provides the adjustment rates at for both h_1 and h_0, and the integrals of those rates. As indicated by the block diagram, the input information to the adjusting mechanism is the error between the output of the reference model and the output of the physical process. It will be assumed that the first and second derivatives of the error are also available. This assumption may be difficult to fulfill in the case of some physical processes, but usually it is possible to obtain derivatives. The objective, then, is to formulate the equations describing the operation of the adjusting mechanism. If no limitations are placed on the values assumed by h_1 and h_0, then, regardless of the values of g_1 and g_0, the corresponding coefficients in Eqs. (6.2-1) and (6.2-2) will be identical whenever Eqs. (6.2-3) and (6.2-4) are satisfied. Two assumptions are made concerning the rate at which g_1 and g_0 vary. They are:

Assumption 1—g_1 and g_0 vary slowly as compared to the basic time constants of the physical process and the reference model.

Assumption 2—g_1 and g_0 vary slowly as compared to the rates at which the adjusting mechanism, which is to be designed, adjusts the parameters h_1 and h_0.

The consequence of violating these two assumptions, as well as a third stated below, is important and may be investigated by analog computer simulation or through analytical stability analysis. Such investigations [12] have shown that the adaptive system to be described still works quite well even when the assumptions are seriously violated. The assumptions are necessary for the derivation of the equations of motion for the adjusting mechanism.

The basis for the operation of the adjusting mechanism is the minimization of a quadratic function of the error and its derivatives. Two functions which may be used are

$$f(\epsilon) = \frac{1}{2}(q_0\epsilon + q_1\dot{\epsilon} + q_2\ddot{\epsilon})^2 \tag{6.2-5}$$

and

$$f_1(\epsilon) = \frac{1}{2}\left(q_0\epsilon^2 + q_1\dot{\epsilon}^2 + q_2\ddot{\epsilon}^2\right) \tag{6.2-6}$$

in which

$$\epsilon = (c - y) \tag{6.2-7}$$

and the parameters q_0, q_1, and q_2 are constants to be specified later. The error function which will be used for this development is that expressed by Eq. (6.2-5). Although the choice of the error function is important in the design of an actual system, it is unimportant in the development of the basic idea for the adjusting mechanism. The technique is equally applicable to error functions of the form of either Eq. (6.2-5) or (6.2-6). Additional discussion of the form of the error function will be given later.

In order to facilitate the discussion, Eq. (6-2.1) will be written as

$$\ddot{c} + \alpha_1 \dot{c} + \alpha_0 c = r(t) \tag{6.2-8}$$

$$\alpha_1 = g_1 + h_1 \tag{6.2-9}$$

and

$$\alpha_0 = g_0 + h_0 \tag{6.2-10}$$

The minimum value of $f(\epsilon)$ is zero. Moreover, the situation in which $f(\epsilon)$ can be caused to vanish for any extended period of time occurs when the error and its derivatives are all zero. The error and its derivatives will vanish only if α_1 and α_0 are equal to a_1 and a_0, respectively, assuming that $r(t)$ is not zero. Therefore, although $f(\epsilon)$ is explicitly a function of the error between the output of the physical process and the output of the reference model and its derivatives, it is also implicitly a function of the error between α_1 and a_1, and α_0 and a_0. Now, defining

$$\delta_1 = \alpha_1 - a_1 \tag{6.2-11}$$

and

$$\delta_0 = \alpha_0 - a_0 \tag{6.2-12}$$

it is possible to think of $f(\epsilon)$ as a surface in the Euclidian space of $f(\epsilon)$, δ_1 and δ_0. If δ_1 and δ_0 are not both zero, the adjusting mechanism is to be designed so that α_1 and α_0 are adjusted by changing h_1 and h_0. In this approach, the relative rates of adjustment will be selected to cause δ_1 and δ_0 to describe an instantaneous steepest descent trajectory along the surface of $f(\epsilon)$ in the $f(\epsilon)$, δ_1, and δ_0 space. The meaning of instantaneous steepest descent will now be explained.

Besides being functions of the values of δ_1 and δ_0, ϵ and its derivatives are also functions of the time-varying input $r(t)$. Therefore, the shape of the $f(\epsilon)$ surface in the $f(\epsilon)$, δ_1, and δ_0 space will change with time in a manner related to the input $r(t)$ and also to the dynamics of the control system and reference model. At this point a third assumption will be stated.

Assumption 3—The adjusting mechanism will be designed so that it adjusts parameters a_1 and a_0 in Eq. (6.2-8) more rapidly than the rate at which $f(\epsilon)$ changes due to the input $r(t)$.

On the basis of Assumption 3, $f(\epsilon)$ will be treated as a function of δ_1 and δ_0 only for the purpose of the following derivation for the trajectory of steepest descent. It is easiest to describe the idea of steepest descent if one considers that δ_1 and δ_0 are changed by finite increments in finite time intervals. Then, the increments and the intervals will be permitted to approach zero as a limit while their ratio is kept constant. This yields the relative rates of adjustments for δ_1 and δ_0. Suppose that the largest increments by which δ_1 and δ_0 may be changed in each time interval, in the Euclidian space of δ_1 and δ_0, are limited by the relation

$$\left(\Delta\delta_1^{\,2} + \Delta\delta_0^{\,2}\right)^{1/2} = d \tag{6.2-13}$$

in which d is a fixed constant, and $\Delta\delta_1$ and $\Delta\delta_0$ are the incremental changes in δ_1 and δ_0 in a given interval. By definition, the path of steepest descent is the one obtained by choosing the relation between $\Delta\delta_1$ and $\Delta\delta_0$ such that the maximum decrease in $f(\epsilon)$ is obtained at each interval. This may be accomplished by choosing $\Delta\delta_1$ and $\Delta\delta_0$ proportional to the negative of the corresponding elements in the gradient vector of $f(\epsilon)$ at that point in the δ_1 and δ_0 space. The reasoning behind this is that the gradient vector at each point is perpendicular to the iso-$f(\epsilon)$ curve passing through that point. Since it is desired to decrease $f(\epsilon)$, the increments are given the sign opposite to the corresponding elements in the gradient vector. A more extensive discussion of this subject may be found in refs. 30 and 31.

The gradient of $f(\epsilon)$ in the δ_1 and δ_0 space is

$$\nabla f(\epsilon) = \mathrm{i}_1 \frac{\partial f(\epsilon)}{\partial \delta_1} + \mathrm{i}_0 \frac{\partial f(\epsilon)}{\partial \delta_0} \tag{6.2-14}$$

in which i_1 is the unit vector in the δ_1 direction and i_0 is the unit vector in the δ_0 direction. Hence, the increments will be chosen as

$$\Delta\delta_1 = -k \frac{\partial f(\epsilon)}{\partial \delta_1} \tag{6.2-15}$$

and

$$\Delta\delta_0 = -k \frac{\partial f(\epsilon)}{\partial \delta_0} \tag{6.2-16}$$

where k is a common proportionality factor or constant. It is necessary to consider how the incremental changes in δ_1 and δ_0 are to be made. In view of Eqs. (6.2-11) and (6.2-12), it is clear that any changes in δ_1 and δ_0 must be made by changes in a_1 and a_0, since a_1 and a_0 are the fixed constants which represent the desired parameter values. Therefore, the changes to be made in a_1 and a_0 may be written

$$\Delta a_1 = \Delta \delta_1 \tag{6.2-17}$$

and

$$\Delta a_0 = \Delta \delta_0 \tag{6.2-18}$$

Also, since a_1 and a_0 are fixed constants, one can treat Eqs. (6.2-11) and (6.2-12) as merely a shift of the coordinate reference system in the plane of δ_1 and δ_0 or a_1 and a_0 in the space of $f(\epsilon)$, and either δ_1 and δ_0 or a_1 and a_0. This being the case, the equivalent of Eqs. (6.2-15) and (6.2-16) may be written in terms of a_1 and a_0. They are

$$\Delta a_1 = -k \frac{\partial f(\epsilon)}{\partial a_1} \tag{6.2-19}$$

and

$$\Delta a_0 = -k \frac{\partial f(\epsilon)}{\partial a_0} \tag{6.2-20}$$

Looking ahead, if the development were to continue to be based upon Eqs. (6.2-19) and (6.2-20), the resulting equations would require explicit knowledge of a_1 and a_0 and consequently g_1 and g_0. Since the objective here is to develop a method of adaptation which does not require explicit knowledge of g_1 and g_0, another approach must be introduced. This will be the subject of the following discussion.

Suppose for the moment that a_1 and a_0 are treated as fixed, and a_1 and a_0 are to be adjusted to cause δ_1 and δ_0 to approach zero. The same arguments used to obtain Eqs. (6.2-19) and (6.2-20) from Eqs. (6.2-15) and (6.2-16) yield the appropriate incremental values for a_1 and a_0. They are

$$\Delta a_1 = -k \frac{\partial f(\epsilon)}{\partial a_1} \tag{6.2-21}$$

and

$$\Delta a_0 = -k \frac{\partial f(\epsilon)}{\partial a_0} \tag{6.2-22}$$

Next, suppose that δ_1 and δ_0 are arbitrarily small compared to a_1 and a_0 in a given interval of time. Then δ_1 and δ_0 will be changed by adding amounts Δa_1 and Δa_0 to the value of a_1 and a_0 respectively.

The objective is not to change a_1 and a_0, however, but to change a_1 and a_0. Notice that the same change in δ_1 and δ_0 can be obtained by subtracting Δa_1 and Δa_0 from a_0 and a_1 respectively, rather than adding them to a_1 and a_0. Now $f(\epsilon)$ is driven to zero when δ_1 and δ_0 are zero, and this may be done by determining the appropriate increments for a_1 and a_0 and applying the negative of these increments to a_1 and a_0. This reasoning is valid at least as long as δ_1 and δ_0 are considered arbitrarily small. The resulting equations for the increments in a_1 and a_0 are then

$$\Delta a_1 = k \frac{\partial f(\epsilon)}{\partial a_1} \tag{6.2-23}$$

and

$$\Delta a_0 = k \frac{\partial f(\epsilon)}{\partial a_0} \tag{6.2-24}$$

The foregoing discussion and observations, which culminated in Eqs. (6.2-23) and (6.2-24), are key points in this method of synthesizing parameter-adjusting model-reference adaptive controls.

It must be recognized that the increments described by Eqs. (6.2-23) and (6.2-24) do not precisely describe the path of steepest descent. However, it may be argued that when δ_1 and δ_0 are arbitrarily close to zero, then the trajectory will be arbitrarily close to a steepest descent. Even if the path is not one of steepest descent, as long as it is one of descent, the objective of adaptation will be fulfilled. The system operation when δ_1 and δ_0 are not arbitrarily small may be investigated by stability analysis and simulation [12].

Since it is desired to establish rates of adjustment for a_1 and a_0, the time interval between incremental adjustments and the size of the increments will be assumed to approach zero as a limit in the same ratio. The instantaneous rates of adjustment for a_1 and a_0 are then

$$\dot{a}_1 = \frac{\partial f(\epsilon)}{\partial a_1} \tag{6.2-25}$$

and

$$\dot{a}_0 = \frac{\partial f(\epsilon)}{\partial a_0} \tag{6.2-26}$$

where the proportionality factor, k, will be absorbed into parameters $q_2, q_1,$ and q_0. Eq. (6.2-5) was substituted into Eqs. (6.2-25) and (6.2-26) and the indicated partial differentiation was carried out to obtain

$$\dot{a}_1 = -(q_0\epsilon + q_1\dot{\epsilon} + q_2\ddot{\epsilon})\left(q_0 \frac{\partial y}{\partial a_1} + q_1 \frac{\partial \dot{y}}{\partial a_1} + q_2 \frac{\partial \ddot{y}}{\partial a_1}\right) \tag{6.2-27}$$

and

$$\dot{a}_0 = -(q_0\epsilon + q_1\dot{\epsilon} + q_2\ddot{\epsilon})\left(q_0\frac{\partial y}{\partial a_0} + q_1\frac{\partial \dot{y}}{\partial a_0} + q_2\frac{\partial \ddot{y}}{\partial a_0}\right) \tag{6.2-28}$$

In view of Assumption 2 and Eqs. (6.2-9) and (6.2-10), it is possible to write

$$\dot{h}_1 \cong \dot{a}_1 \tag{6.2-29}$$

and

$$\dot{h}_0 \cong \dot{a}_0 \tag{6.2-30}$$

It will be convenient to introduce the notation

$$u_1 = \frac{\partial y}{\partial a_1} \tag{6.2-31}$$

and

$$u_0 = \frac{\partial y}{\partial a_0} \tag{6.2-32}$$

The derivative of y with respect to time may be treated as a trivial partial derivative. If the order of differentiation with respect to time and a_1 is interchanged, it is possible to write

$$\frac{\partial \dot{y}}{\partial a_1} = \frac{\partial^2 y}{\partial a_1\,\partial t} = \frac{\partial^2 y}{\partial t\,\partial a_1} = \frac{\partial u_1}{\partial t} = \dot{u}_1 \tag{6.2-33}$$

These operations are mathematically rigorous since y is assumed to be continuous and a_1 does not vary with time. Doing this for all of the partial derivative terms in Eqs. (6.2-27) and (6.2-28) and treating (6.2-29) and (6.2-30) as equalities, it is possible to write

$$\dot{h}_1 = -(q_0\epsilon + q_1\dot{\epsilon} + q_2\ddot{\epsilon})(q_0u_1 + q_1\dot{u}_1 + q_2\ddot{u}_1) \tag{6.2-34}$$

and

$$\dot{h}_0 = -(q_0\epsilon + q_1\dot{\epsilon} + q_2\ddot{\epsilon})(q_0u_0 + q_1\dot{u}_1 + q_2\ddot{u}_0) \tag{6.2-35}$$

These equations yield the rates at which h_1 and h_0 are to be adjusted. They are the basic equations upon which the adjusting mechanism operates. The values of ϵ and its derivatives have been assumed available or obtainable from the output of the physical process and the reference model. The only unknown quantities in Eqs. (6.2-34) and (6.2-35) are u_1, u_0, and their derivatives. These quantities may be easily determined by applying the parameter influence technique of Meissinger [35]. This technique is demonstrated next.

Consider the differential equation for the reference model as given by Eq. (6.2-2), and take the partial derivative of both sides of the equation with respect to the parameter a_1. This yields

$$\frac{\partial \ddot{y}}{\partial a_1} + a_1 \frac{\partial \dot{y}}{\partial a_1} + \dot{y} + a_0 \frac{\partial y}{\partial a_1} = 0 \qquad (6.2\text{-}36)$$

Again, since a_1 is a constant, it is possible to treat the derivative of y with respect to time as a trivial partial derivative, and hence write Eq. (6.2-36) as

$$\frac{\partial^3 y}{\partial a_1 \delta t^2} + a_1 \frac{\partial^2 y}{\partial a_1 \partial t} + a_0 \frac{\partial y}{\partial a_1} = -\dot{y} \qquad (6.2\text{-}37)$$

Interchanging the order of differentiation in Eq. (6.2-37) and employing the notation given by Eqs. (6.2-31) and (6.2-33),

$$\ddot{u}_1 + a_1 \dot{u}_1 + a_0 u_1 = -\dot{y} \qquad (6.2\text{-}38)$$

This is a non-homogeneous differential equation with u_1 as the variable. The forcing function is the negative of the time derivative of the output of the reference model. The form of this equation is precisely the same as that of the reference model. Thus the value of u_1 and its time derivatives are provided for Eq. (6.2-34) by means of a relatively simple differential equation which has a readily available forcing function.

A similar equation may be obtained for u_0 by taking the partial derivative of Eq. (6.2-2) with respect to the parameter a_0 and performing the operations described above. The result is a differential equation of the form

$$\ddot{u}_0 + a_1 \dot{u}_0 + a_0 u_0 = -y \qquad (6.2\text{-}39)$$

This non-homogeneous differential equation is again precisely the same form as the reference model equation, except that the variable is u_0 and the forcing function is the negative of the reference model output. Eq. (6.2-39) then supplies the required values of u_0 and its derivatives with respect to time in Eq. (6.2-35).

The equations used to implement the adjusting mechanism are now available. Notice that these equations do not require explicit knowledge about g_1 and g_0. Some comments about these equations would appear to be in order. Much of the development was based on three somewhat restrictive assumptions about the relative rates at which g_1 and g_0, the outputs of the reference model and physical process, and the parameters h_1 and h_0 change. It was also assumed that the parameters h_1 and h_0 were quite close to the values required to provide the correct compensation for the physical process. The job of the adjusting mechanism was to reduce the samll error that did exist to zero. Assumptions 1 and 2 are reasonable for a wide variety of practical cases. Assumption 3 may be satisfied by the choice of the values of q_2, q_1, and q_0 in the quadratic error function. If these assumptions are actually fulfilled and if the compensation parameters h_1 and h_0 are very close to the correct value when the system is initially set into operation, there is no reason to believe that the system will not adapt to changes occurring in g_1 and g_0. There is, of course, always interest in what might happen if initially h_1 and h_0 are grossly in error, or if g_1 or g_0 should change by a fairly large discontinuous increment. The answers to these questions are taken up in ref. 12. The results presented there show that this adaptive control system will operate quite well even if the assumptions are violated to a considerable degree. The basic idea which has been developed here has been extended to the case of the general linear physical process with a single input and a single output, to cases in which the physical process is described by certain types of nonlinear differential equations, and to multiple input-multiple output control systems [12]. An example of the application of this technique to a system with a nonlinear physical process and a linear reference model is given next. The block diagram for this system is shown in Fig. 6-6. The equation which describes the physical process is

$$\ddot{c} + \left(g_1 - g_2 c^2\right)\dot{c} + g_0 c = \rho \tag{6.2-40}$$

If the output becomes large enough to cause $g_2 c^2$ to be greater than g_1, the physical process may become unstable. The objective in the design of the control system will be to prevent this from occurring. The differential equation which describes the reference model is chosen to be

$$\ddot{y} + a_1 \dot{y} + a_0 y = r \tag{6.2-41}$$

To have the control system assume this form, the feedback compensation is selected as

$$\eta = h_1 \dot{c} + h_2 c^2 \dot{c} + \gamma_0 c \tag{6.2-42}$$

Then the differential equation relating input to output for the control system may be written

$$\ddot{c} + \left(a_1 - a_2 c^2\right)\dot{c} + a_0 c = r \tag{6.2-43}$$

in which

$$a_2 = (g_2 + h_2) \tag{6.2-44}$$

$$a_1 = (g_1 + h_1) \tag{6.2-45}$$

and

$$a_0 = (g_0 + h_0) \tag{6.2-46}$$

As usual, it will be assumed that g_2, g_1, and g_0 vary slowly in some unknown manner with time. The objective of the adjusting mechanism will be to adjust h_2, h_1, and h_0 so that α_2 equals zero, α_1 equals a_1, and α_0 equals a_0 at all times regardless of the values assumed by g_2, g_1, and g_0. The following error function is used as the basis of the adjusting equations:

$$f_1(\epsilon) = \frac{1}{2}\left(q_0\epsilon^2 + q_1\dot{\epsilon}^2 + q_2\ddot{\epsilon}^2\right) \tag{6.2-47}$$

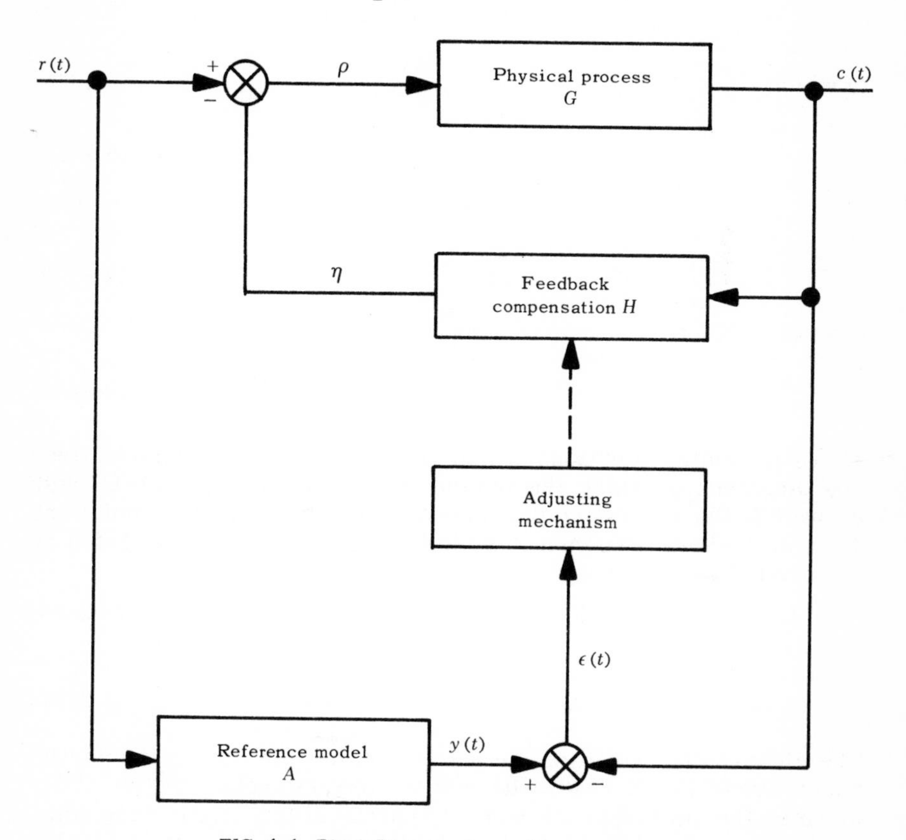

FIG. 6-6. Block Diagram for nonlinear system.

To justify taking partial derivatives with respect to a_2, Eq. (6.2-41) is written as

$$\lim_{a_2 \to 0}\left[\ddot{y} + \left(a_1 - a_2 y^2\right)\dot{y} + a_0 y\right] = r \tag{6.2-48}$$

Now it is possible to obtain the equations for the rates at which h_2, h_1, and h_0 must be adjusted. They are

$$\dot{h}_2 = \dot{a}_2 = \frac{\partial f(\epsilon)}{\partial a_2} = -\left(q_0 \epsilon u_2 + q_1 \dot{\epsilon} \dot{u}_2 + q_2 \ddot{\epsilon} \ddot{u}_2\right) \tag{6.2-49}$$

$$\dot{h}_1 = \dot{a}_1 = \frac{\partial f(\epsilon)}{\partial a_1} = -\left(q_0 \epsilon u_1 + q_1 \dot{\epsilon} \dot{u}_1 + q_2 \ddot{\epsilon} \ddot{u}_1\right) \tag{6.2-50}$$

and

$$\dot{h}_0 = \dot{a}_0 = \frac{\partial f(\epsilon)}{\partial a_0} = -\left(q_0 \epsilon u_0 + q_1 \dot{\epsilon} \dot{u}_0 + q_2 \ddot{\epsilon} \ddot{u}_0\right) \tag{6.2-51}$$

in which

$$u_2 = \frac{\partial y}{\partial a_2} \qquad \text{[see Eq. (6.2-48)]} \tag{6.2-52}$$

$$u_1 = \frac{\partial y}{\partial a_1} \tag{6.2-53}$$

and

$$u_0 = \frac{\partial y}{\partial a_0} \tag{6.2-54}$$

The differential equations which yield u_1, u_0 and their derivatives are obtained by taking the partial derivative of Eq. (6.2-41) with respect to the associated parameter, performing the manipulations described above and then substituting Eq. (6.2-53) or (6.2-54) as required. The result is

$$\ddot{u}_1 + a_1 \dot{u}_1 + a_0 u_1 = -\dot{y} \tag{6.2-55}$$

and

$$\ddot{u}_0 + a_1 \dot{u}_0 + a_0 u_0 = -y \tag{6.2-56}$$

The differential equation for u_2 is obtained by taking the partial derivative of Eq. (6.2-48) with respect to a_2, interchanging the order in which the limiting process and the partial differentiation are done, performing the differentiation, taking the limit, performing the usual manipulations, and finally using Eq. (6.2-52). The result is

$$\ddot{u}_2 + a_1 \dot{u}_2 + a_0 u_2 = y^2 \dot{y} \tag{6.2-57}$$

This then completes all the equations necessary to implement the adaptive portion of the system described in this example.

It is also possible to combine the learning model approach [30, 31] with the model referenced approach to cause the reference model to "learn" certain system parameters while acting as a reference model with respect to other parameters. Such a system might be useful if one wants a control system to have a specified damping characteristic, but it is undesirable, or impossible to control the gain and undamped natural frequency as the dynamics of the physical process change. These generalizations and extensions of the basic concept presented above are contained in ref. 12. In the latter cases, there has not been sufficient analog simulation or analytical stability analysis to evaluate these extensions adequately.

Method 2—Based on Lyapunov's Direct Method [42, 43] Attempts have been made to use the stability considerations of Lyapunov's direct method to derive equations describing the adjusting mechanism for both the learning model approach [42] and the model referenced approach [43]. In the formation of the equations for the adjusting mechanism in a model referenced system, a differential equation which describes the error between the control system and reference model is formulated first. The question then is one of formulating parameter-adjusting equations which assure that the differential equation describing the error is asymptotically stable. To do this, a positive definite Lyapunov function is formulated for the error equation. The adjusting mechanism equations are then selected to cause the time derivative of the Lyapunov function to be negative-definite. By so doing, the error function is guaranteed to be asymptotically stable, and hence the objective of adaptation is achieved. If one considers the geometric proof of Lyapunov's direct method stability theorem (see Chap. 5, Sec. 5.1), considerable similarity appears between this approach and the one described as Method 1. The actual adjusting mechanism equations for the present case are, however, somewhat different. This is due primarily to the manner in which the investigators formulated the adaptive system block diagram.

Although this concept is definitely promising, it will not be presented in detail here since it appears to require further investigation.

Method 3—Based on Partial Derivative of Error Functions This method bears considerable similarity to Method 1 described above; however, there are significant differences. This particular version of the technique was evolved from work that started in 1958 and has been reported in numerous publications [49, 50, 51, 52]. The most recent advances made by these investigators are presented in refs. 38 and 39. This work will be presented here in summary form. The reader interested in the mathematical details is referred to ref. 38. Figure 6-4 supplies the block diagram for the system being considered. An even function of the error $f(\epsilon)$ is the basis for the adjusting mechanism. The only error function considered in detail was

$$f(\epsilon) = \frac{1}{2}\epsilon^2 \tag{6.2-58}$$

After a somewhat lengthy argument, the investigators conclude that each adjustable parameter p_k should be changed continuously at a rate proportional to the negative of the partial derivative of $f(\epsilon)$ with respect to p_k. The equation describing the rate of parameter adjustment is

$$\dot{p}_k = -S_{a_k} f'(\epsilon) \frac{\partial \epsilon}{\partial p_k} \tag{6.2-59}$$

The S_{a_k} are constants to be determined as part of the design process. For the case in which $f(\epsilon)$ is given by Eq. (6.2-58), $f'(\epsilon) = \epsilon$. The function $\epsilon(t)$ is easily determined as the difference between the control system response and the model response. The only problem remaining is that of finding a way to compute $\partial\epsilon/\partial p_k$ in terms of signals available within the system. Referring again to Fig. 6-4. it is seen that $\epsilon(t) = c(t) - y(t)$ and, since $y(t)$ is not a function of the p_k, it follows that $\partial\epsilon/\partial p_k = \partial c/\partial p_k$. It is not possible to determine $\partial c/\partial p_k$ directly. However, an acceptable and useful approximation to it can be obtained as follows.

The differential equation of the control system which relates $c(t)$ to $r(t)$ is of the form

$$\sum_{i=0}^{n} a_i \frac{d^i c}{dt^i} = \sum_{j=0}^{m} b_j \frac{d^j r}{dt^j} \tag{6.2-60}$$

where a_i b_j and $c(t)$ are functions of the parameters, p_k. Partial differentiation of Eq. (6.2-60) with respect to p_k yields

$$\sum_{i=0}^{n} a_i \frac{\partial}{\partial p_k}\left(\frac{d^i c}{dt^i}\right) = \sum_{j=0}^{m} \frac{\partial b_j}{\partial p_k} \frac{d^j r}{dt^j} - \sum_{i=0}^{n} \frac{\partial a_i}{\partial p_k} \frac{d^i c}{dt^i} \tag{6.2-61}$$

If the p_k are changing at a very slow rate, then $dp_k/dt \approx 0$ and the order of the differentiation in the left hand side of Eq. (6.2-61) can be interchanged [38] yielding

$$\sum_{i=0}^{n} a_i \frac{d^i}{dt^i}\left(\frac{\partial c}{\partial p_k}\right) = \sum_{j=0}^{m} \frac{\partial b_j}{\partial p_k} \frac{d^j r}{dt^j} - \sum_{i=0}^{n} \frac{\partial a_i}{\partial p_k} \frac{d^i c}{dt^i} \tag{6.2-62}$$

as the equation to be solved to determine the desired partial derivative, $\partial c/\partial p_k$. Examination of Eq. (6.2-62) reveals that it requires the use of the control system characteristic operator

$\sum_{i=0}^{n} a_i\, d^i/dt^i$ and the coefficient partial derivatives $\partial a_i/\partial p_k$ and $\partial b_j/\partial p_k$. It is precisely the lack of detailed information about this operator and these coefficients which required the use of an adaptive technique in the first place. On the basis of qualitative arguments, it is possible to obtain an approximation to them by considering the basic concept of the operation of the model reference technique. The basic compensation elements are to be designed so that, if the parameters p_k are properly adjusted, the control system response to command inputs will be essentially identical to that of the model response. To achieve this, the dynamic characteristics of the model must therefore be a fairly good approximation to those of the control system whenever the overall adaptive system is working satisfactorily. Therefore, the *known* model characteristic operator and coefficients can be used in Eq. (6.2-62) instead of those of the control system to obtain an equation for use in determining an approximation to $\partial c/\partial p_k$. As long as the difference between the system and model coefficients is not excessive, the approximation is valid. The region of validity is large enough to make this version of the technique practical and effective in applications such as adaptive autopilots for aerospace vehicles.

Discussion and Comparison of the Parameter Adjustment Methods It is now possible to compare Method 1 and Method 3. First of all, Method 1 adjusts the parameters of interest in proportion to the negative of the corresponding elements of the gradient vector of the error function in the parameter space. Method 3 uses somewhat more heuristic arguments to conclude that a given parameter should be adjusted in accordance with the negative of the partial derivative of the error function multiplied by some positive constant. If the constant in the latter technique were equated to unity for all parameters, then the two techniques would differ only by the selected error functions.

In both Method 1 and Method 3, the error function is a very critical factor in the operation of the adaptive portion of the system. Method 1 uses functions of the error and its derivatives and considers functions of the form

$$f(\epsilon) = \left(\sum_{i=0}^{n} q_i \frac{d^i \epsilon}{dt^i} \right)^2$$

and

$$f(\epsilon) = \sum_{i=0}^{n} q_i \left(\frac{d^i \epsilon}{dt^i} \right)^2$$

for nth order systems. Method 3 considers only functions of the form

$$f(\epsilon) = \frac{1}{2}\epsilon^2$$

In comparing the results reported in ref. 38 and ref. 12, it was noted that the parameters adjusted faster in response to a step input for the error functions used in Method 1 than those used in Method 3. This is apparently due to the lead effect produced by including the derivatives of $\epsilon(t)$ in the error function. While this improved speed of adaptability certainly seems desirable, it should be kept in mind that inclusion of such derivative terms complicates the parameter adjustment mechanism somewhat.

It appears that there might be advantages for both of these methods if more general positive-definite quadratic functions of the error and its derivatives were considered. For instance, a function of the form

$$f(\epsilon) = \sum_{i=0}^{n} \sum_{j=0}^{n} a_{ij} \left(\frac{d^i\epsilon}{dt^i}\right)\left(\frac{d^j\epsilon}{dt^j}\right)$$

Although it is certainly true that such an error function would complicate the adaptive system mechanization, it may yield results to justify its use in some applications. At least it would appear to bear further investigation, as does the idea of basing the adjusting mechanism equations on stability concepts as described in Method 2.

In all three approaches, it is necessary to have a non-zero input for parameter adjustments to occur. In the absence of inputs, sizable errors between the actual values of the adjustable parameters and their desired values can result. If the parameter adjustment is slow after such errors build up, then the control system output may deviate severely from the desired output. Simulation of simple systems using Method 1 have shown that parameter adjustment can take place within 10% of the basic time constant of the reference model.

Neither Method 1 nor Method 3 suffers any degradation in performance when the input has noise associated with it. This is because the noise and the signal appear the same to the adaptive loop. For noise introduced inside the control system loop, the problem is quite different and may cause difficulties. At the present time there is insufficient data on this subject to support any conclusion.

A number of additional parameter adjustment techniques are reported in the literature. Although it appears that the continuous-acting systems described above have much in their favor, some applications may call for discrete parameter adjustment techniques. Some apparently promising results have been reported in connection with discrete parameter adjustment based on a truncated integral-of-error-squared error function [10]. In general, however, it appears that error functions which do not involve integration are preferable since they do not introduce additional phase shift into the parameter adjusting loop.

6.2.2 Parameter Perturbation Techniques [32, 33]

The method to be described next could be grouped with parameter adjustment methods; however, it is significantly different from those described above, and hence is treated as a separate approach. Although it will not be described in mathematical detail here, the general concept is presented.

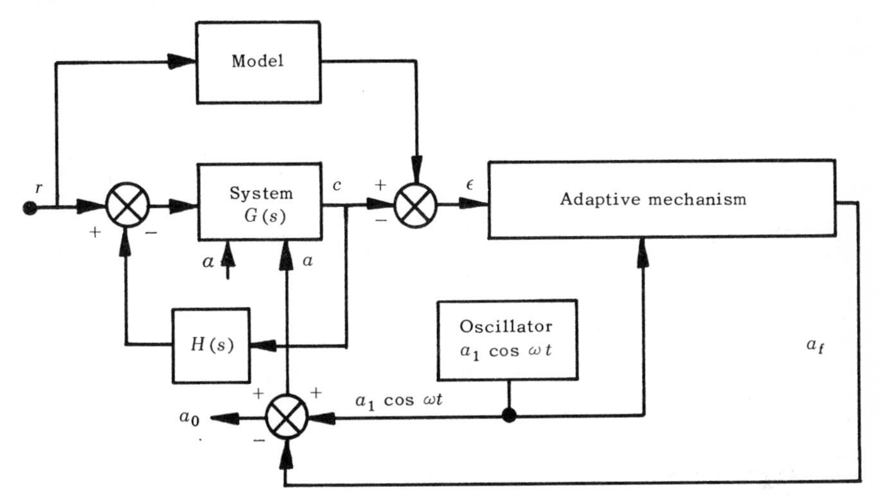

FIG. 6-7. Basic configuration of parameter-perturbation adaptive system.

The block diagram of the basic parameter perturbation approach to model reference adaptive control systems is shown in Fig. 6-7. As in the other approaches the output from the reference model is subtracted from the system output in order to obtain the error, e. A function of the error is generated. Among the functions considered are

$$f(e) = e^2$$

$$f_1(e) = |e|$$

and

$$f_2(e) = |e|^n$$

One parameter a in the system is defined by

$$a = a_0 + a_f + a_1 \cos \omega t$$

where a_0 is the basic system parameter, a_f is the correction applied by the adaptive portion of the system, and $a_1 \cos \omega t$ is a sinusoidal perturbation signal used to create a continuous error. The sinusoidal perturbation signal is also supplied to the adaptive mechanism. Since the parameter "a" is being perturbed in a sinusoidal manner, the error e will also exhibit an oscillatory component of the same frequency. The amplitude and sign of a_f is chosen to reduce the short time average of the error function $f(e)$.

It is possible to synthesize a system which will adjust a number of parameters. This is accomplished by applying perturbation signals of different frequencies to each parameter adjusted. A separate adaptive mechanism loop is required for each parameter. It has been asserted [33] that if each of the loops has the same gain, then the adjustment of the parameters will tend to follow a steepest descent trajectory. To improve performance, it may be desirable to incorporate additional filtering and nonlinear gain terms in the adaptive mechanism.

A primary objection to this approach is that the perturbation signal appears in the output of the system. It is possible to avoid this characteristic by using a model that has either the same or a similar form as the system. In this case, the perturbation signal is applied to the model parameter that corresponds to the system parameter to be controlled. Figure 6-8 shows the block diagram for this approach.

The parameter perturbation technique may be applied to a number of other system configurations. Although the adaptive loop may at first appear to be rather complicated in some cases, it is possible to perform the required computations with a relatively small number of components [32, 33].

6.2.3 Control Signal Augmentation Techniques

In the two approaches to model reference adaptive control systems described above, the adaptation or system adjustment was accomplished by adjusting parameters in the control system. The approach described below considers the problem from a somewhat different basic viewpoint. Again referring to Fig. 6-3, the operation of the model reference adaptive control system can be described in the following manner. The input signal $r(t)$ is applied to both the reference model and the controller. The controller also has as inputs the outputs of both the process and the reference model. On the basis of these inputs, the controller produces an augmented control signal $m(t)$ which is applied to the process. The object of

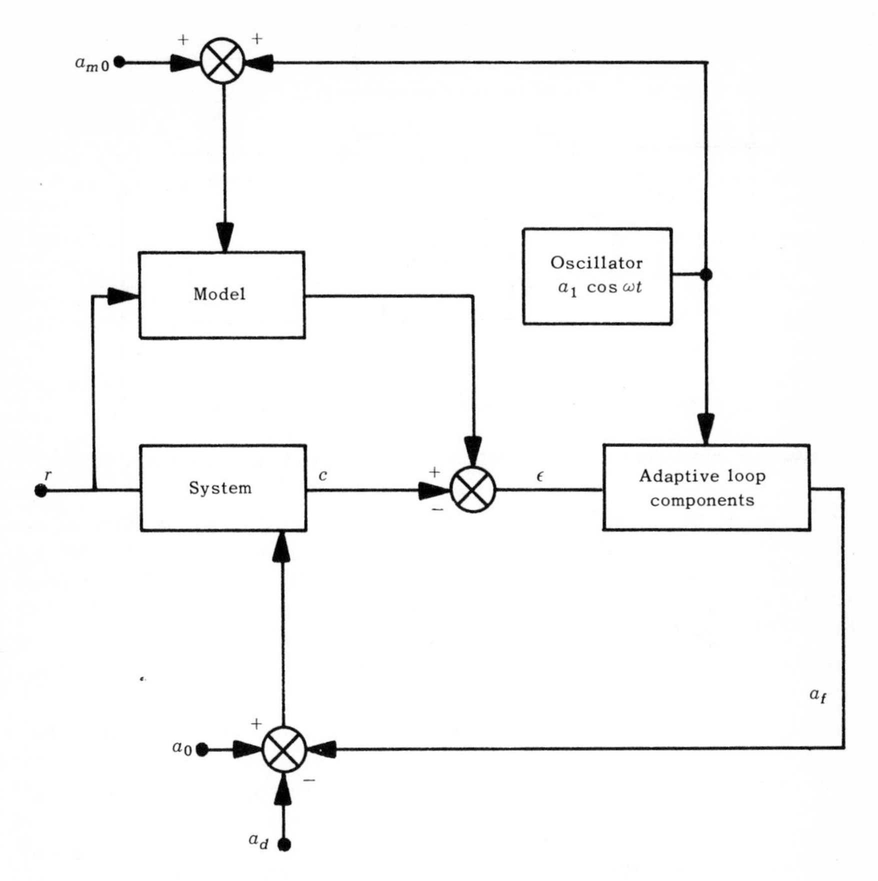

FIG. 6-8. System-adaptive parameter-perturbation system.

the adaptive controller is to produce an $m(t)$ which will cause the output of the process to be the same as the output of the reference model. When described in the degree of generality used thus far in the disucssion, this approach is easy to understand. The difficulty is in the design of the controller and the selection of the model. Although numerous approaches to the controller design might be considered, the discussion here will be limited to one approach that appears to have merit.

This method bases the controller design on stability criteria derived from Lyapunov's direct method [18]. The discussion is based on the block diagram of Fig. 6-9. It should be noted that it is necessary only to regroup the elements in this block diagram to obtain the form shown in Fig. 6-3. With reference to Fig. 6-9, the lower portion of the diagram represents a standard feedback control system, except for an additional summing device in the loop just before the nonlinear gain element. The nonlinear gain

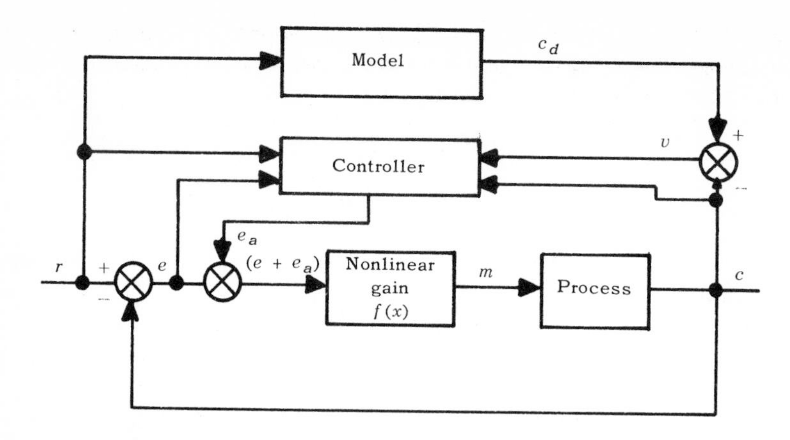

FIG. 6-9. Block diagram for controller design based on Lyapunov's direct method.

element is inclueded merely to make the situation as general as possible. The adaptive characteristic is achieved by generating an additional error signal e_a and injecting this into the system loop through the additional summing device. The nonlinear gain $f(x)$ is restricted to form

$$\text{(a)}\quad f(0) = 0 \tag{6.2-63}$$

$$\text{(b)}\quad xf(x) > 0\,;\; x \neq 0\,. \tag{6.2-64}$$

The output of the process is determined by an equation of the form

$$\frac{d^n c}{dt^n} + \sum_{i=1}^{n-1} a_i(t)\,\frac{d^i c}{dt^i} = k(t) f(e + e_a) \tag{6.2-65}$$

The coefficients in the above equation are assumed to be of the form

$$a_i(t)' = \bar{a}_i + a_i(t)\,,\quad a_{i(\min)} \leq a_i(t) \leq a_{i(\max)}\,,$$

$$(i = 1, 2, \ldots, n-1) \tag{6.2-66}$$

$$k(t) = \bar{k} + K(t)\,,\quad K_{\min} \leq K(t) \leq K_{\max}$$

in which $\bar{a}_i$ and $\bar{k}$ are constants.

The differential equation which describes the model could be either the same order or one order lower than the closed-loop control system that includes the process. For the specific derivation considered below [18], the model will be selected as one order lower. Its input-output relation is described by

$$\frac{d^{n-1} c_d}{dt^{n-1}} + \sum_{i=0}^{n-2} b_i\,\frac{d^i c_d}{dt} = Lr \tag{6.2-67}$$

in which L is a gain term and the b_i are constants. In the investigations to date [18], the following restrictions are assumed:

(a) Input is a step function $R\,u(t)$;

(b) Steady-state gain is unity for both the model and the closed-loop system around the process.

As a consequence of these assumptions, in the steady-state condition $c = R$, $e = 0$, and $e_a = 0$. Also, as a consequence of (b),

$$b_0 = L \tag{6.2-68}$$

The difference between the model output and process output is defined by

$$v = c_d - c. \tag{6.2-69}$$

Next, if:

(a) Equation (6.2-67) is differentiated with respect to time;

(b) Equation (6.2-65) is subtracted from the result of (a);

(c) $\dfrac{d^{n-1}c_d}{dt^{n-1}}$ as obtained from Eq. (6.2-67) is substituted in the result of (b);

(d) The result of (c) is rearranged and Eqs. (6.2-68), (6.2-69), and (6.2-66) are substituted in the result, and it is noted that the input is a step function;

then, by defining

$$v_i = \frac{d^{i-1}v}{dt}; \quad (i = 1, 2, \ldots, n) \tag{6.2-70}$$

it is possible to write the result of (d) as a vector differential equation of the form

$$\dot{\mathbf{v}} = \mathbf{M}\mathbf{v} - \mathbf{y}(t) \tag{6.2-71}$$

in which

$$\mathbf{v} = \begin{bmatrix} v_1 \\ v_2 \\ \cdot \\ \cdot \\ \cdot \\ v_n \end{bmatrix}, \qquad \dot{\mathbf{v}} = \begin{bmatrix} \dot{v}_1 \\ \dot{v}_2 \\ \cdot \\ \cdot \\ \cdot \\ \dot{v}_n \end{bmatrix} \tag{6.2-72}$$

$$\mathbf{M} = \begin{bmatrix} 0 & 1 & 0 & 0 & \cdot & 0 & 0 \\ 0 & 0 & 1 & 0 & \cdot & 0 & 0 \\ \cdot & \cdot & \cdot & \cdot & \cdot & \cdot & \cdot \\ 0 & 0 & 0 & 0 & \cdot & 0 & 1 \\ -m_0 & -m_1 & -m_2 & -m_3 & \cdot & -m_{n-2} & -m_{n-1} \end{bmatrix} \tag{6.2-73}$$

$$
= \begin{bmatrix} 0 & 1 & \cdot & 0 & 0 \\ 0 & 0 & \cdot & 0 & 0 \\ \cdot & \cdot & \cdot & \cdot & \cdot \\ 0 & 0 & \cdot & 0 & 0 \\ -b_0(\bar{a}_{n-1} - b_{n-2}) & -\left[b_1(\bar{a}_{n-1} - b_{n-2}) + b_0\right] & \cdot & -\left[b_{n-2}(\bar{a}_{n-1} - b_{n-2}) + b_{n-3}\right] & -\bar{a}_{n-1} \end{bmatrix}
$$

$$
\mathbf{y}(t) = \begin{bmatrix} 0 \\ 0 \\ \cdot \\ 0 \\ y_0 \end{bmatrix} \tag{6.2-74}
$$

and

$$
\begin{aligned} y_0 = {} & b_0 a_{n-1}(t) v_1 + b_1 a_{n-1}(t) v_2 + \ldots + b_{n-2} a_{n-1}(t) v_{n-1} \\ & + a_{n-1}(t) v_n + b_1\left[a_{n-1}(t) - b_{n-2}\right] - a_1(t) + b_0 \; c^{(1)} \\ & + \ldots + b_{n-2}\left[a_{n-1}(t) - b_{n-2}\right] - a_{n-2}(t) + b_{n-3} \; c^{(n-2)} \\ & - \left[a_{n-1}(t) - b_{n-2}\right](L)e + k(t) f(e + e_a) \end{aligned} \tag{6.2-75}
$$

Equation (6.2-71) represents the differential equation that describes the error between the model and the process. It is a nonautonomous free dynamic system (see definitions in Chapter 5) and by hypothesis [18] $\mathbf{M}\mathbf{v} - \mathbf{y}(t)$ is bounded for all $\mathbf{v}$ and $t > 0$, and further for $\mathbf{v} = 0$ then $\dot{\mathbf{v}} = 0$. If the controller is designed so that this differential equation is asymptotically stable in the large for all step inputs, then the error between the model output and the process ouput will go to zero, i.e., $\mathbf{v} \to 0$. Criteria based on Lyapunov's direct method are used to accomplish the design procedure. A quadratic Lyapanov function $V(\mathbf{v})$ is defined as

$$
V(\mathbf{v}) = \mathbf{v}' \mathbf{P} \mathbf{v} \tag{6.2-76}
$$

in which $\mathbf{P}$ is a positive definite symmetric matrix. A positive definite quadratic form as given by Eq. (6.2-76) will exhibit the properties required by the theorems of Chapter 5. If Eq. (6.2-76) is differentiated with respect to time and Eq. (6.2-71) is substituted in the result, one obtains

$$
\dot{V}(\mathbf{v}) = 2\mathbf{v}' P M \mathbf{v} - 2\mathbf{v}' P \mathbf{y} . \tag{6.2-77}
$$

If all the eigenvalues of M have negative real parts then it is possible to choose the elements of P in conjunction with the elements of a positive definite diagonal matrix Q such that

$$2PM = -Q \tag{6.2-78}$$

Substituting Eq. (6.2-78) in Eq. (6.2-77) yields

$$\dot{V}(\mathbf{v}) = -\mathbf{v}'Q\mathbf{v} - 2\mathbf{v}'\mathbf{P}\mathbf{y} \tag{6.2-79}$$

The first term on the right side of Eq. (6.2-79) is negative-definite and will satisfy the other conditions for $\dot{V}$ in the appropriate theorem of Chapter 5. The second term may be written

$$-2\mathbf{v}'\mathbf{P}\mathbf{y} = -2\left(p_{1n}v_1 + p_{2n}v_2 + \ldots + p_{nn}v_n\right)y_0 \tag{6.2-80}$$

If

$$y_0 = \gamma \operatorname{Sgn}\left(p_{1n}v_1 + p_{2n}v_2 + \ldots + p_{nn}v_n\right), \quad 0 \le \gamma < \infty \tag{6.2-81}$$

where

$$\operatorname{Sgn}(x) = \begin{cases} 1 & \text{if} \quad x > 0 \\ 0 & \text{if} \quad x = 0 \\ -1 & \text{if} \quad x < 0 \end{cases} \tag{6.2-82}$$

then

$$-2\mathbf{v}'\mathbf{P}\mathbf{y} = -2\gamma \le 0 \tag{6.2-83}$$

and consequently, Eq. (6.2-79) will be negative-definite. Eqs. (6.2-76) and (6.2-79) satisfy the conditions of Theorem 5-9 of Chap. 5 if Eq. (6.2-81) holds. Thus, if a controller is designed such that its output e_a determines the sign for y_0 required to satisfy Eq. (6.2-81), then the convergence of the process output to the model output is assured.

Systems of the type described above have been synthesized and simulated, with satisfactory operation obtained [18]. The number of specific applications which may be found for this procedure are probably quite limited, however, due to several basic premises of the design. To be specific, the restriction to step inputs, unity steady-state gain, and the general form of model and process to which it is applicable are in themselves severly limiting. The approach is interesting, however, in terms of developing similar lines of reasoning for other applications. Particularly interesting is the direct application of stability criteria to the design procedure.

6.2.4 Commentary on Model Reference Approach to Adaptive Control Systems

A number of specific schemes have been suggested in the literature which fall into the category of model reference adaptive control systems. In the material presented above, an attempt was made to categorize these techniques further, and select some representative approaches. None of the available techniques is universally applicable, and very few have enjoyed any extensive use. The selection of the proper one from those available depends very strongly on its application. In almost every case, a compromise must be made between complexity, reliability, cost, and system performance. There is no universally applicable set of rules to guide the relative weighting of these factors.

6.3 Optimal-Adaptive Methods

Parameter adjustment techniques are applicable primarily when adjustments of a few parameters will perform the necessary adaptation. More adjustments involve an increase in the amount of computations. As more computations are allowed and as the state of the computer art is advanced, we seek still better methods to improve upon the accuracy of the system. With regard to this, optimal-adaptive methods are investigated. A further incentive for looking into these methods is the possibility of considering amplitude constraints on the control variable.

Basically, optimal-adaptive methods solve some optimization problem of synthesizing the control signal on the assumption that the process parameters and the states are known. Since these quantities are unknown to some extent in an adaptive task, both state estimation and process identification are vital parts in the total scheme. Also, the reference input is required for optimization. In many problems, the reference input is known for the duration of the control operation, as, for example, in a trajectory-following scheme. In this case, the problem reduces to a regulator problem. If the reference input is not known a priori, a predictor must be provided. A schematic diagram for the optimal-adaptive methods is shown in Fig. 6-10. Some of the early contributors in this area of optimal adaptive methods are Kalman [23], Merriam [36], Braun [8], Meditch [34], and Hsieh [20]. Their separate proposals will be reviewed later when an attempt will be made to compare their methods with those in this chapter.

The discussion will be organized in the following manner. First, a general philosophical basis for our optimal-adaptive scheme will be given. Then, methods will be given to solve linear processes and quadratic performance criterion with no inequality constraints on the control variable. The amplitude of the control variable is restrained indirectly by means of a "penalty" function. These methods are then extended to the case with inequality constraints on the control variable. Here, two possible gradient methods are given. Although identification and state estimation are integral parts of optimal-adaptive methods, they are given separately in Sec. 6.4.

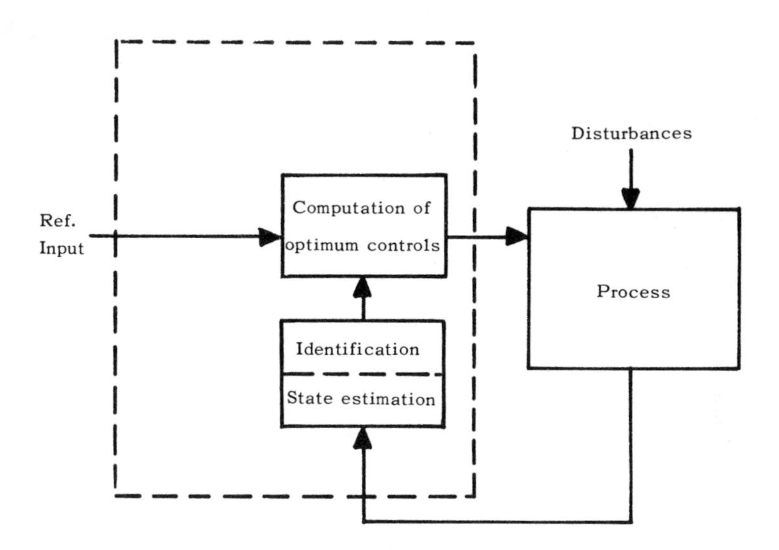

FIG. 6-10. Optimum-adaptive scheme.

To simplify the exposition, the single-input, single-output situation will be discussed. Extensions can be given to the multipole case. Also, the discrete case is discussed because we envision a digital computer to perform the control signal synthesis.

6.3.1 General Philosophy

We envision using an optimal-adaptive control to keep the process output close to some desired trajectory. This operation is to be maintained over some time interval we shall designate as the *operation interval.* In other words, we desire to minimize the performance criterion

$$P = \sum_{k=0}^{N_1} \frac{1}{2} \left[c_d(k) - c(k)\right]^2 \tag{6.3-1}$$

where $c_d(k)$ = desired trajectory or reference input
$c(k)$ = process output
N_1 = number of sampling intervals in the operation interval.

The optimization of Eq. (6.3-1) is impractical for several reasons. First, open-loop control, with its undesirable consequences, ensues. Second, the process is uncertain for time into the future. Also, if a reference input predictor is required, the reference input is known for only a short time into the future. Third, the on-line numerical computational requirements may be too large. As a result, it is more practical to perform the following optimization periodically. We choose a fixed time interval into the future, designated *optimization interval,* and perform a minimization over this interval. Therefore, instead of Eq. (6.3-1) we minimize periodically

$$J = \sum_{j=k+1}^{k+N} \frac{1}{2} \left[c_d(j) - c(j)\right]^2 \tag{6.3-2}$$

where N = number of sampling intervals in the optimization interval
k = present time

The time relation of the intervals under consideration is given in Fig. 6-11.

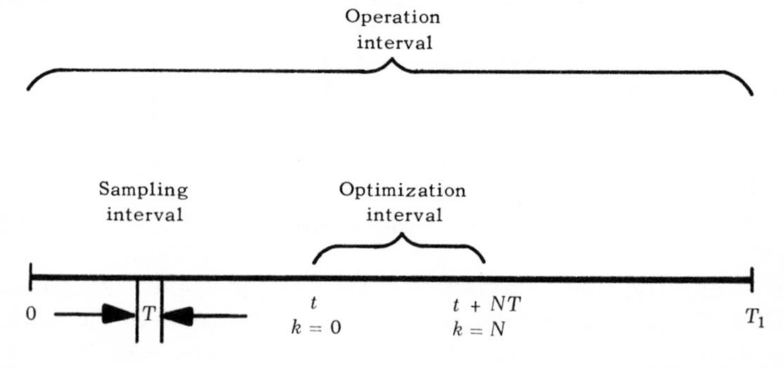

FIG. 6-11. Time relation of intervals under consideration.

The idea of adaptive controls originated from a motive of emulating desirable human characteristics. Therefore as the general philosophy, we offer a human analogy. A similar discussion was first presented by Merriam [36].

A human faced with a control problem, such as driving an automobile, has the problem of optimally selecting the next decision in a multistage decision process. This decision will be based on the present state and the knowledge (maybe intuitive) of the process response (automobile behavior). A human will decide on a particular control on the basis of considerations given over a relatively short time into the future. For example, road conditions may change, and the human will not apply the same control on a rough road as he would on an icy road. With knowledge of the desired path over a short time into the future (optimization interval), and a knowledge of the vehicle response, a human can apply proper control efforts to the steering wheel. The criterion given by Eq. (6.3-2) then replaces the subjective evaluation performed by a human.

Although minimizing Eq. (6.3-2) may lead to suboptimal policies with regards to Eq. (6.3-1), it may be the only proper criterion to apply in a given situation. Inherent in the above discussion were state estimation and process identification. These functions are performed by a human through observation and testing vehicle response. Just as a human can adapt to different vehicles (different responses) and also different changes in the same vehicle (road conditions, tire-blowout, etc.), an adaptive control must be able to perform these tasks if it is to have the finer human capabilities.

6.3.2 Linear Process and Quadratic Criterion Case

In this section an algorithm will be given to treat linear processes and the criterion given by Eq. (6.3-2). The methods which have been widely discussed in the past generally treat this particular important case. The study of this particular case is generally attributed to Merriam [36] and Kalman [24]. Before proceeding to derive the algorithm we shall describe the process in state vector form.

We assume that the changes occurring in the process during an optimization interval is small. This allows us to use constant-coefficient differential equations which in turn will relieve the computational requirements.

The linear process is then described by

$$\begin{aligned} \dot{\mathbf{x}}(t) &= \mathbf{A}\mathbf{x}(t) + \mathbf{b}m(t) \\ c(t) &= \langle \mathbf{h}, \mathbf{x}(t) \rangle \end{aligned} \tag{6.3-3}$$

where $\langle\ ,\ \rangle$ denotes scalar product formed in the usual manner by multiplying the corresponding components of the respective vectors and summing, that is,

$\mathbf{A} = n \times n$ matrix
$\mathbf{b}, \mathbf{h} = n$ vectors

When digital computers are employed as controllers, the control signal will have the appearance of the staircase signal shown in Fig. 6-12. In mathematical notation,

$$m(\tau) = m(k) \qquad (k-1)T \le \tau \le kT \qquad k = 1, 2, \ldots N$$

In this case, the process output is given at discrete instants of time by the following linear difference equation.

$$\begin{aligned} \mathbf{x}(k) &= \phi \mathbf{x}(k-1) + \gamma m(k) \\ (k) &= \langle \mathbf{h}, \mathbf{x} \rangle \end{aligned} \tag{6.3-4}$$

where

$$\begin{aligned} \phi &= n \times n \text{ matrix} \\ \gamma &= n \text{ vector} \end{aligned}$$

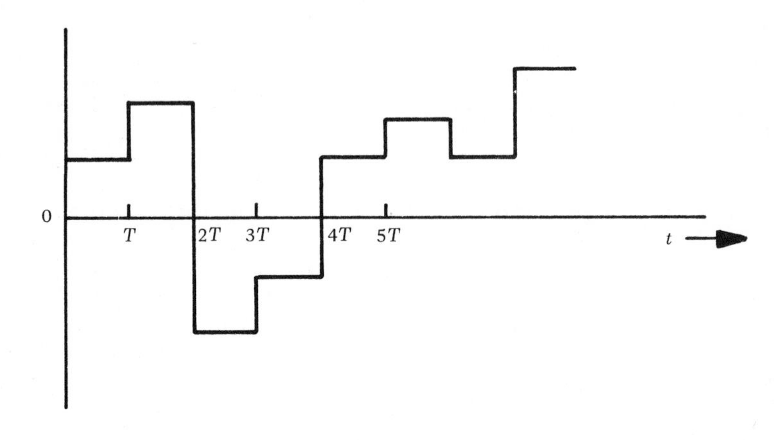

FIG. 6-12. Staircase signal.

In this section, we will constrain the amplitude of $m(k)$ indirectly by using a penalty function on the criterion given by Eq. (6.3-2). Therefore, at every k we minimize the following criterion.

$$J = \sum_{j=k+1}^{k+N} \frac{1}{2}\left(c_d(j) - c(j)\right)^2 + \frac{a}{2} m(j)^2 \tag{6.3-5}$$

where a is the penalty on the use of control.

We will use the calculus of variations and/or the maximum principle approach.* Chang [9] and Katz [25] investigated the maximum principle for the discrete case giving only necessary conditions. The algorithm to be derived is essentially that derived by Kipiniak [27].

The variational problem is to minimize Eq. (6.3-2) subject to Eq. (6.3-4). Using Lagrange multipliers, the constrained functional to be minimized is

$$J = \sum_{j=k+1}^{k+N} \frac{1}{2}\left(c_d(j) - \mathbf{h}^* \mathbf{x}(j)\right)^2 + \frac{a}{2} m(j)^2$$
$$+ \mathbf{p}^*\left(\mathbf{x}(j) - \mathbf{\Phi}\mathbf{x}(j-1) - \gamma m(j)\right)$$

*See Chapter 7.

The necessary condition states that the total differential of J vanishes for independent differentials of $\mathbf{x}(j)$, $\mathfrak{m}(j)$, and $\mathbf{p}(j)$. Taking the differential we get

$$
\begin{aligned}
dJ = & \sum_{j=k+1}^{k+N-1} d\mathbf{x}^*(j)\left\{\left(\mathbf{h}^*\mathbf{x}(j) - c_d(j)\right)\mathbf{h} + \mathbf{p}(j) - \mathbf{\Phi}^*\mathbf{p}(j+1)\right\} \\
& + d\mathbf{x}^*(k+N)\left\{\left(\mathbf{h}^*\mathbf{x}(k+N) - c_d(k+N)\right)\mathbf{h} - \mathbf{p}(k+N)\right\} \\
& + \sum_{j=k+1}^{k+N} d\mathfrak{m}(j)\left\{a\mathfrak{m}(j) - \gamma^*\mathbf{p}(j)\right\} \\
& + d\mathbf{p}^*(j)\left\{\mathbf{x}(j) - \mathbf{\Phi}\mathbf{x}(j-1) - \gamma\mathfrak{m}(j)\right\} = 0
\end{aligned}
$$

Therefore, the following relations must be satisfied*

$$\mathbf{x}(j) = \phi\mathbf{x}(j-1) + \gamma\mathfrak{m}(j) \tag{6.3-6}$$

$$\mathbf{p}(j) = (\phi^*)^{-1}\mathbf{p}(j-1) + (\phi^*)^{-1}\mathbf{h}\mathbf{h}^*\mathbf{x}(j-1) - (\phi^*)^{a1}\mathbf{h}\,c_d(j-1) \tag{6.3-7}$$

$$\mathfrak{m}(j) = \frac{1}{a}\gamma^*\mathbf{p}(j) \tag{6.3-8}$$

with transversality condition

$$\mathbf{p}(k+N) = \mathbf{h}c_d(k+N) - \mathbf{h}\mathbf{h}^*\mathbf{x}(k+N)$$

Substituting Eq. (6.3-8) in Eq. (6.3-6) we have the following two-point boundary-value problem to solve.

$$
\begin{bmatrix} \mathbf{x}(j) \\ \\ \mathbf{p}(j) \end{bmatrix} = \phi \begin{bmatrix} \mathbf{x}(j-1) \\ \\ \mathbf{p}(j-1) \end{bmatrix} + \eta c_d(j-1)
$$

where

$$
\phi = \begin{bmatrix} \phi + \dfrac{1}{a}\gamma\gamma^*(\phi^*)^{-1}\mathbf{h}\mathbf{h}^* & \dfrac{1}{a}\gamma\gamma^*(\phi^*)^{-1} \\ (\phi^*)^{-1}\mathbf{h}\mathbf{h}^* & (\phi^*)^{-1} \end{bmatrix}
$$

*It is noted here that $(\phi^*)^{-1}$ exists since ϕ is a fundamental matrix.

$$\boldsymbol{\eta} = \begin{bmatrix} \dfrac{1}{a} \boldsymbol{\gamma}\boldsymbol{\gamma}^* (\boldsymbol{\phi}^*)^{-1} \mathbf{h} \\ (\boldsymbol{\phi}^*)^{-1} \mathbf{h} \end{bmatrix}$$

with $\mathbf{x}(k)$ given and

$$\mathbf{p}(k + N) = \mathbf{h} c_d(k + N) - \mathbf{h}\mathbf{h}^* \mathbf{x}(k + N)$$

We solve for $\mathbf{x}(k + N)$ in terms of $\mathbf{x}(k)$.

$$\begin{bmatrix} \mathbf{x}(k + N) \\ \mathbf{p}(k + N) \end{bmatrix} = \theta^N \begin{bmatrix} \mathbf{x}(k) \\ \mathbf{p}(k) \end{bmatrix} + \sum_{j=1}^{N} \theta^{N-j} \boldsymbol{\eta}\, c_d(j - 1)$$

The summation can be evaluated. Therefore, we have

$$\begin{bmatrix} \mathbf{x}(k + N) + \boldsymbol{\delta}_1 \\ \mathbf{p}(k + N) + \boldsymbol{\delta}_2 \end{bmatrix} = \boldsymbol{\psi} \begin{bmatrix} \mathbf{x}(k) \\ \mathbf{p}(k) \end{bmatrix}$$

$$\boldsymbol{\psi} = \theta^N = \begin{bmatrix} \boldsymbol{\psi}_{11} & \boldsymbol{\psi}_{12} \\ \boldsymbol{\psi}_{21} & \boldsymbol{\psi}_{22} \end{bmatrix}$$

Thus,

$$\mathbf{x}(k + N) + \boldsymbol{\delta}_1 = \boldsymbol{\psi}_{11} \mathbf{x}(k) + \boldsymbol{\psi}_{12} \mathbf{p}(k)$$

$$\mathbf{h} c_d(k + N) - \mathbf{h}\mathbf{h}^* \mathbf{x}(k + N) + \boldsymbol{\delta}_2 = \boldsymbol{\psi}_{21} \mathbf{x}(k) + \boldsymbol{\psi}_{22} \mathbf{p}(k)$$

Eliminating $\mathbf{x}(k + N)$, we obtain

$$\mathbf{h} c_d(k + N) + \boldsymbol{\delta}_2 + \mathbf{h}\mathbf{h}^* \boldsymbol{\delta}_1 = (\boldsymbol{\psi}_{21} + \mathbf{h}\mathbf{h}^* \boldsymbol{\psi}_{21}) \mathbf{x}(k) + (\boldsymbol{\psi}_{22} + \mathbf{h}\mathbf{h}^* \boldsymbol{\psi}_{12}) \mathbf{p}(k)$$

Therefore, the feedback solution is

$$m(k) = \frac{1}{a} \boldsymbol{\gamma}^* (\boldsymbol{\psi}_{22} + \mathbf{h}\mathbf{h}^* \boldsymbol{\psi}_{12})^{-1} \big(\mathbf{h} c_d(k + N) + \boldsymbol{\delta}_2 + \mathbf{h}\mathbf{h}^* \boldsymbol{\delta}_1 - \boldsymbol{\psi}_{21} + \mathbf{h}\mathbf{h}^* \boldsymbol{\psi}_{11}) \mathbf{x}(k)\big) \tag{6.3-9}$$

Here, it is assumed that the inverse exists. It should be realized that the existence of this inverse is essential in order that we have a unique minimum. In practice, tests should be made to see whether the inverse exists for a wide range of parameter variations.

6.3.3 Inequality Constrains on the Control Variable-Coordinatewise Gradient Method

This section extends considerations given in Sec. 6.3.2 to the case when we impose inequality constraints on the control variable. The optimization problem becomes a problem in quadratic programming for which numerous methods are available. For the control problem, the constraints are simpler than some general nonlinear programming problem. Therefore, we expect some easy method to apply. In turn, a simple method is required if on-line application is to become feasible. Two methods seem to stand out for on-line control* applications which are the coordinatewise gradient method [17] and Ho's simplified gradient projection method [19]. The former will be discussed in this section and the latter will be presented in the next section.

Let us form an array using Eq. (6.3-4), which gives the output at discrete instants of time as a function of the present state and the control to be applied.

$$\begin{bmatrix} c(1) \\ \vdots \\ \vdots \\ c(N) \end{bmatrix} = \begin{bmatrix} \mathbf{h}^*\gamma & 0 & & \\ \mathbf{h}^*\mathbf{\Phi}\gamma & \mathbf{h}^*\gamma & 0 & \\ \vdots & & & \\ \mathbf{h}^*\mathbf{\Phi}^{N-1}\gamma & \cdots & \cdots & \mathbf{h}^*\gamma \end{bmatrix} \begin{bmatrix} m(1) \\ \vdots \\ \vdots \\ m(N) \end{bmatrix} + \begin{bmatrix} \mathbf{h}^*\mathbf{\Phi} \\ \mathbf{h}^*\mathbf{\Phi}^2 \\ \vdots \\ \mathbf{h}^*\mathbf{\Phi}^N \end{bmatrix}$$

$$\underset{\sim}{c} = G\underset{\sim}{m} + \underset{\sim}{c}_0 \tag{6.3-10}$$

$$\text{where } G = \begin{bmatrix} g(1) & 0 & & 0 \\ g(2) & g(1) & 0 & \\ \vdots & & & \\ g(N) & & & g(1) \end{bmatrix} = \begin{bmatrix} \underset{\sim}{g}_1 & \vdots & \underset{\sim}{g}_2 & \vdots & \cdots & \underset{\sim}{g}_N \end{bmatrix}$$

The problem can be restated as

*By on-line control we mean the situation in which certain computations have to be performed by the control computer for the system in order to achieve optimum performance, and that these computations are being performed in real time to provide continuous control.

Problem: Find $\underset{\sim}{m}$ to minimize*

$$J = \frac{1}{2} \left\| \underset{\sim}{c}_d - \underset{\sim}{c} \right\|^2 \tag{6.3-11}$$

Subject to

$$|m(j)| \leq M \, . \tag{6.3-12}$$

Let $\underset{\sim}{d}' = \underset{\sim}{c}_d - \underset{\sim}{c}_0$. Then Eq. (6.3-11) becomes

$$J = \frac{1}{2} \left\| \underset{\sim}{d}' - G \underset{\sim}{m} \right\|^2 \tag{6.3-13}$$

In m-space, Eq. (6.3-13) defines a hyperparabolic surface, and the problem is to find a point in the convex region defined by Eq. (6.3-12) which yields the minimum point on this hyperbolic surface.

In employing the coordinatewise gradient method, we pick a starting point in a m-space, and determine the gradient of J in the direction of one coordinate. A correction to the trial point is made in the negative gradient direction. By considering each coordinate and repeating the process geometrically, we proceed down the hill to the lowest point in the constrained set. More precisely, the derivation is as follows.

The gradient in the $m(j)$ direction is

$$\nabla_j J = \underset{\sim}{g}_j^* G \underset{\sim}{m} - \underset{\sim}{g}_j^* \underset{\sim}{d}' \tag{6.3-14}$$

The corrected value for the $u(j)$ component is

$$m(j)^{(n+1)} = m(j)^{(n)} + \epsilon_n \nabla_j^{(j)} J \tag{6.3-15}$$

The ϵ_n is found by seeking the minimum along the direction of the jth component. Expanding J

$$2J = \langle G^* G \underset{\sim}{m}, \underset{\sim}{m} \rangle - 2 \langle G^* \underset{\sim}{d}', \underset{\sim}{m} \rangle + \| \underset{\sim}{d}' \|^2$$

Let us work with the terms which depend on m

$$Q(\underset{\sim}{m}) = \langle G^* G \underset{\sim}{m}, \underset{\sim}{m} \rangle - 2 \langle G^* \underset{\sim}{d}', \underset{\sim}{m} \rangle \tag{6.3-16}$$

Also,

$$\underset{\sim}{m}^{(n+1)} = \underset{\sim}{m}^{(n)} + \epsilon_n \underset{\sim}{w}^{(n)}$$

*The norm $\| \mathbf{x} \|$ is defined here as $\| \mathbf{x} \| = \sqrt{\sum_{i=1}^{n} x_i^2}$

where $\underset{\sim}{w}^{(n)}$ is zero except for the jth element, which is equal to $\nabla_j^{(n)} J$. Substituting $m^{(n+1)}$ in Eq. (6.3-16), we obtain

$$Q\left(\underset{\sim}{m}^{(n)} + \epsilon_n \underset{\sim}{w}^{(n)}\right) = Q\left(\underset{\sim}{m}^{(n)}\right) + 2\epsilon_n < G^* G \underset{\sim}{m}^{(n)} - G^* \underset{\sim}{d}', \underset{\sim}{w}^{(n)} >$$
$$+ \epsilon_n^2 < G^* G \underset{\sim}{w}^{(n)}, \underset{\sim}{w}^{(n)} >$$

The minimum along a particular direction is then given by

$$\frac{d}{d\epsilon_n} Q(\quad) = 2 < G^* G \underset{\sim}{m}^{(n)} - G^* \underset{\sim}{d}', \underset{\sim}{w}^{(n)} >$$
$$+ 2\epsilon_n < G^* G \underset{\sim}{w}^{(n)}, \underset{\sim}{w}^{(n)} > = 0$$

Or,

$$\epsilon_n = - \frac{||\underset{\sim}{w}^{(n)}||^2}{< G^* G \underset{\sim}{w}^{(n)}, \underset{\sim}{w}^{(n)} >}$$

The vector $\underset{\sim}{w}$ is zero except for the jth element. Therefore, ϵ_n in the jth direction is

$$\epsilon_{n_j} = \frac{-1}{< \underset{\sim}{g}, \underset{\sim}{g}_j >}$$

Therefore, at the nth step, we get the $n+1$ approximation by

$$m(j)^{(n+1)} = m(j)^{(n)} - \frac{\underset{\sim}{\nabla}_j^{(n)} J}{|| \underset{\sim}{g}_j ||^2} \tag{6.3-17}$$

As $m(j)^{(n+1)}$ may possible exceed a bound, we must limit its amplitude

$$\hat{m}(j)^{(n+1)} = \underset{-M,\, M}{\text{sat}} \left[m(j)^{(n+1)}\right] \tag{6.3-18}$$

The quantity on the left is used for the next iteration. Therefore, the vital equations are Eqa. (6.3-14), (6.3-17), and (6.3-18). The simplicity of the equations to be solved is noted. Every iteration requires only $\frac{N^2}{2} + \frac{5N}{2} - 1$ additions, $\frac{N^2}{2} + \frac{5N}{2}$ multiplications, and 1 division.

The proof of convergence has been given by Hildreth [10]. The convergence is assured only for the rectangular type region (in m-space) we have.

Of course, the success of any gradient method depends upon the closeness of the initial trial to the answer. Using the procedure of repeatedly computing optimal controls on the basis of a finite time interval into the future, we have a good initial guess. Although the optimization yields the control force for the entire optimization interval, NT, only the first component is ever used. However, the other components can be used as an initial approximation for the following interval of consideration. If the changes caused by disturbances and process and input variations are small during T, one should be able to compute the optimal controls rapidly since the initial approximations will be very close to the optimal point. In Fig. 6-13, m(2) in interval 1 becomes the first guess for m(1) in interval 2. Only an initial approximation for the last T seconds is missing. For this reason, the interation is initiated from the last T interval, working forward, the repeating this process. In this way, the first iteration will not disturb the initial good approximation of the other intervals.

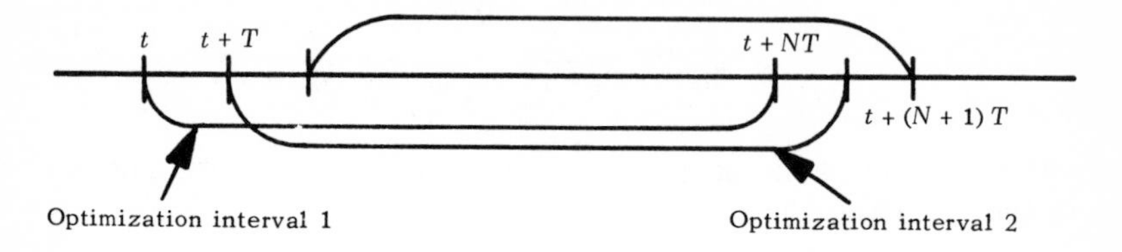

FIG. 6-13. Translation of optimization intervals.

If no initial approximation is available, the unbounded solution can be computed. By simply passing the unbounded solution through a limiter operation, we have a possible initial guess.

Since only one component is initially indeterminate, it is felt that the coordinatewise gradient method may be the most suitable in the optimum-adaptive scheme. Experimental results are reported in ref. 28. Improvement is shown over conventional sampled-data systems.

6.3.4 Inequality Constraints on the Control Variable–Ho's Simplified Gradient Projection Method 19

The other possible method for on-line application of optimal methods is Ho's simplified gradient projection method. This method attempts to move in the steepest direction. Since constraints are present, we limit the components of the steepest gradient vector. As illustrated in Fig. 6-14, this has the effect of projecting the steepest gradient vector on the hypercube. The correction is given by

$$\underset{\sim}{m}^{(n+1)} = \underset{\sim}{m}^{(n)} + \epsilon_n \underset{\sim}{w}^{(n)} \qquad (6.3\text{-}19)$$

where $\underset{\sim}{w}^{(n)}$ is the gradient vector. The ϵ_n is given as described previously by

$$\epsilon_n = -\frac{\|\underset{\sim}{w}^{(n)}\|^2}{\langle G^* G\, \underset{\sim}{w}^{(n)}, \underset{\sim}{w}^{(n)} \rangle} \tag{6.3-20}$$

As we may exceed the constraints, we must limit the corrections to be applied. Therefore

$$\hat{m}_i^{(n+1)} = \begin{cases} M & \text{if } m_i^{(n+1)} > M \\ m_i^{(n+1)} & \text{if } M \le m_i^{(n+1)} \le M \\ -M & \text{if } m_i^{(n+1)} < -M \end{cases} \tag{6.3-21}$$

It should be emphasized that this procedure, like the coordinate-wise gradient method, is restricted to the rectangular-type constraint. Ho gives convergence considerations for this method.

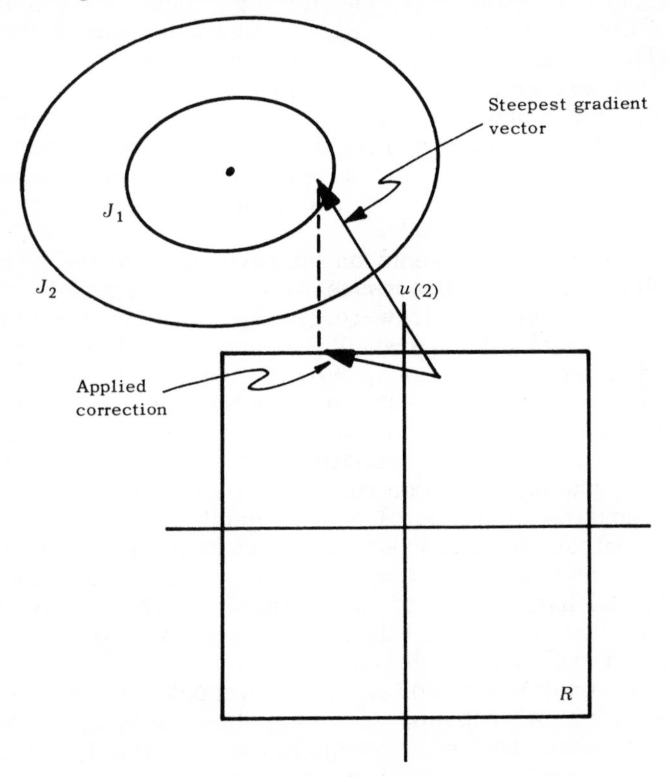

FIG. 6-14. Ho's simplified gradient projection method.

This method requires $2N^2 + 3N - 4$ additions, $2N^2 + 5N$ multiplications, and 1 division per iteration. Of course, although the computations are more numerous, the movement per iteration may be greater than in the coordinatewise gradient method. Up to this time, no direct comparison of the two methods has been made.

6.3.5 Historical Background of Optimal-Adaptive Methods

Some representative contributors to optimal-adaptive methods are Kalman [23], Merriam [36], Braun [8], Meditch [34], and Hsieh [20]. Briefly, their methods will be described.

One of the first contributions in the optimal-adaptive area was given by Kalman. The discrete controller used by Kalman reduced the error to a step in minimum time. He used a property of sampled-data theory which allows eliminating error at a finite number of discrete steps. Kalman's later work [24] parallels the philosophy given in Sec. 6.3.2. He gives an alternate derivation using dynamic programming and his algorithms solve a Ricatti equation. Computationally, the Ricatti equation algorithm and the method given in Sec. 6.3.2 appear to be equivalent.

Merriam was also an early contributor. His work is also confined to the linear process, quadratic criterion case, and parallels Sec. 6.3.2. Merriam also used the dynamic programming approach. He described a criterion which performed optimization over a finite time into the future. No experimental verification of this philosophy was given, however.

Braun's method represents an approximation to the optimization problem for linear processes with a resulting system which is simpler in terms of hardware. Restricting the control signal to the staircase function, Braun's performance criterion involves matching the actual response, $c(t)$, to the desired response at periodic instants of time. These sampling instants, T seconds apart, represent the time instants when the level of the control effort changes. Close approximations by $c(t)$ of $c_d(t)$ can be obtained by reducing the control interval T wth, however, the attendant possible increase of control amplitude, $m(i)$. Thus, the available control effort (bounds on the control signal) can dictate the control interval to be used, with the accuracy of the system limited by the narrowness of these bounds on the control signal. It is noted that by this method there is no indication of the accuracy during the control interval.

Meditch's on-line controller also represents an approximation to the optimization problem. Instead of the "floating" optimization interval described in this chapter, the time axis is subdivided into equal intervals. In other words, the following optimization is performed periodically.

$$J_N = \int_{NT}^{(N+1)T} \frac{1}{2}\,[c_d(t) - c(t)]^2 + \frac{a}{2}\,m(t)^2\,dt$$

Instead of the constant signal employed by Braun, Meditch employs a finite sum of orthonormal functions to perform the optimization.

The method of Hsieh, like that of Sec. 6.3.2, uses the linear-process, quadratic-criterion case. He shows, however, that when the process id identified in terms of the weighing function, it is more convenient to employ the method of functional steepest descent for the optimization.

6.4 Parameter and State Estimation

The problems of estimation of parameters (identification) and state estimation are closely coupled. That is, the states can be estimated if the process parameters are known and visa versa. We will employ the following philosophy to get out of this predicament. If identification methods which can operate with inaccurate knowledge of the state variables are available, the identified process can be used in the state variable estimation. A possible reason for taking this route is that the state variables generally change faster than the process parameters. Subsequent discussions will be made using these assumptions.

6.4.1 Preliminary Remarks

Many methods have been proposed for the identification problem.

Our discussion will be restricted to those methods which have the following characteristics. First, the process is assumed linear and stationary. The stationarity is assumed for the time interval beginning with determination of the identification data. Second, the identification should be performed without inserting externally generated test signals. It should depend only on the normal signals present in the system. Finally, because noise is inevitable in the systems, smoothing should be provided.

For linear processes, either the weighting function or the coefficients of the difference equation (discrete case) are identified. We confine ourselves to the determination of the coefficients. Discussions on the determination of the weighting function are given by Levin [29], Kerr and Surber [26], Balakrishnan [3], and Hsieh [21].

Restricting ouselves to determination of the corfficients of the difference equation, essentially two different approaches are available: 1) the explicit mathematical-relation method, and 2) the learning-model method. The explicit mathematical-relation method requires knowledge of the exact form of the difference equation. This restriction is somewhat relaxed for the learning-model method in the sense that a lower-order model can be made to approximate a higher-order process. Section 6.4.2 discusses the explicit mathematical-relation method and Sec. 6.4.3 discusses the learning-model method.

The methods described in Secs. 6.3.3 and 6.3.4 require, in addition to the determination of the coefficients of the difference equation, the pulse response at several sampling instants. The two methods to be described can also be employed for the determination of the pulse response. The learning-model approach may, however, become unwieldy if the number of parameters is large.

6.4.2 The Explicit Mathematical-Relation Method

The explicit mathematical-relation method was used by Kalman[1] but the basic philosophy dates as far back as 1951 when Greenberg discussed methods for determining stability derivatives of an airplane [14]. Subsequent work on this method was performed by Bigelow and Ruge [6].

Briefly, the method reconstructs the equation of the process by measuring the output and input and their previous values (enough so that all of the terms in the equation are accounted for). By taking redundant measurements, filtering is provided.

The method can best be described by an example. Let us determine the coefficients of the difference equation

$$c(k) = a_1 c(k-1) + a_2 m(k) \tag{6.4-1}$$

The problem is to determine a_1 and a_2. These parameters can be constant but unknown or changing due to changes in environment. Usually, $c(k)$ will not be directly observed but with a contaminating noise quantity as depicted in Fig. 6-15. Thus,

$$z(k) = c(k) + v(k) \tag{6.4-2}$$

The values of $z(k)$ and $m(k)$ will be stored for some interval of time into the past, and throughout this interval the parameters a_1 and a_2 are assumed to be constant. Since $c(k)$ cannot be directly measured, Eq. (6.4-1) is rewritten in terms of $z(k)$

$$z(k) - v(k) = a_1[z(k-1) - v(k-1)] + a_2 m(k)$$

Or,

$$z(k) = a_1 z(k-1) + a_2 m(k) + v_1(k) \tag{6.4-3}$$

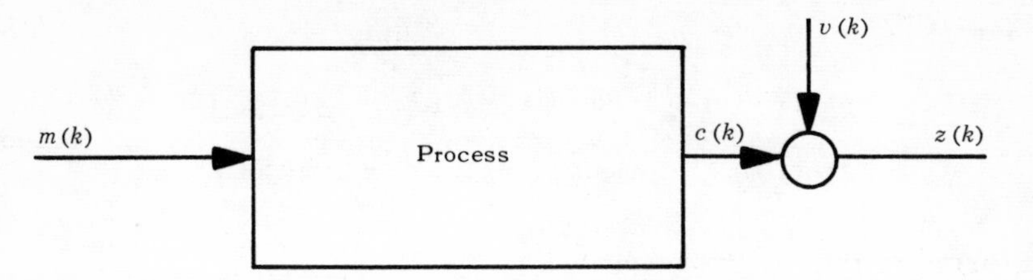

FIG. 6-15. Configuration of problem.

where $v_1(k) = v(k) - a_1 v(k-1)$.

Taking a set of measurements, Eq. (6.4-3) can be rewritten in vector form.*

$$\underset{\sim}{z}_k = a_1 \underset{\sim}{z}_{k-1} + a_2 \underset{\sim}{m}_k + \underset{\sim}{v}_k$$

where

$$\underset{\sim}{z}_k = \begin{bmatrix} z(k-N+1) \\ \vdots \\ z(k) \end{bmatrix}, \quad \text{etc.}$$

or,

$$\underset{\sim}{z}_k = \mathbf{A}\boldsymbol{\alpha} + \underset{\sim}{v}_k \tag{6.4-4}$$

where

$$\mathbf{A} = \left[\underset{\sim}{z}_{k-1} \vdots \underset{\sim}{m}_k\right]$$

Let

$$\check{\underset{\sim}{z}}_k = \mathbf{A}\boldsymbol{\alpha}$$

The $\check{\underset{\sim}{z}}_k$ is in the manifold [54] of $\underset{\sim}{z}_{k-1}$ and $\underset{\sim}{m}_k$. The quantity $\underset{\sim}{z}_k$ is not necessarily in the linear manifold because of $\underset{\sim}{v}_{1k}$. Since $\underset{\sim}{v}_{1k}$ is unknown, a reasonable estimate of the parameters would be those values which result from the projection of $\underset{\sim}{z}_k$ on the manifold of $\underset{\sim}{z}_{k-1}$ and $\underset{\sim}{m}_k$. The projection yields

$$\langle \underset{\sim}{z}_k - \underset{\sim}{z}_k, \underset{\sim}{z}_{k-1} \rangle = 0$$

$$\langle \underset{\sim}{z}_k - \underset{\sim}{z}_k, \underset{\sim}{m}_k \rangle = 0$$

*The k signifies that N data points going into the past from time k are considered.

or,

$$\begin{aligned} a_1 < \underset{\sim}{z}_{k-1}, \underset{\sim}{z}_{k-1} > + a_2 < \underset{\sim}{m}_k, \underset{\sim}{z}_{k-1} > &= < \underset{\sim}{z}_k, \underset{\sim}{z}_{k-1} > \\ a_1 < \underset{\sim}{z}_{k-1}, \underset{\sim}{m}_k > + a_2 < \underset{\sim}{m}_k, \underset{\sim}{m}_k > &= < \underset{\sim}{z}_k, \underset{\sim}{m}_k > \end{aligned} \tag{6.4-5}$$

In terms of the matrix equation,

$$A^* A \alpha = A^* \underset{\sim}{z}_k \tag{6.4-6}$$

Equations (6.26) or (6.27) are known as normal equations; and if $\underset{\sim}{z}_{k-1}$ and $\underset{\sim}{m}_k$ are linearly independent, then the solution is given by

$$\alpha = (A^* A)^{-1} A^* \underset{\sim}{z}_N \tag{6.4-7}$$

If $\underset{\sim}{z}_{N-1}$ and $\underset{\sim}{m}_N$ are not linearly independent, there are many solutions. A unique solution is provided by employing the pseudo-inverse studied extensively by Penrose [40, 41].

$$\hat{a} = A^+ \underset{\sim}{z}_k \tag{6.4-8}$$

where A^+ = pseudo-inverse of A.

Recursive methods to evaluate the pseudo-inverse have been given by Greville [16].

6.4.3 The Learning-Model Approach

The other approach available for estimation of coefficients of a difference equation is the learning-model method. It is felt that if some a priori estimate of the unknown parameters is available then we should be able to use this information to advantage. This is probably the motivation for the learning-model method. This method was originally studied by Margolis [31] using the sensitivity function. The sensitivity function is also used by Staffanson [45], who was concerned with parameter determination from flight-test data. There are two disconcerting characteristics of the sensitivity function approach.

1. One must choose the gain in the steepest descent procedure.
2. The use of sensitivity functions is, generally, valid only for a small region about a trial point.

To overcome these problem areas, a modified Newton's procedure is described in this section. The philosophy for the learning model approach is depicted in Fig. 6-16.

Two other possibilities of the learning-model method should be mentioned. First, the quasi-linearization approach described by Bellman et al. [4]. This method was found to be very cumbersome for the discrete case. The other method is the orthogonal-function approach used by Elkind et al. [13]. Fixing the model time-constants a priori seems to be a crude method.

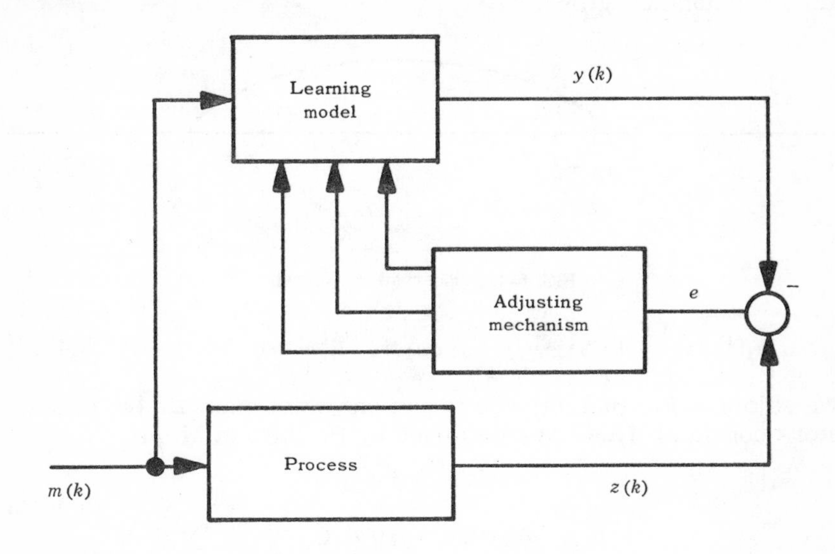

FIG. 6–16. Margolis' learning model approach.

Let us describe the method through an example. Instead of operating on the error, as shown in Fig. 6-16, the stability problem can possibly be alleviated by solving the problem,

Problem 6.1: Find the parameters (a_i) of the model which minimizes

$$J = \sum_{j=1}^{N} [z(j) - y(j)]^2 \tag{6.4-9}$$

where $y(j)$ is subject to the dynamical constraint

$$y(j) = a_1 y(j-1) + a_2 m(j) \tag{6.4-10}$$

The time indices are shown in Fig. 6-17.* In our case, the model, Eq. (6.4-10), could be of lower order than the actual process (model-fitting problem). We start form an initial trial or estimate of the parameters $a_i^{(1)}$ and the initial conditions for the interval

*To simplify the notation, the index k is dropped. Thus, at the time of computation, $j = 0 \to j = k - N$ and $j = N \to j = k$.

of observation $y(0)^{(1)}$. With these initial trials, Eq. (6.4-10) is solved to obtain a nominal solution, $y(j)^{(1)}$, $j = 0, 1, \ldots, N$. Next, the perturbation equations of Eq. (6.4-10) are written, evaluated along the nominal $y(j)^{(1)}$.

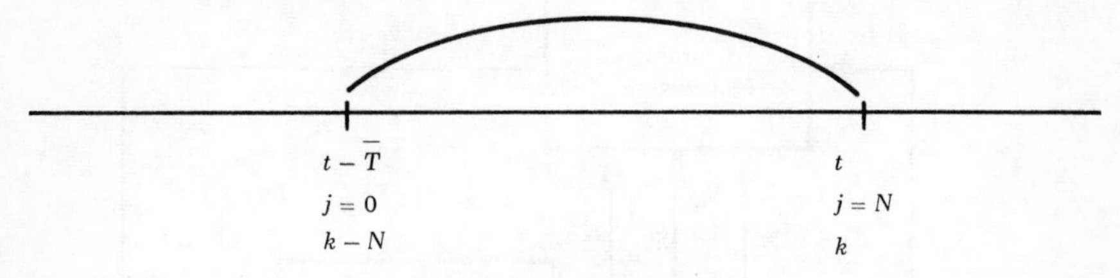

FIG. 6-17. Observation interval.

$$\delta y(j) = a_1^{(1)} \delta y(j-1) + y^{(1)}(j-1)\, \delta a_1(j-1) + m(j)\, \delta a_2(j-1) \quad (6.4\text{-}11)$$

We adjoin to Eq. (6.4-11) other equations which maintain the parameters constant. This trick was used by Bellman et al. [4].

$$\begin{aligned} \delta a_1(j) &= \delta a_1(j-1) \\ \delta a_2(j) &= \delta a_2(j-1) \end{aligned} \quad (6.4\text{-}12)$$

Let

$$\zeta(j) = \begin{bmatrix} \delta y(j) \\ \delta a_1(j) \\ \delta a_2(j) \end{bmatrix}$$

then

$$\zeta(j) = \mathbf{\Phi}(j-1)\, \zeta(j-1) \quad (6.4\text{-}13)$$

where

$$\mathbf{\Phi}(j-1) = \begin{bmatrix} a_1^{(1)} & y^{(1)}(j-1) & m(j) \\ 0 & 1 & 0 \\ 0 & 0 & 1 \end{bmatrix} \quad (6.4\text{-}14)$$

At this stage, instead of solving the optimization problem, Problem 6.1, we solve the following problem.

Problem 6.2: Find the initial conditions of Eq. (6.4-10) which minimizes

$$J = \sum_{j=1}^{N} z(j) - y^{(1)}(j) - \delta y^{(1)}(j)^{\;2} \quad (6.4\text{-}15)$$

where $\delta y^{(1)}(j)$ is subject to the constraint (6.4-13).
We have converted a nonlinear problem into a linear problem. By repeatedly solving this last problem, we hope to approach the solution to the first problem.

Problem 6.2 is solved by using the least-square curve-fitting procedure. It is noted that

$$y^{(1)}(j) + \delta y^{(1)}(j) = z(j) + n(j) \tag{6.4-16}$$

where $n(j)$ is the discrepancy caused by noise and error in parameter adjustment. Let

$$\delta y^{(1)}(j) = z(j) - y^{(1)}(j) \tag{6.4-17}$$

The right-hand side of Eq. (6.4-17) is known and is desired to determine $\delta y^{(1)}(j)$, subject to Eq. (6.4-13), which best approximates $z(j) - y(j)^{(1)}$. Equation (6.4-17) can be rewritten

$$\mathbf{h}^* \boldsymbol{\zeta}(j) \doteq z(j) - y^{(1)}(j) \tag{6.4-18}$$

where $\mathbf{h}^* = (1 \quad 0 \quad 0)$.

The N equations represented by Eq. (6.4-18) can all be rewritten in terms of $\zeta(0)$ by using Eq. (6.4-14)

$$\begin{aligned}
\mathbf{h}^* \boldsymbol{\zeta}(0) &= z(0) - y^{(1)}(0) \\
\mathbf{h}^* \boldsymbol{\Phi}(1,0) \boldsymbol{\zeta}(0) &= z(1) - y^{(1)}(1) \\
&\vdots \\
\mathbf{h}^* \boldsymbol{\Phi}(N,0) \boldsymbol{\zeta}(0) &= z(N) - y^{(1)}(N)
\end{aligned}$$

Or, in matrix form

$$\mathbf{A} \boldsymbol{\zeta}(0) = \boldsymbol{\xi} \tag{6.4-19}$$

where

$$\mathbf{A} = \begin{bmatrix} \mathbf{h} \\ \mathbf{h}\boldsymbol{\Phi}(1,0) \\ \vdots \\ \mathbf{h}\boldsymbol{\Phi}(N,0) \end{bmatrix} \qquad N+1 \times 3 \text{ matrix}$$

$$\boldsymbol{\xi} = \begin{bmatrix} z(N) - y^{(1)}(0) \\ \vdots \\ z(N) - y^{(1)}(N) \end{bmatrix} \qquad N+1 \times 1 \text{ vector}$$

The pseudo-inverse routine is used to solve Eq. (6.4-19).

$$\zeta(0)^{(1)} = A^{+}\xi^{(1)} \tag{6.4-19}$$

From Eq. (6.4-19) we can make corrections to the initial trial of the parameters and initial conditions.

$$\begin{aligned} a_i^{(2)} &= a_i^{(1)} + \delta a_i^{(1)}(0) \\ y(0)^{(2)} &= y(0)^{(1)} + \delta y^{(1)}(0) \end{aligned} \tag{6.4-20}$$

The procedure can now be repeated.

The procedure just outlined may well be divergent. Procedures using the digital computer can, however, be used to give monotone convergence. An algorithm which assures this important property will be described.

From the initial trial and solution we can compute the error index.

$$J_1 = \sum \left(z(j) - y^{(1)}(j) \right)^2 = \| \underset{\sim}{z} - \underset{\sim}{y}^{(1)} \|^2$$

The problem is to find a $\delta y(j)$ such that J_2 given by

$$J_2 = \sum \left(z(j) - y^{(1)}(j) - \delta y(j) \right)^2$$

is less than J_1.

The difference $J_1 - J_2$ must be greater than zero.

$$\begin{aligned} J_1 - J_2 = {} & \cancel{\| \underset{\sim}{z} - \underset{\sim}{y}^{(1)} \|^2} - \cancel{\| \underset{\sim}{z} - \underset{\sim}{y}^{(1)} \|^2} \\ & + 2 < \delta \underset{\sim}{y}, \underset{\sim}{z} - \underset{\sim}{y}^{(1)} > \\ & - \| \delta \underset{\sim}{y} \|^2 \geq 0 \end{aligned}$$

Or,

$$2 < \delta \underset{\sim}{y}, \underset{\sim}{z} - \underset{\sim}{y}^{(1)} > - \| \delta \underset{\sim}{y} \|^2 \geq 0 \tag{6.4-21}$$

Equation (6.4-21) is the condition for convergence. If

$$< \delta \underset{\sim}{y}, \underset{\sim}{z} - \underset{\sim}{y}^{(1)} > \neq 0$$

then, for $\delta \underset{\sim}{y}$ sufficiently small, Eq. (6.4-21) can be satisfied since the first term is linear in $\delta \underset{\sim}{y}$ while the second term is quadratic. It is noted that the first term in Eq. (6.4-21) is positive since it is a scalar product between the error and the projection of the error in the linear manifold.

The condition

$$< \delta \underset{\sim}{y}, \underset{\sim}{z} - \underset{\sim}{y}^{(1)} > = 0 \tag{6.4-22}$$

requires that $\underset{\sim}{y}^{(1)}$ be closer to $\underset{\sim}{z}$ than any nearby point obtained through linear perturbation. In other words, the gradient is zero and we have a local minimum.

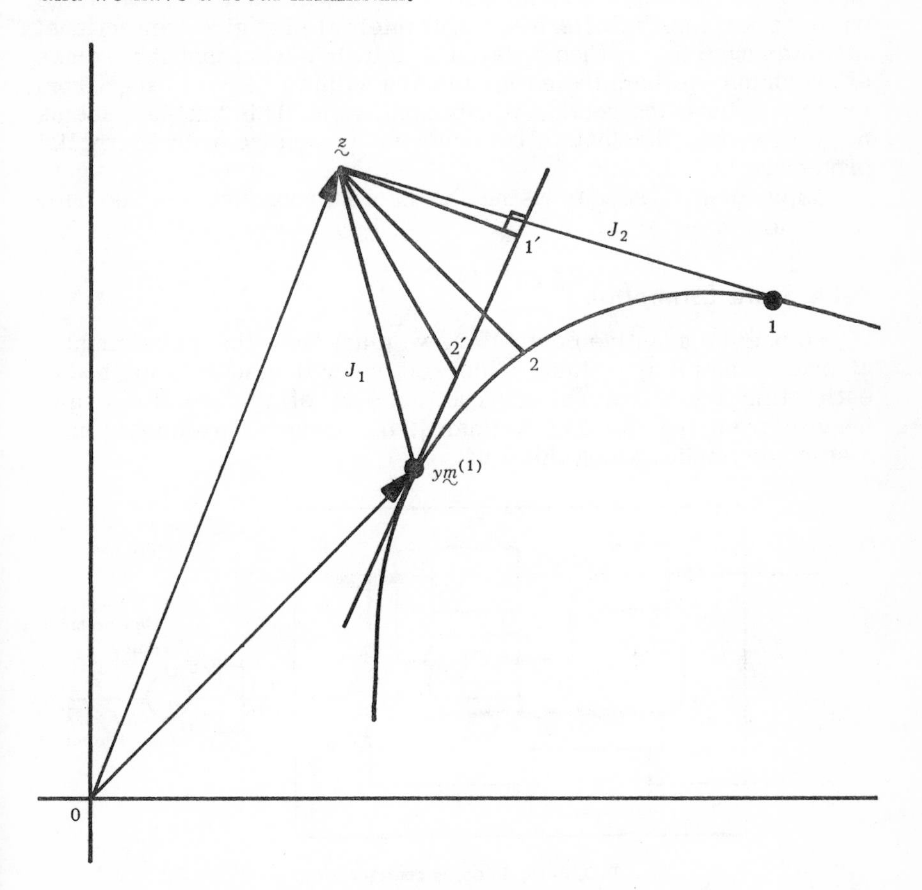

FIG. 6-18. Two-dimensional picture of correction scheme.

The situation is shown in Fig. 6-18. The first linear correction is $1'$. Upon solving Eq. (6.4-10), point 1 is obtained which may well give a J greater than J_1. If $J_2 > J_1$, then we cut the correction, $\delta y(k)$, by a half yielding point 2. If the J at point 2 is less than J_1, then we keep the correction given by $\delta y(k)^{(1)}/2$. If not, we cut $\delta y(k)^{(1)}/2$ by a half and repeat this process. By using this cutting procedure, we have monotone convergence until condition of Eq. (6.4-22) is reached.

In an on-line task, we are limited in the number of iterations we can make at a given time. The requirement is not as stringent, however, as the control synthesis problem because the estimation can be make at wider time intervals for slowly varying processes. If we limit the number of cutting procedures described in the last paragraph, we may never find the correction which will give a smaller J. In this case, no corrections will be made, and we go on to the next interval. Here again, no interval may give corrections, in which case the method fails. It is felt, however, that for a class of problems in which the estimates are within a certain range from the true values, the routine will be applicable. This problem seems no worse than the instability problem associated with Margolis' procedure.

Experimental results using the above procedure are reported in ref. 12.

6.4.4 State Estimation

To use the adaptive controller, we must know the state variable at every sampling instant. This section will discuss a method of estimating these variables. The contens of this section draws heavily from the work of Kalman [24]. Joseph and Tou [22] have also made studies along this line.

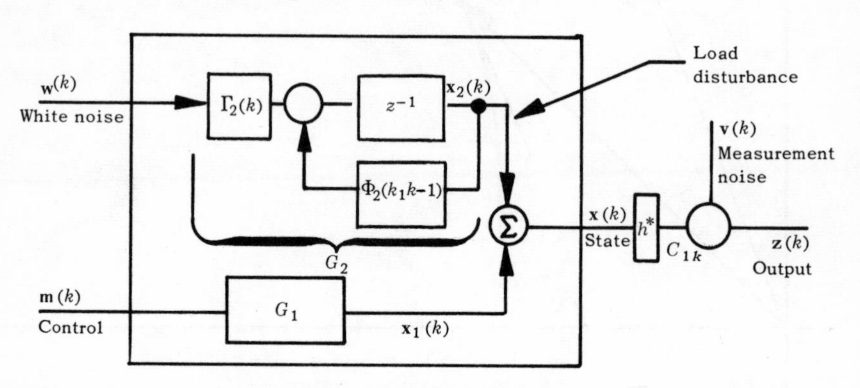

FIG. 6-19. Process configuration.

Let us refer to the process configuration shown in Fig. 6-19. From knowledge of $z(k)$ and $m(k)$, it is required to estimate the state $\mathrm{x}(k)$ at the present time. The past values of $\mathrm{z}(k)$ and $\mathrm{m}(k)$ are known from some initial time. The process characteristics, G_1 and G_2, are known, the former through identification. In an adaptive task, the transfer characteristics are time-varying. As new parameter values are obtained, the corresponding values used in the estimation will be changed. The covariance matrices of $v(k)$

and $w(k)$ are also known. These noise sources can be taken to be white noise. It is noted that because of G_2 the load disturbance can have a nonwhite spectrum.

We note

$$\mathrm{x}(k) = \mathrm{x}_1(k) + \mathrm{x}_2(k)$$

where $\mathrm{x}_1(k)$ is known. Let

$$\nu(k) = z(k) - \mathrm{h}^* \mathrm{x}(k)$$

$$\mathrm{x}_2(k) = \mathrm{x}(k) - \mathrm{x}_1(k)$$

The problem is now simply the determination of $\hat{\mathrm{x}}_2(k)$, the conditioned expectation given $\nu(k)$, $k = 0, 1, \ldots, k$. From $\hat{\mathrm{x}}_2(k)$, the estimate of the state is

$$\hat{\mathrm{x}}(k) = \hat{\mathrm{x}}_2(k) + \mathrm{x}_1(k)$$

Therefore, it can be seen that Kalman's filtering algorithm [24], which can treat time-varying processes, is applicable here.

REFERENCES

1. Aseltine, J. A., Mancini, A. R., and Sarture, C. W., "A Survey of Adaptive Control Systems," *IRE Transactions on Automatic Control*, PGAC-6:102-108, Dec., 1958.
2. Ashby, W. R., *Design for a Brain*, John Wiley and Sons, Inc., New York, 1952 (Second Edition, 1960).
3. Balakrishnan, A. V., "*Determination of Nonlinear System from Input-Output Data*," Presented at the Conference on Identification and Representation Problems, Princeton University, March, 1963.
4. Bellman, R., Kagiwada, H., and Kalaba, R., "*A Computational Procedure for Optimal System Design and Utilization*," Proc. Nat. Acad. of Sci., 48:1524-1528, Sept., 1962.
5. Benner, A. H., and Drenick, R., "An Adaptive Servo System," *IRE 1955 Convention Record*, Pt. 4, pp. 8-14.
6. Bigelow, S. C., and Ruge, H., "An Adaptive System Using Periodic Estimation of the Pulse Transfer Function," *IRE National Convention Record*, Part 4, pp. 25-38, 1961.
7. Boskovich, B., Cole, G. H., Mellen, D. L., "*Advanced Flight Vehicle Self-Adaptive Flight Control System*," Part I Study, Part II Design, WADD TR 50-651 (CONFIDENTIAL).
8. Braun, Jr., L., *On Adaptive Control Systems*, Doctoral Dissertation, Polytechnic Institute of Brooklyn, 1959.
9. Chang, S. S. L., *Synthesis of Optimal Control Systems*, McGraw-Hill Book Co., New York, 1961.
10. Clark, R. N., and Wheeler, P. C., "A Self-Adjusting Control System with Large Initial Error," *IRE Transactions on Automatic Controls, PGAC*, January, 1962.

11. Cooper, G. R., and Gibson, J. E., "A Survey of Philosophy and State of the Art of Adaptive Systems," *Technical Report No. 1,* Contract AF33(616)-6890, Project 8225, Task 82181, PRF 2358, School of Electrical Engineering, Purdue University, Lafayette, Indiana, July 1, 1960.
12. Donalson, D. D., "*The Theory and Stability Analysis of a Model-Referenced Parameter Tracking Technique for Adaptive Automatic Control Systems,*" Doctoral Dissertation, University of California, Los Angeles, May, 1961.
13. Elkind, J. I., Green, D. M., and Starr, E. A., "Application of Multiple Regression Analysis to Identification of Time-Varying Linear Dynamic Systems," Correspondence, *IRE Trans. on Automatic Controls,* AC-8; 163-166, April, 1963.
14. Greenberg, H., "A Survey of Methods for Determining Stability Parameters of an Airplane from Dynamic Flight Measurements," *NACA* TN-2340, April, 1951.
15. Gregory, P. C., Davis, H. M., "*Future Control Systems for Vehicles with Large Thermoelastic Effects,*" ASD TR 61-645, October, 1961.
16. Greville, T. N. E., "Some Applications of the Pseudo-Inverse of a Matrix," *SIAM Review,* 2:15-2 , 1960.
17. Hildreth, C., "Quadratic Programming Procedure," *Naval Research Logistics Quarterly,* 4:79-85, 1957.
18. Hiza, J. C. and Li, C. C., "On Analytical Synthesis of a Class of Model-Reference Time-Varying Control Systems," *IEEE Paper 63-123,* presented at the IEEE 1963 Winter General Meeting, New York, Jan. 27-Feb. 1, 1963.
19. Ho, Y. C. and Brentain, P. B., "On Computing Optimal Control with Inequality Constraints," Paper presented at the Symposium on Multivariable Control Systems, Boston, Mass., November, 1962.
20. Hsieh, H. C., "On the Synthesis of Adaptive Controls by the Hilbert Space Approach," *Report No. 62-19,* Dept. of Engineering, University of California, Los Angeles, June, 1962.
21. Hsieh, H. C., *Synthesis of Adaptive Control Systems by the Function Space Methods,* Doctoral Dissertation, Dept. of Engineering, University of California, Los Angeles, Calif., 1963.
22. Joseph, R. D. and Tou, J. D., "On Linear Control Theory," *AIEE Trans.,* Part II, Applications and Industry, 80:193-196, September, 1961.
23. Kalman, R. E., "Design of a Self-Optimizing Control System," *ASME Transactions,* 80:468-478, February, 1958.
24. Kalman, R. E., Englar, T. S., and Bucy, R. S., "Fundamental Study of Adaptive Control Systems," *Tech. Report No. ASD-TR-61-27,* Vol. 1, Aeronautical Systems Division, Air Force System Command, Wright-Patterson Air Force Base, Ohio, April, 1962.

25. Katz, S., "A Discrete Version of Pontryagin's Maximum Principle," *Jour. of Electronics and Control*, 1962.
26. Kerr, R. P., and Surber, W. H., "Precision of Impulse Response Identification Based on Short Normal Operation Records," *IRE Trans.* PGAC, AC-6:173-182, May, 1961.
27. Kipiniak, W., *Dynamic Optimization and Control*, MIT Press and John Wiley and Sons, Inc., New York, 1961.
28. Kishi, F. H., "On Line Computer Control Techniques and Their Application to Re-entry Aerospace Vehicle Control," Advances in Control Systems Theory and Applications, Volume I, Academic Press, 1964.
29. Levin, M. J., "Optimum Estimation of Impulse Response in the Presence of Noise," *IRE Trans.—PGCT*, CT-7:50-56, March, 1960.
30. Margolis, M., and Leondes, C. T., "A Parameter Tracking Servo for Adaptive Control Systems," *IRE Transactions on Automatic Control*, PGAC-4, No. 2:100-111, November, 1959.
31. Margolis, M., "*On the Theory of Process Adaptive Control Systems, the Learning-Model Approach,*" Doctoral Dissertation, Department of Engineering, University of California, Los Angeles, 1959.
32. McGrath, R. J., Rideout, V. C., "A Simulator Study of a Two-Parameter Adaptive System," *IRE Transactions on Automatic Control*, Vol. AC-6, No. 1, February, 1961.
33. McGrath, R. J., Rajaraman, V., and Rideout, V. C., "A Parameter Perturbation Adaptive Control System," *IRE Transactions on Automatic Control*, Vol. AC-6, No. 2, May, 1961.
34. Meditch, J. S. and Gibson, J. E., "On the Real-Time Control of Time-Varying Linear Systems," *IRE Trans. on Automatic Controls*, AC-7, No. 4:3-10, July, 1962.
35. Meissinger, H. R., "The Use of Parameter Influence Coefficients in Computer Analysis of Dynamic Systems," Unpublished Notes, Hughes Aircraft Company, System Development Laboratories, December, 1957.
36. Merriam, C. W., "Use of a Mathematical Error Criterion in the Design of Adaptive Control System," *AIEE Transactions*, Part II, 79:506-512, January, 1960.
37. Mishkin, E. and Braun, Jr., L., *Adaptive Control Systems*, McGraw-Hill Book Co., New York, 1961.
38. Osburn, P. V., "*Investigation of a Method of Adaptive Control,*" Doctor of Science Thesis, Massachusetts Institute of Technology, 1961, MIT Instrumentation Laboratory Report No. T 266.
39. Osburn, P. V., Whitaker, H. P., Kezer, A., "*New Developments in the Design of Model Reference Adaptive Systems,*" IAS Report No. 61-39, January, 1961.
40. Penrose, R., "A Generalized Inverse of Matrices," *Proc. Cambridge Phil. Soc.*, 51:406-413, 1955.

41. Penrose, R., "On Best Approximate Solutions of Linear Matrix Equations," *Proc. Cambridge Phil. Soc.*, 52:17-19, 1956.
42. Rang, E. R., and Johnson, C. W., "*The Learning Model Approach to Self-Evaluation and the Method of Lyapunov,*" AIEE Paper No. DP 61-603, presented at the AIEE Great Lakes District Meeting, Minneapolis, Minnesota, April 19-21, 1961.
43. Rang, E. R., and Stone, C. R., "Adaptive State Vector Control Adaptive Controllers Derived by Stability Considerations," Minneapolis-Honeywell Regulator Company, *Military Products Group Report 1529-TR 9*, March 15, 1962.
44. Rutman, R. S., "Adaptive Systems with Dynamic Characteristic Adjustment," *Automation and Remote Control*, 23:602-625, 1962.
45. Staffanson, Forrest L., "Determining Parameter Corrections According to System Performance—A Method and its Application to Real-Time Missile Testing," Army Missile Test Center, White Sands Missile Range, *Lab. Res. Rpt. 20*, July, 1960.
46. Stear, E. B., and Gregory, P. C., "Capabilities and Limitations of Some Adaptive Techniques," Report, Flight Control Laboratory, Aeronautical Systems Division, Wright Air Development Division.
47. Stromer, Peter R., "Adaptive or Self-Optimizing Control Systems—A Bibliography," *IRE Transactions on Automatic Control*, PGAC-4; No. 1:65-68, May, 1959.
48. Tsien, H. S., *Engineering Cybernetics*, McGraw-Hill Book Co., New York, 1954.
49. Whitaker, H. P., Yarmon, J., and Kezer, A., "*Design of Model Reference Adaptive Control Systems for Aircraft,*" MIT Instrumentation Laboratory Report R-164, September, 1958.
50. Whitaker, H. P., "*An Adaptive System Control for of the Dynamic Performance of Aircraft and Spacecraft,*" IAS Paper No. 59-100, June, 1959.
51. Whitaker, H. P., "*Model Reference Adaptive Control Systems for Large Flexible Boosters,*" MIT Instrumentation Laboratory Report E-1036, May, 1960, presented to SAE-18 Committee on Aerospace Vehicle Flight Control Systems on June 16, 1961.
52. Whitaker, H. P., and Kezer, A., "*Use of Model Reference Adaptive Systems to Improve Reliability,*" Paper 1936-61, presented to ARS Guidance, Control, and Navigation Conference, August 7-9, 1961.
53. Zadeh, L. A., "On the Definition of Adaptivity," *Proceedings of the IEEE*, 51:469-470, March, 1963.
54. James, G. and James, R., Editors, "Mathematics Dictionary," van Nostrand, Princeton, New Jersey, 1959.

7

An Introduction to the Pontryagin Maximum Principle

JAMES S. MEDITCH
MEMBER OF THE TECHNICAL STAFF
AEROSPACE CORPORATION, LOS ANGELES 45, CALINFORNIA

In this chapter, we shall present the salient features of the Pontryagin maximum principle [1-4]. Our interest will be to familiarize ourselves with the method, its development, and its utility in practice. Wherever possible, we shall give physical and geometrical interpretations of the results. Our presentation will proceed along the following lines. First, we shall formulate the optimal control problem in fairly general terms, and then proceed to a statement and discussion of the first theorem of the maximum principle. Second, we shall outline the derivation of the first theorem and present some of its extensions which will permit us to treat more general optimization problems. We shall conclude our presentation by applying the maximum principle to three different classes of optimization problems.

7.1 Problem Formulation

We shall consider dynamic systems whose state at any instant of time is characterized by n variables $x_1, \ldots, x_n$. For example, these variables might represent the position and velocity coordinates of a space vehicle. The vector space X of the vector variable $x = (x_1, \ldots, x_n)$ is termed the phase space of the system. The behavior of the system is simply the time-varying nature of the vector variable x, commonly referred to as the *state vector*. The x_i, $i = 1, \ldots, n$, are termed the *state variables* of the system.

We assume that the system's state can be *controlled*, that is, that we have access to a set of system inputs (controls) whose manipulation governs the system's state. We shall assume that we have r such controls and that they are characterized by a point u

in a region U of r-dimensional Euclidean space where U is independent of x and t. We shall term u the *control vector*, and note that it is given by $u = (u_1, \ldots, u_r) \epsilon U$. We shall call every piecewise continuous function $u = u(t)$, defined on some time inverval $t_0 \leq t \leq t_1$ whose range is in U, a *control*.*

Admissible Controls. We shall define a control $u(t)$ to be *admissible* if each component $u_i(t)$, $i = 1, \ldots, r$, of $u(t)$ is a bounded, piecewise continuous function such that $u(t) \epsilon U$ for all t, $t_0 \leq t \leq t_1$, i.e., the period of time over which we are concerned with system operation. The initial time t_0 is assumed fixed, but the terminal time t_1 may be either free or fixed.

To illustrate the notion of admissible controls, let us consider a system having two control inputs $u_1(t)$ and $u_2(t)$ on which we impose the constraint

$$[u_1(t)]^2 + [u_2(t)]^2 \leq M_0^2 \qquad t_0 \leq t \leq t_1$$

where M_0 is a positive constant. The set U becomes the interior and boundary of a circle of radius M_0 as shown in Fig. 7-1. Thus, for this example, an admissible control vector can assume any orientation, but its length is constrained to be $\leq M_0$.

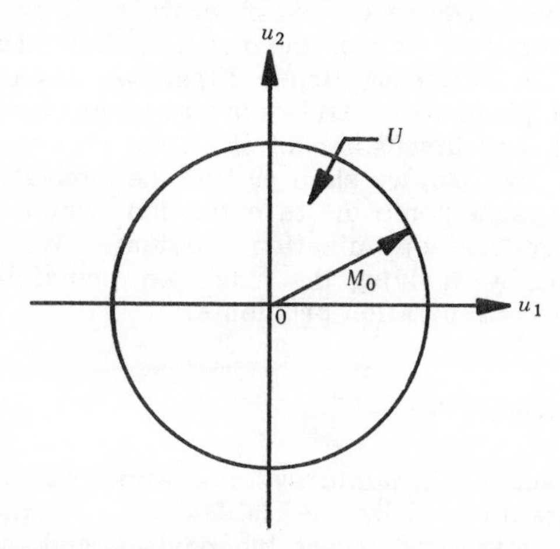

FIG. 7-1. Set of admissible controls U for the chosen example.

Statement of the Problem. We shall assume that the dynamic system's behavior is characterized by the system of differential equations:

$$\frac{dx_i}{dt} = f_i(x_1, \ldots, x_n; u_1, \ldots, u_r) = f_i(x, u), \; i = 1, \ldots, n \tag{7.1}$$

*The more general results as developed by Pontryagin et al [1, 2] assumed U to be any topological Hausdorf space, and $u = u(t)$, $t_0 \leq t \leq t_1$, to be a measurable and bounded vector function. However, for practical applications, the assumption of piecewise continuous controls is of most interest and will be considered here.

For the present, we shall consider only those dynamical systems for which the right-hand side of Eq. (7.1) does not depend explicitly on time t. Such systems are termed *autonomous*. In Section 7.4, we shall discuss the maximum principle for nonautonomous systems, that is, systems for which the right-hand side of Eq. (7.1) depends explicitly on time.

In vector form, Eq. (7.1) is expressed by

$$\frac{dx}{dt} = f(x, u) \tag{7.2}$$

where $f(x, u)$ is a vector-valued function of x and u. The functions f_i, for every $x \epsilon X$ and $u \epsilon U$, are assumed to be continuous with respect to all variables $x_1, \ldots, x_n$, and $u_1, \ldots, u_r$. We also assume that the f_i are continuously differentiable with respect to $x_1, \ldots, x_n$. That is,

$$f_i(x, u) \quad \text{and} \quad \frac{\partial f_i(x, u)}{\partial x_j} \qquad i, j = 1, \ldots, n$$

are defined and continuous for all $x \epsilon X$ and $u \epsilon U$.

A symbolic representation of the system of Eqs. (7.1) is given in Fig. 7-2.

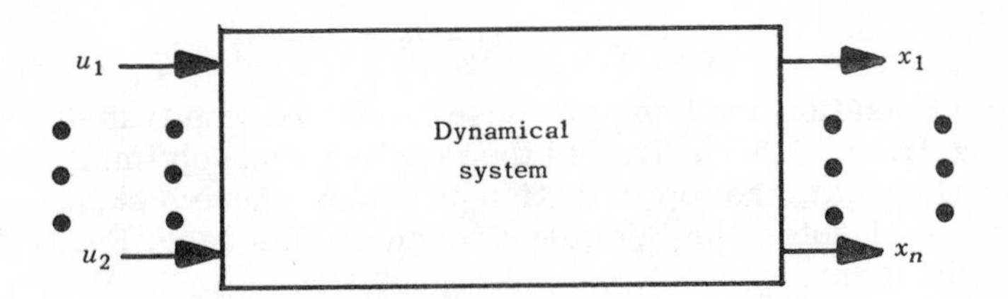

FIG. 7-2. Representation of dynamical system.

If an admissible control $u(t)$ is chosen along with a given initial condition $x(t_0) = x^0$, Eq. (7.2) takes the form

$$\frac{dx}{dt} = f[x, u(t)], \quad x(t_0) = x^0 \tag{7.3}$$

The solution of Eq. (7.3) then uniquely defines the system's state vector (response for $t \geq t_0$ starting at the initial state $x(t_0) = x^0$). It is well known that this solution is an absolutely continuous vector function which satisfies Eq. (7.3) everywhere [except at the points of discontinuity of $u(t)$] on its segment of definition [5].

We say that the admissible control $u(t)$ transfer the point x^0 to some point x^1 if the solution $x(t)$ of Eq. (7.3) passes through the point x^1 at a certain instant of time t_1, i.e., $x(t_1) = x^1$.

Let us now suppose that the quality of system performance is to be measured in terms of some functional

$$J = \int_{t_0}^{t_1} f_0\left[x(t), u(t)\right] dt \tag{7.4}$$

which will hereafter be termed the cost function. The value of J for a given admissible control will be called the *cost* for that control. We assume that the function $f_0\left[x(t), u(t)\right] = f_0(x_1, \ldots, x_n; u_1, \ldots, u_r)$ is defined and continuous, and differentiable with respect to x_i, $i = 1, \ldots, n$, for $x \epsilon X$ and $u \epsilon U$.

We now state the fundamental problem to be considered here: Given the dynamic system of Eq. (7.2) with $x(t_0) = x^0$ and a point $x^1 \epsilon X$, find an admissible control $u = u(t)$ which transfers x^0 to x^1, if any exist, such that the cost function, Eq. (7.4), takes on the least possible value.

For the present, it will be assumed that t_1 is not fixed *a priori*.

An admissible control which gives the solution to the above problem is termed an *optimal control*. The corresponding trajectory connecting x^0 and x^1 is termed an *optimal trajectory*.

Examples. (a) Consider the optimization problem where $f_0(x, u) = 1$. Equation (7.4) becomes

$$J = t_1 - t_0$$

The minimization of J in this case minimizes the transition time in going from x^0 to x^1. This is the familiar time-optimal problem.

(b) Consider the optimization problem where $f_0(x, u) = \| u(t) \|$ where $\| \ \|$ denotes the Euclidean norm. In this case, Eq. (7.4) assumes the form

$$J = \int_{t_0}^{t_1} \| u(t) \| \, dt$$

The minimization of J leads to a control which transfers x^0 to x^1 with a minimum of integrated control effort.

Equivalent Problem Statement. To facilitate the development of the maximum principle, we shall reformulate the above problem. We begin by adding an additional coordinate x_0, defined by the relation

$$\frac{dx_0}{dt} = f_0\left(x^1, \ldots, x^n; u^1, \ldots, u^r\right)$$

to the phase coordinates $x_1, \ldots, x_n$. The function f_0 is the one which appears in the definition of the cost function J above. If

$x_0(t_0) = 0$, then it is clear from Eq. (7.4) that $x_0(t_1) = J$. Thus, we shall now consider the system of ordinary differential equations

$$\begin{aligned} \frac{dx_0}{dt} &= f_0(x, u) & x_0(t_0) &= 0 \\ \frac{dx_1}{dt} &= f_1(x, u) & x_1(t_0) &= x_1^0 \\ &\vdots \\ \frac{dx_n}{dt} &= f_n(x, u) & x_n(t_0) &= x_n^0 \end{aligned} \tag{7.5}$$

We observe that the right-hand sides of Eqs. (7.5) do not depend on x_0.

We now introduce the vector

$$\mathbf{x} = (x_0, x_1, \ldots, x_n) = (x_0, x)$$

of the $(n+1)$-dimensional space $\mathbf{X}$ and rewrite Eqs. (7.5) in the vector form

$$\frac{d\mathbf{x}}{dt} = \mathbf{f}(x, u), \quad \mathbf{x}(t_0) = (0, x^0) \tag{7.6}$$

(Boldfaced letters will be used throughout to indicate $(n+1)$-dimensional vectors.) It follows from Eqs. (7.5) that the right-hand side of Eq. (7.6) does not depend on the coordinate x_0 of the vector $\mathbf{x}$.

Let $u(t)$ be an admissible control which transfers x^0 to x^1, and let $x = x(t)$ be the solution of Eq. (7.3) with $x(t_0) = x^0$. We denote the point $(0, x^0)$ by $\mathbf{x}^0$. That is, $\mathbf{x}^0$ is the point whose coordinates are $\left(0, x_1^0, \ldots, x_n^0\right)$, where $\left(x_1^0, \ldots, x_n^0\right)$ are the coordinates of x^0 in X. It is clear that $\mathbf{x}^0$ is the state of the system in Eq. (7.6) at $t = t_0$. It then follows that the solution of Eq. (7.6) assumes the form:

$$x_0 = \int_{t_0}^{t} f_0[x(s), u(s)]\, ds$$

$$x = x(t)$$

In particular, for $t = t_1$, we obtain

$$x_0 = \int_{t_0}^{t_1} f_0[x(t), u(t)]\, dt = J$$

$$x = x^1$$

Thus, the solution of Eq. (7.6) passes through the point $\mathbf{x}^1 = (J, x^1)$ at $t = t_1$. Now, if we let ℓ be a straight line in X passing through the point $\mathbf{x} = (0, x^1)$ and parallel to the x_0 axis, we can say that the solution $\mathbf{x}(t)$ passes through a point on ℓ, with coordinate $x_0 = J$ at the time $t = t_1$. Conversely, if $u(t)$ is an admissible control such that the solution of Eq. (7.6) passes through the point $x^1 \in \ell$ at some time $t_1 > t_0$ with coordinate $x_0 = J$, then $u(t)$ transfers x^0 to x^1 in X. Moreover, the functional, Eq. (7.4), takes on the value J. We give a three-dimensional geometric interpretation of these results for a second-order dynamical system by letting x_0 be one coordinate and $x = (x_1, x_2)$ be the other two coordinates as shown in Fig. 7-3. The space X in this example is simply Euclidean three-space in which the coordinates are (x_0, x), where the first coordinate is the *cost* and the second coordinate is the *state* of the dynamical system. A typical trajectory joining $(0, x^0)$ with a point on the line ℓ is also shown in the figure. Since the coordinate x_0 is the value of the cost function, it is desired to find an admissible control whose corresponding trajectory intersects the line ℓ at the smallest possible value of x_0.

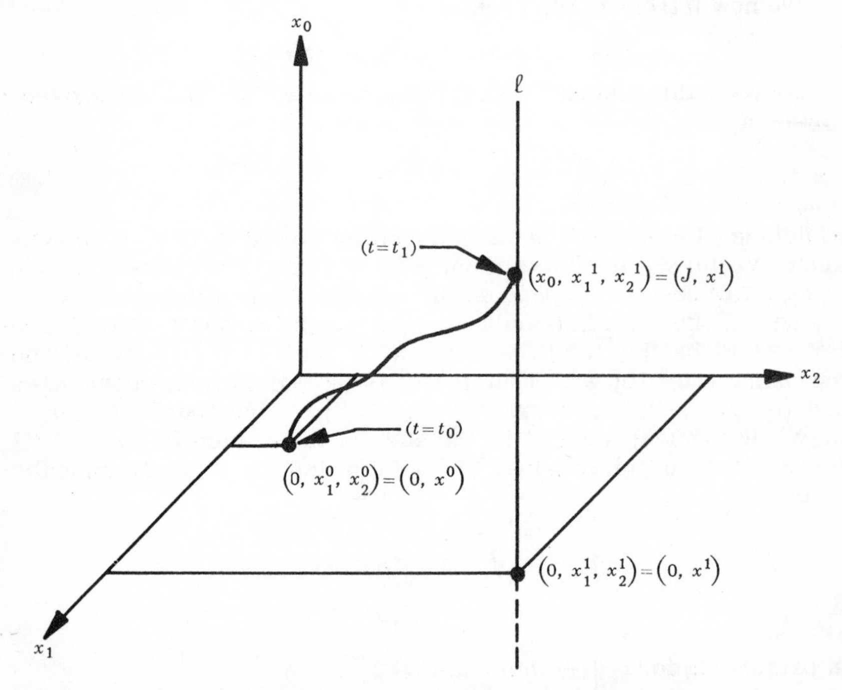

FIG. 7-3. Geometric interpretation of optimization problem for second order dynamical system.

We restate the optimization problem in terms of this new formulation: The point $\mathbf{x}^0 = (0, x^0)$ and the line ℓ are given in the

$(n+1)$-dimensional phase space X. The line ℓ passes through the point $(0, x^1)$ and is parallel to the x_0 axis. From among all the admissible controls $u = u(t)$ for which the solution of Eq. (7.6) intersects ℓ, find one whose point of intersection with ℓ gives the smallest value of the coordinate x_0.

7.2 The Maximum Principle

We now state and discuss the maximum principle, and illustrate its use with a particular example. We begin by writing Eq. (7.6) in the form

$$\frac{dx_i}{dt} = f_i(x, u) \qquad i = 0, 1, \ldots, n \tag{7.7}$$

and adjoining to this latter system, the set of equations

$$\frac{d\psi_i}{dt} = -\sum_{a=0}^{n} \frac{\partial f_a[x(t), u(t)]}{\partial x_i} \psi_a(t) \qquad i = 0, 1, \ldots, n \tag{7.8}$$

It will be expedient for us to write Eqs. (7.7) and (7.8) in a more convenient form. To do this, we consider the function H defined by the relation

$$H(\psi, x, u) = \psi' \mathbf{f}(x, u) = \sum_{a=0}^{n} \psi_a f_a(x, u) \tag{7.9}$$

where ψ is an $(n+1)$-dimensional column vector, $\psi = (\psi_0, \psi_1, \ldots, \psi_n)$, and the prime denotes the transpose. We observe that H is a continuous function of the $2n + 1 + r$ variables $x_1, \ldots, x_n$; $\psi_0, \ldots, \psi_n$; and $u_1, \ldots, u_r$. The function H is termed the *Hamiltonian.*

From Eq. (7.9), we see that Eqs. (7.7) and (7.8) can be written in the form

$$\frac{dx_i}{dt} = \frac{\partial H}{\partial \psi_i}, \qquad \frac{d\psi_i}{dt} = -\frac{\partial H}{\partial x_i} \qquad i = 0, \ldots, n \tag{7.10}$$

respectively. Equations (7.10) are termed the *Hamiltonian system.*

For given fixed values of ψ and x, H is a function of u only. Let us denote the least upper bound of the values of H by $M(\psi, x)$ so that

$$M(\psi, x) = \sup_{u \in U} H(\psi, x, u) \tag{7.11}$$

If H assumes its upper bound on U, then $M(\psi, x)$ is the maximum of H, for fixed ψ and x.

We now state the maximum principle for autonomous systems.

Theorem 1. Let $u(t)$ be an admissible control and let $x(t)$ be the corresponding trajectory of Eq. (7.6) which passes through a point on the line ℓ at some time $t_1 > t_0$. So that $u(t)$ and $x(t)$ will be optimal for $t_0 \leq t \leq t_1$, it is *necessary* that there exist a nontrivial continuous vector function $\psi(t) = \left[\psi_0(t), \psi_1(t), \ldots, \psi_n(t)\right]$ which satisfies Eq. (7.8) such that at all points of continuity of $u(t)$, the Hamiltonian $H(\psi, x, u)$ of the variable $u \epsilon U$ achieves a maximum at $u = u(t)$; that is,

$$H\left[\psi(t), x(t), u(t)\right] = M\left[\psi(t)\ x(t)\right] \tag{7.12}$$

In addition,

$$\psi_0(t) = \text{constant} \leq 0 \quad \text{and} \quad M\left[\psi(t), x(t)\right] = 0 \tag{7.13}$$

The principle result of Theorem 1 lies in Eq. (7.12). We observe, however, that Theorem 1 is only a necessary condition for optimality. That is, the conditions of the theorem must be satisfied by an optimal control. On the other hand, the satisfaction of the conditions by a given admissible control does not mean that the latter is an optimal control.

For some problems (e.g., minimal time and minimal effort control of linear dynamical systems), it has been shown [6-10] that if there exists an admissible control which transfers x^0 to x^1, then there also exists an optimal (not necessarily unique) control.

We can gain some insight into the usefulness of the maximum principle, as well as some of the difficulties encountered, by considering the time-optimal control of a particular class of dynamical systems.

Let us consider dynamical systems characterized by the system of equations

$$\frac{dx}{dt} = f(x) + Bu(t) \tag{7.14}$$

where x is an n-dimensional column vector, u is an r-dimensional column vector, f is an n-dimensional vector-valued function of x, and B is a constant $n \times r$ matrix. We assume f continuous and continuosuly differentiable in x. We constrain the controls so that $|u_i| \leq 1, i = 1, \ldots, r$. We require that the system of Eq. (7.14) be transferred from a given initial state $x(t_0) = x^0$ to a specified terminal state $x(t_1) = x^1$ in minimal time. Note that t_1 is one of the unknown parameters.

As noted earlier, we have $f_0(x, u) = 1$ for time optimality. Hence, the Hamiltonian for our problem is

$$H = \psi_0 + \psi' f(x) + \psi' Bu \tag{7.15}$$

where the notation has been defined earlier.

The Hamiltonian, Eq. (7.15), is maximized if we choose

$$u(t) = \operatorname{sgn}\left[\psi'(t)B\right]' \qquad t_0 \le t \le t_1 \tag{7.16}$$

where the prime denotes the transpose and the function sgn is defined by the relation

$$\operatorname{sgn} y = \begin{cases} +1 & y > 0 \\ -1 & y < 0 \end{cases}$$

and is undefined for $y \equiv 0$.

Assuming no component of $\psi'(t), B$ vanishes on any subinterval of the closed interval $t_0 \le t \le t_1$; and recalling that the maximum principle is a necessary condition for optimality, we may assert that if an optimal control exists for our problem, it must be of the form given in Eq. (7.16).

For our problem, the Hamiltonian system is

$$\frac{dx}{dt} = f(x) + Bu(t) \tag{7.17}$$

and

$$\frac{d\psi_i}{dt} = -\sum_{a=1}^{n} \frac{\partial f_a[x(t)]}{\partial x_i} \psi_a(t), \qquad i = 1, \ldots, n \tag{7.18}$$

with the boundary conditions $x(t_0) = x^0$ and $x(t_1) = x^1$. We note that since x_0 does not appear in $f(x)$, the indices i and a run from 1 to n instead of 0 to n.

From Eqs. (7.12), (7.13), (7.15), and (7.16), we obtain

$$M[\psi(t), x(t)] = \psi_0 + \psi'(t)f[x(t)] + \psi'(t)B \operatorname{sgn}\left[\psi'(t)B\right]'$$

Since ψ_0 = constant ≤ 0, it follows that

$$\psi'(t)\left\{f[x(t)] + B \operatorname{sgn}\left[\psi'(t)B\right]'\right\} = -\psi_0 = \text{constant} \ge 0 \tag{7.19}$$

Hence, we now have $2n + 1$ relations, Eqs. (7.17), (7.18), and (7.19), for $2n + 1$ variables, x, ψ, and t_1.

From this example, we observe two important aspects of the maximum principle which usually arise in applications. First, we note that the form of the optimal control, Eq. (7.16), follows readily from the Hamiltonian. This feature is obviously a desirable one since it specifies the nature of the controller. Our second observation is, unfortunately, an unpleasant one. We note that the switching times for the optimal control are governed by $\psi(t)$. If the initial conditions on $\psi(t)$ were known, Eq. (7.18) could be integrated directly, and the switching times would be known. Instead, we observe that only the boundary conditions on $x(t)$ are known. Hence, we are confronted with a nonlinear, two-point boundary value problem.

If we assume the initial conditions on Eq. (7.18) are known, we can obtain the structural form of the optimal system for our problem by considering Eqs. (7.16), (7.17), and (7.18). This structural form is given in Fig. 7-4.

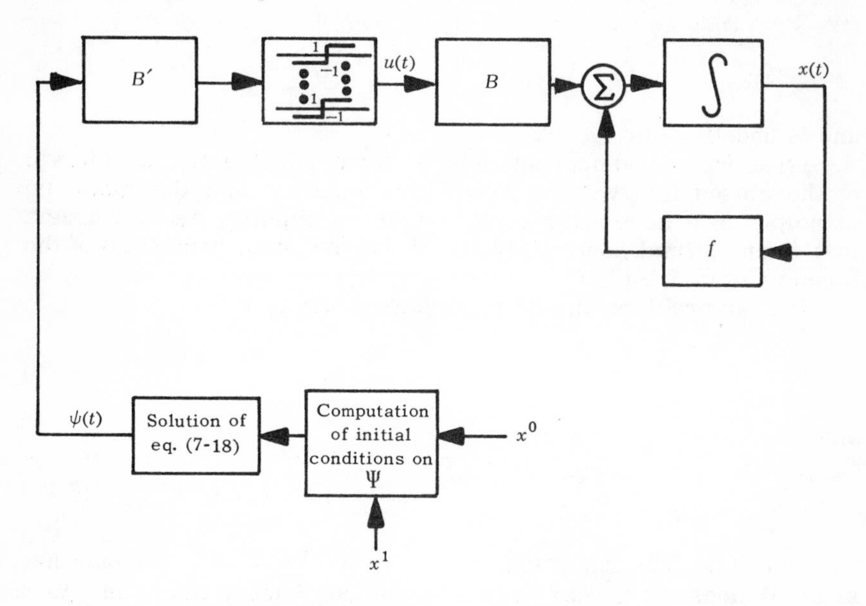

FIG. 7-4. Structural form of optimal system.

7.3 Derivation of the Maximum Principle

In this section, we shall outline the development of the principal results stated in Theorem 1. (A detailed mathematical derivation of these results is available [1, 2].) We shall develop the maximum principle by considering perturbations in the optimal control. By examining certain properties of these perturbations, we shall be able to formulate the *necessary* conditions which an optimal control must satisfy.

Before proceeding with the derivation, it will be necessary for us to introduce some new definitions and concepts.

Variational Trajectories and Controls. Let $\mathrm{x}(t)$ be an optimal trajectory (starting at x^0) which corresponds to an optimal control $u(t)$ for a given problem. Let us assume that the optimal control $u(t)$ is perturbed and that this gives a corresponding perturbed trajectory. We shall denote the perturbed trajectory (which still starts at x^0) by $\mathrm{x}^*(t)$, and the perturbed control (which must also be admissible) by $u^*(t)$. We shall denote the last n coordinates of $\mathbf{x}^*(t)$ by $x^*(t)$.

We shall consider perturbed controls of a particular kind, as described below.

Let τ, $t_0 \leq \tau \leq t_1$, be a point of continuity of $u(t)$. Let $\delta t_1, \ldots, \delta t_s$ be arbitrary nonnegative numbers, and let $v_1, \ldots, v_s$ be arbitrary, not necessarily distinct, points of U. Also, let ϵ be a small positive number. Let us consider the series of adjoining intervals defined by

$$
\begin{aligned}
I_1 &: \tau - \epsilon\delta t_1 \leq t < \tau \\
I_2 &: \tau - \epsilon(\delta t_1 + \delta t_2) \leq t < \tau - \epsilon\delta t_1 \\
&\quad\cdot \qquad\qquad \cdot \\
&\quad\cdot \qquad\qquad \cdot \\
&\quad\cdot \qquad\qquad \cdot \\
I_s &: \tau - \epsilon(\delta t_1 + \ldots + \delta t_s) \leq t < \tau - \epsilon(\delta t_1 + \ldots + \delta t_{s-1})
\end{aligned}
$$

The particular class of perturbed controls we shall consider is given by

$$
u^*(t) = \begin{cases} v_i & t \epsilon I_i, \quad i = 1, \ldots, s \\ u(t) & t \not\epsilon I_i, \quad i = 1, \ldots, s \end{cases}
$$

We assume that the choice of ϵ and the δt_i, $i = 1, \ldots, s$ is such that $\tau - \epsilon \sum_{i=1}^{s} \delta t_i > t_0$. We indicate a typical perturbed control in Fig. 7-5. We point out that the δt_i and $|v_i - u(\tau)|$ for $i = 1, \ldots, s$ are not necessarily small, but that ϵ is.

Let $\delta t = \sum_{i=1}^{s} \delta t_i$, and denote the interval $(\tau - \epsilon\delta t) \leq t < \tau$ by I. Since the perturbation in $u(t)$ begins at the left-hand endpoint of I, we have

$$
\mathbf{x}(\tau - \epsilon\delta t) = \mathbf{x}^*(\tau - \epsilon\delta t) \tag{7.20}
$$

For the unperturbed trajectory, then

$$
\mathbf{x}(\tau) = \mathbf{x}(\tau - \epsilon\delta t) + \int_{\tau-\epsilon\delta t}^{\tau} \dot{\mathbf{x}}(t)\,dt \tag{7.21}
$$

For the perturbed trajectory, we have likewise

$$
\mathbf{x}^*(\tau) = \mathbf{x}(\tau - \epsilon\delta t) + \int_{\tau-\epsilon\delta t}^{\tau} \dot{\mathbf{x}}^*(t)\,dt \tag{7.22}
$$

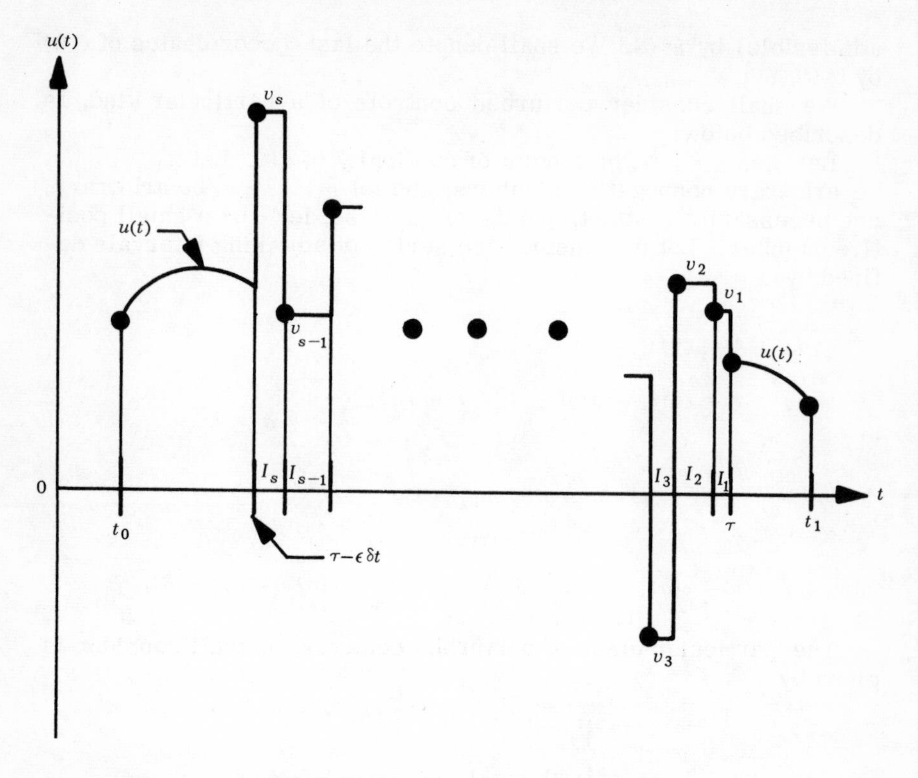

FIG. 7-5. Example of perturbed control.

by utilizing Eq. (7.20). Subtracting Eq. (7.21) from Eq. (7.22)

$$\delta \mathbf{x}(\tau) \triangleq \mathbf{x}^*(\tau) - \mathbf{x}(\tau) = \int_{\tau-\epsilon\,\delta t}^{\tau} \left[\dot{\mathbf{x}}^*(t) - \dot{\mathbf{x}}(t)\right] dt \qquad (7.23)$$

Substituting $\dot{\mathbf{x}}^*(t) = \mathbf{f}\left[x^*(t), u^*(t)\right]$ and $\dot{\mathbf{x}}(t) = \mathbf{f}\left[x(t), u(t)\right]$ in Eq. (7.23)

$$\delta \mathbf{x}(\tau) = \int_{\tau-\epsilon\,\delta t}^{\tau} \left[\mathbf{f}\left[x^*(t), u^*(t)\right] - \mathbf{f}\left[x(t), u(t)\right]\right] dt \qquad (7.24)$$

Using the definition of $u^*(t)$, we rewrite Eq. (7.24) as a summation of integrals:

$$\delta \mathbf{x}(\tau) = \sum_{i=1}^{s} \int_{I_i} \left[\mathbf{f}\left(x^*(t), v_i\right) - \mathbf{f}\left(x(t), u(t)\right)\right] dt \qquad (7.25)$$

As $\epsilon \to 0$, the length of the interval $I \to 0$ so that

$$x(t) \xrightarrow[\epsilon \to 0]{} x(\tau) \qquad \text{on } I$$

Moreover, since the effect of the perturbed control becomes arbitrarily small as $\epsilon \to 0$, we have $x^*(t) \to x(t)$ as $\epsilon \to 0$, so that

$$x^*(t) \xrightarrow[\epsilon \to 0]{} x(t) \qquad \text{on } I$$

Combining the last two relations, we obtain

$$x^*(t) \xrightarrow[\epsilon \to 0]{} x(\tau)$$

Also, since u is continuous at τ, we have

$$u(t) \xrightarrow[\epsilon \to 0]{} u(\tau)$$

on I. Therefore, since f is continuous in x and u,

$$\mathrm{f}\left[x^*(t), v_i\right] = \mathrm{f}\left[x(\tau), v_i\right] + \mathrm{g}(t) \qquad \text{for } t \epsilon I_i \tag{7.26}$$

where $\mathrm{g}(t) \xrightarrow[\epsilon \to 0]{} 0$, and

$$\mathrm{f}\ x(t), u(t)\big] = \mathrm{f}\left[x(\tau), u(\tau)\right] + \mathrm{h}(t) \qquad \text{for } t \epsilon I \tag{7.27}$$

where $\mathrm{h}(t) \xrightarrow[\epsilon \to 0]{} 0$.

Subtracting Eq. (7.27) from Eq. (7.26), and substituting the result in Eq. (7.25), we obtain

$$\delta \mathrm{x}(\tau) = \sum_{i=1}^{s} \int_{I_i} \left\{ \mathrm{f}\left[x(\tau), v_i\right] - \mathrm{f}\left[x(\tau), u(\tau)\right] \right\} dt + \int_{\tau - \epsilon \delta t}^{\tau} \left[\mathrm{g}(t) - \mathrm{h}(t)\right] dt \tag{7.28}$$

The last term in Eq. (7.28) can be approximated as follows:

$$\int_{\tau - \epsilon \delta t}^{\tau} \left[\mathrm{g}(t) - \mathrm{h}(t)\right] dt \le \max |\mathrm{g}(t) - \mathrm{h}(t)| \cdot \epsilon \delta t, \ t \epsilon I$$

Since $\mathrm{g}(t) \to 0$ and $\mathrm{h}(t) \to 0$ as $\epsilon \to 0$, this term is of order ϵ^2.

In addition, each of the integrands in the first term of Eq. (7.28) is constant in its respective subinterval I_i, $i = 1, \ldots, s$. Thus, for ϵ sufficiently small, Eq. (7.28) becomes

$$\delta \mathrm{x}(\tau) = \epsilon \sum_{i=1}^{s} \left\{ \mathrm{f}\left[x(\tau), v_i\right] - \mathrm{f}\left[x(\tau), u(\tau)\right] \right\} \delta t_i + \cdots \tag{7.29}$$

where the dots denote higher order terms in ϵ.

For the work which follows, it will be convenient for us to rewrite Eq. (7.29) in the form

$$\delta \mathrm{x}(\tau) = \mathrm{x}^*(\tau) - \mathrm{x}(\tau) = \epsilon \Delta \mathrm{x}(\tau) + \cdots \tag{7.30}$$

where

$$\Delta x(\tau) = \sum_{i=1}^{s} \left\{ f[x(\tau), v_i] - f[x(\tau), u(\tau)] \right\} \delta t_i \tag{7.31}$$

Equation (7.30) gives us the perturbation in the optimal trajectory at time τ for the perturbation we have introduced into the optimal control. This perturbation will "propagate" along the trajectory as indicated in Fig. 7-6. In particular, we are interested in determining how the perturbation in the optimal control affects the terminal state of the system. That is, given $\delta x(\tau)$, we wish to determine $\delta x(t_1)$, $t_1 > \tau$, which is given by

$$\delta x(t_1) = x^*(t_1) - x(t_1) = \epsilon \Delta x(t_1) + \ldots$$

We emphasize here that we are only considering perturbations in the optimal control on the interval $\tau - \epsilon \delta t \leq t < \tau$. Hence, for $t_1 \geq t \geq \tau$, the control is once again the optimal one, i.e., $u^*(t) = u(t)$ for $t \geq \tau$.

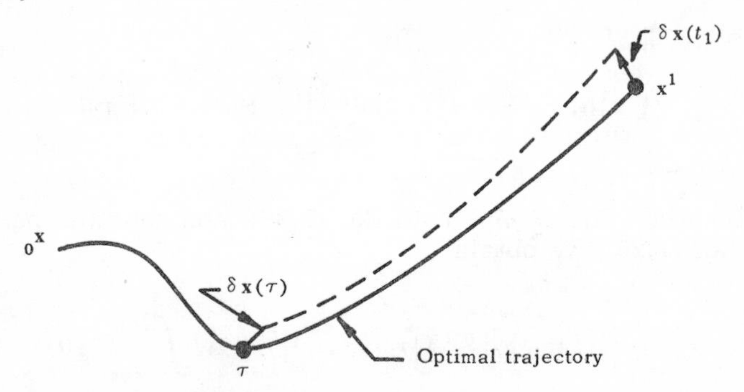

FIG. 7-6. Propagation of a perturbation in the optimal trajectory.

To determine $\Delta x(t_1)$ from $\Delta x(\tau)$, we must solve the linearized perturbation equations

$$\frac{d[\Delta x_i(t)]}{dt} = \sum_{\beta=0}^{n} \frac{\partial f_i[x(t), u(t)]}{\partial x_\beta} [\Delta x_\beta(t)] \qquad i = 0, 1, \ldots, n \tag{7.32}$$

on the interval $\tau \leq t \leq t_1$ with the initial conditions $\Delta x_i(\tau)$, $i = 0, 1, \ldots, n$. In Eq. (7.32), the $\Delta x_i(t)$ and $\Delta x_\beta(t)$ are the components of the vector $\Delta x(t)$, and the partial derivatives are evaluated along the optimal trajectory. Evaluation of the solution of Eq. (7.32) at t_1 then gives us $\Delta x(t_1)$.

We remark that the systems of Eqs. (7.8) and (7.32) are linear and homogeneous, and that the matrix

$$\left[- \frac{\partial f_a[x(t), u(t)]}{\partial x_i} \right]$$

of Eq. (7.8) is the negative of the transpose of the matrix of Eq. (7.32). Hence, the two systems are *adjoint*. For convenience, we shall hereafter refer to Eq. (7.8) as the adjoint system of equations.

Since the system of Eq. (7.32) is linear, $\Delta x(t_1)$ is obtained from $\Delta x(\tau)$ by a linear transformation which we shall denote by A_{t_1,τ_j}. Thus, we have

$$\Delta x(t_1) = A_{t_1,\tau} \cdot \Delta x(\tau) \tag{7.33}$$

Up to now, we have only considered a single set of perturbations in the optimal control near one fixed time τ, and have computed its effect on the terminal state of the system. Let us generalize this result by considering a finite number of points along the optimal trajectory which occur at distinct instants of time τ_j, $j = 1, \ldots, p$. We assume that each of these τ_j is a point of continuity of $u(t)$. Let us introduce perturbations in the optimal control of the type shown in Fig. 7-5 at each of the τ_j, and assume that we have chosen ϵ small enough so that none of the perturbations overlap. We shall use the same ϵ for every τ_j.

We note, first of all, that these perturbations are independent, and that Eq. (7.33) is a linear transformation. Hence, by employing arguments similar to those given above, we can show that the net effect of these perturbations on the terminal state of the system is obtained by superposition of the effects of the individual perturbations. Hence,

$$\Delta x(t_1) = \sum_{j=1}^{p} A_{t_1,\tau_j} \Delta x(\tau_j) \tag{7.34}$$

where each $\Delta x(\tau_j)$ has the form given in Eq. (7.31).

In addition to perturbations in the optimal control, we allow perturbations $\epsilon\delta t$ in the terminal time t_1. Here δt may be either positive or negative (not necessarily small), and ϵ is the same as that used above.

Since we are now considering only perturbations in the terminal time, we have $u^*(t) = u(t)$. Hence,

$$\delta x(t_1)\Big|_{\text{due to } \epsilon\delta t} = \int_{t_1}^{t_1+\epsilon\delta t} f[x^*(t), u(t)]\, dt \tag{7.35}$$

As before, $x^*(t) \xrightarrow[\epsilon\to 0]{} x(t)$, and for $\delta t \le 0$, Eq. (7.35) may be written as

$$\delta x(t_1)\Big|_{\text{due to } \epsilon\delta t} = \epsilon f[x(t_1), u(t_1)]\,\delta t + \ldots \tag{7.36}$$

where the dots denote higher order terms in ϵ. For $\delta t \ge 0$, i.e., $t_1 \le t \le t_1 + \epsilon\delta t$, we let $u(t) = u(t_1)$. Then, Eq. (7.36) holds for either positive or negative δt. Hence, to within first-order terms in ϵ, we have

$$\delta x(t_1)\Big|_{\text{due to } \epsilon\delta t} = \epsilon \Delta x(t_1)\Big|_{\text{due to } \epsilon\delta t} = \epsilon f\left[x(t_1), u(t_1)\right]\delta t \qquad (7.37)$$

From Eq. (7.37), we obtain

$$\Delta x(t_1)\Big|_{\text{due to } \epsilon\delta t} = f\left[x(t_1), u(t_1)\right]\delta t \qquad (7.38)$$

Combining Eqs. (7.34) and (7.38), we have

$$\Delta x(t_1) = f\left[x(t_1), u(t_1)\right]\delta t + \sum_{j=1}^{p} A_{t_1, \tau_j} \Delta x(\tau_j) \qquad (7.39)$$

Substitution of Eq. (7.31) into Eq. (7.39) gives us

$$\Delta x(t_1) = f\left[x(t_1), u(t_1)\right]\delta t + \sum_{j=1}^{p} A_{t_1, \tau_j} \sum_{i=1}^{s} \left\{ f\left[x(\tau_j), v_i^j\right] - f\left[x(\tau_j), u(\tau_j)\right]\right\} \delta t_i^j \qquad (7.40)$$

where the j superscripts are used to indicate that the v_i and δt_i are different for the different points τ_j.

Linear Combinations of Variations. If any of the δt_i^j vanish, we can eliminate them in the definition of the perturbed controls $u^*(t)$, along with the corresponding v_i^j. Conversely, the addition of new v_i^j, whose corresponding δt_i^j equal zero, will not change $u^*(t)$. We shall make use of this in the sequel. Indeed, if we consider a finite number of perturbed controls $u_1^*(t), \ldots, u^*_q(t)$, we can suppose that the same v_i^j occur in the definition of every one of the controls. The difference in these controls then lies in the fact that the δt_i^j and δt differ for the different controls.

The vector $\Delta x(t_1)$ in Eq. (7.40) is independent of ϵ, but does depend on τ_j, v_i^j, δt_i^j, and δt. We shall denote the totality of these quantities by $\gamma = \left\{\tau_j, v_i^j, \delta t_i^j, \delta t\right\}$. To emphasize the fact that $\Delta x(t_1)$ depends on these quantities, we shall adopt the notation Δx_γ.

Let us consider a finite number of these quantities γ of the form:

$$\gamma' = \left\{\tau_j, v_i^j, \overline{\delta t}_i^j, \overline{\delta t}\right\}$$

$$\gamma'' = \left\{\tau_j, v_i^j, \overline{\overline{\delta t}}_i^j, \overline{\overline{\delta t}}\right\}$$

$$\cdots\cdots\cdots$$

We define linear combinations of the γ's by

$$\lambda'\gamma' + \lambda''\gamma'' + \dots = \left\{\tau_j,\ v_i^j,\ \lambda'\overline{\delta t}_i^j + \lambda''\overline{\overline{\delta t}}_i^j + \dots,\ \lambda'\overline{\delta t} + \lambda''\overline{\overline{\delta t}} + \dots, \dots\right\}$$

where $\lambda' \geq 0,\ \lambda'' \geq 0, \dots$.

The Cone of Attainability. We shall now consider the vectors $\Delta \mathbf{x}_\gamma$ for various γ.

From the fact that Eq. (7.40) is linear in δt_i^j and δt, it follows that if $\gamma = \lambda'\gamma' + \lambda''\gamma'' + \dots,\ \lambda' \geq 0,\ \lambda'' \geq 0, \dots$, the corresponding vectors $\Delta \mathbf{x}$ are related by

$$\Delta \mathbf{x}_\gamma = \lambda'\Delta \mathbf{x}_{\gamma'} + \lambda''\Delta \mathbf{x}_\gamma + \dots \tag{7.41}$$

We shall consider $\Delta \mathbf{x}_\gamma$ to be a vector connected to, and originating from the point $\mathbf{x}(t_1) = \mathbf{x}^1$, the endpoint of the optimal trajectory, in the $(n+1)$-dimensional space X. If we consider all the possible γ as described above, the vectors $\Delta \mathbf{x}_\gamma$ generate a set K in X. In particular, K is a convex cone* in X.

To show that K is a convex cone, we let $\mathbf{b}'$ and $\mathbf{b}''$ be two distinct points of X which also lie in K. That is, we assume there exist two quantities γ' and γ'' such that $\mathbf{b}' = \Delta \mathbf{x}_{\gamma'}$ and $\mathbf{b}'' = \Delta \mathbf{x}_{\gamma''}$. Then, from Eq. (7.41), we have

$$\lambda'\mathbf{b}' + \lambda''\mathbf{b}'' = \lambda'\Delta \mathbf{x}_{\gamma'} + \lambda''\Delta \mathbf{x}_{\gamma''} = \Delta \mathbf{x}_{(\lambda'\gamma' + \lambda''\gamma'')}$$

which means the point $\lambda'\mathbf{b}' + \lambda''\mathbf{b}''$ also lies in K. Hence, K is a convex cone in X with its vertex at $\mathbf{x}(t_1) = \mathbf{x}_1$.

We shall call K the *cone of attainability.*

A simple geometric interpretation of K is given in Fig. 7-7 for a second-order system. In the figure, $\mathbf{x}(t)$ is the optimal trajectory, ℓ is the line through the point $(0, x_1^0, x_2^0)$ parallel to the x_0 axis, and J_0 is the optimal (minimum) cost associated with the optimal control $u(t)$.

Hyperplanes of Support. To clarify the arguments which follow, we shall employ a two-dimensional model of the $(n+1)$-dimensional phase space X as shown in Fig. 7-8. We assume $\mathbf{x}(t),\ t_0 \leq t \leq t_1$, is an optimal trajectory joining the point $\mathbf{x}^0 = (0, x^0)$ with the point (J_0, x^1) on the line ℓ, that $u(t)$ is the corresponding optimal control, and that K is the cone of attainability. Since $\mathbf{x}(t)$ is an optimal trajectory, J_0 is the minimum cost associated with transferring the dynamical system of Eq. (7.2) from the initial state x^0 to the terminal state x^1.

It is clear that any ray originating at (J_0, x^1), the vertex of K, parallel to the x_0 axis and directed in the negative x_0 direction,

*A set K, in a vector space X is called a convex cone if: (1) it has a vertex O such that for any point P in the set distinct from O, the ray OP lies entirely within the set, and (2) for any two points in the set, the line segment joining them lies entirely within the set.

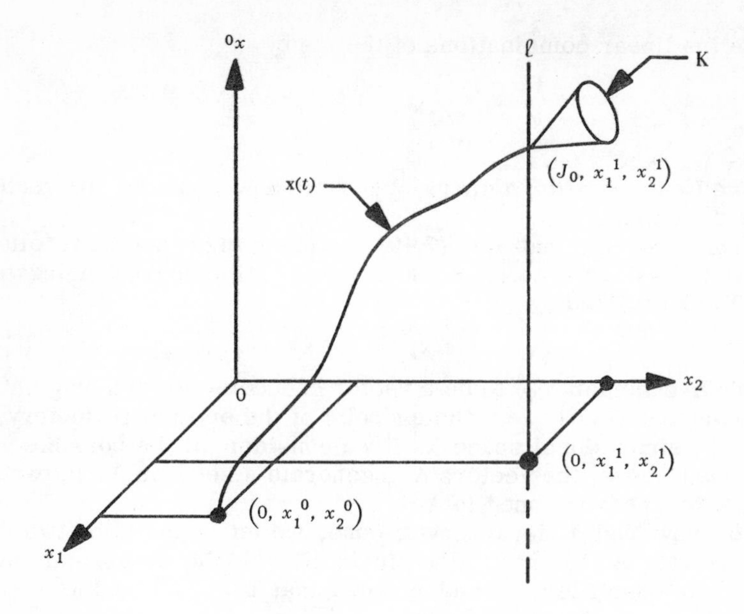

FIG. 7-7. Geometric interpretation of cone of attainability for second-order system.

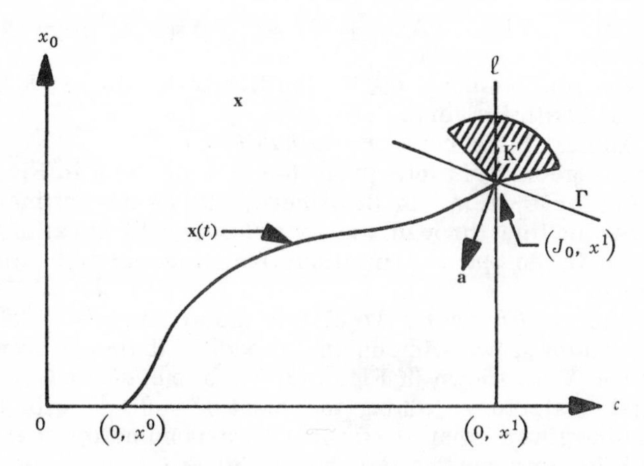

FIG. 7-8. Model of phase space X.

cannot lie in K. If this were not so, we could find an admissible perturbation in $u(t)$ which would transfer the system of Eq. (7.2) from x^0 to x^1 with a cost less than J_0, thereby contradicting the assumption that $\mathrm{x}(t)$ is an optimal trajectory. Hence, K does not fill out the entire space X. This means that we can construct a hyperplane of support for K at its vertex, that is, a hyperplane Γ, as in Fig. 7-8, such that K lies in one of the two closed half-spaces defined by Γ. We note that Γ need not be unique.

Now let a be any normal to Γ at (J_0, x^1) directed away from K as shown in the figure. The normal a is an $(n+1)$-dimensional vector in X having components $a_0, a_1, \ldots, a_n$. Since a "points away from" K, i.e., "downward" as in Fig. 7-8, $a_0 \leq 0$. By our choice of the direction of a, the cone of attainability K lies in the negative half-space. In other words, for any $\Delta x \epsilon K$

$$a'\Delta x \leq 0 \tag{7.42}$$

where the prime denotes the transpose.

Equation (7.42) is the fundamental relationship from which we shall develop the maximum principle stated in Theorem 1. Before we can complete this development, we must establish one more auxiliary relation: that if $\psi(t) = [\psi_0(t), \psi_1(t), \ldots, \psi_n(t)]$ is a solution of Eq. (7.8), then

$$\psi'(t)\, A_{t,\tau}\, y(\tau) = \text{constant} \quad t \geq \tau \tag{7.43}$$

where $y(\tau)$ is any $(n+1)$ dimensional column vector.

To establish Eq. (7.43), we shall show that the derivative of the left-hand side vanishes. Taking this derivative, we obtain

$$\frac{d}{dt}\left[\psi'(t)\, A_{t,\tau}\, y(\tau)\right] = \frac{d\psi'(t)}{dt} A_{t,\tau}\, y(\tau) + \psi'(\tau)\frac{d}{dt}\left[A_{t,\tau}\, y(\tau)\right] \tag{7.44}$$

In vector-matrix form, Eq. (7.8) may be written as

$$\frac{d\psi(t)}{dt} = -\left[\frac{\partial f[x(t), u(t)]}{\partial x}\right]' \psi(t)$$

and its transpose as

$$\frac{d\psi'(t)}{dt} = -\psi'(t)\left[\frac{\partial f[x(t), u(t)]}{\partial x}\right] \tag{7.45}$$

where the partial derivatives are evaluated along the optimal trajectory and its corresponding optimal control.

Substituting Eq. (7.45) in Eq. (7.44), the right-hand side of Eq. (7.44) becomes

$$-\psi'(t)\left[\frac{\partial f[x(t), u(t)]}{\partial x}\right] A_{t,\tau}\, y(\tau) + \psi'(t)\frac{d}{dt}\left[A_{t,\tau}\, y(\tau)\right] \tag{7.46}$$

From Eq. (7.32) and the definition of $A_{t,\tau}$,

$$\frac{d}{dt}\left[A_{t,\tau}\, y(\tau)\right] = \left[\frac{\partial f[x(t), u(t)]}{\partial x}\right] A_{t,\tau}\, y(\tau) \tag{7.47}$$

Upon substitution of Eq. (7.47) in Eq. (7.46), the latter vanishes, and we obtain

$$\frac{d}{dt}\,\psi'(t)\left[A_{t,\tau}\,\mathrm{y}(\tau)\right] = 0 \quad t \geq \tau$$

Hence, our assertion is proved.

The Maximum Principle. Now, Eq. (7.42) must be satisfied for every perturbation in the optimal control and the terminal time.

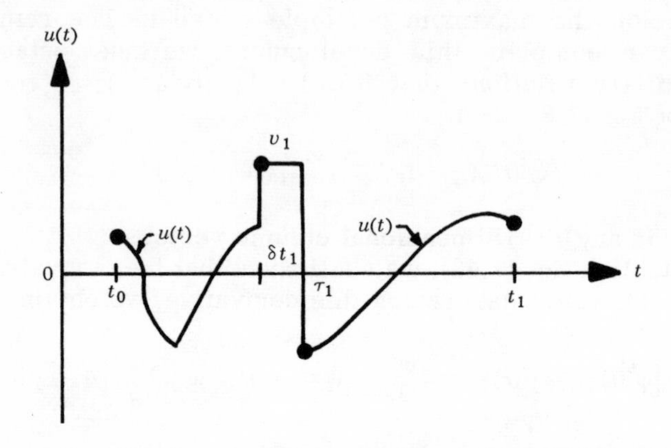

FIG. 7-9. Single perturbation in optimal control.

First, let us consider a single perturbation in the optimal control as shown in Fig. 7-9. We let τ_1 be a point of continuity of $u(t)$, and let v_1 be an arbitrary point of U. Since we are considering only a single perturbation in $u(t)$, we set $p = s = 1$ in Eq. (7.40). In addition, since we are not perturbing the terminal time, we set $\delta t = 0$. Thus, we are considering the symbol γ for which

$$\gamma = \{\tau_1, v_1, \delta t_1 > 0, 0\}$$

Then the vector $\Delta\mathrm{x}(t_1)$, given by Eq. (7.40) for this γ, is

$$\Delta\mathrm{x}(t_1) = A_{t_1,\tau_1}\left\{\mathrm{f}\left[x(\tau_1), v_1\right] - \mathrm{f}\left[x(\tau_1), u(\tau_1)\right]\right\}\delta t_1$$

Substituting this value into Eq. (7.42), we obtain

$$\mathrm{a}' \cdot A_{t_1,\tau}\left\{\mathrm{f}\left[x(\tau_1), v_1\right] - \mathrm{f}\left[x(\tau_1), u(\tau_1)\right]\right\}\delta t_1 \leq 0 \tag{7.48}$$

To simplify the notation, we shall rewrite Eq. (7.48) in the form

$$\mathrm{a}' \cdot A_{t_1,\tau_1}\left[\Delta\mathrm{f}(v_1, \tau_1)\right]\delta t_1 \leq 0 \tag{7.49}$$

Now let $\psi(t) = [\psi_0(t), \psi_1(t), \ldots, \psi_n(t)]$ be the solution of Eq. (7.8) with the boundary conditions

$$\psi(t_1) = a \tag{7.50}$$

Substituting Eq. (7.50) into Eq. (7.49), we get

$$\psi'(t_1) \cdot A_{t_1,\tau_1}[\Delta f(v_i, \tau_1)]\,\delta t_1 \leq 0 \tag{7.51}$$

From Eq. (7.43)

$$\psi'(t_1) \cdot A_{t_1,\tau_1}[\Delta f(v_1, \tau_1)] = \psi'(\tau_1) \cdot A_{\tau_1,\tau_1}[\Delta f(v_1, \tau_1)] \tag{7.52}$$

Since $A_{t_1,\tau_1}[\Delta f(v_1, \tau_1)]$ is the solution of Eq. (7.32) evaluated at the initial time τ_1, Eq. (7.52) becomes

$$\psi'(t_1) \cdot A_{t_1,\tau_1}[\Delta f(v_1, \tau_1)] = \psi'(\tau_1) \cdot [\Delta f(v_1, \tau_1)] \tag{7.53}$$

Substituting Eq. (7.53) in Eq. (7.51), we have

$$\psi'(\tau_1) \cdot \Delta f(v_1, \tau_1)\,\delta t_1 \leq 0 \tag{7.54}$$

From the definition of $\Delta f(v_1, \tau_1)$, Eq. (7.54) becomes

$$\psi'(\tau_1) \cdot \left\{ f[x(\tau_1), v_1] - f[x(\tau_1), u(\tau_1)] \right\} \delta t_1 \leq 0$$

Since δt_1 is positive,

$$\psi'(\tau_1) \cdot \left\{ f[x(\tau_1), v_1] - f[x(\tau_1, u(\tau_1)] \right\} \leq 0 \tag{7.55}$$

From the definition of the Hamiltonian, Eq. (7.55) may be written in the form

$$H[\psi(\tau_1), x(\tau_1), v_1] \leq H[\psi(\tau_1), x(\tau_1), u(\tau_1)] \tag{7.56}$$

Since Eq. (7.56) is valid for any point $v_1 \epsilon U$, we obtain

$$H[\psi(\tau_1), x(\tau_1), u(\tau_1)] = \max_{v_1 \epsilon U} H[\psi(\tau_1), x(\tau_1), v_1] = M[\psi(\tau_1), x(\tau_1)] \tag{7.57}$$

Moreover, since Eqs. (7.56) and (7.57) hold at any τ_1 which is a point of continuity of $u(t)$, we have, finally

$$H[\psi(t), x(t), u(t)] = M[\psi(t), x(t)] \qquad t_0 \leq t \leq t_1 \tag{7.58}$$

along an optimal trajectory at all points of continuity of $u(t)$. Thus, Eq. (7.12) of Theorem 1 is proved.

Now we shall consider a perturbation in the terminal time only. Thus, we consider the symbol γ for which

$$\gamma = \{0, 0, 0, \delta t\}$$

That is, we have no points τ_j. For this γ, we obtain, from Eqs. (7.40) and (7.42),

$$\mathbf{a}' \cdot \mathbf{f}[x(t_1), u(t_1)]\,\delta t \leq 0 \tag{7.59}$$

Substituting Eq. (7.50) into Eq. (7.59), and utilizing the definition of the Hamiltonian, we obtain

$$H[\psi(t_1), x(t_1), u(t_1)]\,\delta t \leq 0 \tag{7.60}$$

Since δt may be either positive or negative, Eq. (7.60) can only be satisfied if

$$H[\psi(t_1), x(t_1), u(t_1)] = 0 \tag{7.61}$$

That is, the Hamiltonian vanishes at the endpoint of an optimal trajectory. To emphasize that Eq. (7.61) is valid for an optimal trajectory, i.e., one along which the Hamiltonian is a maximum, we shall write it as

$$M[\psi(t_1), x(t_1)] = 0 \tag{7.62}$$

We shall now show that Eq. (7.62) holds for all t, $t_0 \leq t \leq t_1$. Differentiating Eq. (7.58) with respect to t, we obtain

$$\frac{dH}{dt} = \sum_{a=0}^{n}\left(\frac{\partial H}{\partial \psi_a}\frac{d\psi_a}{dt} + \frac{\partial H}{\partial x_a}\frac{dx_a}{dt}\right) + \sum_{a=1}^{r}\frac{\partial H}{\partial u_a}\frac{du_a}{dt} \tag{7.63}$$

If we substitute the Hamiltonian system Eq. (7.10) into Eq. (7.63), we obtain

$$\frac{dH}{dt} = \sum_{a=0}^{n}\left(\frac{dx_a}{dt}\frac{d\psi_a}{dt} - \frac{d\psi_a}{dt}\frac{dx_a}{dt}\right) + \sum_{a=1}^{r}\frac{\partial H}{\partial u_a}\frac{du_a}{dt}$$

which simplifies to

$$\frac{dH}{dt} = \sum_{a=1}^{r}\frac{\partial H}{\partial u_a}\frac{du_a}{dt} \tag{7.64}$$

If U is a closed region in r-dimensional Euclidean space, the values of u in the interior of U which maximize the Hamiltonian are determined by setting $\partial H/\partial u_a = 0$, $a = 1, \ldots, r$. Equation (7.64) then becomes $dH/dt = 0$.

Now, if the Hamiltonian can only be maximized by selecting values of u on the boundary of U, we no longer have $\partial H/\partial u_a = 0$, $a = 1, \ldots, r$. However, it can be shown in this case that the two vectors comprising the elements $\partial H/\partial u_a$ and du_a/dt, $a = 1, \ldots, r$, respectively, are orthogonal. Hence, the right-hand side of Eq. (7.64) will also be zero in this case. Therefore, we have

$$M[\psi(t), x(t)] = 0 \qquad t_0 \leq t \leq t_1 \tag{7.65}$$

along an optimal trajectory.

From Eq. (7.50) and the fact that $a_0 \leq 0$, we have

$$\psi_0(t_1) \leq 0 \tag{7.66}$$

As observed earlier, none of the f_a, $a = 0, 1, \ldots, n$, in Eqs. (7.5) depend on x_0. Therefore, for $i = 0$ in Eq. (7.8), we have

$$\frac{d\psi_0}{dt} = 0 \qquad t_0 \leq t \leq t_1$$

It then follows from Eq. (7.66) that

$$\psi_0(t) = \text{constant} \leq 0 \qquad t_0 \leq t \leq t_1 \tag{7.67}$$

Equations (7.65) and (7.67) give us Eq. (7.13) of Theorem 1. This completes our derivation.

Discussion of Results—The Synthesis Problem. We conclude this section by considering certain aspects of the problem of implementing or synthesizing optimal controls. We observe from the preceding section that the boundary conditions on the adjoint system, Eq. (7.8), are $\psi(t_1) = a$ where a is a vector normal to the hyperplane of support Γ. To determine these boundary conditions, we must know a hyperplane of support Γ for the cone of attainability K and determine an appropriate vector normal to this hyperplane. The computational difficulties involved are apparent. Even if an a could be obtained, Eq. (7.8) would have to be integrated backward in time from $t = t_1$ to $t = t_0$ before the optimal control could be determined. Since t_1 is free, i.e., unknown *a priori*, the problem is further complicated. In its general form, we see that the synthesis problem is a nonlinear, two-point boundary value problem. This was seen to be the case for the particular optimization problem considered in Section 7.2.

We shall comment further on the synthesis problem at the end of this chapter.

7.4 Extensions of the Maximum Principle

In this section we shall consider two control optimization problems whose solutions are obtained as extensions of the preceding results.

The Maximum Principle for Nonautonomous Systems. Let us consider the class of problems in which the dynamic system's equations are

$$\frac{dx_i}{dt} = f_i(x, u, t) \qquad i = 1, 2, \ldots, n$$

and we wish to minimize the functional

$$J = \int_{t_0}^{t_1} f_0[x(t), u(t), t]\,dt$$

We assume t_1 is not fixed and that $u \epsilon U$.

As before, we introduce the new coordinate

$$x_0 = \int_{t_0}^{t} f_0[x(s), u(s), s]\,ds$$

or equivalently,

$$\frac{dx_0}{dt} = f_0[x(t), u(t), t] \qquad x_0(t_0) = 0$$

We now state the optimization problem we wish to solve. The point $(0, x^0)$ and the line ℓ, parallel to the x_0 axis and passing through the point $(0, x^1)$, are given in the $(n+1)$-dimensional phase space X. From among all the admissible controls $u = u(t)$ having the property that the solution $\mathrm{x}(t)$ of the system

$$\frac{dx_i}{dt} = f_i(x, u, t) \qquad \mathrm{x}(t_0) = (0, x^0) \qquad i = 0, 1, \ldots, n \tag{7.68}$$

intersects ℓ, find one whose point of intersection with ℓ gives the smallest value of the coordinate x_0.

We observe that the problem is of the same form as in Eq. (7.6). The difference is that we now permit the functions f_α to depend explicitly on time.

We approach this problem by introducing yet another coordinate x_{n+1}, defined by

$$\frac{dx_{n+1}}{dt} = 1 \qquad x_{n+1}(t_0) = t_0$$

It is clear from the definition that $x_{n+1} \equiv t$. Hence, if we make this substitution in Eq. (7.68), we obtain the *autonomous* system

$$\frac{dx_i}{dt} = f_i(x, u, x_{n+1}) \qquad i = 0, 1, \ldots, n \qquad \frac{dx_{n+1}}{dt} = 1 \tag{7.69}$$

Let us now outline the development of the maximum principle for the problem at hand. Analogous to the adjoint system of Eqs. (7.8) for the autonomous case, we write

$$\frac{d\psi_i}{dt} = -\sum_{a=0}^{n} \frac{\partial f_a}{\partial x_i} \psi_a \qquad i = 0, 1, \ldots, n \tag{7.70}$$

and

$$\frac{d\psi_{n+1}}{dt} = -\sum_{a=0}^{n} \frac{\partial f_a}{\partial x_{n+1}} \psi_a$$

Since $x_{n+1}(t) \equiv t$, the latter equation can be written as

$$\frac{d\psi_{n+1}}{dt} = -\sum_{a=0}^{n} \frac{\partial f_a}{\partial t} \psi_a \tag{7.71}$$

Let us now consider the function

$$\sum_{a=0}^{n} \psi_a f_a(x, u, t) + \psi_{n+1} \frac{dx_{n+1}}{dt} \tag{7.72}$$

We denote this function by H^* and define

$$H(\psi, x, u, t) = \sum_{a=0}^{n} \psi_a f_a(x, u, t) \tag{7.73}$$

Let us denote the maxima of H^* and H with respect to u, for fixed x_i and ψ_i, by M^* and M respectively. From Eqs. (7.72) and (7.73) and the fact that

$$\frac{dx_{n+1}}{dt} = 1$$

we have the relations

$$H^* = H + \psi_{n+1} \quad \text{and} \quad M^* = M + \psi_{n+1} \tag{7.74}$$

Since Eqs. (7.69) represent an autonomous system, it follows from Theorem 1 that the relation

$$H^*[\psi(t), x(t), u(t), t] = M^*[\psi(t), x(t), u(t), t] = 0 \tag{7.75}$$

is satisfied at all values of t which are points of continuity of $u(t)$. Substituting in Eq. (7.75) from Eq. (7.74), we obtain

$$H[\psi(t), x(t), u(t), t] = M[\psi(t), x(t), u(t), t] = -\psi_{n+1}(t)$$

It is shown [1, 2] that $\psi_{n+1}(t_1) = 0$, so that, by Eq. (7.71)

$$-\psi_{n+1}(t) = \int_{t_1}^{t} \sum_{a=0}^{n} \frac{\partial f_a[x(t), u(t), t]}{\partial t} \psi_a(t)\, dt$$

From Theorem 1, we have

$$\psi_0(t) = \text{constant} \leq 0$$

Summarizing these results, we obtain the maximum principle for nonautonomous systems.

Theorem 2. Let $u(t)$ be an admissible control and let $x(t)$ be the corresponding trajectory of Eq. (7.68) which passes through a point on the line ℓ at some time $t_1 > t_0$. In order that $u(t)$ and $x(t)$ be optimal for $t_0 \leq t \leq t_1$, it is necessary that there exist a nontrivial continuous vector function $\psi(t) = [\psi_0(t), \ldots, \psi_n(t)]$ which satisfies Eq. (7.70) such that at all points of continuity of $u(t)$, the Hamiltonian $H(\psi, x, u, t)$ on the variable $u \epsilon U$ achieves a maximum at $u = u(t)$; that is,

$$H[\psi(t), x(t), u(t), t] = M[\psi(t), x(t), x(t), t] \tag{7.76}$$

In addition,

$$\psi_0(t) = \text{constant} \leq 0$$

and

$$M[\psi(t), x(t), t] = -\psi_{n+1}(t) = \int_{t_1}^{t} \sum_{a=0}^{n} \frac{\partial f_a[x(t), u(t), t]}{\partial t} \psi_a(t)\, dt \tag{7.77}$$

The Maximum Principle for the Fixed-Time Problem. We shall now consider the optimization problem for the autonomous case when the terminal time t_1 is specified *a priori*. That is, when the interval of system operation $(t_1 - t_0)$ is a known constant.

We have the system of equations

$$\frac{dx_i}{dt} = f_i(x, u) \qquad x(t_0) = (0, x^0) \qquad i = 0, 1, \ldots, n \tag{7.78}$$

We wish to transfer the system to a point on the line ℓ in a fixed time $(t_1 - t_0)$ such that the value of the coordinate $x_0(t_1)$ is minimized.

Since t_1 = constant, we cannot consider perturbations in the terminal time t_1; i.e., we cannot consider δt as part of γ. However, all of the other results developed in Section 7.3 still hold. We are thus led to the maximum principle for the fixed time problem.

Theorem 3. Let $u(t)$ be an admissible control and let $x(t)$ be the corresponding trajectory of Eq. (7.78) which passes through a point on the line ℓ at a specified time t_1. So that $u(t)$ and $x(t)$ be optimal for $t_0 \leq t \leq t_1$, it is necessary that there exist a nontrivial continuous vector function $\psi(t) = [\psi_0(t), \psi_1(t), \ldots, \psi_n(t)]$, which satisfies Eq. (7.8), such that at all points of continuity of $u(t)$, the Hamiltonian

$H(\psi, x, u)$ of the variable $u \epsilon U$ achieves a maximum at $u = u(t)$; that is

$$H[\psi(t), x(t), u(t)] = M[\psi(t), x(t)] \tag{7.79}$$

In addition,

$$\psi_0(t) = \text{constant} \leq 0 \tag{7.80}$$

Other Extensions. In conclusion, we shall mention some other extensions of the maximum principle, the details of which are given elsewhere [1, 2].

First of all, the extension of Theorem 3 to the nonautonomous case follows readily by utilizing the techniques and results introduced to obtain Theorem 2.

By introducing the so-called transversality conditions [1, 2], Theorems 1 through 3 can be extended to include those cases where one desires to transfer a dynamical system (a) from a fixed initial state to a point in a specified subset of phase space, (b) from an initial state in a specified subset of phase space to a point in a second specified subset of phase space, (c) from an initial state in a specified subset of phase space to a fixed terminal point in phase space. In addition, it is possible to treat the case where the terminal point is moving, i.e., $x^1 = x^1(t)$.

The basic results developed above can also be extended to treat optimization problems where the functional to be minimized is of the form

$$J = \int_{t_0}^{\infty} f_0[x(t), u(t)]\, dt$$

Finally, the maximum principle can also be developed to deal with optimization problems where the system dynamical equations contain parameters, i.e.,

$$\frac{dx_i}{dt} = f_i(x, u, w) \qquad i = 1, \ldots, n$$

where the vector w represents a set of adjustable system parameters. The problem is essentially the following. Prior to the initial motion, select a value w^0 for the parameter w and an admissible control $u(t)$, such that the corresponding trajectory $x(t)$ starts at $x(t_0) = x^0$ and passes through $x(t_1) = x^1$ at some time $t_1 > t_0$, and such that the functional

$$J = \int_{t_0}^{t_1} f_0[x(t), u(t), w^0]\, dt$$

is minimized.

7.5 Some Applications of the Maximum Principle

We shall now apply the results of the preceding sections to three different classes of control optimization problems. Since actual synthesis of optimal controls will, in general, require digital computation, we shall concern ourselves with establishing properties of the optimal controls. Indeed, we shall find that a very significant feature of the maximum principle is the relative ease with which properties of optimal controls follow.

Time-Optimal Control with a Constraint on the Length of the Control Vector. In Section 7.2, we investigated time-optimal control of a class of dynamical systems where the elements of the control vector u were constrained by $|u_i| \leq 1$, $i = 1, 2, \ldots, r$. Here, we shall consider the constraint to be $\| u(t) \| \leq 1$, $t_0 \leq t \leq t_1$, where $\| \;\|$ is the Euclidean norm defined by

$$\| u(t) \| = \sqrt{[u_1(t)]^2 + \cdots + [u_r(t)]^2}$$

The dynamical system is characterized by the system of equations

$$\frac{dx}{dt} = f(x) + Bu(t)$$

where all the terms were defined in Section 7.2. We wish to transfer the system from a given initial state $x(t_0) = x^0$ to a specified terminal state $x(t_1) = x^1$ in minimal time subject to the constraint $\| u(t) \| \leq 1$, $t_0 \leq t \leq t_1$. For this problem, U is the unit sphere centered at the origin in r-dimensional Euclidean space.

As before $f_0(x, u) = 1$, and the Hamiltonian for our problem is

$$H(\psi, x, u) = \psi_0 + \psi' f(x) + \psi' B u$$

where ψ_0 is a scalar function of time, ψ is an n-dimensional column vector $\psi = (\psi_1, \psi_2, \ldots, \psi_n)$, and the prime denotes the transpose.

From Theorem 1, $\psi_0(t) = \text{constant} \leq 0$, $t_0 \leq t \leq t_1$, and the Hamiltonian becomes

$$H(\psi, x, u) = \psi' f(x) + \psi' B u - 1 \tag{7.81}$$

where we have arbitrarily chosen $\psi_0(t) = -1$.

We observe that Eq. (7.81) is a scalar equation we wish to maximize with respect to u, subject to $\| u(t) \| \leq 1$, $t_0 \leq t \leq t_1$. In addition, we note that we need consider only the second term of Eq. (7.81) in this maximization since it is the only one which depends upon u. To make this term as large as possible subject to the constraint, we choose

$$u(t) = \frac{[\psi'(t) B]'}{\| \psi'(t) B \|}, \qquad t_0 \leq t \leq t_1 \tag{7.82}$$

From Eq. (7.82) we observe that $\|u(t)\| = 1$, $t_0 \le t \le t_1$, so that the length of the control vector is always at the constraint limit whereas its "direction" is time-varying in accordance with Eq. (7.82). Geometrically, this tells us that the tip of the control vector always lies on the surface of the unit sphere in r-dimensional Euclidean space. In maximizing Eq. (7.81) with respect to u, we have chosen the length of u to be the maximum possible and the direction of u so that the product $\psi'(t)\, B\, u(t)$ is always positive.

The Hamiltonian system for our problem is

$$\frac{dx}{dt} = f(x) + B\,u(t)$$

$$\frac{d\psi_i}{dt} = -\sum_{a=1}^{n} \frac{\partial f_a[x(t)]}{\partial x_i}\psi_a(t) \qquad i = 1, \ldots, n \tag{7.83}$$

with the boundary conditions $x(t_0) = x^0$ and $x(t_1) = x^1$.

The extra equation needed to establish a condition on t_1 is obtained by substituting Eq. (7.82) into Eq. (7.81) and utilizing Eq. (7.13) of Theorem 1. We then have

$$\psi'(t)\, f(x) + \|\psi'(t)\, B\| = 1 \qquad t_0 \le t \le t_1 \tag{7.84}$$

We thus have $2n+1$ relations, Eqs. (7.83) and (7.84), in as many unknowns, $x(t)$, $\psi(t)$, t_1, where $u(t)$ is given by Eq. (7.82).

The Fixed Time–Minimum Energy Problem. Let us now consider the problem where we wish to transfer the system

$$\frac{dx}{dt} = A(t)\, x(t) + B(t)\, u(t) \tag{7.85}$$

from the initial state $x(t_0) = x^0$ to the terminal state $x(t_1) = x^1$ in a fixed time interval $(t_1 - t_0)$. In Eq. (7.85), x is an n-dimensional column vector, u is an r-dimensional column vector, and $A(t)$ and $B(t)$ are continuous $n \times n$ and $n \times r$ matrices, respectively. We wish to effect the transfer with a minimum amount of control energy. We place a measure on the control energy by specifying the cost function

$$J = \frac{1}{2}\int_{t_0}^{t_1} u'(t)\, u(t)\, dt \tag{7.86}$$

where the prime denotes the transpose. We shall permit the set of admissible controls U to be the entire r-dimensional Euclidean space.

Our system of equations becomes

$$\frac{dx_0}{dt} = \frac{1}{2} u'(t)u(t) = \frac{1}{2}\left\{[u_1(t)]^2 + \cdots + [u_r(t)]^2\right\}$$

$$\frac{dx}{dt} = A(t)x(t) + B(t)u(t)$$

Setting $\psi_0(t) = -1$, $t_0 \leq t \leq t_1$, we write the Hamiltonian as

$$H(\psi, x, u) = \psi' Ax + \psi' Bu - \frac{1}{2} u'u \tag{7.87}$$

Since u can be any point in r-dimensional Euclidean space, the value of u which maximizes the Hamiltonian can be found by setting the derivative of H with respect to u equal to zero. Since only the last two terms of Eq. (7.87) depend on u, we shall define a "new" Hamiltonian for our problem by

$$G(\psi, x, u) = \psi' Bu - \frac{1}{2} u'u$$

Letting $c_j(t)$ be the jth element of the $1 \times r$ row vector $\psi'(t)B(t)$, we express G by the relation

$$G(\psi, x, u) = \sum_{j=1}^{r} c_j(t)u_j(t) - \frac{1}{2}\sum_{j=1}^{r}[u_j(t)]^2$$

Taking the derivative of G with respect to u_j, $j = 1, 2, \ldots, r$ and setting the result equal to zero, we obtain

$$u_j(t) = c_j(t) \qquad j = 1, 2, \ldots, r \qquad t_0 \leq t \leq t_1$$

In vector form, this equation becomes

$$u(t) = [\psi'(t)B(t)]' \qquad t_0 \leq t \leq t_1 \tag{7.88}$$

Substituting Eq. (7.88) in Eq. (7.86), we obtain the following expression for the minimum energy:

$$J_0 = \int_{t_0}^{t_1} \psi'(t)B(t)B'(t)\psi(t)\,dt$$

We note, in passing, that the optimal control Eq. (7.88) is linear and continuous.

The Hamiltonian system for our problem is

$$\frac{dx}{dt} = A(t)\,x(t) + B(t)\,u(t)$$
$$\frac{d\psi}{dt} = -A'(t)\,\psi(t) \tag{7.89}$$

with the boundary conditions $x(t_0) = x^0$ and $x(t_1) = x^1$. From Eqs. (7.87) and (7.88), and the extension of Theorem 3 to nonautonomous systems, we obtain

$$M\left[\psi(t), x(t)\right] = \psi'(t)\,A(t)\,x(t) + \psi'(t)\,B(t)\,B'(t)\,\psi(t) = \text{constant}, \qquad t_0 \le t \le t_1$$

In conclusion, we observe that since t_1 is fixed, Eqs. (7.89) constitute $2n$ relations in $2n$ unknowns with the $2n$ boundary conditions $x(t_0) = x^0$ and $x(t_1) = x^1$. Since the optimal control is linear, we have a linear, two-point boundary value problem whose solution will give the optimal trajectory and control.

The Fixed Time—Minimum Energy—Constrained Control Vector Problem. We consider the same problem as above, but we now impose the constraint $|u_i(t)| \le 1$, $i = 1, 2, \ldots, r$, $t_0 \le t \le t_1$, on the elements of the control vector u.

The Hamiltonian for the constrained problem is the same as above

$$H(\psi, x, u) = \psi' Ax + \psi' Bu - \frac{1}{2} u' u$$

However, the maximization is now subject to the above constraint.

As in the preceding example, we let $c_j(t)$ be the jth element of the $1 \times r$ row vector $\psi'(t)B(t)$. The control dependent portion of the Hamiltonian is then expressed by

$$G(\psi, x, u) = \sum_{j=1}^{r} c_j(t)\,u_j(t) - \frac{1}{2}\sum_{j=1}^{r}\left[u_j(t)\right]^2$$

Differentiating G with respect to u_j, $j = 1, 2, \ldots, r$, and setting the result equal to zero, we obtain

$$u_j(t) = c_j(t) \qquad j = 1, 2, \ldots, r \qquad r_0 \le t \le t_1 \tag{7.90}$$

Because of the constraints, however, Eq. (7.90) is a valid expression for the elements of u only for those values of t, $t_0 \le t \le t_1$ at which $|c_j(t)| \le 1$. For those values of t at which $|c_j(t)| \ge 1$, we maximize G by choosing

$$u_j(t) = \operatorname{sgn} c_j(t) \qquad k = 1, 2, \ldots, r \qquad t_0 \le t \le t_1 \tag{7.91}$$

From Eqs. (7.90) and (7.91), we obtain the control characteristics shown in Fig. 7.10 for the components of $u(t)$. We observe that the control is linear over a fixed range and saturates whenever $|c_j(t)| \geq 1$.

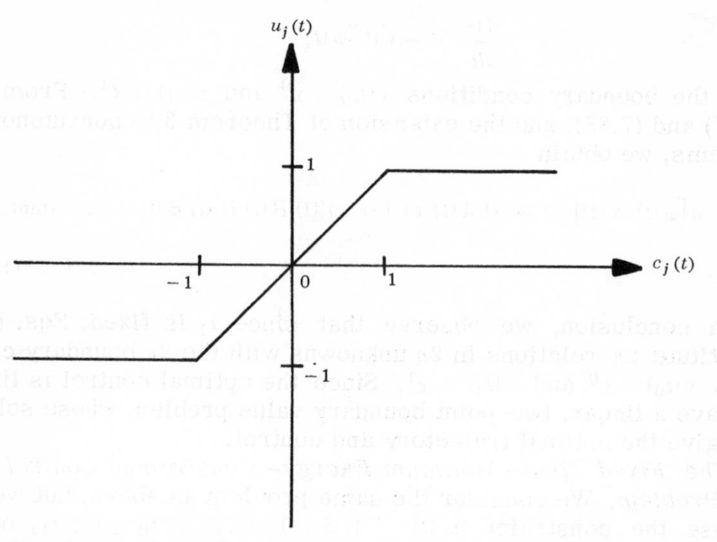

FIG. 7-10. Control characteristics for the minimum energy problem with amplitude constrained controls.

In vector matrix form, Eqs. (7.90) and (7.91) become

$$u(t) = \begin{cases} [\psi'(t)B(t)]' & \text{when } [\psi'(t)B(t)] \text{ sgn } [\psi'(t)B(t)]' \leq 1 \\ \text{sgn } [\psi'(t)B(t)]' & \text{when } [\psi'(t)B(t)] \text{ sgn } [\psi'(t)B(t)]' \geq 1 \end{cases} \tag{7.92}$$

for $t_0 \leq t \leq t_1$. We observe that the linear portion of Eq. (7.92) is the same as Eq. (7.88) for the minimum energy problem with no constraint on u.

The Hamiltonian system for this problem is given in Eq. (7.89) along with the boundary conditions $x(t_0) = x^0$ and $x(t_1) = x^1$.

7.6 Conclusion

While there is no known method for solving the general two-point boundary value problem which arises in applications of the maximum principle, numerous results have been obtained for certain classes of optimization problems. For the most part, these results are restricted to problems in which the dynamic system is linear or the control enters linearly. In conclusion, we shall indicate some of the types of problems treated.

The time-optimal problem for linear dynamic systems has been treated extensively [7, 11-14], as has the problem of minimal effort controls [14-17]. The techniques in some of these references provide a formal iterative procedure for obtaining the initial conditions on the adjoint system of equations and, therefore, the optimal control as a function of time. Some results have also been obtained for the time-optimal and minimal effort problems which lead to the synthesis of optimal feedback controls [18].

For certain restricted dynamical systems, closed-form solutions for the optimal feedback control as a continuous function of the state of the system have also been obtained [19].

In addition, a steepest-descent procedure for the minimum terminal miss problem for linear dynamical systems has been obtained [20].

In many of these cases, the authors were led to the methods of solution either by direct application of the maximum principle or by utilization of some of the concepts and constructions used in its proof. Thus, while the maximum principle does not provide a formal synthesis procedure for optimal controls, its utility as a research tool has already been established.

REFERENCES

1. Boltyanskiy, V. G., R. V. Gamkrelidze, and L. S. Pontryagin, The Theory of Optimal Processes, I, The Maximum Principle, *Am. Math. Soc. Translations, Series 2*, Vol. 18, pp. 341-382, 1961.
2. Pontryagin, L. S., V. G. Boltyanskiy, R. V. Gamkrelidze, and E. F. Mishchenko, "The Mathematical Theory of Optimal Processes," Intersci. Div. of J. Wiley and Sons, Inc., New York, N. Y., 1962 (An English translation).
3. Boltyanskiy, V. G., The Maximum Principle in the Theory of Optimal Processes, *Dokl. Akad. Nauk. SSSR*, Vol. 119, No. 6, pp. 1070-1073, 1958.
4. Gamkrelidze, R. V., On the Theory of Optimal Processes in Linear Systems, *Dokl. Akad. Nauk. SSSR*, Vol. 116, No. 1, pp. 9-11, 1957.
5. Coddington, E. A., and N. Levinson, "Theory of Ordinary Differential Equations," McGraw-Hill Book Company, Inc., New York, 1955.
6. Krasovskiy, N. N., On the Theory of Optimal Control, *Avtomat. i. Telemekh.*, Vol. 18, No. 11, pp. 960-970, 1957 (English translation in *Automation and Remote Control*, Vol. 18, pp. 1005-1016, 1957).
7. LaSalle, J. P., Time Optimal Control Systems, *Proc. Natl. Acad. Sci. U. S.*, Vol. 45, pp. 573-577, 1959.
8. Lee, E. B., and L. Markus, Optimal Control for Nonlinear Processes, *Arch. Rational Mech. Anal.*, Vol. 8, No. 1, pp. 36-58, 1961.

9. Roxin, E., The Existence of Optimal Controls, *Mich. Math. J.*, Vol. 9, pp. 109-119, 1962.
10. Neustadt, L. W., The Existence of Optimal Controls in the Absence of Convexity Conditions, *J. Math. Anal. and Appl.*, Vol. 7, 1963.
11. Neustadt, L. W., Synthesizing Time Optimal Controls, *J. Math. Anal. and Appl.*, Vol. 1, No. 3, pp. 484-493, 1960.
12. Gamkrelidze, R. V., The Theory of Time Optimal Processes in Linear Systems, *Izvest. Akad. Nauk. SSSR, Ser. Mat.*, Vol. 22, pp. 449-474, 1958 (English translation in Report No. 61-7, University of California, Dept. of Engineering, Los Angeles 24, California).
13. Paiewonsky, B. H., The Synthesis of Optimal Controllers, *Proc. of the Optimum System Synthesis Conf.*, ASD-TDR-63-119, Wright-Patterson Air Force Base, Ohio, pp. 76-99, February, 1963.
14. Neustadt, L. W., Minimum Effort Control Systems, *J. SIAM on Control*, Vol. 1, No. 1, pp. 16-31, 1962.
15. Athanassiades, M., Optimal Control for Linear Time-Invariant Plants with Time, Fuel, and Energy Constraints, *Trans. AIEE—Pt. II. Applications and Industry*, Vol. 81, No. 64, pp. 321-325, January, 1963.
16. Meditch, J. S., and L. W. Neustadt, An Application of Optimal Control to Midcourse Guidance, *Proc. of Second Congress of the International Federation of Automatic Control*, Butterworths, London, 1963 (to appear).
17. Meditch, J. S., On Minimal Fuel Satellite Attitude Control, Paper No. XIX-5-IEEE (TOC), 1963 Joint Automatic Control Conference, Minneapolis, Minn., June, 1963.
18. Neustadt, L. W., A Synthesis Method for Optimal Controls, *Proc. of the Optimum System Synthesis Conf.*, ASD-TDR-63-119, Wright-Patterson Air Force Base, Ohio, pp. 374-381, February, 1963.
19. Athans, M., P. L. Falb, and R. T. Lacoss, On Optimal Control of Self-Adjoint Systems, Paper No. V-5-IEEE (TOC), 1963 Joint Automatic Control Conference, Minneapolis, Minn., June, 1963.
20. Balakrishnan, A. V., An Operator Theoretic Formulation of a Class of Control Problems and a Steepest Descent Method of Solution, *Jour. of SIAM, Series A on Control*, Vol. 1, No. 2, 1963.

8

Minimum Norm Problems and Some Other Control System Optimization Techniques

MASANAO AOKI
ASSOCIATE PROFESSOR OF ENGINEERING
UNIVERSITY OF CALIFORNIA, LOS ANGELES, CALIFORNIA

The theory of control has seen rapid development in recent years, and many new tools have become available for control system optimization. It is the purpose of this chapter to discuss a few such techniques not treated elsewhere in the book. For example, it has become fairly well known that certain problems in optimal control theory and trajectory optimizations can be formulated as problems of finding linear functionals with minimal norms, and that theories developed for Krein's L-problem are applicable [2, 33, 39].* The L-problems and related topics are discussed in the first five sections. The L-problems are formulated mathematically in Sec. 8.2. The theory developed in Sec. 8.2 is applied in Sec. 8.3 to the discussion of problems of approximately solving a set of incompatible systems. In Sec. 8.4, the topic of approximation in normed linear spaces is introduced by reformulating the L-problems as problems in approximation. The expression for error is derived in Sec. 8.5 when the space is Hilbert space, and the problem of uniqueness is discussed in Sec. 8.6. The remaining four sections discuss some successive approximation methods in solving functional equations.

PART I The L-Problem and Approximation in Normed Linear Space

8.1 Introduction

Let us illustrate the connection between the minimal norm problem in functional analysis and certain control problems by

*For another approach, see for example ref. 4.

briefly considering a time-optimal problem with scalar control variable $u(t)$.

Since there exists an extensive set of references on the subject, the reader is referred to them for more complete treatment [28].

The problem is usually formulated as the problem of finding, for given scalar functions of time $z_i(t)$ and $h_i(t)$, a control variable $u(t)$ from an admissible set of control variables Ω such that

$$\int_0^t h_i(\tau)u(\tau)\,d\tau = z_i(t) \qquad 1 \le i \le n \tag{8.1}$$

at the minimal time $t = T$.

The functions $h_i(t)$ are taken to be such that

$$\int_0^t |h_i(\tau)|^q\,d\tau < \infty \qquad 1 \le i \le n, \quad q > 1$$

for any finite t.

The set of all points in the n-dimensional Euclidean space E_n that can be reached by admissible u over $[0, t]$ is called the reachable or attainable region [28, 35] $R(t)$

$$R(t) = \left\{ x(t) : x(t) = \int_0^t h(\tau)u(\tau)\,d\tau, u \in \Omega \right\}$$

where $h = (h_1, \ldots, h_n)$.

Under suitable assumptions on h and Ω, the set $R(t)$ turns out to be a bounded closed convex body in E_n, and the minimal time T occurs at the time t when $z(t)$ is on the boundary of $R(t)$. The set R plays an important role in many modern control problems.

When u is considered elements of $L_p(0, t)$* such that

$$\Omega = \left\{ u(t) : \left(\int_0^t |u(t)|^p\,dt \right)^{1/p} \le c_p \right\} \qquad p > 1$$

then the optimal control variable u^0 satisfies

$$\left(\int_0^t |u^0(t)|^p\,dt \right)^{1/p} = c_p$$

Eq. (8.1) may be regarded as a linear operator on $L_q(0, T)$ which maps L_q into the space of real numbers and satisfies n constraints.

$$Ah_i = \int_0^T h_i(t)u(t)\,dt = z_i(T) \qquad 1 \le i \le n \tag{8.2}$$

*$L_p(0, t) = \left\{ x : x \text{ measurable}, \int_0^t |x(s)|^p\,ds < \infty \right\}$

Then the norm of the operator (called a linear functional on L_q) [26] is defined by

$$\|A\| = \sup_{x \neq 0} \frac{|Ax|}{\|x\|}$$

is given by

$$\left(\int_0^T |u(t)|^p \, dt \right)^{1/p}$$

The time-optimal operator A^* has the norm

$$\|A^*\| = \left(\int_0^T |u^0(t)|^p \, dt \right)^{1/p} = c_p$$

Loosely speaking, the norm of A^* expresses the amount of control or control effort (energy) needed to reach the desired point in minimal time.

When T is regarded as fixed, and the original problem is rephrased as the minimal control effort problem, [5, 27, 29, 32] i.e., the problem of reaching the desired point at time T, $z(T)$, with minimal control effort [27, 28] then the problem becomes one of minimizing $\|A\|$ over all A which satisfy Eq. (8.2).

This corresponds to the problem of minimal norm [33, 34] in a normed linear space [26].

8.2 Mathematical Formulation

After this introduction, the L-problem is now formulated. The exposition in part follows that of ref. 2. Let E be a normed linear space, and let x be an arbitrary element of E.

Problem A: Given n linearly independent elements $x_1, \ldots, x_n$ of E, given $c = (c_1, \ldots, c_n)$, a non-zero vector in E_n (Euclidean n space), and given a positive real number L, find necessary and sufficient conditions for the existence of a linear functional in E^*, the space of all linear functionals defined on E [37], satisfying

$$f(x_i) = c_i \qquad i = 1,2,\ldots,n \tag{8.3}$$

and

$$\|f\| \leq L \tag{8.4}$$

The space E^* is called the conjugate space of E. The problem can be put in a slightly different form.

Problem A$'$: Find a functional $f(x)$ of minimal norm satisfying

$$f(x_i) = c_i$$

where $x_1, \ldots, x_n$ are linearly independent elements of E and $c = (c_1, \ldots, c_n)$ is a vector in E_n. An element x such that $|f(x)| = ||f|| \cdot ||x||$ is called an extremal element of f. Later, we will have occasion to discuss the uniqueness of extremal elements in Sec. 8.6. There is still another problem formulation closely related to Problems A and A$'$.

Problem B: Given n linearly independent elements $x_1, \ldots, x_n$ of E, find λ such that

$$\frac{1}{\lambda} = \inf_{\xi \cdot c = 1} ||\xi_1 x_1 + \ldots + \xi_n x_n|| \tag{8.5}$$

where $\xi = \xi_1, \ldots, \xi_n$ is a vector in E_n. The dot in Eq. (8.5) denotes the usual inner product. Since

$$\begin{aligned} \frac{1}{\lambda} &= \inf_{\xi \cdot c = 1} ||\xi_1 x_1 + \ldots + \xi_n x_n|| = \inf_{\xi \epsilon E_n} \frac{||\sum_i \xi_i x_i||}{|\xi \cdot c|} \\ &= \left[\sup_{\xi \epsilon E_n} \frac{|\xi \cdot c|}{||\sum_i \xi_i x_i||} \right]^{-1} \end{aligned} \tag{8.6}$$

the number λ in Eq. (8.5) can also be given as

$$\lambda = \sup_{\xi \epsilon E_n} \frac{|\xi \cdot c|}{||\sum_i \xi_i x_i||} = \sup_{\xi \epsilon H} |\xi \cdot c| \tag{8.7}$$

where

$$H = \left\{ \xi : \xi \epsilon E_n , \; ||\sum_i \xi_i x_i|| = 1 \right\} \tag{8.8}$$

From now on λ is written explicitly as a function of $c_1, \ldots, c_n$. At this point, the close relation between Problems A and B becomes apparent.

If a linear functional $f(x)$ satisfies Eq. (8.3) and (8.4), then

$$\begin{aligned} |\sum_i \xi_i c_i| &= |\sum_i \xi_i f(x_i)| = |f \sum_i \xi_i x_i| \\ &\leq ||f|| \cdot ||\sum_i \xi_i x_i|| \leq L \; ||\sum_i \xi_i x_i||, \end{aligned}$$

therefore

$$L \geq \frac{|\sum_i \xi_i c_i|}{||\sum_i \xi_i x_i||}$$

taking the supremum on the right, one has, from Eq. (8.7),

$$L \geq \lambda \tag{8.9}$$

Next, suppose $\lambda(c_1, \ldots, c_n) \leq L$ and define a functional $\phi(x)$ on G by

$$\phi(x) = \phi \sum_{i=1}^{n} \xi_i x_i \triangleq \sum_{i=1}^{n} \xi_i c_i$$

Then, the norm of $\phi(x)$ in G where G is the subspace of E spanned by x_i, $1 \leq i \leq n$ is given by

$$||\phi||_G = \sup_{x \epsilon G} \frac{|\phi(x)|}{||x||} = \sup_{\xi \epsilon E_n} \frac{|\sum_i \xi_i c_i|}{||\sum_i \xi_i x_i||} = (c_1, \ldots, c_n)$$

By the Hahn-Banach Theorem [26, 37], $\phi(x)$ can be extended to E without increasing the norm. Let $f(x)$ be the extension of $\phi(x)$ to E, then

$$||f|| = ||\phi||_G = \lambda \leq L$$

and

$$f(x_i) = \phi(x_i) = c_i \qquad i = 1,2,\ldots, n$$

Thus λ is the minimal norm of a functional satisfying Eq. (8.3).

This is a basic result which establishes a relation between the two problem formulations, A and B.

Proposition: Problem A has a solution if and only if $L \geq \lambda(c_1, \ldots, c_n)$.

Since a finite-dimensional vector space [22] is one of the most familiar spaces to us, let us first re-state Problem A in the m-dimensional Euclidean space E_m, and indicate sets in E_m which may be regarded as reachable sets R. The conjugate (dual) space E_m^* has the same dimension as E_m, and there exists a unique linear functional such that $f(x_i) = c_i$, $i = 1,2,\ldots, m$, where c_i are given constants and $\{x_1, \ldots, x_m\}$ is a basis in E_m. If, instead of imposing m conditions on f, only n conditions ($n < m$) are imposed on f, then there are more than one linear functionals which satisfy the imposed conditions.

In terms of the adjoint space E_m^*, only n of the m elements in the dual basis have been determined. Without loss of generality, one can assume that $f(x_i) = c_i$, $1 \le i \le n$.

Such linear functionals form an $(m - n)$-dimensional subspace in E_m^*.

As an additional constraint on f, let us require that $||f||$ be less than a specified number L. This will further restrict the set of linear functionals in the $(m - n)$-dimensional subspace.

Let K be the set of such linear functionals, i.e.,

$$K = \{f : f \epsilon E_m^*, \ ||f|| \le L, \ f(x_i) = c_i, \ 1 \le i \le n\}$$

Let us first consider the set K for the Euclidean norm, i.e.,

$$||x|| \triangleq \left(\sum_i x_i^2\right)^{1/2}$$

Then,

$$|f(x)| = |\sum_i f_i x_i| \le \left(\sum_i f_i^2\right)^{1/2} \left(\sum_i x_i^2\right)^{1/2} = \left(\sum_i f_i^2\right)^{1/2} \cdot ||x||$$

where

$$f_i = f(x_i), \ 1 \le i \le m$$

therefore

$$||f|| = \sup_{x \ne 0} \frac{|f(x)|}{||x||} \le \left(\sum_i f_i^2\right)^{1/2} \tag{8.10}$$

Consider the particular vector given by

$$x_i = f_i, \ 1 \le i \le m$$

then

$$|f(x)| = |\sum_i f_i x_i| = \left(\sum_i f_i^2\right)^{1/2} \left(\sum_i x_i^2\right)^{1/2} = \left(\sum_i f_i^2\right)^{1/2} ||x||$$

Therefore the equality in Eq. (8.10) actually holds.

$$||f|| = \left(\sum_i f_i^2\right)^{1/2} \tag{8.11}$$

As a second case, let us next consider the set K when the norm in E_m is defined by

$$||x|| = \operatorname*{Max}_i |x_i|$$

where

$$x = (x_1, \ldots, x_m).$$

Then

$$|f(x)| = |\sum_i f_i x_i| \leq \left(\sum_i |f_i|\right) \operatorname*{Max}_i x_i = \left(\sum_i |f_i|\right) ||x||$$

Therefore

$$||f|| = \sup_{x \neq 0} \frac{|f(x)|}{||x||} \leq \sum_i |f_i|$$

The equality actually holds since by setting

$$x_i = \begin{cases} \operatorname{sgn} f_i \ , & f_i \neq 0 \\ 0 \ , & f_i = 0 \end{cases}$$

one has

$$|f(x)| = \sum_i |f_i| = \left(\sum |f_i|\right) ||x||$$

thus

$$||f|| = \sum_{i=1}^{m} |f_i|$$

In the first case, where the norm of f is given by Eq. (8.11), the set K is nonvacuous if and only if

$$\left(\sum_{i=1}^{n} c_i^2\right)^{1/2} \leq L$$

The set K is defined by

$$K = \left\{ f : f_i = c_i, \ 1 \leq i \leq n, \ \sum_{n+1}^{m} f_i^2 \leq L^2 - \left(\sum_1^n c_i^2\right) \right\} \tag{8.12}$$

In the second case, K is nonvacuous if and only if

$$\sum_{i=1}^{n} |c_i| \le L$$

and the set K is given by

$$K = \left\{ f : f_i = c_i,\ 1 \le i \le n,\ \sum_{n+1}^{m} |f_i| \le L - \sum_{i=1}^{n} |c_i| \right\} \tag{8.13}$$

If the norm $\|f\|$ is specified, then the set K is restricted further by f such that

$$\sum_{i=n+1}^{m} f_i^2 = \|f\|^2 - \sum_{i=1}^{n} c_i^2 \tag{8.14}$$

and

$$\sum_{i=n+1}^{m} |f_i| = \|f\| - \sum_{i=1}^{n} |c_i| \tag{8.15}$$

Equations (8.14) and (8.15) define the subsets K' and K respectively of Eqs. (8.12) and (8.13). The sets K and K' may be regarded as the reachable set R in the previous section.

Let us next consider the question whether it is possible to attain the value y for some $x \in E_m$ with

$$f(x_i) = c_i \qquad 1 \le i \le n$$

$\|f\|$ given. This has obvious implications for minimum-energy problems. If x is in the subspace spanned by $\{x_1, \ldots, x_n\}$ then

$$x = \sum_{1}^{n} \xi_i x_i$$

$$f(x) = \sum_{1}^{n} \xi_i c_i$$

therefore, by choosing ξ_i to satisfy

$$\sum_{i=1}^{n} \xi_i c_i = y$$

any one of the f in K' will do. If x is independent of $x_1, \ldots, x_n$, then let it be x_{n+1}. Then the question becomes: For what value of y is

$$f(x_{n+1}) = y$$

satisfied with f in K'?

This question will be answered later when linear functionals defined on a general normed linear space (not necessarily finite-dimensional) are discussed. To do this, it is convenient to discuss problem B first.

First, one notes that in Eq. (8.5) the infimum is attained. Hence it can be replaced by the minimum, and consequently the supermum in Eq. (8.7) can be replaced by the maximum.

To see this, denote by $\{\xi^{(i)}\}$ a sequence of vectors in E_n, $\xi^{(i)} = \left(\xi_1^{(i)}, \ldots, \xi_n^{(i)}\right)$ such that

$$\frac{1}{\lambda} = \lim_{i \to \infty} \left|\left| \sum_{j=1}^{n} \xi_j^{(i)} x_j \right|\right| \tag{8.16}$$

with

$$\xi^{(i)} \cdot c = 1, \quad i = 1,2,\ldots$$

holds, and by $\{x^{(i)}\}$ a sequence of elements in E where

$$x^{(i)} = \xi_1^{(i)} x_1 + \xi_2^{(i)} x_2 + \cdots + \xi_n^{(i)} x_n \qquad i = 1,2,\ldots \tag{8.17}$$

where $x_1, \ldots, x_n$ are given. Because of the linear independence of $x_1, \ldots, x_n$, to every vector x in E_n there corresponds a unique vector x in G, an n-dimensional subspace of E, and vice versa. It is easily seen that this is a one-to-one mapping of G onto E_n, continuous both ways, i.e., G and E_n are homeomorphic [37]. Therefore compactness [26, 37] is preserved and any bounded set in G is compact. The sequence $\{x^{(i)}\}$ is obviously bounded hence there exists a subsequence $\{x^{(i')}\}$ which converges to some x^0

$$\lim_{i' \to \infty} \left(\xi_1^{(i')} x_1 + \cdots + \xi_n^{(i')} x_n\right) = \xi_1^0 x_1 + \cdots + \xi_n^0 x_n \tag{8.18}$$

From Eqs. (8.16) and (8.18), however,

$$\frac{1}{\lambda} = \lim_{i' \to \infty} || \xi^{(i')} \cdot x || = || \xi^0 \cdot x || \tag{8.19}$$

From linear independence of $x_1, \ldots, x_n$, and from Eq. (8.18),

$$\left.\begin{aligned} &\xi_j^0 = \lim_{i' \to \infty} \xi_j^{(i')}, \quad j = 1,2\ldots, n \\ &\text{and} \\ &\lim_{i' \to \infty} \xi^{(i')} \cdot c = \xi^0 \cdot c = 1 \end{aligned}\right\} \tag{8.20}$$

Thus

$$\frac{1}{\lambda} = ||\xi^0 \cdot x|| \quad \text{and} \quad \xi^0 \cdot c = 1$$

From Eq. (8.7), it is easily seen that the function $\lambda(c_1, \ldots, c_n)$ satisfies:

$$\begin{aligned} &\text{a.} \quad \lambda(c_1, \ldots, c_n) > 0 \quad \text{if} \quad \sum_{i=1}^{n} c_i^2 > 0 \\ &\text{b.} \quad \lambda(tc_1, \ldots, tc_n) = |t|\, \lambda(c_1, \ldots, c_n) \\ &\text{c.} \quad \lambda(c_1 + c_1', \ldots, c_n + c_n') \le \lambda(c_1, \ldots, c_n) + \lambda(c_1', \ldots, c_n') \end{aligned} \tag{8.21}$$

When one remembers the way time-optimal control problems are disucssed, it is clear that the concept of reachable set, i.e., the range of some functionals satisfying given constraints, is one of the basic concepts we utilize in deriving optimal control policies.

It is therefore of great interest to investigate the question: When one more linearly independent element of y of E is added to $x_1, \ldots, x_n$, what is the range of $f(y)$, with $||f|| \le L$, provided $L > \lambda(c_1, \ldots, c_n)$? One can treat y as x_{n+1} in the original problem formulation and discuss the $(n+1)$-dimensional problem

$$\left.\begin{aligned} f(x_i) &= c_i, \quad i = 1, 2, \ldots, n \\ f(y) &= \eta \\ ||f|| &\le L \end{aligned}\right\} \tag{8.22}$$

This is possible if and only if

$$\lambda(c_1, \ldots, c_n, \eta) \le L$$

Thus, we are led to consider the set K_{n+1}'

$$K_{n+1}' = \left\{(c_i, \ldots, c_n, \eta) \; : \; \lambda(c_1, \ldots, c_n, \eta) \le L\right\}$$

or, more generally,

$$K_m = \left\{(c_1, \ldots, c_m) \; : \; \lambda(c_1, \ldots, c_m) \le 1\right\}$$

Because of Eq. (8.21), the set of c such that $\lambda(c_1, \ldots, c_m) \le t$ can be generated from K_m in its expansion by a factor t. Thus, in this section, $\lambda(c_1, \ldots, c_n) = 1$ and $L > 1$ are assumed. From Eqs. (8.7) and (8.8),

$$\operatorname*{Max}_{\xi \in H} \left|\sum_{i=1}^{n} \xi_i c_i\right| \le M\left(\sum_{i=1}^{n} c_i^2\right)^{1/2}$$

where

$$M = \operatorname*{Max}_{\xi \epsilon H} \sum_{i=1}^{n} \xi_i^2{}^{1/2} \tag{8.23}$$

The constant M is independent of c. One has

$$\lambda(c_1, \ldots, c_n) \leq M \cdot \left(\sum_1^n c_i^2\right)^{1/2}$$

hence λ is a continuous function of the arguments, $c_1, \ldots, c_n$, and K_n contains a Euclidean sphere of radius $\frac{1}{M}$.

By defining another constant μ as

$$\mu = \operatorname{Min} \left| \sum_{i=1}^{n} \xi_i c_i \right| \tag{8.24}$$

$$\sum_{i=1}^{n} c_i^2 = 1$$

from Eq. (8.24),

$$\left(\sum_i c_i^2\right)^{1/2} \leq \frac{1}{\mu} \cdot \lambda(c_1, \ldots, c_n) = \frac{1}{\mu}$$

Thus, K_n is contained in the Euclidean sphere of radius $\frac{1}{\mu}$.

From Eq. (8.21), K_n is convex. Therefore, K_n is a bounded closed convex body, i.e., it has interior points.

By identifying every point $(c_1, \ldots, c_n) \epsilon E_n$ with the point $(c_1, \ldots, c_n, 0, \ldots, 0) \epsilon E_m \ (m > n)$, one has

$$\lambda(c_1, \ldots, c_n) \leq \lambda(c_1, \ldots, c_m) \qquad m > n$$

If $(c_1, \ldots, c_m) \epsilon K_m$, then $\lambda(c_1, \ldots, c_n) \leq 1$, therefore the projection of K_m on E_n is contained in K_n. To see that every point K_n is the projection of some point K_m, one notes from the basic proposition that there exists a functional

$$f(x_i) = c_i, \quad i = 1, 2, \ldots, n$$

$$\|f\| = \lambda(c_1, \ldots, c_n)$$

Define c_k' by

$$c_k' = f(x_k), \quad k = n+1, \ldots, m.$$

Then, since $\lambda(c_1, \ldots, c_n, c'_{n+1}, \ldots, c'_m)$ is the minimal norm,

$$1 \geq \lambda(c_1, \ldots, c_n) = ||f|| \geq \lambda(c_1, \ldots, c_n, c'_{n+1}, \ldots, c'_m)$$

This proves that the projection of K_m on $E_n\,(m > n)$ coincides with K_n. One can therefore determine a point $P = (c_1, \ldots, c_n, c^0_{n+1})$ such that

$$\lambda(c_1, \ldots, c_n) = \lambda(c_1, \ldots, c_n, c^0_{n+1}) = 1$$

This is a point in K'_{n+1}. Thus, the straight line $\eta_i = c_i$, $i = 1,2, \ldots, n$, passing through P, will have as its intersection with K'_{n+1} some closed interval

$$\eta_i = c_i, \quad i \neq 1, \ldots, n$$

$$\eta' \leq \eta \leq \eta''$$

If $\eta = \eta'$ or $\eta = \eta''$, Eq. (8.22) is satisfied with $||f|| = L$. Thus, the inequality

$$\lambda(c_1, \ldots, c_n, \eta) \leq L$$

is satisfied only for η in $[\eta', \eta'']$. From the basic proposition, for those η which satisfies the above inequality, there exists f in E^* satisfying Eq. (8.22).

By similar arguments, one has the proposition: If $x_1, \ldots, x_n$, $y_1, \ldots, y_p$ are linearly independent elements of E and if $\lambda(c_1, \ldots, c_n) < L$, then the set of points $(\eta_1, \ldots, \eta_p)$ for which

$$f(x_i) = c_i, \quad i = 1,2, \ldots, n$$

$$f(y_k) = \eta_k, \quad k = 1,2, \ldots, p$$

are satisfied, with $||f|| \leq L$, forms a p-dimensional convex body in E^*.

As an application of this proposition, one has: If the dimension of the space E is not less than $n+2$, then for any $L > \lambda(c_1, \ldots, c_n)$

$$f(x_i) = c_i, \quad i + 1,2, \ldots, n$$

$$||f|| = L$$

has a continum of different solutions. This is because the boundary of a p-dimensional convex body $(p > 2)$ of the solutions of the related $(n+p)$-dimensional problem has a continum of different points. This question of uniqueness has important implications in the theory of approximations which will be discussed briefly later.

Now, the connection between the solutions of problems A and B can be stated concisely:

The minimizing element of problem B is the extremal element of problem A′ (or the extremal element of the linear functional of minimal norm satisfying Eq. (8.3) in Problem A) and conversely.

Proof: Let f_0 be a solution of the problem A with minimal norm, i.e.,

$$f_0(x_i) = c_i, \qquad i = 1,2,\ldots,n$$

$$\|f_0\| = \lambda(c_1,\ldots,c_n)$$

and let x_0 be the minimizing solution of the problem B, i.e.,

$$x_0 = \sum_{i=1}^{n} \xi_i^0 x_i, \quad \|x_0\| = \frac{1}{\lambda(c_1,\ldots,c_n)}$$

$$1 = \sum_i \xi_i c_i$$

then

$$1 = \left|\sum_{i=1}^{n} \xi_i c_i\right| = |f_0(x)| = \|f_0\| \cdot \|x\|$$

Therefore x_0 is an extremal element of f_0, the extremal element being that element which achieves the equality in the defining equation for the norm. Conversely if x_0 is an extremal element for f_0, then reading the equation above backward, one sees that x_0 is the minimizing element of the problem B.

8.3 Application to Chebychev Approximation Problems

Let us discuss another application of the L problem to solve approximately a system of incompatible linear algebraic equations in the Chebychev sense. The problem of Chebychev approximation has been studied extensively and there is a large body of references on the subject [1, 9, 30, 31]. It would require a separate treatment to discuss the problem adequately. Therefore we will be content merely to point out the connection of the Chebychev approximation problem to the L problems discussed in the previous section. This problem can be stated as follows. Given

$$a_1^i \xi_1 + a_2^i \xi_2 + \cdots + a_n^i \xi_n = 1^i, \quad i = 1,2,\ldots,m \tag{8.25}$$

find $\xi = \xi_1, \xi_2, \ldots, \xi_n$ such that

$$\underset{\substack{\xi_j \\ 1\le j\le n}}{\text{Min}} \; \underset{1\le i\le m}{\text{Max}} \left|\sum_{j=1}^{n} a_j^i \xi_j - 1^i\right| = \epsilon \tag{8.26}$$

The quantity ϵ gives the minimum of the absolute maximum error in Eq. (8.25). In this problem formulation, it can be assumed that vectors $(a_i^1, a_i^2, \ldots, a_i^m)$ and $(1^1, \ldots, 1^m)$ are to be linearly independent without loss of generality.

To see how the theories developed in the previous section can be applied, one begins by defining

$$a_{n+1}^i = -1^i$$

Eq. (8.26) can then be rewritten as

$$\underset{\eta}{\text{Min}} \quad \underset{1 \leq i \leq m}{\text{Max}} \left| \sum_{j=1}^{n+1} a_j^i \eta_j \right| \tag{8.27}$$

with

$$\sum_{i=1}^{n+1} c_i \eta_i = 1 \tag{8.28}$$

where

$$\left.\begin{aligned} \eta &= (\eta_1, \ldots, \eta_{n+1}) \\ \eta_i &= \xi_i, \qquad i = 1,2,\ldots, n \\ \eta_{n+1} &= 1 \\ c_i &= 0 \qquad i = 1,2,\ldots, n \\ c_{n+1} &= 1 \end{aligned}\right\} \tag{8.29}$$

Consider $(n + 1)$ linearly independent elements of E_m

$$\left.\begin{aligned} y_1 &= (a_1^1, a_1^2, \ldots, a_1^m) \\ &\;\vdots \\ y_{n+1} &= (a_{n+1}^1, a_{n+1}^2, \ldots, a_{n+1}^m) \end{aligned}\right\} \tag{8.30}$$

Let y be the linear combination of $y_1, \ldots, y_{n+1}$

$$y = \eta_1 y_1 + \cdots + \eta_{n+1} y_{n+1} \tag{8.31}$$

such that Eq. (8.28) is satisfied. Then, the kth element of y, y^k is given by

$$y^k = \sum_{i=1}^{n+1} a_i^k \eta_i \qquad k = 1,2,\ldots, m \tag{8.32}$$

Note that Eq. (8.32) has the same form as Eq. (8.27). Define the norm in E_m by

$$||y|| = \underset{1 \leq i \leq m}{\text{Max}} |y^i| \tag{8.33}$$

Since linear functionsal defined on E_m are elements of the conjugate space E_m^* which is also m-dimensional [22], linear functionals are defined uniquely by specifying m numbers $f^1, \ldots, f^m$

$$f(y) = \sum_{i=1}^{m} f^i y^i \tag{8.34}$$

Thus,

$$||f|| = \sup_{\substack{y \in E_m \\ y \neq 0}} \frac{|f(y)|}{||y||} = \sum_{i=1}^{m} |f^i| \tag{8.35}$$

Then, Eq. (8.26) can be formulated as problem A, i.e., that of finding a functional f of minimal norm with the conditions $f(y_i) = c_i$, $i = 1, 2, \ldots, n+1$,

$$\epsilon = \text{Min} \sum_{i=1}^{m} |f^i| \tag{8.36}$$

with

$$\left.\begin{array}{l} a_1^1 f^1 + a_1^2 f^2 + \cdots + a_1^m f^m = c_1 \\ - - - - - \\ a_{n+1}^1 f^1 + a_{n+1}^2 f^2 + a_{n+1}^m f^m = c_{n+1} \end{array}\right\} \tag{8.37}$$

Thus, if the rank of the matrix

$$\begin{bmatrix} a_1^1 & a_{n-1}^1 & 1^1 \\ \cdot & & \cdot \\ \cdot & & \cdot \\ \cdot & & \cdot \\ a_1^m \ldots & a_{n-1}^m & 1^m \end{bmatrix}$$

is n ($n < m$), then from Eq. (8.28) and (8.37),

$$\epsilon = \underset{\eta}{\text{Min}} \underset{1<i<n}{\text{Max}} |a_1^i \eta_1 + \cdots + a_n^i \eta_n - 1^i|$$

is solved by solving

$$\epsilon = \operatorname{Min} \sum_{i=1}^{m} |f^i| \tag{8.38}$$

with

$$\sum_{i=1}^{m} a_j^i f^i = 0 \qquad j = 1,2,\ldots,n$$

$$\sum_{i=1}^{m} 1^i f^i = -1$$

To obtain the minimizing ξ in Eq. (8.26), let $f_0 = f_0^1, f_0^2, \ldots, f_0^m$ be the vector for which $\|f_0\| = \epsilon$ is attained, and let $f_0^i \neq 0$, $i = 1,2,\ldots,m$, and let z be the extremal element. Then

Then

$$|f_0(z)| = |\sum_{i=1}^{m} f_0^i z^i| = \left(\sum_i |f_0^i|\right)\left(\operatorname{Max}_i |z^i|\right) = \|f_0\| \cdot \|z\|$$

hence

$$z^i = k \operatorname{sgn} f_0^i \qquad i = 1,2,\ldots,m \tag{8.39}$$

Thus, for some constant $k > 0$

$$f_0(z) = k \sum_{i=1}^{m} |f_0^i| \tag{8.40}$$

On the other hand, since the extremal element z is given by

$$z = \sum_{i=1}^{n+1} y_i \eta_i$$

with

$$\sum_{i=1}^{n+1} c_i \eta_i = 1$$

Thus,

$$f_0(z) = f_0\left(\sum_i y_i \eta_i\right) = \sum_i c_i \eta_i = 1 \tag{8.41}$$

Therefore from Eqs. (8.40), (8.41) and (8.38)

$$k = \frac{1}{\sum_{i=1}^{m} |f_0^i|} = \frac{1}{\epsilon}$$

Hence from Eqs. (8.39) and (8.41), $\eta_1, \ldots, \eta_{n+1}$ are obtained from

$$\frac{1}{\epsilon} \operatorname{sgn} f_0^i = \sum_{j=1}^{n+1} a_j^i \eta_j \qquad i = 1,2,\ldots, m \tag{8.42}$$

From Eq. (8.29), Eq. (8.42) is really

$$a_1^i \xi_1 + a_2^1 \xi_2 + \cdots + a_n^i \xi_n - 1^i = \frac{\operatorname{sgn} f_0^i}{\epsilon} \qquad i = 1,2,\ldots, m \tag{8.43}$$

when $n = m$, $\xi_1, \ldots, \xi_n$ will be determined uniquely by Eq. (8.43).

8.4 *L*-Problem as an Approximation Problem in Normed Linear Space

In this section we shall show the connection between certain optimization problems formulated as the L-problem and the approximation problems in normed linear space [1]. Consider the L-problem at some t in $[0, T]$

$$\min_{\substack{\xi_i \\ 0 \le i \le n}} \left\| \sum_i \xi_i h_i(t) \right\| \tag{8.44}$$

with

$$\sum_{i=0}^{n} \xi_i c_i(T) = 1 \tag{8.45}$$

where $h_i(t)$ are the system weighting functions and $c_i(T)$ are given, $i = 0,1,\ldots, n$. Since not all of $c_i(T)$'s are zero, we may assume without loss of generality that $c_0(T) \neq 0$. Then from Eq. (8.45),

$$\xi_0 = \frac{1}{c_0(T)} - \sum_{i=1}^{n} \xi_i \frac{c_i(T)}{c_0(T)} \tag{8.46}$$

Substituting Eq. (8.46) in Eq. (8.44),

$$\sum_{i=0}^{n} \xi_i h_i(t) = \phi_0(t) - \sum_{i=1}^{n} \xi_i \phi_i(t)$$

where

$$\phi_0(t) = \frac{h_0(t)}{c_0(T)}$$

and

$$\phi_i(t) = \frac{c_i(T)}{c_0(T)} h_0(t) - h_i(t) \qquad i = 1,2,\ldots,n$$

Thus, the constrained minimization problem of Eq. (8.44) and Eq. (8.45) becomes the unconstrained minimization over $\xi_1, \ldots, \xi_n$ of

$$\underset{\substack{\xi_i \\ 1 \leq i \leq n}}{\text{Min}} \; \|\phi_0 - \sum_{i=1}^{n} \xi_i \phi_i\| \tag{8.47}$$

Since all the ϕ_i are known Eq. (8.47) means that the ξ_i are to be chosen so that the metric between the point ϕ_0 and the point in the linear manifold spanned by $\phi_1, \ldots, \phi_n$ is minimized; in other words, ξ_i, $i = 1,2,\ldots,n$ will be determined by the point nearest ϕ_0 in the linear manifold spanned by $\phi_1, \ldots, \phi_n$. This is a general statement of approximation problems in normed linear spaces. Since the ϕ are functions of time in Eq. (8.47) the ξ become functions of time.

8.5 Approximation in Hilbert Space and Gram's Determinant

When the concept of inner product is introduced and the norm is defined in terms of the inner product, i.e., when the space is a Hilbert space [37], it is possible to give an explicit expression for the error of the approximation.

Let ϕ be the best approximation in

$$\rho = \underset{\substack{\lambda_i \\ 1 \leq i \leq n}}{\text{Min}} \; \|\phi_0 - \phi\| \tag{8.48}$$

where

$$\phi = \sum_{i=1}^{n} \lambda_i \phi_i \tag{8.49}$$

and $\phi, \phi_1, \ldots, \phi_n$ are given.

In this case, the space is strictly normed [1] (see Sec. 8.6) and the unique best approximation exists and is given by the orthogonal projection of ϕ_0 onto the linear manifold spanned by $\phi_1, \ldots, \phi_n, H$.

Since $\phi_0 - \phi$ is orthogonal to ϕ_i, $i = 1,2,\ldots,n$.

$$(\phi_0 - \phi, \phi_i) = 0, \qquad i = 1,1,\ldots,n \tag{8.50}$$

From Eqs. (8.49) and (8.50)

$$\sum_{i=1}^{n} \lambda_i (\phi_i, \phi_j) = (\phi_0, \phi_j) \qquad j = 1,2,\ldots, n \tag{8.51}$$

a system of linear algebraic equations for the λ_i. Let

$$G(\phi_1, \ldots, \phi_n) = \det[(\phi_i, \phi_j)]$$

The determinant G is called the Gram determinant [1] and is not zero because of the linear independence of ϕ_i, $i = 1,2,\ldots, n$. From Eq. (8.48)

$$\begin{aligned} \rho^2 &= (\phi_0 - \phi, \phi_0 - \phi) \\ &= (\phi_0 - \phi, \phi_0) - (\phi_0 - \phi, \phi) \\ &= (\phi_0, \phi_0) - (\phi, \phi_0) \end{aligned}$$

Therefore

$$(\phi, \phi_0) = (\phi_0, \phi_0) - \rho^2 \tag{8.52}$$

Eqs. (8.51) and (8.52) constitute $(n + 1)$ equations needed to solve $(n + 1)$ unknowns $\lambda_1, \ldots, \lambda_n$ and ρ.

The result is

$$\rho^2 = \frac{G(\phi_1, \ldots, \phi_n, \phi_0)}{G(\phi_1, \ldots, \phi_n)}$$

When, instead of the subspace spanned by n linearly independent elements of H, we consider a set $H_n \subset H$

$$H_n = \left\{ \phi : \phi = \sum_{i=1}^{n} a_i \phi_i , \ |a_i| \le a_i \right\}$$

for some given $a_1, \ldots, a_n$. Then the problem becomes one of restricted approximation

$$\min_{\phi \in H_n} \| \phi_0 - \phi \|$$

and it is no longer true that best approximations are given by orthogonal projections of ϕ on H.

The class of approximation problems in control system optimizations is usually of the restricted type [6, 7, 8, 10, 12, 13].

8.6 Uniqueness of External Elements and Best Approximation

Because of the close connection between the L-problem and approximation problems, it may be expected that the condition for the unique existence up to the scalar constraint of the extremal elements in the L-problem is the same as the condition of uniqueness of the best approximation in approximation problems. The necessary condition for the uniqueness of the best approximation is that the space be strictly normed, i.e., the normed space E is such that for two arbitrary non-zero elements x, y of E, the equality holds in

$$||x + y|| \leq ||x|| + ||y||$$

if and only if

$$x = ay \text{ or } y = ax, \quad a \geq 0$$

To see this, let

$$\sum_{i=1}^{n} \lambda_i \phi_i \quad \text{and} \quad \sum_{i=1}^{n} \lambda_i' \phi_i$$

be the two best approximations to ϕ_0 in a strictly normed space E, where ϕ_i are linearly independent, i.e.,

$$\rho = ||\phi_0 - \sum_{i=1}^{n} \lambda_i \phi_i|| = ||\phi_0 - \sum_{i=1}^{n} \lambda_i' \phi_i||$$

One can assume $\rho \neq 0$ without loss of generality, since otherwise, from the linear independence of ϕ_i, one has

$$\lambda_i = \lambda_i', \qquad i = 1,2,\ldots,n$$

Consider the error of approximation ϕ_0 by

$$\sum_{i=1}^{n} \frac{\lambda_i + \lambda_i'}{2} \phi_i$$

On the one hand

$$\rho \leq ||\phi_0 - \sum_{i=1}^{n} \frac{\lambda_i + \lambda_i'}{2} \phi_i||$$

and on the other

$$||\phi_0 - \sum_{i} \frac{\lambda_i + \lambda_i'}{2} \phi_i|| \leq \frac{1}{2} ||\phi_0 - \sum_{i} \lambda_i \phi_i|| + \frac{1}{2} ||\phi_0 - \sum_{i} \lambda_i' \phi_i||$$

Therefore

$$\rho = ||\phi_0 - \sum_{i} \frac{\lambda_i + \lambda_i'}{2} \phi_i||$$

and from the strict normality of E

$$\left(\phi_0 - \sum_i \lambda_i \phi_i\right) = a\left(\phi_0 - \sum_i \lambda_i' \phi_i\right) a \geq 0$$

The constant is $a = 1$, otherwise ϕ_0 is a linear combination of ϕ_i and $\rho = 0$, contrary to the assumption.

Since $a = 1$,

$$\sum_i \left(\lambda_i - \lambda_i'\right)\phi_i = 0$$

therefore

$$\lambda_i = \lambda_i', \qquad i = 1,2,\ldots,n.$$

In L-problems, let x_1 and x_2 be linearly independent elements such that

$$||x_1 + x_2|| = ||x_1|| + ||x_2||$$

Let f be a linear functional with $(x_1 + x_2)$ as one of its extremal elements. Then

$$|f(x_1 + x_2)| = ||f||\,||x_1 + x_2|| = ||f||\,||x_1|| + ||f||\,||x_2||$$

Since

$$|f(x_1 + x_2)| \leq |f(x_1)| + |f(x_2)| \leq ||f||\,||x_1|| + ||f||\,||x_2||$$

one obtains

$$|f(x_1)| = ||f|| \cdot ||x_1|| \quad \text{and} \quad |f(x_2)| = ||f|| \cdot ||x_2||$$

Thus if E is strictly normed, there are no two linearly independent extremal elements for any linear functional on E.

What spaces are the strictly normed spaces? From Hölder's inequality, one sees that $L_p[a,b]$ spaces $(p>1)$ are all strictly normed.

However, $L[a,b]$ and $c[a,b]^*$ are not strictly normed as the following examples [1] show:

Take $x(t)$ and $y(t)$ in $c[a,b]$ such that $x(t) \not\equiv y(t)$ but

$$\max_t |x(t)| = x(a) = \max_t |y(t)| = y(a)$$

then

$$||x + y|| = ||x|| + ||y||$$

* $c[a,b] = x : x(t)$ continuous for all t in $[a,b]$.

In $L[a, b]$, take x and y to be

$$x(t) = \begin{cases} 1, & b \geq t \geq \dfrac{a+b}{2} \\ 0, & a \leq t < \dfrac{a+b}{2} \end{cases}$$

and

$$y(t) = 1 - x(t)$$

then

$$||x + y|| = \int_a^b |x + y| \, dt = b - a$$

$$||x|| = \int_a^b |x| \, dt = \frac{b-a}{2}$$

and

$$||y|| = \int_a^b |y| \, dt = \frac{b-a}{2}$$

Hence

$$||x + y|| = ||x|| + ||y||$$

but

$$x(t) \neq k\, y(t)$$

Since the space is not strictly normed, the approximation problem in these spaces generally do not have unique solutions. By sufficiently restricting the class of functions to be considered, one can again restore uniqueness in approximations. One such example is discussed in ref. 10.

PART II Approximate Solution of Functional Equations

8.7 Introduction

It often happens that equations which define optimal system behaviors and/or optimal control variables are too complex to admit of ready analytical answers in closed forms. Even some iterative or successive approximation schemes are too difficult. In many such cases, equations which approximate the given equation in some sense are used to derive near optimal system behaviors and/or control variables. This section is devoted to considerations of this type of approximation.

8.8 Inversion of Operators

As an example let us consider the approximate inversion of linear operators [3].

Let A be a nonsingular bounded linear operator with its domain and range in a complete normed linear space X.

We will discuss a procedure to construct a successive approximation to A^{-1}, and will evaluate the error of each approximation.

Let R_1 be its initial approximation such that

$$||I - AR_1|| = e < 1 \tag{8.53}$$

where I is the identity operator in X.

Construct the sequence $\{R_n\}$ where each element is defined by

$$R_{n+1} = R_n\left(I + T_n + \cdots + T_n^{p-1}\right) \tag{8.54}$$

where

$$T_n = (I - AR_n), \quad n = 1,2,\ldots$$
$$p \geq 2$$

then

$$I - AR_{n+1} = T_n^p = (I - AR_n)^p \tag{8.55}$$

and iterating Eq. (8.55)

$$I - AR_{n+1} = (I - AR_1)^{p^n} \tag{8.56}$$

thus, the accuracy improves with power p.

Defining the error for each R_n by

$$e_n = ||A^{-1} - R_n|| \quad n \geq 1 \tag{8.57}$$

then, from Eq. (8.56),

$$A^{-1} - R_{n+1} = A^{-1}(I - AR_1)^{p^n} \tag{8.58}$$

Thus

$$e_{n+1} = ||A^{-1} - R_{n+1}|| \leq ||A^{-1}||\, e^{p^n} \tag{8.59}$$

Since A^{-1} may not be known exactly, it is better to obtain error estimates involving R_1. To do this, we obtain from Eq. (8.58) and from the definition of T_1

$$A^{-1} = R_1(I - T_1)^{-1} \tag{8.60}$$

Thus

$$||A^{-1}|| \leq ||R_1|| \cdot ||I - T_1||^{-1} \leq \frac{||R_1||}{1 - e} \tag{8.61}$$

From Eqs. (8.59) and (8.61), the error estimate is given by

$$e_{n+1} \leq ||R_1|| \cdot \frac{e^{p^n}}{1 - e} \qquad n = 1,2,\ldots \tag{8.62}$$

As an application, let X be a Hilbert space H and let A be a positive definite self-adjoint operator in H.

Then A has positive eigenvalues m and M such that

$$m(x,x) \leq (Ax,x) \leq M(x,x) \tag{8.63}$$

for every $x \epsilon H$. Consider

$$R_1 = aI \qquad 0 < a < 2/M \tag{8.64}$$

then

$$e = ||I - aA|| \tag{8.65}$$

To see $e < 1$, define an operator T_a by

$$T_a = I - aA \tag{8.66}$$

then

$$e = ||T_a|| \tag{8.67}$$

From Eqs. (8.63) and (8.66)

$$1 - aM \leq \frac{(T_a x, x)}{(x,x)} \leq (1 - am)$$

hence

$$\frac{M - m}{M + m} \leq ||Ta|| < 1 \qquad \text{for } 0 < a < 2/M$$

and the minimal norm is attained by $a = \dfrac{2}{M + m}$. From Eq. (8.62)

$$e_{n+1} \leq \frac{a}{1 - e} e^{p^n} \qquad n = 1,2,\ldots$$

To accelerate the convergence, the smallest value of e is given by

$$e = \frac{M - m}{M + m}$$

and is achieved by

$$R_1 = \frac{2}{M + m} I$$

8.9 Convergence of Approximate Solutions (20, 38)

Let T be a given operator in a metric space X and let T^* be its approximation. For example, if T is an inversion operator, T^* could be its power series approximation. More precisely, it is assumed that T and T^* have a common domain D in X, and there is a positive constant ϵ such that

$$||Tx - T^*x|| < \epsilon \tag{8.68}$$

for arbitrary x in D.

Suppose the problem is to solve

$$Tx = x \tag{8.69}$$

approximately by solving

$$T^*x = x \tag{8.70}$$

Let us assume the mapping T is a contraction mapping

$$||Tx_1 - Tx_2|| \leq k\,||x_1 - x_2||, \qquad 0 < k < 1 \tag{8.71}$$

so that an iterative method can be used to solve Eq. (8.69)

$$x_{n+1} = Tx_n, \qquad n = 0,1,\ldots \tag{8.72}$$

Then it is necessary to infer the behaviors of $\{x_n\}$ from that of $\{x_n^*\}$,

$$x_{n+1}^* = T^*x_n^*, \quad x_0^* = x_0 \tag{8.73}$$

the error $\|x_n - x_n^*\|$ at each stage of iteration being of particular interest.

Define

$$d_n = \|x_n - x_{n+1}\|, \quad d_n^* = \|x_n^* - x_{n+1}^*\| \tag{8.74}$$

and

$$E_n = \|x_n - x_n^*\|, \qquad E_0 = 0 \tag{8.75}$$

From Eq. (8.75) and (8.68)

$$E_{n+1} = \|Tx_n - T^* x_n^*\| \leq \|Tx_n - Tx_n^*\| + \|Tx_n^* - T^*x_n^*\| \leq kE_n + \epsilon \tag{8.76}$$

From Eq. (8.76)

$$E_{n+1} \leq k^{n+1} E_0 + (k^n + k^{n-1} + \cdots + 1)\epsilon \leq \frac{\epsilon}{1-k} \triangleq \delta \tag{8.77}$$

for all n. Thus, the errors remain bounded.

Let

$$K = \left\{x : \|x - x_1\| \leq \frac{k}{1-k} d_0\right\} \tag{8.78}$$

Then, since

$$\|x_n - x_1\| \leq \frac{k}{1-k} d_0 \tag{8.79}$$

for all n, x_n^* belongs to δ neighborhood of K for all n. For arbitrary $x \epsilon K$,

$$\|x - x_1^*\| \leq \|x - x_1\| + \|x_1 - x_1^*\| \leq \frac{k}{1-k} d_0 + E_1,$$

$$d_0 = \|x_0 - x_1\| \leq \|x_0^* - x_1^*\| + \|x_1^* - x_1\| = d_0^* + E_1$$

and

$$E_1 = \|x_1^* - x_1\| = \|T^*x_0 - Tx_0\| < \epsilon,$$

Therefore

$$\|x - x_1^*\| \leq \frac{k}{1-k}\left(d_0^* + \epsilon\right) + \epsilon = \frac{k}{1-k} d_0^* + \delta \tag{8.80}$$

Define K^* by

$$K^* = \left\{x : ||x - x_1^*|| \leq \frac{k}{1-k} d_0^* + \delta\right\} \qquad (8.81)$$

Since K^* contains K from Eq. (8.80), if D contains a δ-neighborhood of K^*, then $x_n^* \epsilon D$ for all n. Let

Let

$$e_n = ||x_n^* - x|| \qquad (8.82)$$

where $Tx = x$. Proceeding as in the case of Eq. (8.76), one has

$$e_{N+1} = k\, e_N + ||Tx - Tx_N^*|| \leq k\, e_N + \epsilon \qquad (8.83)$$

Iterating Eq. (8.83)

$$e_{N+m} \leq k^m e_N + \epsilon\left(1 + k + \cdots + k^{m-1}\right)$$

but

$$e_N = ||x_N^* - x|| \leq ||x_{N+m}^* - x|| + ||x_{N+m}^* - x_N^*||$$

$$= e_{N+m} + ||x_{N+m}^* - x_N^*||$$

hence

$$\left(1 - k^m\right)e_{N+m} \leq k^m ||x_{N+m}^* - x_N^*|| + \frac{\epsilon}{1-k}$$

or

$$e_{N+m} \leq \frac{k^m}{1-k^m} ||x_{N+m}^* - x_N^*|| + \frac{\delta}{1-k^m} \qquad (8.84)$$

From Eq. (8.84)

$$d_n^* = ||x_n^* - x_{n+1}^*|| \leq ||T^* x_{n-1}^* - Tx_{n-1}^*|| + ||Tx_{n-1}^* - Tx_n^*||$$
$$+ ||Tx_n^* - T^* x_n^*|| \leq kd_{n-1}^* + 2\epsilon \qquad (8.85)$$

Iterating Eq. (8.85)

$$d_N^* \leq k^N d_0^* + 2\delta \qquad (8.86)$$

Since $k < 1$, for any $c > 0$, there exists an integer N such that

$$k^N d_0^* < 2c \quad \text{for} \quad n \geq N \qquad (8.87)$$

therefore from Eq. (8.86)

$$d_n^* \leq 2\delta^* \tag{8.88}$$

where

$$\delta^* = \delta + c$$

Applying Eq. (8.88) to Eq. (8.84)

$$e_{N+m} \leq \frac{k^m}{1 - k^m} m \, 2\, \delta^* + \frac{\delta}{1 - k^m}$$

Therefore, for every $c < 0$,

$$e_n \leq c + \delta$$

for sufficiently large n.

From Eq. (8.82), one sees that the sequence $\{x_n^*\}$ is eventually in a δ-neighborhood of x and therefore it has a limit point in it.

8.10 Quasi-Linearization

In this section, we will investigate the possibility of solving some nonlinear equations by considering related linear equations. This is only one example of the idea of solving a problem by first transforming it into another problem whose solution is modified to obtain the desired solution. The idea is obviously of great importance and has been the subject of many extensive and scholarly investigations. Since every aspect of the topic cannot be covered even very briefly, we shall focus our attention on the topic of quasi-linearization [15, 23] and its related concept of positivity of operators [15]. Even in this restricted sense, it is impossible to present all relevant facts in a short exposition of this type.

One well-known successive approximation technique is the contraction mapping [26] used, among many other applications, to prove the theorem of the existence and uniqueness of solutions for some types of algebraic, differential, integral, and other equations.

There is another type of successive approximation procedure known as quasi-linearization.

Let us consider an analog of Newton's method of approximation to the differential equation [17]

$$\frac{du}{dt} = g(u, t) \qquad u(0) = c \tag{8.89}$$

Assuming the derivative of $g(u, t)$ exists,

$$h(u,t) \triangleq \frac{\partial g}{\partial u}(u,t) \tag{8.90}$$

and is continuous and monotonically increasing in u for $0 \le t \le t_0$, the successive approximation scheme is given by

$$\frac{du_0}{dt} = g(v_0,t) + (u_0 - v_0)h(v_0,t), \quad u_0(0) = c$$

$$\frac{du_{n+1}}{dt} = g(u_n,t) + (u_{n+1} - u_n)h(u_n,t), \quad u_{n+1}(0) = c \quad n = 0,1,2,\ldots \tag{8.91}$$

where $v_C(t)$ is a known function of time and is continuous over $[0, t_0]$.

Let us investigate one typical stage of induction

$$\frac{dw}{dt} = g(v,t) + (w - v)h(v,t), \; w(0) = c \tag{8.92}$$

From the relation

$$g(v,t) + (w - v)h(v,t) \le g(V,t) + (w - V)h(V,t) \tag{8.93}$$

where V is the function which maximizes the left-hand side of Eq. (8.93), the w of Eq. (8.92) and the W defined by

$$\frac{dW}{dt} = g(V,t) + (W - V)h(V,t), \; W(0) = c \tag{8.94}$$

are compared.

To establish an order relation between the solutions of Eq. (8.93) and Eq. (8.94), an essential use is made of this fact:

If

$$\frac{dx}{dt} = a(t)x + b(t) \quad x(0) = c \tag{8.95}$$

$$\frac{dy}{dt} \ge a(t)y + b(t) \quad y(0) = c \tag{8.96}$$

then

$$x(t) \le y(t) \tag{8.97}$$

This follows from the positivity of the kernal $\exp\left[\int_s^t a(r)\,dr\right]$ which appears in the expression for $y(t)$ when Eq. (8.96) is rewritten as

$$\frac{dy}{dt} = a(t)y + b(t) + f(t) \;,\; f(t) \geq 0 \tag{8.98}$$

Because of Eq. (8.93), the solution of Eq. (8.92) is majorized by the solution of Eq. (8.94), and we obtain from Eq. (8.97)

$$w(t) \leq W(t) \tag{8.99}$$

Since by assumption $g(u,t)$ is strictly convex in u

$$g(u,t) = \underset{v}{\text{Max}} \left[g(v,t) + (u - v)h(v,t) \right] \tag{8.100}$$

Identifying w in Eq. (8.93) with u_{n+1} of Eq. (8.91), V is identified with u_{n+1}. Thus, W in Eq. (8.94) is identified with u_{n+2} in Eq. (8.91). From Eq. (8.99),

$$u_{n+2} \geq u_{n+1} \;,\; n = 0,1,2,\ldots$$

also

$$u_1 \geq u_0$$

Thus,

$$u_0(t) \leq u_1(t) \leq \cdots$$

within a common interval of existence. The convergence, therefore, follows once the uniform boundedness of $\{u_n(t)\}$ is established.

In the scalar case just discussed, in establishing the key relation of Eq. (8.97), the kernel is always positive regardless of $a(t)$. In vector cases, however, this is no longer true. Consider

$$\frac{dx}{df} = Ax \;,\; x(0) = c$$

$$\frac{dy}{dt} \geq Ay \;,\; y(0) = c$$

where x, y and c are n-vectors and A is a constant $n \times n$ matrix.

When does $y(t) \geq x(t)$ follow? Or equivalently, when does $x(t) \geq 0$ follow from $(dx/df) \geq Ax$, $x(0) = 0$?

Since

$$x(t) = \int_0^t e^{A(t-s)} f(s) us$$

where

$$dx/dt = Ax + f(t) \;,\; x(0) = 0$$

the inequality $x(t) \geq 0, t \geq 0$ depends on the positivity of e^{At}, which is no longer always positive when A is a matrix. A sufficient condition for the positivity of e^{At} is that $a_{ij} \geq 0, i \neq j$ and a_{ii} real.

The connection between the positivity of an operator and the quasi-linearization technique is now clear. Stated in general terms:

Let u be a solution of a nonlinear equation defined on D

$$L(u) = f(u, x) \quad , \quad x \epsilon D \tag{8.101}$$

where L is a linear operator defined on D, and D is a complete subset of E_n, n-dimensional Euclidean space. The ranges of L and f are in a metrical, partially ordered space. Let f be continuous in (u, x), twice differentiable in u, and a strictly convex (i.e., $f'' > 0$ function of u for all finite values of u and for all $x \epsilon D$. The function $f(u, x)$, therefore, satisfies

$$f(u, x) = \underset{v}{\text{Max}} \left[f(v, x) + (u - v) f_u(v, x) \right] \tag{8.102}$$

for all $x \epsilon D$.

Appropriate boundary conditions are assumed to exist such that Eq. (8.101) has a unique solution. Without loss of generality, the boundary condition can be taken to be

$$u = 0 \quad \text{for} \quad x \epsilon \partial D$$

where ∂D denotes the boundary of D.

Since the maximum is actually attained for $v = u$ in Eq. (8.102), the class of admissible functions v may be restircted to some suitable class S including u. The set S is to be chosen so that v is sufficiently smooth to insure that a unique solution exists for each admissible v in

$$L(w) = f(v, x) + (w - v) f_u(v, x) \tag{8.103}$$

Let us denote the solution as

$$w = w[x; v]$$

$$\text{For } v = u \text{ note that } w[x; u] = u \tag{8.104}$$

Let the positivity property be satisfied for the operator of Eq. (8.103) for all x in D anf for all v in S, i.e., if $L(z) - f_u(v, x) z \geq 0$ for every $x \epsilon D$ and for every $v \epsilon S$ and $z = 0$ for every $x \epsilon \partial D$, then $z \geq 0$. Then comparing solutions of Eq. (8.101) and Eq. (8.103), because of the positivity property

$$u \geq w[x; v], \; x \epsilon D$$

and

$$u = w[x;v], \ x \epsilon \partial D$$

Thus

$$u \geq \underset{v \epsilon S}{\text{Max}} \, w[x;v]$$

from Eq. (8.104)

$$u = \underset{v \epsilon S}{\text{Max}} \, w[x;v], \ x \epsilon D \tag{8.105}$$

As before, Eq. (8.103) can be used to generate a sequence $u_n(x)$ which converges monotonically to u of Eq. (8.105).

If f is concave (i.e., $f'' < 0$) and the condition $L(z) - f_u(v,x)z \leq 0$ on D implies $z < 0$ on D, then the maximization operation in Eq. (8.105) is replaced by the minimization operation.

Iterative procedures based on the quasi-linearization, under the stated assumptions, thus have a very desirable property of monotone convergence, i.e., steady improvement of accuracy. The convergence can be shown to be quadratic under suitable conditions on f and its derivatives. This means that a positive constant k exists such that for sufficiently large n

$$||u_{n+1} - u_n|| \leq k||u_n - u_{n-1}||^2$$

This is a definite advantage over the usual Picard method of iteration which converges only linearly.

We have seen that the quasi-linearization technique can be used to give unusual representations to functions of interest, thus providing a new approach to solving problems.

As an example of this unusual representation, an application of the quasi-linearization in stochastic differential equations is discussed by Bellman [18]. He considers

$$\frac{du}{dt} = g(u) + r(t) \qquad u(0) = c \tag{8.106}$$

where $r(t)$ is a random function with given distribution function.

Let $g(u)$ be twice differentiable and strictly convex in u so that

$$g(u) = \underset{u}{\text{Max}} \, [g(u) + (u - v)g'(v)] \tag{8.107}$$

is true.

By considering two associated equations

$$\frac{du}{dt} = \operatorname{Max}\,[g(v) + (u - v)g'(u)] + r(t) \qquad u(0) = c \tag{8.108}$$

and

$$\frac{dU}{dt} = g(v) + (U - v)g'(v) + r(t) \qquad U(0) = c \tag{8.109}$$

one obtains

$$U(t\,;v) \le u(t)$$

for all v. Therefore

$$\operatorname{Prob}\,[u \ge x] \ge \operatorname{Prob}\,[U(t\,;v) \ge x]$$

for all random functions v and $-\infty < x < \infty$, where $u(t)$ is the random function obtained by the solution of the stochastic differential Eq. (8.106), which is assumed to exist over $[0, T]$.

Then, for $-\infty < x < \infty$ and for $0 \le t \le T$

$$\operatorname{Prob}\,[u \ge x] = \operatorname*{Max}_{v} \operatorname{Prob}\,[U(t\,;v) \ge x] \tag{8.110}$$

where the maximization is over all random functions defined over $[0, T]$.

A sequence $u_n(t)$ can again be constructed, the distribution of which converges monotonically to that of u.

Let us remember that there are many other techniques for successive approximations such as applications of fixed-point theorems in pseudo-metric spaces [20]* (contraction mapping is one such special case), the rules of false position, and many others.

Dvoretsky [21] considered a fixed-point problem disturbed by random variables, and investigated its convergence with probability one and in mean square. This type of iteration scheme is known as stochastic approximation [25] and is useful in parameter estimations of certain nonlinear control problems and other system optimization problems [19, 36]. The subject matter is discussed in more detail elsewhere in this book.

Before leaving the subject of this section, let us note that the application of quasi-linearization is not limited to successive approximations.

It can be used effectively to represent some nonlinear functions [16]. The simplest example is perhaps furnished by

*A space for any two elements of which pseudo-metric $\rho(x, y)$ is defined satisfying
1) $\rho(x, y) \le \rho(x, z) + \rho(z, y)$
2) $x = y \implies \rho(x, y) = 0$

$$|x| = \max_{-1 \leq y \leq 1} xy \tag{8.111}$$

Here the nonlinear function $|x|$ is expressed as an envelope of linear functions xy, $-1 \leq y \leq 1$. Another simple example [16] is given by

$$\text{Min}\,(x, y) = \min_{0 \leq \phi < 1} [\phi x + (1 - \phi) y]$$

Here again, the nonlinear function is expressed as an envelope of linear functions.

The apparent complication of introducing a maximization operation is justified in that it can lead easily to solutions which may otherwise be rather difficult to obtain, especially in problems where the minimax theorem [7, 11, 24] can be applied effectively.

Eq. (8.111), in its somewhat more general form [15], appears as

$$\phi(x) = \max_{R(y)} L(x, y) \tag{8.112}$$

Where $L(x, y)$ is a function of two variables and $\phi(x)$ is a function of x alone defined by Eq. (8.112). Whatever properties $L(x, y)$ has as a function of x, for all y in $R(y)$, is reflected by $\phi(x)$. If these properties are easier to establish in $L(x, y)$ than to prove the corresponding properties in $\phi(x)$ directly, then the representation of Eq. (8.112) can be quite useful.

REFERENCES

1. Ahieser, N. I., *Theory of Approximation*, Ungar, New York, 1956.
2. Ahieser, N. I., and M. Krein, *Some Questions in the Theory of Moments*, Translation of Mathematical Monographs, Vol. 2, American Mathematical Society, Providence, R.I., 1962.
3. Altman, M., *Approximation Methods in Functional Analysis*, Lecture Note given at the California Institute of Technology, 1959.
4. Antosiewicz, H. A., "Linear Control Systems," *Arch. Rat. Mech. and Analysis* 12, No. 4, pp. 313–324, 1963.
5. Athans, M., "Minimum Fuel Control of Second-Order Systems with Real Poles," *Proc. JACC 1964*, pp. 232–239, June 1963.
6. Aoki, M., "Minimizing Integrals of Absolute Deviations in Linear Control Systems," *AIEE Trans. Application and Industry*, No. 61, 125–128, July 1962.
7. Aoki, M., "Synthesis of Optimum Controllers for a Class of Maximization Problems," *Automatica 1*, No. 1, 69–80, 1963.

8. Aoki, M., "On Realization of Given Trajectories in Control Systems," to appear in *J. Math. Anal. and Application.*
9. Aoki, M., "Successive Generation of Chebychev Approximation Solution," to appear in *A.S.M.E. Transaction J. of Basic Engineering.*
10. Aoki, M., P. L. Elliot and L. A. Lopes, "Correction and Addendum to Minimize Integrals of Absolute Deviation in Linear Control Systems." To appear.
11. Arrow, J. J., M. J. Beckmann, and S. Karlin, "The Optimal Expansion of the Capacity of a Firm," *Studies in the Mathematical Theory of Inventory and Production*, Stanford University Press, Stanford, Calif., 1958.
12. Barbashin, E. A., "Estimating the Mean-Square Deviation From a Given Trajectory," *Aut. and Remote Control,* English Translation *21*, pp. 661-667, February 1961.
13. Barbashin, E. A., "An Estimate of the Maximum Deviation From a Given Trajectory," *Aut. and Remote Control*, English Translation *21*, pp. 945-952, April 1961.
14. Barbashin, E. A., "On the Realization of Motion Along a Given Trajectory," *Aut. and Remote Control*, English Translation *27*, pp. 681-687, June 1961.
15. Beckenbach, E. F., and R. Bellman, *Inequalities*, Ergebnise der Mathematik and ihre Grenzgebiete neue Folge, heft 30, Springer-Verlag, Berlin, 1961.
16. Bellman, R., Glicksberg, I., and O. Gross, "Some Nonclassical Problems in the Calculus of Variations," *Proc. Amer. Math. Soc. 7*, No. 1, pp. 87-94, February 1956.
17. Bellman, R., "On Monotone Convergence to Solutions of $u' = g(u, t)$," *Proc. Am. Math. Soc. 8*, pp. 1007-1009, 1957.
18. Bellman, R., "On the Representation of the Solution of a Class of Stochastic Differential Equations," *Proc. Am. Math. Soc. 9*, pp. 326-7, 1958.
19. Bertram, J. E., "Control by Stochastic Adjustment," *AIEE Application and Industry*, 1-7, January 1960.
20. Collatz, L., "Application of Functional Analysis to Numerical Analysis," *Tech. Report No. 5,* Department of Mathematics, University of California, Berkeley, Calif., May 1960.
21. Dvoretzky, A., "On Stochastic Approximation," *Proc. Third Berkeley Symposium on Math. Stat. and Prob.*, Vol. I, 39-55, University of California Press, 1956.
22. Halmos, P. R., *Finite Dimensional Vector Spaces,* D. Van Nostrand Company, Inc., Princeton, New Jersey, 1958.
23. Kalaba, R., "On Nonlinear Differential Equation, the Maximum Operator, and Monotone Convergence," *J. Math. and Mech. 8*, No. 4, pp. 519-574, July 1959.
24. Karlin, S., "The Theory of Infinite Games," *Ann. Math. 58*, 371-401, 1953.

25. Kiefer, J., and J. Wolfowitz, "Stochastic Estimation of the Maximum of a Regression Function," *Ann. Math. Stat. 23*, pp. 462-466, 1952.
26. Kolmogorov, A. N., and S. V. Fomin, *Elements of the Theory of Functions and Functional Analysis, Vol. 1, Metric Space and Normed Spaces* (English Translation), Graylock Press, Rochester, N.Y., 1957.
27. Krasovskiy, N. N., "On the Theory of Optimum Regulation," *Aut. and Remote Control 18*, pp. 1005-1016, 1957. (English Translation)
28. Kreindler, E., "Contributions to the Theory of Time Optimal Control" *J. Franklin Inst. 275*, No. 4, pp. 314-344, April 1963.
29. Kulikowski, R., "On Optimal Control with Constraints," Bull. de l'Academie Polonaise de Sciences. Serie des sciences Technique 7, No. 4, pp. 285-294, 1959.
30. Lanczos, D., *Applied Analysis*, Prentice-Hall, Inc., Englewood Cliffs, N.J., 1956.
31. Langer, R. (Editor), *On Numerical Approximation*, The University of Wisconsin Press, Madison, Wis., 1959.
32. Neustadt, L. W., "Minimum Effort Control Systems," *J. Soc. Ind. Appl. Math. Ser. A on Control 1*, 16-31, 1962.
33. Neustadt, L. W., "Optimization, a Moment Problem, and Nonlinear Programming," *Report No. TDR-169-(3540-10) TN-1*, Aerospace Corporation, El Segundo, Calif., July 1963.
34. Reid, W., "Ordinary Linear Differential Operators of Minimum Norm," *Duke Math. J. 29*, pp. 591-606, 1962.
35. Roxin, E., "A Geometric Interpretation of Pontryagin's Maximum Principle," *RIAS Tech. Report 61-15*, Baltimore, Md., December 1961.
36. Sakrison, D., "Application of Stochastic Approximation Methods to System Optimization," *Tech. Report 391*, Massachusetts Institute of Technology, Research Laboratory of Electronics, Cambridge, Mass., July 1962.
37. Taylor, A. E., *Introduction to Functional Analysis*, John Wiley & Sons, Inc., New York, 1958.
38. Urabe, M., "Convergence of Numerical Iteration in Solution of Equations," J. of Science of the Hiroshima Uni. Ser. A, *19*, No. 3, pp. 479-489, January 1956.
39. Zadeh, Lotfi A., *Computer Control Systems Technology*, Chapter 14, McGraw-Hill Book Co., New York, N.Y., 1961.

9

Analytical Design Techniques for an Optimal Control Problem

P. R. SCHULTZ
HEAD, INJECTION GUIDANCE SECTION
SYSTEMS RESEARCH AND PLANNING DIVISION
AEROSPACE CORPORATION, LOS ANGELES, CALIFORNIA

9.1 Introduction

This chapter presents a discussion of the problem of optimizing continuous linear systems whose performance is evaluated by a quadratic criterion. Two approaches are used. One is a dynamic programming approach which uses the principle of optimality and treats this problem as the optimization of a continuous decision process. This was developed by Bellman and others. The other approach involves the use of the maximum principle developed by Pontryagin and his colleagues. It involves an extension of the classical calculus of variations.

This kind of problem has been discussed extensively in the literature. Some (but by no means all) of this literature is listed in the references at the end of these notes. The solution of the problem is well known. There are three reasons for considering it in a discussion of modern control theory. The first is that it is one of the few problems for which analytical techniques provide satisfactory results. The second is that it can be applied to some practical problems. The third is that it provides a pedagogical tool for illustrating and comparing the use of dynamic programming techniques and the maximum principle in the optimization of control systems.

This chapter is primarily concerned with the first and third reasons. The solutions obtained by the two different techniques will be compared and their differences and similarities emphasized.

9.2 Definition of the Problem

It is desired to control the behavior of a physical system so that a quadratic performance index is optimized. The motion of system in an n-dimensional phase space is described by the following differential equations written in vector-matrix form:

$$\dot{\mathrm{x}}(t) = A(t)\mathrm{x}(t) + \mathrm{B}(t)u(t) + \mathrm{F}(t), \quad \mathrm{x}(t_0) = \mathrm{x}_0 \tag{9.1}$$

In Eq. (9.1), $A(t)$ is an $n \times n$ matrix whose elements are piecewise-continuous functions of the independent variable t (time); $\mathrm{x}(t)$ and $\mathrm{B}(t)$ are $n \times 1$ (column) matrices; and $\dot{\mathrm{x}}(t)$ is a column matrix whose elements are the derivatives of the corresponding elements of $\mathrm{x}(t)$. The elements of $\mathrm{B}(t)$ are also piecewise-continuous. The scalar $u(t)$ represents the input which can be adjusted to optimize the system's behavior. Function $\mathrm{F}(t)$ is an $n \times 1$ matrix or vector whose elements represent inputs to the system and are known to the controller (the computer in which $u(t)$ is calculated). However, the element of $\mathrm{F}(t)$ cannot be varied by the controller.

For a given initial condition $\mathrm{x}(t_0)$, it is desired to pick $u(t)$ for $t_0 \le t \le T$ so that the following performance index or payoff function is minimized:

$$\text{payoff} = \mathrm{x}'(T)Qx(T) + \int_{t_0}^{T} \left[\mathrm{x}'(s)P(s)\mathrm{x}(s) + \lambda u^2(s)\right] ds \tag{9.2a}$$

$$= \rho(T) + \int_{t_0}^{T} \left[\mathrm{x}'(s)P(s)\mathrm{x}(s) + \lambda u^2(s)\right] ds \tag{9.2b}$$

Here $P(\tau)$ is a positive-semidefinite, symmetric $n \times n$ matrix whose elements can vary with time; Q is a positive-semidefinite symmetric matrix whose elements are constant; and λ is a positive scalar. The prime associated with a matrix indicates the transpose of this matrix. Now we can write

$$\begin{aligned} \dot{\rho}(t) &= \frac{d\rho}{dt} = \mathrm{x}'Q\dot{\mathrm{x}} + \dot{\mathrm{x}}'Q\mathrm{x} \\ &= \mathrm{x}'[QA + A'Q]\mathrm{x} + \mathrm{x}'Q[\mathrm{B}u + \mathrm{F}] + [\mathrm{F}' + u\mathrm{B}']Q\mathrm{x} \end{aligned} \tag{9.3}$$

where the independent variable t has not been denoted on the right-hand side. Thus,

$$\rho(T) = \rho(t_0) + \int_{t_0}^{T} \dot{\rho}(s)\,ds \tag{9.4}$$

Consequently, the substitution of Eq. (9.4) in Eq. (9.2) gives

$$\begin{aligned}\text{payoff} &= \mathbf{x}'(t_0)Q\mathbf{x}(t_0) + \int_{t_0}^{T} \left[\mathbf{x}'(s)P(s)\mathbf{x}(s) + \lambda u^2(s)\right] ds \\ &+ \int_{t_0}^{T} \{\mathbf{x}'(s)[A'(s)Q + QA(s)]\mathbf{x}(s) + \mathbf{x}'(s)Q[\mathbf{B}(s)u(s) \\ &+ \mathbf{F}(s)] + [\mathbf{F}'(s) + \mathbf{B}'(s)u(s)]Q\mathbf{x}(s)\} ds\end{aligned} \tag{9.5}$$

Since the first term on the right-hand side of Eq. (9.5) is specified (because the initial conditions $\mathbf{x}(t_0)$ are specified), the payoff will be minimized if the second and third terms on the right-hand side are minimized. Therefore, the subsequent discussions will deal with the choice of $u(\tau)$, $t_0 \le \tau \le T$, which minimizes the second and third terms on the right-hand side of Eq. (9.5).

The reason for reformulating the problem so that the payoff or performance criterion is defined by Eq. (9.5) rather than Eq. (9.2) is to show that this can be done. The solution can be achieved just as easily without the reformulation. The difference between the two approaches lies in the boundary conditions and form of the adjoint equations [Eqs. (9.47) and (9.51)] when the maximum principle is used to obtain a solution, and in the boundary conditions and form of certain matrix differential equations [Eqs. (9.23) and (9.26)] when dynamic programming is used.

9.3 The Dynamic Programming Treatment

Those familiar with the calculus of variations will recognize that the problem of finding a $u(\tau)$, $t_0 \le \tau \le T$, which minimizes the payoff function given in Eq. (9.2) subject to the constraint that the state vector $\mathbf{x}(t)$ satisfy the differential Eq. (9.1) is a version of the problem of Bolza in the classical calculus of variations. Kalman has shown that this version and similar versions do have unique solutions, and that the optimal payoff function (the payoff function when the optimal control $u(t)$ is implemented) is a twice-differentiable function of the initial state variables. The reader who is interested in these questions of existence and other similar questions is referred to Kalman's papers [1, 2]. The subsequent development assumes these results to be true. Under these conditions, it is possible to use dynamic programming to obtain an analytical solution to the problem, i.e., the optimal policy and payoff function. The results obtained by dynamic programming will be given in this section and the next two sections.

Let the optimal payoff function be defined as follows:

$$f\ \mathbf{x}(t_0), t_0, T = \min_{\substack{u(s)\\ t_0 \le s \le T}} \left\{ \int_{t_0}^{T} \left[\mathbf{x}'P\mathbf{x} + \lambda u^2 + \mathbf{x}'(A'Q + QA)\mathbf{x} + \mathbf{x}'Q(\mathbf{B}u + \mathbf{F}) + (\mathbf{F}' + \mathbf{B}'u)Q\mathbf{x}\right] ds \right\} \tag{9.6}$$

Note that the independent variable s has been suppressed in the integrand in Eq. (9.6). Note also that $\mathbf{x}(t)$ must satisfy Eq. (9.1). To simplify the subsequent discussion, define

$$\Delta(\mathbf{x}, u, s) = \mathbf{x}'P(s)\mathbf{x} + \lambda u^2 + \mathbf{x}'[A'(s)Q + QA(s)]\mathbf{x} + \mathbf{x}'Q[\mathbf{B}(s)u + \mathbf{F}(s)] + [\mathbf{F}'(s) + \mathbf{B}'(s)u]Q\mathbf{x} \tag{9.7}$$

Then

$$f(\mathbf{x}_0, t_0, T) = \min_{\substack{u(s)\\ t_0 \le s \le T}} \left\{ \int_{t_0}^{T} \Delta[\mathbf{x}(s), u(s), s]\, ds \right\} \tag{9.8}$$

where

$$\dot{\mathbf{x}}(t) = A(t)\mathbf{x}(t) + \mathbf{B}(t)u(t) + \mathbf{F}(t)$$

$$\mathbf{x}(t_0) = \mathbf{x}_0$$

Now we can rewrite the right-hand side of Eq. (9.8) as the sum of the following two integrals

$$f[\mathbf{x}(t_0), t_0, T] = \min_{\substack{u(s)\\ t_0 \le s \le T}} \left\{ \int_{t_0}^{t_0+\delta} \Delta\, ds + \int_{t_0+\delta}^{T} \Delta\, ds \right\} \tag{9.8a}$$

For small δ, we can write

$$f[\mathbf{x}(t_0), t_0, T] = \min_{\substack{u(s)\\ t_0 \le s \le T}} \left\{ \Delta[\mathbf{x}(t_0), u(t_0), t_0]\delta + \int_{t_0+\delta}^{T} \Delta\, ds + 0(\delta^2) \right\} \tag{9.9}$$

where $0(\delta^2)$ signifies terms which go to zero as fast as δ^2 goes to zero, i.e., there exists an m such that $m < \infty$, and the terms denoted by $0(\delta^2)$ are bounded by $M\delta^2$ as $\delta \to 0$. Since $\Delta[\mathbf{x}(t_0), u(t_0), t_0]$ is independent of $u(\tau)$, $t_0 + \delta < \tau \le T$, then Eq. (9.9) becomes

$$f[\mathbf{x}(t_0), t_0] = \min_{u(t_0)} \Bigg\{ \Delta[\mathbf{x}(t_0), u(t_0), t_0]\delta + \min_{\substack{u(s) \\ t_0 + \delta \le s \le T}} \int_{t_0}^{T} \Delta[\mathbf{x}(s), u(s), s]\,ds + 0(\delta^2) \Bigg\}$$

$$= \min_{\mathbf{u}(t_0)} \left\{ \Delta[\mathbf{x}_0, u(t_0), t_0]\delta + f(\mathbf{x}_0 + \delta\mathbf{x}_0, t_0 + \delta, T) + 0(\delta^2) \right\} \tag{9.10}$$

by the principle of optimality. Quantity $\delta\mathbf{x}_0$ is given by

$$\delta\mathbf{x}_0 = \dot{\mathbf{x}}(t_0)\delta + 0(\delta^2) = A(t_0)\mathbf{x}_0\delta + \mathbf{B}(t_0)u(t_0)\delta + \mathbf{F}(t_0)\delta + 0(\delta^2) \tag{9.11}$$

Let $\nabla\mathbf{f}$ be a column matrix whose ith element (the one element composing the ith row) is the partial derivative of f with respect to x_i. Since the second term in the brackets on the right-hand side of Eq. (9.10) can be expanded in a Taylor's series about the point $\mathbf{x}_0$, t_0,

$$f(\mathbf{x}_0, t_0, T) = \min_{u(t_0)} \left\{ \Delta[\mathbf{x}_0, u(t_0), t_0]\delta + f(\mathbf{x}_0, t_0, T) + \nabla\mathbf{f}'(\mathbf{x}_0, t_0, T)\delta_0 + \frac{\partial f}{\partial t_0}(\mathbf{x}_0, t_0, T)\delta + 0(\delta^2) \right\} \tag{9.12}$$

Since $f(\mathbf{x}_0, t_0, T)$ is independent of $u(t_0)$, it can be removed from the brackets and canceled with the corresponding term on the left-hand side. Likewise, the fourth term in the brackets on the right-hand side of Eq. (9.12) can be transposed to the left-hand side because it is independent of $u(t_0)$. Therefore, using Eq. (9.11):

$$\frac{\partial f}{\partial t_0}\delta = -\min_{u(t_0)} \left\{ \Delta[\mathbf{x}_0, u(t_0), t_0]\delta + \nabla\mathbf{f}(x_0, t_0, T)'[A(t_0)\mathbf{x}_0 + \mathbf{B}(t_0)u(t_0) + \mathbf{F}(t_0)]\delta + 0(\delta^2) \right\} \tag{9.13}$$

Letting $\delta \to 0$ and canceling δ on both sides of the equation

$$\frac{\partial f}{\partial t_0}(\mathrm{x}_0, t_0, T) = -\min_{u(t_0)} \Big\{ \Delta(\mathrm{x}_0, u_0, t_0) + \nabla \mathrm{f}(\mathrm{x}_0, t_0, T)' [A(t_0)\mathrm{x}_0 + \mathrm{B}(t_0) u(t_0) + \mathrm{F}(t_0)] \Big\} \tag{9.14}$$

$$f(\mathrm{x}_0, T, T) = 0 \text{ -- the boundary condition}$$

The reader is reminded that $\Delta[\mathrm{x}_0, u(t_0), t_0]$ is defined by Eq. (9.7). Equation (9.14) must be solved to obtain the optimal control policy and the corresponding payoff function.

Note that the value of $u(t_0)$ obtained by minimizing the right-hand side of Eq. (9.14) is the system's input or control policy which should be used when the system is in state x_0 at time t_0. Consequently, the solution of Eq. (9.14) will provide the optimal policy for all states $\mathrm{x}(t_0)$ at a time t_0. Thus the optimal policy is determined as a function of $\mathrm{x}(t)$ and t rather than as a function of t for $t_0 \le t \le T$ and a given x_0. This is highly desirable for control problems. Thus it is proper to represent the optimal policy $u^*(t_0)$ by $u^*(x_0, t_0)$ or $u^*(x, t)$. This will be done in the subsequent sections concerning the dynamic programming approach to emphasize that the optimal policy is determined as a function of the current state variables.

9.4 The Solution Via Dynamic Programming

In this section, the fact that $f(\mathrm{x}_0, t_0, T)$ is twice differentiable will be used to obtain a solution to Eq. (9.14). From this solution the optimal policy can be obtained.

Since $\Delta(\mathrm{x}_0, u_0, t_0)$ is a quadratic function of u_0 and since $\lambda > 0$, the expression in brackets on the right-hand side of Eq. (9.14) will have a minimum determined by

$$\frac{\partial \Delta}{\partial u_0}(\mathrm{x}_0, u_0, t_0) + \nabla \mathrm{f}'(\mathrm{x}_0, t_0, T)\mathrm{B}(t_0) = 0 \tag{9.15}$$

Substituting Eq. (9.7) in Eq. (9.15) gives

$$\begin{aligned} u_0^* &= -\frac{1}{2\lambda}[\mathrm{x}_0' Q \mathrm{B} + \mathrm{B}' Q \mathrm{x}_0 + \mathrm{B}' \nabla \mathrm{f}] \\ &= -\frac{1}{2\lambda}[\mathrm{B}'(t_0) \nabla \mathrm{f}(\mathrm{x}_0, t_0, T) + 2\mathrm{B}'(t_0) Q \mathrm{x}(t_0)] \end{aligned} \tag{9.16}$$

Equation (9.16) gives the expression for the optimal policy $u^*(\mathrm{x}_0, t_0)$ when the system is in state x_0 at time t_0. Substituting this result in Eq. (9.14) eliminates u_0 from the right-hand side of that equation and gives (after some manipulations)

$$\begin{aligned}\frac{\partial f}{\partial t_0} &= \frac{1}{4\lambda}\nabla f' BB' \nabla f + \frac{1}{2\lambda} x_0' QBB' \nabla f + \frac{1}{2\lambda} \nabla f' BB' Q x_0 \\ &+ \frac{1}{\lambda} x_0' QBB' Q x_0 - \frac{1}{2}\nabla f'(Ax_0 + F) - \frac{1}{2}(x_0' A' + F')\nabla f \\ &- 2F' Q x_0 - x_0' [P + QA + A'Q] x_0\end{aligned} \tag{9.17}$$

Now let us assume that the solution to Eq. (9.17) has the form

$$f(x_0, t_0, T) = k_0(t_0, T) + k_1'(t_0, T)x_0 + x_0' k_2(t_0, T)x_0 \tag{9.18}$$

In other words, $f(x_0, t_0, T)$ is assumed to be a quadratic function of the state variables. If higher order terms are assumed to exist on the right-hand side of Eq. (9.18), then the subsequent steps leading to a separation of variables will show that these terms are identically zero.

Now

$$\begin{aligned}\nabla f(x_0, t_0, T) &= k_1(t_0, T) + [k_2(t_0, T) + k_2'(t_0, T)]x_0 \\ &= k_1(t_0, T) + K(t_0, T)x_0\end{aligned} \tag{9.19}$$

where

$$K(t_0, T) = K'(t_0, T)$$

If we denote dk/dt_0 by $\dot{k}$ (where k may be any element of K, k_1 or k_0), then the substitution of Eq. (9.19) in Eq. (9.17) gives

$$\begin{aligned}\dot{k}_0 + \dot{k}_1' x_0 + \frac{1}{2} x_0' \dot{K} x_0 &= \frac{1}{4\lambda}(k_1' + x_0' K) BB' (k_1 + Kx_0) \\ &+ \frac{1}{2\lambda}[x_0' QBB'(k_1 + Kx_0) + (k_1' + x_0' K) BB' Q x_0] \\ &+ \frac{1}{\lambda} x_0' QBB' Q x_0 - x_0' [P + QA + A'Q] x_0 \\ &- \frac{1}{2}(F' + x_0' A')(k_1 + Kx_0) - \frac{1}{2}(k_1' + x_0' K)(Ax_0 + F) \\ &- 2F' Q x_0\end{aligned} \tag{9.20}$$

Equating, on both sides of Eq. (9.20), the terms independent of x_0, those linear in x_0, and those quadratic in x_0 requires that the following differential equations must be satisfied (since Eq. (9.20) must hold for all x_0):

$$\frac{dk_0}{dt_0} = \frac{1}{4\lambda} k_1' BB' k_1 - F' k_1 \tag{9.21}$$

$$\begin{aligned} \frac{dk_1}{dt_0} &= \frac{1}{2\lambda} KBB'k_1 + \frac{1}{\lambda} QBB'k_1 - A'k_1 - KF - 2QF \\ &= -A' - \frac{1}{2\lambda} K + \frac{1}{\lambda} Q \; BB' \; k_1 - (K + 2Q)F \end{aligned} \tag{9.22}$$

$$\begin{aligned} \frac{dK}{dt_0} &= \frac{1}{2\lambda} KBB'K + \frac{1}{\lambda} [QBB'K + KBB'Q] \\ &\quad - A'K - KA + 2\left[\frac{QBB'Q}{\lambda} - P - QA - A'Q\right] \end{aligned} \tag{9.23}$$

The boundary condition in Eq. (9.14) requires that Eqs. (9.21), (9.22) and (9.23) satisfy the following boundary conditions:

$$k_0(T, T) = 0 \tag{9.24}$$

$$k_1(T, T) = 0 \tag{9.25}$$

$$K(T, T) = 0 \tag{9.26}$$

Note that if $F(t)$ is identically zero, then k_1 and k_0 are also identically zero. Then, the optimal payoff function is a quadratic form in the state variables (containing no linear or constant terms). The coefficients of this quadratic form are the elements of the K matrix. Note also that the K matrix is independent of F and consequently is invariant to changes in F.

When a solution to Eqs. (9.21), (9.22) and (9.23) is obtained, subject to the boundary conditions given in Eqs. (9.24), (9.25) and (9.26), then the optimal policy given by Eq. (9.16) becomes

$$u_0^*(x_0, t_0) = -\frac{1}{2\lambda} B'(t_0)[k_1(t_0, T) + K(t_0, T)x_0 + 2Qx_0] \tag{9.27}$$

Thus the optimal policy is a linear function of the state variables. This result is characteristic of optimization problems involving quadratic performance indices and linear physical systems whose behavior can be described by the set of equations in Eq. (9.1).

Equations (9.21) and (9.22) are linear in k_0 and k_1. Consequently, solving them should present no conceptual difficulties.

However, in the general case where the elements of the matrices A and B are time-varying, computational solutions of one kind or another will probably be necessary. Equation (9.23) is nonlinear due to the presence of the first term on the right-hand side. However, equations of this form are called matrix Ricatti equations, and their characteristics have been studied extensively by mathematicians. A discussion of some of their properties and a bibliography of previous work is given in Levin's paper. Some of Levin's result which indicate what must be done to solve Eq. (9.23) are outlined below.

The following definitions will simplify the discussions:

$$G_1 = \frac{1}{\lambda} QBB' - A' \tag{9.28}$$

$$G_2 = 2\left[P + QA + A'Q - \frac{1}{\lambda} QBB'Q\right] \tag{9.29}$$

$$G_3 = -\frac{1}{2\lambda} BB' \tag{9.30}$$

$$G_4 = -\frac{1}{\lambda} BB'Q + A \tag{9.31}$$

Then Eq. (9.23) can be rewritten as

$$\begin{aligned} \frac{dK}{dt_0}(t_0, T) &= -K(t_0, T)G_3(t_0)K(t_0, T) + G_1(t_0)K(t_0, T) \\ &\quad + G_2(t_0) - K(t_0, T)G_4(t_0) \end{aligned} \tag{9.32}$$

Let z be an $2n \times 2n$ matrix defined by the following system of linear differential equations:

$$\frac{dz}{dt} = \begin{pmatrix} G_1(t) & G_2(t) \\ G_3(t) & G_4(t) \end{pmatrix} z(t) \tag{9.33}$$

Let $M(t,s)$ be a fundamental solution to the system of differential equations in (9.33); that is, let $M(t,s)$ satisfy (9.33) with $M(s,s) = I_{2n}$, where I_{2n} is the $2n \times 2n$ identity matrix. Let $M_1(t,s)$, $M_2(t,s)$, $M_3(t,s)$, and $M_4(t,s)$ be $n \times n$ matrices which result from the following partition of $M(t,s)$:

$$M(t,s) = \begin{pmatrix} M_1(t,s) & M_2(t,s) \\ M_3(t,s) & M_4(t,s) \end{pmatrix} \tag{9.34}$$

Then the boundary condition given in Eq. (9.26) and the result given in Theorem 1 in Levin's paper imply that

$$K(t_0, T) = [M_1(t_0, T)K(T,T) + M_2(t_0, T)][M_3(t_0, T)K(T,T) + M_4(t_0, T)]^{-1}$$
$$= M_2(t_0, T)M_4^{-1}(t_0, T) \quad (9.35)$$

Thus the solution to the nonlinear matrix Ricatti equation can be obtained from the fundamental solutions to the linear differential Eq. (9.33).

Also, it is easy to show that Eqs. (9.18) and (9.19) imply that

$$x'K_2x = \frac{1}{2}x'(k_2 + k_2')x = \frac{1}{2}x'Kx = x'k_2'x \quad (9.36)$$

This results from the fact that an $n \times n$ matrix can be decomposed into symmetric and antisymmetric parts, and the antisymmetric parts cancel when the quadratic form in (9.36) is composed. Thus the determination of $K(t_0, T)$ is adequate for the solution of the optimization problem.

It is worth noting here that the solution to this problem with an infinite operating time can be obtained (if it exists) by letting $t_0 \to -\infty$. One condition under which a solution to this problem will exist is the case where A is a constant matrix whose eigenvalues have negative real parts and B is constant. Kalman [2] shows that the solution to Eqs. (9.21), (9.22) and (9.23) does indeed exist under these circumstances. Letov [7] also discusses this problem.

At this point, it is worthwhile to mention that $F(t)$ can represent other things besides inputs to the system. For example, consider the case where it is desired that the system follow a specified trajectory. Then a reasonable quadratic cost criterion might be

$$\text{cost} = \int_{t_0}^{T} [\rho_1(s) + \lambda u^2(s)] \quad (9.37)$$

where

$$\rho_1(s) = [x(s) - x_D(s)]' P[x(s) - x_D(s)] \quad (9.38)$$

$$x_D(s) = \text{desired trajectory} \quad t_0 \le s \le T$$

If $N(t, t_0)$ is a fundamental solution of the homogeneous form of Eq. (9.1) with $N(t_0, t_0)$ equal to the identity matrix, then the complete solution to Eq. (9.1) is given by

$$x(t) = N(t, t_0)\left\{x(t_0) + \int_{t_0}^{t} N^{-1}(s, t_0)[B(s)u(s) + F(s)]ds\right\} \quad (9.39)$$

$$= N(t, t_0)\left\{x(t_0) + \int_{t_0}^{t} N^{-1}(s, t_0)B(s)u(s)ds\right\} + x_D(t) \qquad (9.40)$$

Consequently, if an $F(t)$ exists such that the desired trajectory can be described by

$$x_D(t) = N(t, t_0)\int_{t_0}^{t} N^{-1}(s, t_0)F(s)ds \qquad (9.41)$$

then $F(t)$ can be used to represent the effect of the desired trajectory on the optimization problem. This and other such possibilities result from the fact that the differential equations in (9.1) are linear.

9.5 The Treatment of the Problem Via the Maximum Principle

The treatment of this optimization problem by the maximum principle will now be discussed. This will involve the use of modifications of concepts discussed in the chapter by Dr. Meditch on the maximum principle. The modifications consist in replacing the maximization of the Hamiltonian function by the minimization of this function. To maintain the validity of the maximum principle (as it is developed, for example, by Rozenoer [13]) it is necessary to make the vector boundary conditions on the adjoint equations equal to the gradient of the cost function evaluated at the terminal time T. In the usual development [13] the Hamiltonian function is maximized and the (vector) boundary conditions on the adjoint equations are set equal to minus the gradient of the optimal cost function. This modification is analogous to replacing the maximization of a function $f(x)$ by the minimization of the function $-f(x)$. The proofs [13] require no nontrivial changes when this modification is introduced. The results for introducing this modification is that it facilitates the comparison of the results that are obtained here with the results of the dynamic programming approach. This comparison could also be made without the modification. The choice of whether or not to use the above modification of the maximum principle is essentially a matter of taste.

The plant equations are given by Eq. (9.1) and are given below for reference:

$$\dot{x} = g(x, u, t) = A(t)x + B(t)u(t) + F(t) \qquad (9.1)$$

The following equation is added to the nth order system defined by Eq. (9.1):

$$\begin{aligned}\dot{x}_0 = g_0(x, u, t) &= x'[P + A'Q + QA]x + \lambda u^2 \\ &+ x'Q[Bu + F] + [F' + uB']Qx = f_0(x, u, t)\end{aligned} \qquad (9.42)$$

Equation (9.42) (defining the rate of change of the performance criterion) is added to the nth order system given by Eq. (9.1) to obtain an $(n+1)$-st order system which is denoted by

$$\dot{\mathrm{x}} = \mathrm{g}(\mathrm{x}, u, t) \tag{9.43}$$

Here x denotes an $n+1$ dimensional vector which is often called the state vector of the augmented plant. The system of equations adjoint to Eq. (9.43) is given by

$$\dot{\mathrm{z}} = -\left[\frac{\partial \mathrm{g}}{\partial \mathrm{x}}\right]' \mathrm{z} \tag{9.44}$$

The matrix on the right-hand side of Eq. (9.43) is the transpose of the Jacobian of g with respect to x, i.e., its ijth element is $\partial g_i/\partial x_j$. It is an $n+1$ by $n+1$ matrix in this case. Since $\partial g_i/\partial x_0 = 0$ as a result of the definition of g, we have

$$\dot{z}_0 = 0 \tag{9.45a}$$

$$z_0 = \text{constant} \tag{9.45b}$$

The other n equations defined by Eq. (9.44) are given by

$$z_i = -\sum_{j=1}^{n} a_{ji}(t) z_j - z_0 \frac{\partial g_0}{\partial x_i}(\mathrm{x}, u, t), \quad i = 1, \ldots, n \tag{9.46}$$

$$\left[\text{since } \frac{\partial g_i}{\partial x_j} = a_{ij}(t)\right]$$

The equations in (9.46) can be written in vector matrix form as

$$\dot{\mathrm{z}} = -A'\mathrm{z} - 2z_0\left[(P + A'Q + QA)\mathrm{x} + Q\mathrm{B}u + Q\mathrm{F}\right] \tag{9.47}$$

Now Pontryagin's Maximum Principle, with the previous modifications, states that a necessary condition that a system input or control be optimum is that $H(z,x,u,t)$ be minimized with respect to the allowable u. $H(\mathrm{z},\mathrm{x},u,t)$ is defined as

$$\begin{aligned} H(\mathrm{z},\mathrm{x},u,t) &= \mathrm{z}'g(\mathrm{x},u,t) \\ &= z_0\big[\mathrm{x}'(P + A'Q + QA)\mathrm{x} + \lambda u^2 \\ &\quad + \mathrm{x}'Q(\mathrm{B}u + \mathrm{F}) + (\mathrm{F}' + u\mathrm{B}')Q\mathrm{x}\big] \\ &\quad + \mathrm{z}'A\mathrm{x} + \mathrm{z}'\mathrm{B}u + \mathrm{z}'F \end{aligned} \tag{9.48}$$

Minimizing the right-hand side of Eq. (9.48) with respect to u gives (by partial differentiation)

$$u = -\frac{1}{2\lambda z_0}(2z_0 B'Q\mathrm{x} + B'\mathrm{z}) \tag{9.49}$$
$$= -u(\mathrm{x},\mathrm{z},t)$$

Equation (9.49) must be satisfied by the optimal control since Pontryagin's Maximum Principle (and the modification introduced here) is a necessary condition for optimality.

It will be shown in the next section that it is of great interest to determine the behavior of the plant and the adjoint systems of differential equations along an optimal trajectory. This information can be obtained by substituting Eq. (9.49) in Eq. (9.47) and Eq. (9.1). The result is

$$\dot{\mathrm{x}} = \left(A - \frac{1}{\lambda}BB'Q\right)\mathrm{x} - \frac{1}{2\lambda z_0}BB'\mathrm{z} + F \tag{9.50}$$

$$\dot{\mathrm{z}} = -\left(A' - \frac{1}{\lambda}QBB'\right)\mathrm{z} - 2z_0\left(P + A'Q + QA - \frac{1}{\lambda}QBB'Q\right)\mathrm{x} - 2z_0 QF \tag{9.51}$$

It is illuminating to write Eqs. (9.51) and (9.50) as the set of the following second first-order equations:

$$\begin{bmatrix}\dot{\mathrm{x}} \\ \dot{\mathrm{z}}\end{bmatrix} = \begin{bmatrix} A - \frac{1}{\lambda}BB'Q & -\frac{1}{2\lambda z_0}BB' \\ -2z_0\ P + A'Q + QA - \frac{1}{\lambda}QBB'Q & -A' - \frac{1}{\lambda}QBB' \end{bmatrix}\begin{bmatrix}\mathrm{x} \\ \mathrm{z}\end{bmatrix} + \begin{bmatrix} F \\ -2QF \end{bmatrix} \qquad \begin{aligned} \mathrm{x}(t_0) &= \mathrm{x}_0 \\ \mathrm{z}(T) &= 0 \end{aligned} \tag{9.52}$$

This is the set of equations which an optimal trajectory must satisfy. The boundary condition $\mathrm{z}(T) = 0$ is required in that the cost of operating the system in the interval $t_0 \leq t \leq T$ is

$$x_0(t_0) = \int_{t_0}^{T} \dot{x}_0(t)\,dt = \text{cost}$$

Hence $x_0(T) = 0$ for all $\mathrm{x}(T)$. The appropriate modifications to [13] show that $\mathrm{z}(T)$ is given by the gradient of the cost function at time T. Therefore, $\mathrm{z}(T) = 0$ and $z_0(T) = 1$. Hence, with $\mathrm{x}(t_0) = \mathrm{x}_0$ and $\mathrm{z}(T) = 0$, the solution to this optimization problem involves a two-point boundary value problem. However, the differential equations for this particular case are linear. Consequently, the solution of the two-point boundary value problem is not as difficult as in other cases (due to the fact that the differential equations are linear).

Note that the adjoint equations correspond to the Euler-Lagrange equations in the classical treatment of the appropriate form of the problem of Bolza. The application of the classical variational theory (in the form of the problem of Bolza) to the control optimization problem is discussed by Berkovitz, Kalman, Letov and others.

9.6 Comparison of the Maximum Principle Appproach with the Dynamic Programming Approach

To make a comparison between the solution obtained by the application of the maximum principle and that obtained by dynamic programming, it is instructive to derive the differential equation which the gradient of the optimum performance criterion satisfies as the system moves through phase or state space under the influence of the optimal control. The gradient of the optimum performance criterion is defined by Eq. (9.19). Removing the subscript 0 in Eq. (9.19) and differentiating with respect to t (denoted by a dot over the symbol)

$$\frac{d}{dt}\nabla \mathrm{f}(\mathrm{x},t,T) = \dot{\mathrm{k}}_1(t,T) + \dot{K}(t,T)\mathrm{x} + K(t,T)\dot{\mathrm{x}} \tag{9.53}$$

Substituting Eq. (9.22), (9.23) and (9.1) in Eq. (9.53) gives, after some algebraic manipulations which involve the use of Eq. (9.19)

$$\begin{aligned}\frac{d}{dt}\nabla f = &-\left(A' - \frac{1}{\lambda}QBB'\right)\nabla f - 2Q\mathrm{F}\\ &- 2\left(P + A'Q + QA - \frac{1}{\lambda}QBB'\right)\mathrm{x}\end{aligned} \tag{9.54}$$

This differential equation is identical to Eq. (9.51) since $z_0 = 1$.

Note that the boundary conditions in Eqs. (9.14), (9.25) and (9.26) imply that $\nabla \mathrm{f}(\mathrm{x},T,T) = 0$. But

$$\mathrm{z}(T) = 0 \tag{9.55}$$

With these boundary conditions and $z_0 = 1$, the adjoint equations satisfy the same differential equations and boundary conditions that ∇f must satisfy. Therefore, the solution to the adjoint equations as defined in Eq. (9.51) [obtained by eliminating $u(t)$ from Eq. (9.48)] gives ∇f. Equation (9.51) shows how ∇f, the gradient of the optimal performance criterion, varies along an optimal trajectory. With this interpretation of the adjoint solutions, the modified maximum principle can then be interpreted as meaning that to optimize the system's performance, one chooses the control (from the allowable set of controls) to minimize the scalar product of the $n+1$ dimensional vectors $[1, \nabla f(\mathrm{x}, t)]$ and $(\dot{x}_0, \dot{\mathrm{x}})$. Likewise, equation (9.14) can be rewritten as

$$0 = -\min_{u(t_0)} \left\{ \Delta[\mathrm{x}_0, u(t), t_0] + \frac{\partial f}{\partial t_0} + \nabla \mathrm{f}' [A\mathrm{x}_0 + \mathrm{B}u(t_0) + \mathrm{F}] \right\} \quad (9.56)$$

If the minimizing value of $u(t)$, given by Eq. (9.16) is substituted in Eq. (9.56), then we have the result that the following relation should hold along an optimal trajectory:

$$\frac{d}{dt} f(\mathrm{x}, t, T) = -\Delta[\mathrm{x}, u^*(t), t] \quad (9.57)$$

The $*$ associated with $u(t)$ indicates that the optimal control given by Eq. (9.16) is substituted in Δ. In view of the definition of Δ and the performance criterion [see Eqs. (9.7) and (9.5)], this result should not be unexpected. It was first discussed in connection with a more general problem by Krassovskiy and Desoer.

Note again that the solution of Eqs. (9.50) and (9.51) or, equivalently, the system of equations given in (9.52), involves the solution of a two-point boundary value problem. This occurs because the initial conditions on the state variables are specified at the initial time t_0 [see Eq. (9.1)] while the boundary conditions on the adjoint equations are given at the terminal time T.

The boundary conditions on the differential equations (9.21), (9.22) and (9.23) which determine the optimal performance index are also specified at the terminal time [see Eqs. (9.24), (9.25) and (9.26)]. However, there is no coupling between these equations and the state variables as there is in the adjoint equations. Thus the solution of these equations would appear to be considerably simpler than the solution of the Eq. (9.52) because of the fact that the boundary conditions on the state variables are specified at the initial time t_0 is irrelevant in Eqs. (9.21), (9.22) and (9.23). However, one must specify the terminal time T in order to obtain suitable boundary conditions for Eqs. (9.21), (9.22) and (9.23).

9.7 Related Subjects

Many people have discussed this control system optimization problem in the literature [10, 11, 17, 18, 1, 2, 6].

This and other control system optimization problems can be studied with the classical calculus of variations. Berkovitz discusses some of the mathematical aspects of this topic in ref. 8. Dreyfus illustrates heuristically some relations between the calculus of variations and dynamic programming. Friedland discusses the canonical structure of optimal controllers. Kelley [21] and Breakwell, Speyer and Bryson [22] consider "neighboring optimum" trajectories and use variational methods involving the second variation of the payoff to obtain a control law. This work can be used to derive control laws for some nonlinear problems. Finally, Letov shows how certain analytical results canbe obtained for this and some other problems with the classical calculus of variations. These results are similar in some ways to the ones presented here. They have not been included because a suitable presentation in this chapter would require lengthy discussion of some topics in the calculus of variations.

Finally, a word of caution is in order. The type of problem discussed in this chapter is one of a few which can be studied in terms of the solutions to a set of *linear* equations [see Eqs. (9.52) and Eqs. (9.21), (9.22), (9.23), (9.34) and (9.35)]. If the plant contains nonlinearities or the performance criterion has a different form, the results which one can obtain with these approaches are less complete and elegant. Usually, numerical methods must be used to obtain quantitative solutions. Reference 9 discusses this type of problem in detail.

In addition to variational methods, Balakrishnan and Hsieh have used techniques from functional analysis to study optimization problems related to this one [23, 24]. They involve the use of a steepset-descent technique to obtain the optimal control as a function of time for a given initial state vector, and use concepts from the theory of Hilbert space.

REFERENCES

1. Kalman, R. E., "The Theory of Optimal Control and the Calculus of Variations," RIAS Technical Report 61-3.
2. Kalman, R. E., "Contributions to the Theory of Optimal Control Systems," Boletín de la Sociedad Matemática Mexicana, 1960, pp. 102-119.
3. Levin, J. J., "On the Matrix Riccati Equation," Proceedings of American Mathematical Society, Vol. 10 (1959), pp. 519-524.
4. Dreyfus, S. E., "Dynamic Programming and the Calculus of Variations," Journal of Mathematical Analysis and Applications, Vol. 1, No. 2 (Sept. 1960).
5. Desoer, C. A., "Pontryagin's Maximum Principle and the Principle of Optimality," Journal of the Franklin Institute, Vol. 271, No. 5 (May, 1961).

6. Friedland, B., "The Structure of Optimum Control Systems," Transactions of the ASME, Journal of Basic Engineering, Vol. 84, Series D, No. 1 (March, 1962), pp. 1-12.
7. Letov, A. M., "Analytical Controller Design I-IV," Automation and Remote Control, Vol. 21, Nos. 4, 5 and 6 and Vol. 22, No. 4.
8. Berkovitz, L. D., "Variational Methods in Problems of Control and Programming," Journal of Mathematical Analysis and Applications, Vol. 3, No. 1 (August, 1961), pp. 145-169.
9. Kipiniak, W., "Dynamic Optimization and Control, A Variational Approach," MIT Press—John Wiley and Sons, 1961.
10. Merriam, C. W., "An Optimization Theory for Feedback Control System Design," Journal of Information and Control, Vol. 3, No. 1 (March, 1960).
11. Merriam, C. W., "A Class of Optimum Control Systems," Journal of the Franklin Institute, Vol. 267, No. 4 (April, 1959), pp. 267-281.
12. Breakwell, J. V., "The Optimization of Trajectories," Journal of Society for Industrial and Applied Mathematics, Vol. 7, No. 2 (June, 1959), pp. 215-247.
13. Rozenoer, L. T., "The Maximum Principle of L. S. Pontryagin in the Theory of Optimal Systems," Automation and Remote Control, Vol. 20, Nos. 10, 11 and 12 (October, November and December, 1959).
14. Boltyanskiy, V. G., R. V. Gamkrelidze, and L. S. Pontryagin, "The Theory of Optimal Processes, I: The Maximum Principle," American Mathematical Society Translations, Series 2, Vol. 18 (1961), pp. 341-382.
15. Pontryagin, L. S., V. G. Boltyanskiy, R. V. Gamkrelidze and E. F. Mishchenko, "The Mathematical Theory of Optimal Processes," Interscience (a division of John Wiley and Sons, New York, N. Y., 1962) (an English translation by L. W. Neustadt and K. N. Trirogoff).
16. Krasovskiy, N. N., "On the Theory of Optimum Control," Applied Mathematics and Mechanics, Vol. 23, No. 4 (1959), pp. 899-919.
17. Bellman, R. E., "Dynamic Programming," Princeton University Press, 1957.
18. Bellman, R. E., "Adaptive Control Processes: A Guided Tour," Princeton University Press, 1961.
19. Bryson, A. E. and W. F. Denham, "Multivariable Terminal Control for Minimum Mean-Square Deviation from a Nominal Path," Proceedings of the IAS Symposium on Vehicle Systems Optimization; Garden City, N. Y., Nov. 28-29, 1961.
20. Ellert, F. J. and C. W. Merriam III, "Synthesis of Feedback Controls Using Optimization Theory," IEEE Transactions on Automatic Control, Vol. AC-8, No. 2 (April, 1963), pp. 89-103.

21. Kelley, H. J., "Guidance Theory and External Fields," IRE Transactions on Automatic Control, Vol. AC-7, No. 5 (Oct., 1962), pp. 75-82.
22. Breakwell, J. V., J. L. Speyer and A. E. Bryson, "Optimization of Nonlinear Systems Using the Second Variation," Journal of Society of Industrial and Applied Math on Control, Series A, Vol. 1, No. 2, pp. 193-223.
23. Hsieh, H. C., "On the Synthesis of Adaptive Controls by the Hilbert Space Approach," *University of California, Los Angeles: Department of Engineering Report* No. 62-19, June, 1962.
24. Balakrishnan, A. V., "An Operator Theoretic Formulation of a Class of Control Problems and a Steepest-Descent Method of Solution," Journal of SIAM on Control, Series A, Vol. 1, No. 2, pp. 109-127.

10

Some Elements of Stochastic Approximation Theory and its Application to a Control Problem

P. R. SCHULTZ
HEAD, INJECTION GUIDANCE SECTION
SYSTEMS RESEARCH AND PLANNING DIVISION
AEROSPACE CORPORATION, LOS ANGELES, CALIFORNIA

10.1 Introduction

This chapter presents an introduction to the subject of stochastic approximation theory and shows how it can be applied to the study of certain types of control system problems. It is worth mentioning at this point that stochastic approximation theory is one approach to stochastic convergence problems. There are other such problems and approaches [2, 4, 5, 12]. The special cases discussed here are of interest because their mean convergence proofs are much simpler than those for more general cases and because they are of some interest as far as control system problems are concerned. It is worth noting that it is also possible to obtain convergence with probability 1 in certain rather general cases. This aspect of stochastic approximation theory is not discussed in this chapter due to its mathematical complexity.

In a qualitative and heuristic sense, stochastic approximation theory can be considered to deal with questions regarding the convergence of certain sequences of random variables. This subject has attracted the attention in recent years of pure mathematicians, statisticians, and engineers. The reasons for this are: (a) that it presents many interesting mathematical problems and (b) the concepts can be applied fruitfully to problems in feedback control, filtering and statistical estimation.

10.2 An Example of Stochastic Approximation

In this section we will prove that under certain conditions a sequence of scalar-valued random variables converges in the mean-square sense to a real number θ. Since this sequence is obtained from the transformation defined below in Eq. (10.1), the mean convergence of this sequence or lack thereof can be used to discuss questions of "stability" of a system whose behavior can be described by Eq. (10.1).

Let $\{X_n(\omega)\}$ and $\{Y_n(\omega)\}$ be sequences of real scalar random variables defined over the points $\{\omega\}$ of an approximate sample space Ω. Let x_i and y_i denote values assumed by $X_i(\omega)$ and $Y_i(\omega)$. Let $T_n[X_1(\omega), X_2(\omega), \ldots, X_n(\omega)]$ be a measurable transformation of its arguments. Then, we define the sequence of random variables by

$$X_{n+1}(\omega) = T_n[X_1(\omega), X_2(\omega), \ldots, X_n(\omega)] + Y_n(\omega) \tag{10.1}$$

where the transformations T_n are assumed to satisfy the inequality

$$|T_n(x_1, \ldots, x_n) - \theta| \leq F_n|x_n - \theta| \tag{10.2}$$

The sequence $\{F_n\}$ of positive real numbers is assumed to satisfy the following conditions:

$$\prod_{n=r}^{\infty} F_n = 0 \tag{10.3a}$$

$$\prod_{n=r}^{s} F_n < A, \quad \text{all } s > r, \ A \text{ finite} \tag{10.3b}$$

with these definitions we are now ready to state and prove the following theorem.

Theorem 1. If the conditions given in Eqs. (10.5) hold with probability 1 for all n, then

$$\lim_{n \to \infty} E(X_n - \theta)^2 = 0 \tag{10.4}$$

The required conditions are as follows:

$$E(X_1^2) < \infty \tag{10.5a}$$

$$\sum_{n=1}^{\infty} E(Y_n^2) < \infty \tag{10.5b}$$

$$E(Y_n | x_1, \ldots, x_n) = 0 \tag{10.5c}$$

The proof of Theorem 1 follows. Let $V_n^2 = E(X_n - \theta)^2$ and $\sigma_n^2 = EY_n^2$. The use of Eqs. (10.2), (10.5c) and (10.1) gives

$$V_{n+1}^2 \leq F_n^2 V_n^2 + \sigma_n^2 \tag{10.6}$$

Iterating Eqs. (10.6) gives

$$\begin{aligned} V_{n+1}^2 \leq \sigma_n^2 + \sigma_{n-1}^2 F_n^2 + \dots + \sigma_m^2 F_{m+1}^2 F_{m+2}^2 \dots F_n^2 + \\ \dots + \sigma_1^2 F_2^2 F_3^2 \dots F_n^2 + V_1^2 F_2^2 \dots F_n^2 \end{aligned} \tag{10.7}$$

To prove that $\lim_{n \to \infty} V_n^2 = 0$, note that Eq. (10.7) gives

$$\begin{aligned} V_{n+1}^2 \leq \sum_{j=m}^{n} \sigma_j^2 \left(\max_{m \leq k \leq n} \prod_{j=k+1}^{n} F_j^2 \right) \\ + \left(V_1^2 + \sum_{j=1}^{m-1} \sigma_j^2 \right) \max_{m < k \leq m} \prod_{j=k+1}^{n} F_j^2 \end{aligned} \tag{10.8}$$

Now, the assumption in Eq. (10.3b) is that the partial procucts πF_j^2 are bounded, and the condition in Eq. (10.5b) enables us to choose an m large enough so that for any given $\epsilon > 0$

$$A^2 \sum_{j=m}^{\infty} \sigma_j^2 < \epsilon/2 \tag{10.9}$$

Recall that the condition in Eq. (10.56) requires that

$$\sum_{j=1}^{\infty} \sigma_j^2 < \infty$$

Consequently,

$$V_{n+1}^2 \leq \frac{\epsilon}{2} + \left(V_1^2 + \sum_{j=1}^{\infty} \sigma_j^2 \right) \max_{1 \leq k \leq m} \left(\prod_{j=k+1}^{n} F_j^2 \right) \tag{10.10}$$

Now, since m is fixed by the inequality in Eq. (10.9), we have

$$\lim_{n \to \infty} \max_{1 \leq k \leq m} \prod_{j=k+1}^{n} F_j^2 = 0 \tag{10.11}$$

This follows from Eq. (10.3a). Consequently, there exists an N such that for all $n > N$, $V_n^2 < \epsilon$. But ϵ can be chosen arbitrarily small. Therefore, $V_n^2 \to 0$ as $n \to \infty$. This completes the proof of Theorem 1.

Dvoretzky derives this and analogous results [2] and shows that $X_n(\omega)$ converges to θ with probability one for less restrictive conditions on the T_n than that given in Eq. (10.2). The interested reader is referred to these results. Also note that minor modifications in the proof (e.g., replacing $(X_{n+1} - \theta)^2$ by the norm) enable one to obtain analogous results for vector-valued random variables (such as the state vector in a control system). This fact will be demonstrated and used later in this chapter.

10.3 An Example: The Robbins-Munro Process

In this section, the Robbins-Munro process [6] will be discussed from the viewpoint of the previous section. This process is classic in stochastic approximation theory and involves a method of experimental estimation of the root θ of a regression function.

Let Z_x be a stochastic process, i.e., a set of random variables for which x is an index. We define the expected value of Z_x (which will be a function of x) as

$$f(x) = E(Z_x) \tag{10.12}$$

Here $f(x)$ is called the regression function of the random variables Z_x, and is assumed to exist for all x.

Consider the sequence of numbers $\{a_n\}$ defined on the positive integers which satisfy the following conditions:

$$\sum_{n=1}^{\infty} a_n = \infty \tag{10.13a}$$

$$\sum_{n=1}^{\infty} a_n^2 < \infty \tag{10.13b}$$

Let X_n^* be an observation of the random variable Z_n and define the following sequence of random variables:

$$X_{n+1} = X_n - a_n X_n^* \tag{10.14}$$

This sequence of random variables defines the Robbins-Munro process. It will be shown that $\{X_n\}$ converges to θ, the root of the regression function, if $f(x)$ and $\{Z_x\}$ satisfy certain conditions. Dvoretzky shows that this convergence can also be obtained under less stringent conditions. With the restrictions given below, the convergence of the Robbins-Munro process follows as a corollary to Theorem 1.

Corollary: Let the following conditions be satisfied:

1. $f(x)$ is a measurable function of x
2. $E\ Z_x^2 \leq \sigma^2 < \infty$ for all t

3. $0 < A \leq \dfrac{f(x)}{x - \theta} \leq B < \infty$

4. $\sum_{n=1}^{\infty} a_n = \infty, \; \sum_{n=1}^{\infty} a_n^2 < \infty$

Then the Robbins-Munro procedure coverges to θ, the root of $f(x)$. This proof is presented by Dvoretzky [2].

Proof: Define the random variable $X_{n+1}(\omega)$, $Y_n(\omega)$ and the transformation $T_n(X_n)$ by

$$X_{n+1}(\omega) = X_n - a_n f(X_n) + Y_n(\omega) \tag{10.15}$$

$$Y_n(\omega) = -a_n \left[Z_{X_n} - f(X_n)\right] \tag{10.16}$$

$$T_n(X_n) = X_n - a_n f(X_n) \tag{10.17}$$

Condition 3 of the corollary gives

$$\begin{aligned} |T_n(x_n) - \theta| &\leq |x_n - \theta| \sup \left(1 - a_n \frac{f(x)}{x - \theta}\right) \\ &\leq |x_n - \theta| \max(1 - A a_n, B a_n - 1) \end{aligned} \tag{10.18}$$

Since condition 4 requires $a_n \to 0$ as $n \to \infty$, we can define a sequence $\{F_n\}$ which satisfies the conditions given in Eqs. (10.2) and (10.3a). Equation (10.2) can be satisfied by setting $F_n = \max(1 - A a_n, B a_n - 1)$. The requirement that $\sum_{n=1}^{\infty} a_n$ diverge is sufficient to insure that the infinite product $\prod_{i=1}^{\infty} F_i = 0$, as required by Eq. (10.3a) (see, for example, ref. 8, pp. 13-15). An example of a suitable $\{a_n\}$ sequence is $a_n = 1/n$. Furthermore, we choose $X_1(\omega)$ so that $EX_1^2 < \infty$. Since the regression function is measurable and satisfies condition 3 for all X, and since $E(Z_x^2) \leq \sigma^2$ by condition 2,

$$E\, Y_n^2 = a_n^2 E\left[Z_{x_n} - f(X_n)\right]^2 \leq a_n^2 M \tag{10.19}$$

where M is finite. Thus inequality (10.5b) is satisfied. Finally, the definition of the regression function given in Eq. (10.12) insures that condition (10.5c) of Theorem 1 is satisfied. Hence, all the conditions of Theorem 1 are satisfied and $E(X_n - \theta)^2 \to 0$ as $n \to \infty$. Consequently, X_n converges in the mean, to θ, the root of the regression function. The comparison of Eq. (10.14) with Eqs. (10.15) and (10.16) shows that the Robbins-Munro process converges to the root of the regression function. For more general results, see ref. 2.

10.4 An Example: The Kiefer-Wolfowitz Process

The Kiefer-Wolfowitz [7] process and its generalizations have received much attention in the literature. This process is used for estimating the maximum of the regression function $f(x)$. Sakrison has used these ideas to obtain the optimum parameter settings for a filter. His work is discussed in more detail in the conclusions of this chapter.

Let Z_x be a stochastic process with $E\{Z_x\}$ as before. We assume $f(x)$ to be a measurable function of x. If $z(x)$ is an observation of the Z_x process, the Kiefer-Wolfowitz process is defined as follows:

$$X_{n+1} = X_n + \frac{b_n}{c_n}\left[z(x_n + c_n) - z(x_n - c_n)\right] \tag{10.20}$$

For purposes of simplifying the proof, the problem of estimating the maximum θ of the regression function $f(x)$ of the random process Z_x will be replaced by the problem of estimating the minimum θ of the regression function $F(x) = -f(x)$ of the random process $-Z_x = Z_x^*$. We define the following sequence of random variables:

$$X_{n+1}^* = X_n^* - b_n\left[\frac{F(X_n^* + c_n) - F(X_n^* - c_n)}{c_n}\right] + Y_n^* \tag{10.21}$$

$$Y_n^* = -\frac{b_n}{c_n}\left[Z_{X_n+c_n}^* - F(X_n + c_n) - Z_{X_n-c_n}^* + F(X_n + c_n)\right] \tag{10.22}$$

This amounts to a modification of the form of Eq. (10.20). Now the following corollary can be proved:

Corollary: let $\{X_n^*(\omega)\}$ and $\{Y_n^*(\omega)\}$ be sequences of random variables defined by Eqs. (10.21) and (10.22). Furthermore, let

a. $\sum_{n=1}^{\infty} b_n = \infty$

b. $\lim_{n\to\infty} c_n = 0$

c. $\sum_{n=1}^{\infty}\left(\frac{b_n}{c_n}\right)^2 < \infty$

d. $E\{Z_n^{*2}\} \le \sigma^2 < \infty$ for all x

e. $0 < A \le \left|\frac{F(X_n^* + c_n) - F(X_n^* - c_n)}{c_n(X_n^* - \theta)}\right| \le B < \infty$

f. $E(X_1^{*2}) < \sigma_1^2 < \infty$

Then $\lim E(X_n^* - \theta)^2 = 0$.

Proof: Define

$$T_n = X_n^* - b_n \left[\frac{F(X_n^* + c_n) - F(X_n^* - c_n)}{c_n} \right] \tag{10.23}$$

Then

$$\begin{aligned} |T_n - \theta| &\leq \sup_{r \neq \theta} \left| 1 - b_n \left[\frac{F(r + c_n) - F(r - c_n)}{c_n (r_n - \theta)} \right] \right| |r - \theta| \\ &\leq \left\{ \max\left((1 - b_n A), (B\, b_n - 1)\right) \right\} |r - \theta| \end{aligned} \tag{10.24}$$

From here it is easy to show that conditions (a) through (f) imply that the requirements of Theorem 1 are satisfied. The steps are almost identical to those used previously to prove convergence of the Robbins-Munro process. Thus X_n^* converges in the mean-square sense to θ, the minimum of the regression function $F(x)$.

As was the case with the Robbins-Munro process, mean-square convergence and convergence with probability 1 can be proven under more general conditions.

10.5 A Control Problem

In this and the following sections, some results presented by J. E. Bertram [1] will be discussed. Bertram is interested in designing a suitable sampled data controller for the following type of plant. It is assumed that the plant is linear and time-varying with a single input, and that the differential equations governing the behavior of the state variables are of the form

$$\dot{x}_i(t) = \sum_{j=2}^{n} a_{ij}(t) x_j(t) + d_i(t) m(t) \qquad i = 1, \ldots, n \tag{10.25}$$

The fact that j runs from 2 to n rather than 1 to n means that there is an integration in the system's equations. The requirements on the $a_{ij}(t)$ terms are given later in terms of requirements on the transition matrix and the measurement noise. It is assumed that all the state variables can be measured by the controller, but that these measurements are corrupted by additive noise. It is also assumed that the control $m(t)$ can change only at discrete time due to the presence of periodic sampling in the system and therefore $m(t)$ is constant over a given sampling interval, i.e.,

$$\dot{m}(t) = m(kT), \; kT < t \leq (k+1)T \tag{10.26}$$

With noise or errors and sampling present in the measurement of the state variables, the information available to the controller for the computation of $m(t)$ (the input to the plant) has the following form:

$$\mathbf{y}(kT) = \mathbf{x}(kT) + \mathbf{n}(kT) \quad k = 0, 1, \ldots \tag{10.27}$$

where $\mathbf{n}(kT)$ represents the vector error in the measurement of the state vector $\mathbf{x}(kT)$ and $\mathbf{y}(kT)$ is the quantity available to the controller. Finally, it is assumed that the elements of the vector $\mathbf{d}(t)$ can vary between known limits (due to the designer's imperfect knowledge of the plant).

The task which the controller is supposed to perform is to return the system from a perturbed condition $\mathbf{x}(0)$ to an equilibrium condition $\mathbf{x}^r$, where $\mathbf{x}^r$ is a vector such that

$$\mathbf{x}^r = \begin{pmatrix} x_1^r \\ 0 \\ 0 \\ \cdot \\ \cdot \\ 0 \end{pmatrix} \tag{10.28}$$

i.e., only its first component is unequal to zero. Also, it is assumed that the control $m(t)$ is of the form

$$m^*(t) = \mathbf{a}_k^T \left[\mathbf{x}^r - \mathbf{y}(k)\right] \quad , \; kT < t \le (k+1)T \tag{10.29}$$

i.e., the control is piecewise constant and is a linear combination of the components of the difference between the equilibrium (or desired) state vector and the measured state vector. The gains in this controller are represented by the vector $\mathbf{a}_k$. The subscript k indicates that these gains are time-varying. In the following part of these notes we will use the stochastic approximation theory approach to derive conditions under which the state vector $\mathbf{x}(k)$ will converge in the mean to the desired state vector $\mathbf{x}^r$.

10.6 The Application of Stochastic Approximation Methods to the Control Problem

In the following discussion, we are interested in the state vector only at the sampling instants (i.e., $t = kT$, $k = 1, 2, \ldots$). Thus the behavior of the system can be described by the following difference equation:

$$\begin{aligned} \mathbf{x}(k+1) &= \Phi_k \mathbf{x}(k) + \mathbf{d}_k m(k) \\ &= \Phi_k \mathbf{x}(k) + \mathbf{d}_k \mathbf{a}_k^T \left[\mathbf{x}^r - \mathbf{x}(k) - \mathbf{n}(k)\right] \end{aligned} \tag{10.30}$$

where $\mathbf{x}(k)$ is used to indicate the value of the state vector $\mathbf{x}(t)$ for $t = kT$. The subscript k on $\mathbf{d}_k$, $\mathbf{a}_k$ and the matrix Φ_k have the same meaning. This Φ_k is the transition matrix of the system given by the solution to the following matrix differential equation:

$$\frac{d\Phi}{dt} = [a_{ij}(t)]\,\Phi(t)$$

$$\Phi_k = \Phi[(k+1)T] \quad \text{with} \quad \Phi[(k)T] = I \text{ (identity matrix)}$$

$$[a_{ij}(t)] = \text{matrix coefficients defined in Eq. (10.25)} \tag{10.31}$$

Since we are interested in the convergence of $\mathbf{x}(k)$ to $\mathbf{x}^r$, it is expedient to define the error state vector $\mathbf{e}(k)$ as $\mathbf{e}(k) = \mathbf{x}(k) - \mathbf{x}^r$, and study the convergence of $\mathbf{e}(k)$ to the zero vector in norm. With this change of variable, the equations of motion become

$$\mathbf{e}(k+1) = \left(\Phi_k - \mathbf{d}_k \mathbf{a}_k^T\right)\mathbf{e}(k) + (\Phi_k - I)\mathbf{x}^r - \mathbf{d}_k \mathbf{a}_k^T \mathbf{n}(k) \tag{10.32}$$

But, from the assumptions involved in Eq. (10.25), $a_{i1}(t) = 0$ for all i. Consequently, it is easy to show that $(\Phi_k)_{11} = 1$ and $(\Phi_k)_{i1} = 0$ for $i \neq 1$, i.e., all terms in the first column of Φ_k are zero except the first (or diagonal term) which is equal to one. This and the fact that only the first component of $\mathbf{x}^r$ is nonzero [see Eq. (10.28)] implies that

$$(\Phi_k - I)\mathbf{x}^r = 0 \tag{10.33}$$

Therefore Eq. (10.32) becomes

$$\mathbf{e}(k+1) = \left(\Phi_k - \mathbf{d}_k \mathbf{a}_k^T\right)\mathbf{e}_k - \mathbf{d}_k \mathbf{a}_k^T \mathbf{n}(k) \tag{10.34}$$

Now, it will be shown that under conditions A, B, and C (given below) that

$$\lim_{k \to \infty} E\,\|\mathbf{e}(k)\| = 0 \tag{10.35}$$

where $\|\mathbf{e}(k)\|$ denotes the norm of the vector $\mathbf{e}(k)$.

In addition to the norm of a vector, it is also necessary to define the norm of a matrix. One satisfactory definition of the norm of a matrix is

$$\|A\| = \max_{\mathbf{x} \neq 0} \left[\frac{\|A\mathbf{x}\|}{\|\mathbf{x}\|}\right] \tag{10.36}$$

The maximizing operation on the right-hand side of Eq. (10.36) is done over all nonzero vectors $\mathbf{x}$. Note that the norm of the matrix

$||A||$ depends on the definition of the vector norm used. Consequently, it is possible to define many norms of a given matrix [1, 19]. Note that Eq. (10.36) gives for any $\mathbf{x}$

$$||A\mathbf{x}|| \leq ||A|| \, ||\mathbf{x}|| \tag{10.36a}$$

The conditions which imply the result stated in Eq. (10.35) are summarized in the following theorem.

Theorem 2. If the following conditions are satisfied:

A. $||\Phi_k - \mathbf{d}_k \mathbf{a}_k^T|| < \mu_k$, where μ_k is a sequence of positive numbers such that

i. $\prod_{i=0}^{\infty} \mu_i = 0$

ii. $\prod_{i=r}^{s} \mu_i < A$ for all r, s such that $s > r$ and A finite

B. $E||\mathbf{n}(k)|| < \infty$ and $\mathbf{n}(k)$ is statistically independent of the state of the system

C. $\sum_{k=0}^{\infty} E\left(||\mathbf{d}_k \mathbf{a}_k^T||\right) E\left(||\mathbf{n}(k)||\right) < \infty$

then Eq. (10.35) holds, i.e.,

$$\lim_{k \to \infty} E\left(||\mathbf{e}(k)||\right) = 0$$

The proof of Theorem 2 follows.

From Eq. (10.36)

$$||\mathbf{e}(k+1)|| \leq ||\Phi_k - d_k a_k^T|| \, ||\mathbf{e}(k)|| + ||\mathbf{d}_k \mathbf{a}_k^T|| \, ||\mathbf{n}(k)|| \tag{10.37}$$

Define $V_k = E\left(||\mathbf{e}(k)||\right)$ and $\sigma_k = E\left(||\mathbf{d}_k \mathbf{a}_k^T||\right) E\left(||\mathbf{n}(k)||\right)$. Then Eq. (10.37) and condition A of the theorem imply that

$$V_{k+1} < \mu_k V_k + \sigma_k \tag{10.38}$$

Define

$$b_{k,m} = \prod_{j=m}^{k} \mu_j \tag{10.39}$$

Then, "iterating" Eq. (10.38) gives

$$V_{k+1} < b_{k,0} V_0 + \sum_{i=0}^{k} b_{k,i+1} \sigma_i \tag{10.40}$$

where $b_{k,k+1} = 1$, by definition.

Now we fix m. As $k \to \infty$, $b_{k,m} \to 0$ due to condition A. Consequently, $b_{k,0} V_0 \to 0$ as $k \to \infty$. Also,

$$\sum_{i=0}^{k} b_{k,i+1} \sigma_i = \sum_{i=0}^{m} b_{k,i+1} \sigma_i + \sum_{i=m+1}^{k} b_{k,i+1} \sigma_i \tag{10.41}$$

Now, the first term on the right-hand side goes to zero as k becomes infinite since $b_{k,i+1} \to 0$ as $k \to \infty$ for all i less than m from condition A. Since condition A implies that $b_{r,s}$ is bounded for all r and s such that $r > s$, condition C gives

$$\sum_{k=0}^{\infty} \sigma_k < \infty \tag{10.42}$$

Consequently, we can choose m large enough so that

$$\sum_{i=m+1}^{\infty} b_{k,i+1}\, \sigma_i < \epsilon \tag{10.43}$$

For any $\epsilon > 0$. Since we can choose an arbitrarily small $\epsilon > 0$ before specifying m, the second term on the right-hand side of Eq. (10.41) approaches zero as k becomes infinite. Thus

$$\lim_{k \to \infty} V_k = 0 \tag{10.44}$$

and Theorem 2 is proved.

Thus we have established a set of sufficient conditions for the convergence of $E\|\mathbf{e}(k+1)\|$ to zero. This is another way of saying that the control system is stable (in the sense that the expected value of the norm of the error converges to zero) in the presence of measurement errors in the state vector. Note that condition C requires that

$$E\left(\|\mathbf{d}_k \mathbf{a}_k^T\|\right) E\left(\|\mathbf{n}(k)\|\right) \to 0 \quad \text{as} \quad k \to \infty \tag{10.45}$$

This is a rather strong requirement on the characteristics of the measurement errors. However, the conditions of Theorem 2 [which imply Eq. (10.45)] are sufficient to insure that the system will reach the desired equilibrium point (in the sense that $\lim_{k \to \infty} E\left(\|\mathbf{e}(k)\|\right) = 0$).

Note that the assumptions that $a_{i1} = 0$ and $\mathbf{x}_r$ have only one nonzero component [see Eqs. (10.25) and (10.28)] that can be modified under certain conditions. The results obtained in this section can also be obtained if Eq. (10.33) and the assumptions used to obtain it are replaced by the following condition:

$$\sum_{k=0}^{\infty} \|(\Phi_k - I)\mathbf{x}_r\| < \infty \tag{10.46}$$

Any $\mathbf{x}_r$ for which Eq. (10.46) is satisfied will permit convergence of the analog of Eq. (10.37). In addition, it is unnecessary that $(\Phi_k)_i = 0$ for $i \neq 1$ and $(\Phi_k)_{11} = 1$ to satisfy Eq. (10.46) and thus obtain the proof of Theorem 2. Finally, it should be possible to make $\mathbf{x}_r$ a function of time and find some suitable sequences $\{\mathbf{x}_r(k)\}$ for which Eq. (10.46) is satisfied.

10.7 Conclusions

In the preceding section, a special case of the concept of stochastic approximation has been discussed. This concept has been applied to the problem of determining sufficient conditions for stability of a linear, time-varying controller subject to measurement errors and variations in certain parameters. There are many more general approaches to stochastic approximation [2, 4, 5, 12]. These alternate and more general approaches may also be useful in other control problems.

Techniques of this type have been applied to other problems. Sakrison [3, 9, 10] describes the application of a stochastic approximation theoretic technique to optimal filter design. He assumes that a finite number of parameters can be adjusted to optimize the performance of the system, and that the inputs are stationary ergodic random processes. A gradient procedure for adjusting the parameters so that the performance of the system is improved is given. The performance index takes the form of the expected value of a convex function of the difference between the total output and the desired output of the system. This optimization procedure is an experimental one. The major part of the paper consists of the application of stochastic approximation theoretic methods to prove that this experimental optimization procedure converges to the optimum. The conditions the performance criterion, the system, and the inputs must satisfy to obtain convergence are discussed. This is an extension of the procedures described in refs. 2 and 7.

Kushner [16, 17, 18] describes methods for experimental system optimization and for obtaining the peak of an unknown function. Some of this work involves simulation studies in addition to analytical results. These techniques have much in common with stochastic approximation theory.

Some other papers of interest are listed in the references. Schmetterer [12] summarizes many of the recent results in this field. Such topics as asymptotic distributions of the random variables in the sequences, stopping rules, rates of convergence and the higher moments are discussed in ref. 12. Work by Czechoslovakians on this subject is presented in refs. 4 and 5 of this chapter. Much of this work involves the theory of random functions or random variables which range over a function (Banach) space.

REFERENCES

1. Bertram, J. E., "Control by Stochastic Adjustment," AIEE Transactions—Part II, Applications and Industry, January, 1960.
2. Dvoretzky, A., "On Stochastic Approximation," Proceedings of the Third Berkeley Symposium on Mathematical Statistics

and Probability, Vol. 1, University of Calinfornia Press, 1956.

3. Sakrison, D. J., "Application of Stochastic Approximation Methods to Optimum Filter Design," IRE 1961 International Convention Record, Part 4.
4. "Transactions of the First Prague Conference on Information Theory, Statistical Decision Functions and Random Processes," Publishing House of the Czechoslovak Academy of Sciences, Prague, 1957.
5. "Transactions of the Second Prague Conference on Information Theory, Statistical Decision Functions, and Random Processes," Publishing House of the Czechoslovak Academy of Sciences, Prague, 1960.
6. Robbins, H. and S. Munro, "A Stochastic Approximation Method," Annals of Mathematical Statistics, Vol. 22 (1951) pp. 400-407.
7. Kiefer, J. and J. Wolfowitz, "Stochastic Estimation of the Maximum of a Regression Function." Annals of Mathematical Statistics, Vol. 23 (1952), pp. 462-466.
8. Titchmarsh, E. C., *Theory of Functions*, 2nd ed. Oxford University Press, 1939.
9. Sakrison, D. J., "Iterative Design of Optimum Filters for Nonmean-Square-Error Performance Criteria," IEEE Transactions on Information Theory, Vol. IT-9, No. 3, July, 1963, pp. 161-167.
10. Sakrison, D. J., "Application of Stochastic Approximation Methods to System Optimization," MIT Research Laboratory of Electronics Technical Report No. 391, July 10, 1962.
11. Gardner, L. A. Jr., "Stochastic Approximation and its Application to Problems of Prediction and Control Synthesis," in *Proceedings of the International Symposium on Nonlinear Differential Equations and Nonlinear Mechanics*, edited by J. P. LaSalle and S. Lefschetz. Academic Press, NYC., 1963.
12. Schmetterer, L., "Stochastic Approximation," *Proceedings of the Fourth Berkeley Symposium on Mathematical Statistics and Probability*, Vol. 1, edited by J. Neyman. University of California Press, 1961.
13. Chung, K. L., "On a Stochastic Approximation Method," Annals of Mathematical Statistics, Vol. 25 (1963), pp. 463-483.
14. Kesten, H., "Accelerated Stochastic Approximation," Annals of Mathematical Statistics, Vol. 29 (1958), pp. 41-49.
15. Ho, Y. C., "On the Stochastic Approximation Method and Optimal Filtering Theory," Journal of Mathematical Analysis and Applications, Vol. 6 (1962), pp. 152-154.
16. Kushner, H. J., "A Versatile Stochastic Model of a Function of Unknown and Time-Varying Form," Journal of Mathematical Analysis and Applications, Vol. 5 (1962), pp. 150-167.

17. Kushner, H. J., "Hill-Climbing Methods for the Optimization of Multiparameter Noise Disturbed Systems," Proceedings of the 1962 Joint Automatic Control Conference, New York City, Paper No. 8-4.
18. Kushner, H. J., "A New Method of Locating the Maximum Point of an Arbitrary Multipeak Curve in the Presence of Noise," Proceedings of the 1963 Joint Automatic Control Conference, Minneapolis, Minn., pp. 69-80.
19. Bellman, R. E., Introduction to Matrix Analysis, McGraw-Hill, NYC, 1960.

11

Analysis and Synthesis of Discrete-Time Systems

GEORGE A. BEKEY
ASSOCIATE PROFESSOR
ELECTRICAL ENGINEERING DEPARTMENT
UNIVERSITY OF SOUTHERN CALIFORNIA
LOS ANGELES, CALIFORNIA

11.1 Introduction

If the signals at one or more points of a system can change only at discrete values of time, the system is known as a "discrete" or "sampled-data" system. Such systems generally contain elements operating on continuous signals, elements operating on discrete signals, devices for transforming continuous to discrete information (usually known as samplers), and devices for transforming discrete to continuous information (usually known as hold circuits). The sampling operations are often periodic, but may be arbitrary.

The analysis of sampled-data systems with linear elements has been accomplished in the past largely by the use of a special form of the Laplace transform known as the "z transform." The use of such transforms is generally limited to linear systems with periodic sampling and negligible sampling times, but can be extended to certain other cases.

In keeping with current trends in control system theory, discrete systems have recently been studied by means of state-space concepts. As with continuous systems, the concepts of *state* and *state transformation* make possible the systematic formulation of a large class of problems, including those with arbitrary sampling patterns and nonlinear operations.

The purpose of this chapter is to review some of the characteristics and methods of analysis of linear sampled-data systems, present the state-variable formulation of sampled-data problems, and discuss the solution of certain nonlinear problems. The use of Lyapunov's second method for the study of asymptotic stability of discrete systems is also presented.

The chapter is not intended to be a complete survey of analysis and synthesis techniques. Rather, it is a discussion of a few selected problems which illustrate the use of state-space concepts in the analysis and synthesis of sampled-data systems.

11.2 The Sampling Process

The transition from continuous to discrete information is performed by means of a sampling switch or "sampler." The operation of a sampler may be viewed as the modulation of a train of pulses $p(t)$ by a continuous, information-carrying signal $x(t)$, as indicated in Fig. 11-1. If $x(t)$ is used to modulate the amplitude of the pulses,

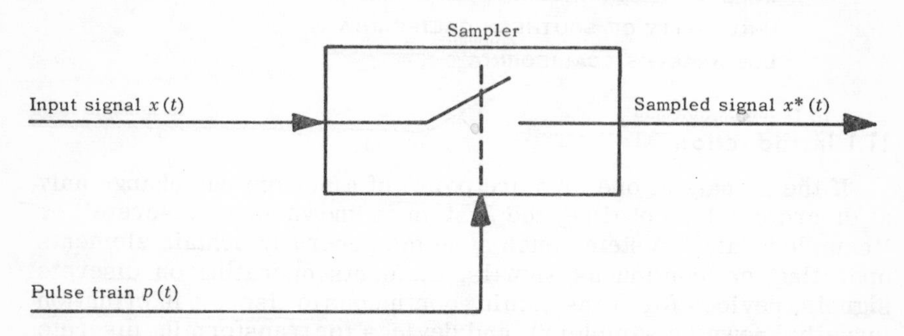

FIG. 11-1. Operation of a sampler.

the process is termed *pulse amplitude modulation (PAM)*. If each pulse has equal width h and unit amplitude, and the pulses are periodic with period T, the pulse train can be described by

$$p(t) = \sum_{k=-\infty}^{+\infty} [u(t - kT) - u(t - kT - h)] \tag{11.1}$$

where $u(t)$ is the unit step function. The output pulse train in Fig. 11.1 is then given by

$$x^*(t) = x(t)\, p(t) \tag{11.2}$$

If the sampling pulse width is small, i.e. $h \ll T$, the output of the sampler may simply be considered the number sequence $\{x(kT^+)\}$. We shall designate this type of periodic sampling with negligible pulse width as *ordinary* or *conventional sampling*. It can be seen that the ordinary sampler is a time-varying amplifier, and that ordinary sampling is a linear operation, i.e.

$$[a_1 x_1(t) + a_2 x_2(t)]^* = a_1 x_1^*(t) + a_2 x_2^*(t) \tag{11.3}$$

where $x_1(t)$ and $x_2(t)$ are continuous signals, and a_1 and a_2 are arbitrary constants. In accordance with convention, the asterisk (*) denotes sampled signals.

If the modulation pattern is signal-dependent, the sampler becomes a nonlinear device. Pulse-width modulation (PWM) and pulse-frequency modulation (PFM) are examples of nonlinear sampling. We shall consider the analysis and stability of such systems in later sections of this chapter.

Data reconstruction. Reconstruction of sampled signals is performed by clamping and extrapolation devices. The simplest data reconstruction device is the *zero-order hold*, which produces an output

$$x_R(t) = x(kT), \quad kT \leq t < (k+1)T \tag{11.4}$$

i.e., the output is held constant between samples at the last sampled value. More complex devices can be used to extrapolate from sampled values with an arbitrary polynomial.

Quantization. If the sampled signal is to be used in a digital computer, it cannot assume any arbitrary value but can only take on a finite sequence of amplitudes dependent on the register length in the computer. Thus, the sampled signal must also be quantized in amplitude. Quantization is a nonlinear operation which may or may not be negligible depending on the resolution available in the computer. In this chapter we shall ignore the effect of quantization.

Frequency characteristics. The unit pulse train $p(t)$ of Eq. (11.1) can be represented by a Fourier series in the form

$$p(t) = \sum_{k=-\infty}^{+\infty} c_k e^{jk\omega_s t} \tag{11.5}$$

where $\omega_s = 2\pi/T$ is the sampling frequency and the coefficients c_k are

$$c_k = \frac{1 - e^{-jk\omega_s h}}{jk\omega_s T} \tag{11.6}$$

where, again, h is pulse width. With the aid of Eq. (11.5) the sampler output (11.2) can be written

$$x^*(t) = x(t)\left\{\frac{h}{T} + \frac{2h}{T}\sum_{k=1}^{\infty} \frac{\sin(k\pi h/T)}{k\pi h/T}\cos[k(\omega_s t - \phi)]\right\} \tag{11.7}$$

where $\phi = h\omega_s/2$. Equation (11.7) shows that pulse-amplitude modulation involves multiplication by a function which contains an infinite number of harmonics of the sampling frequency. Consequently, the output of the sampler contains not only the original signal frequencies ω_i, but also an infinite number of sideband frequencies, $\omega_i \pm k\omega_s$, $k = 1, 2, \cdots$.

Impulse modulation. The mathematical representation of the ordinary sampling process can be simplified if the sampler is replaced by an idealized sampler called an *impulse modulator.* The sampling pulse train $p(t)$ is then replaced by

$$p(t) = \sum_{k=-\infty}^{+\infty} \delta(t - kT) \tag{11.8}$$

which is a train of impulses. If the input to the idealized sampler is a continuous signal $x(t)$ defined for $t \geq 0$, then the sampler output is

$$x^*(t) = \sum_{k=0}^{\infty} x(kT)\,\delta(t - kT) \tag{11.9}$$

which is a modulated impulse train. Equation (11.9) can be viewed as a limiting case of Eq. (11.2) if the finite pulse-width sampler is assumed to have a gain of $1/h$ so that a particular output pulse is given by

$$h^{-1}[u(t - kT) - u(t - kT - h)]\,x(kT) \tag{11.10}$$

for $h \ll T$, and we allow h to approach zero. Unfortunately, the problems raised by the use of generalized functions such as $\delta(t)$ have only been treated heuristically in the literature on ordinary sampled-data systems [1-3], but the process can be defined rigorously [7].

Stationarity. Assume that signals in a continuous time-invariant linear system are sampled. Since the sampler is a time-varying amplifier, the resulting signals will be nonstationary. The sequence $\{x(kT)\}$ however will be stationary.

11.3 Time-Domain Response of Conventional Sampled Systems

Consider the system of Fig. 11-2 where the output of an idealized sampler (impulse modulator) is used as the input to a linear time-invariant system described by its weighting function $g(t)$. Since the input to the plant is a train of impulses, the output $y(t)$ can be obtained by summing the impulse responses, i.e.

$$y(t) = \sum_{k=0}^{\infty} g(t - kT)\,x(kT) \tag{11.11}$$

where $x(t) = 0$ for $t < 0$. At the nth sampling instant

$$y(nT) = \sum_{k=0}^{n} g(nT - kT)\,x(kT) \tag{11.12}$$

which is known as the *convolution summation*. The sequence $\{g(kT)\}$, by its analogy with the weighting function $g(t)$ is called the *weighting sequence*. Equation (11.12) is a linear relation between the input sequence $\{x(kT)\}$ and the output sequence $\{y(kT)\}$ and consequently characterizes a discrete system. It can be noted that such a relationship among discrete values of continuous signals does not require the actual presence of samplers in the system. The summation of Eq. (11.11) gives the output or response of the system for all values of time. If sampling instants are denoted by t_k, $(0 \le k \le n)$, then Eq. (11.12) can be written as

$$y(t_n) = \sum_{k=0}^{n} g(t_n - t_k)x(t_k) \tag{11.13}$$

which is valid even if the sampling is not periodic.

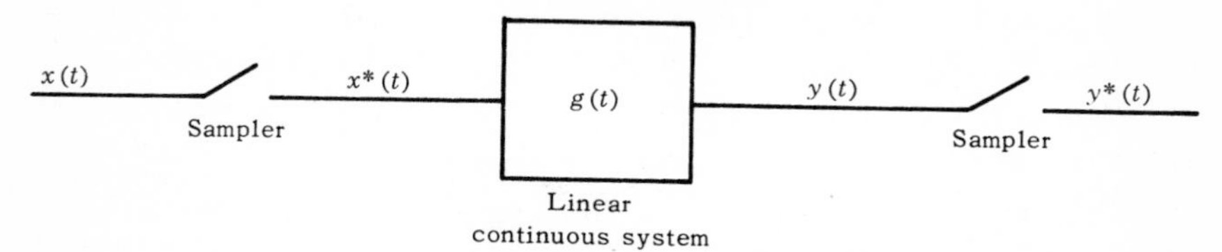

FIG. 11-2. Open-loop discrete system.

Difference equations. The input and output sequences of a discrete dynamic system are related by difference equations, analogous to the differential equations relating continuous dynamic systems. The difference equations will have constant coefficients for periodic sampling. An nth order system is described by an equation which expresses the output at any sampling instant in terms of input and output values at the n past sampling instants.

Example 11.1. Let the system of Fig. 11.2 be a double integration, defined by

$$g(t) = t \tag{11.14}$$

Then, applying Eq. (11.13),

$$y(t_n) = \sum_{k=0}^{n} (n - k)T\, x(t_k) \tag{11.15}$$

where the sampling is periodic. Consequently,

$$y(t_{n+1}) = \sum_{k=0}^{n} (n - k)T\, x(t_k) + T \sum_{k=0}^{n} x(t_k) \tag{11.16}$$

and

$$y(t_{n+2}) = \sum_{k=0}^{n} (n - k)T\, x(t_k) + 2T \sum_{k=0}^{n} x(t_k) + T\,x(t_{n+1}) \tag{11.17}$$

Combining Eqs. (11.15) through (11.17):

$$y(t_{n+2}) - 2y(t_{n+1}) + y(t_n) = T\,x(t_{n+1}) \tag{11.18}$$

This is a second-order linear difference equation which relates the input and output of the system at the sampling instants. Relationships of the form (11.18) can be generalized most conveniently using the state variable formulation of discrete systems to be introduced later in this chapter. Note that since the system is of order 2, two past samples are required in this difference equation.

11.4 Frequency Domain Analysis of Conventional Linear Sampled-Data Systems

Consider the simple system of Fig. 11-2 again. In accordance with the definition of the impulse modulator, the sampled signal $x^*(t)$ is given by

$$x^*(t) = \sum_{k=0}^{\infty} x(kT)\,\delta(t - kT) \tag{11.19}$$

The Laplace transform of this signal is given by

$$X^*(s) = \sum_{k=0}^{\infty} x(kT)\,e^{-kTs} \tag{11.20}$$

It can be shown that an equivalent representation is

$$X^*(s) = \frac{1}{T}\sum_{n=-\infty}^{+\infty} X(s + nj\omega_s) + \frac{1}{2}\,x(0) \tag{11.21}$$

where $x(0)$ is the initial value of the time function and $X(s)$ is its Laplace transform. Equations (11.20) and (11.21) show that the sampled function is periodic in the frequency domain, i.e.

$$X^*(j\omega) = X^*(j\omega + nj\omega_s) \tag{11.22}$$

The Laplace transform of the system output is obtained from Eq. (11.20):

$$Y(s) = X^*(s)G(s) \tag{11.23}$$

where

$$G(s) = \mathcal{L}[g(t)] \tag{11.24}$$

By direct application of the above relationships, the transform of sampled output is

$$Y^*(s) = X^*(s)G^*(s) \tag{11.25}$$

This expression is periodic in ω_s. To avoid the difficulties connected with the evaluation of the infinite series of Eq. (11.25), the s plane is commonly mapped into a new complex plane, called the z plane, by the transformation

$$z = e^{sT} \tag{11.26}$$

Periodicity is eliminated by this transformation since horizontal strips of the left half of the s plane, ω_s rad/sec in height, overlie the inside of the unit circle in the z plane [1, 2].

Applying the transformation of Eq. (11.26) to Eq. (11.20), we have the following definition:

Definition: The z transform of the function $x(t)$ is the function X of the complex variable z defined by

$$X(z) = \sum_{k=0}^{\infty} x(kT)z^{-k} \tag{11.27}$$

If we now define $G(z)$ as the z transform of the weighting sequence $\{g(nT)\}$, Eq. (11.25) becomes

$$Y(z) = X(z)G(z) \tag{11.28}$$

$G(z)$ is known as the "pulse transfer function." Inversion of (11.28) gives information on the behavior of the output signal only at the sampling instants. It should be noted that the definition (11.27) is ambiguous if $x(t)$ has discontinuities at any of the sampling instants. We therefore require that if $X(z)$ is to exist, and $x(t)$ has any discontinuities at the sampling instants, then $x(nT^-)$ and $x(nT^+)$ must exist, and Eq. (11.27) is written

$$X(z) = \sum_{k=0}^{\infty} x(kT^+)z^{-k} \tag{11.27$'$}$$

The above relationships are the basis of the so-called "z-transform method" of analysis of conventional sampled-data systems. The method was introduced by Hurewicz [6] in the U.S., Barker [5] in England and Tsypkin [4] in the U.S.S.R. The application of z transforms to the analysis and design of sampled-data systems is treated extensively in several texts [1-3, 23].

Other applications of z transforms. While the z transform method is primarily applicable to systems with conventional sampling and with all sampling operations synchronous, it can be extended to certain other problems. By appropriately delaying or advancing the functions to be sampled, response between sampling instants can be obtained. This variation is called the "modified z transform" [1-3, 5]. Systems with several samplers where the sampling periods are staggered but equal, and systems where certain samplers operate at different frequencies (generally multiples of one another) can also be analyzed by this method, but the labor involved can be considerable. If the sampling period varies in length periodically, z transforms can still be applied [2, 8, 9]. However, even specialized techniques become inapplicable in the nonlinear case.

Relation of z transforms and difference equations. The z transform equation of (11.28) is a relationship between the sequences $\{x(nT)\}$ and $\{y(nT)\}$ and consequently it may be expected that a very close relationship exists between this equation and the corresponding difference equation. It is easy to show from the definition, Eq. (11.27), that if

$$Z\{x(nT)\} = X(z)$$

then

$$Z\{x(n+1)T\} = zX(z) \tag{11.29}$$

This relationship immediately establishes a relationship between corresponding terms in a z transform expression and in a difference equation.

Example 11.2. Consider the system of Example 11.1 again. By applying the definition (or consulting tables), if $g(t) = t$,

$$G(z) = \frac{Tz}{(z-1)^2} = \frac{Y(z)}{X(z)} \tag{11.30}$$

This expression can be expanded to yield

$$z^2Y(z) - 2zY(z) + Y(z) = Tz\,X(z) \tag{11.31}$$

Applying Eq. (11.29) to each term in Eq. (11.31), one obtains

$$y(n+2)T - 2y(n+1)T + y(nT) = Tx(n+1)T \tag{11.32}$$

which is identical with Eq. (11.18). Thus, z transforms are a useful way of obtaining the difference equation in a linear discrete system.

11.5 The State-Space Formulation of Discrete-Time Problems

The concepts of *state* and *state transformation* [24], introduced in earlier chapters in connection with continuous-time dynamic systems, also provide a unifying framework for the description of discrete-time systems. The use of state-space concepts in the analysis of discrete-time systems in this country is due in large part to the work of Kalman and Bertram [11-14] and Bellman [15]. The state-space approach makes it possible to formulate both linear and nonlinear problems in a uniform and concise manner without restriction to conventional sampling methods. The resulting equations are well suited to digital computer solution. The concepts of vector spaces will be used in this chapter in a formal and heuristic manner; for more rigorous treatment, the reader is urged to consult the references.

Intuitive definition of the state of a dynamic system. Assume that the initial conditions which describe a dynamic system at time t_0 are known. Then, the *state* of the system at any time $t > t_0$ is a minimum set of quantities (called the *state variables*) sufficient to describe the present and future outputs of a system, provided the inputs to the system are known. (For a more rigorous definition see Zadeh [10] and Zadeh and Desoer [24].) For a continuous system characterized by an nth order differential equation, it is clear that such a set of quantities is the system output $y(t)$ and its $(n-1)$ time derivatives

$$y^{(1)}(t),\ y^{(2)}(t),\ \ldots\ y^{(n-1)}(t)$$

Thus, these quantities determine the state of the system at time t, and can be considered the elements of a vector $\mathbf{y}(t)$, the *state vector.*

In Sec. 11.3, it was shown that an unforced nth order discrete system can be represented by means of a difference equation which relates the value of the output signal at the kth sampling instant and at n past sampling instants. Thus, the quantities $y(t_k), y(t_{k-1}), \cdots y(t_{k-n})$ represent the state of the discrete system.

State transition equations of continuous dynamic systems. To illustrate these concepts, consider first a linear continuous system characterized by the vector differential equation

$$\dot{\mathbf{x}} = \mathbf{A}\mathbf{x} + \mathbf{B}\mathbf{u}; \quad \mathbf{x}(0) = \mathbf{x}_0 \tag{11.33}$$

where $\mathbf{x}$ is an n vector (the state vector), $\mathbf{u}$ is an m-dimensional input vector, $\mathbf{A}$ is an $n \times n$ matrix, and $\mathbf{B}$ is an $n \times m$ matrix. The solution of this equation is given by [16, 17]

$$\mathbf{x}(t) = \mathbf{\Phi}(t - t_0)\mathbf{x}(t_0) + \int_{t_0}^{t} \Phi(t - \tau)\mathbf{B}\,\mathbf{u}(\tau)\,d\tau \tag{11.34}$$

where $\mathbf{\Phi}$ is the *state transition matrix* of the system (11.33), defined by

$$\mathbf{\Phi}(t) = e^{t\mathbf{A}} = \sum_{k=0}^{\infty} \mathbf{A}^k \frac{t^k}{k!} \tag{11.35}$$

The solution can also be obtained by taking Laplace transforms of each term in Eq. (11.33). This procedure yields

$$\mathbf{X}(s) = (s\mathbf{I} - \mathbf{A})^{-1}\mathbf{x}(t_0) + (s\mathbf{I} - \mathbf{A})^{-1}\mathbf{B}\,\mathbf{U}(s) \tag{11.36}$$

where I is the unit matrix and $\mathbf{X}(s) = \mathcal{L}[\mathbf{x}(t)]$. By comparison of Eqs. (11.34) and (11.36), the state transition matrix for this system can be defined as

$$\Phi(t - t_0) = \mathcal{L}^{-1}\left[(s\mathbf{I} - \mathbf{A})^{-1}\right] \tag{11.37}$$

Example 11.3. Let the linear constant-coefficient system described by the differential equation

$$\begin{aligned} \ddot{y} + a_1\dot{y} + a_0 y &= b\,u(t) \\ y(0) = c_0,\ \dot{y}(0) &= c_1 \end{aligned} \tag{11.38}$$

where the constants a_1, a_0 and b are not zero. Then, if we let

$$\begin{aligned} x_1 &= y \\ x_2 &= \dot{y} \end{aligned} \tag{11.39}$$

Equation (11.38) can be written in the form of the vector differential equation (11.33), with the matrices of coefficients A and B being given by

$$\mathbf{A} = \begin{bmatrix} 0 & 1 \\ -a_0 & -a_1 \end{bmatrix}, \quad \mathbf{B} = \begin{bmatrix} 0 \\ b \end{bmatrix} \tag{11.40}$$

and the state of the system is given by

$$\mathbf{x} = \begin{bmatrix} x_1 \\ x_2 \end{bmatrix}, \quad \mathbf{x}(0) = \begin{bmatrix} c_0 \\ c_1 \end{bmatrix} \tag{11.41}$$

The solution can be obtained by using Eq. (11.34). Let us turn now to the discrete case.

State transition equations of linear discrete systems. We begin by considering linear sampled-data systems with conventional sampling and assume that we are interested only in the system behavior at sampling instants. In many systems the sampling process can be idealized sufficiently so that the system can be described by a finite set of quantities, e.g., the values of the input and output at time t_k and at n past sampling instants. (The difference equation formulation of Eq. (11.27) was an example.) Thus, by the definition above, these quantities can be considered the *state variables* and they constitute the components of a *state vector* $\mathbf{x}(t_k)$. The dynamic behavior of such systems can then be described by *vector difference equations* of the form

$$\mathbf{x}(t_{k+1}) = \mathbf{A}\,\mathbf{x}(t_k) + \mathbf{B}\,\mathbf{u}(t_k)$$
$$\mathbf{x}(t_0) = \mathbf{x}_0 \tag{11.42}$$

for $k = 0,1,2, \cdots$. As before, $\mathbf{x}_0$ represents the initial state of the system. Equation (11.42) corresponds to the vector differential equation (11.33) for a linear continuous system. (It should be noted that a system need not in fact be sampled for Eq. (11.42) to apply; if the behavior of a continuous system is observed or measured once every T seconds, it can be considered a discrete system.)

Example 11.4. Let a linear discrete system be described by the difference equation

$$x(t_{k+2}) + a_1 x(t_{k+1}) + a_0 x(t_k) = u(t_k) \tag{11.43}$$

(where $u(t_k)$ is the input, and initial conditions are given) or by the z transform relationship

$$\frac{X(z)}{U(z)} = \frac{1}{z^2 + a_1 z + a_0} \tag{11.44}$$

If we let

$$x_1(t_k) = x(t_k)$$
$$x_2(t_k) = x_1(t_{k+1}) \tag{11.45}$$

Then Eq. (11.43) can be written as the equivalent set

$$x_1(t_{k+1}) = x_2(t_k)$$
$$x_2(t_{k+1}) = -a_0 x_1(t_k) - a_1 x_2(t_k) + u(t_k) \tag{11.46}$$

This set of first-order scalar difference equations can be written as the *vector difference equation*

$$\mathbf{x}(t_{k+1}) = \mathbf{A}\,\mathbf{x}(t_k) + \mathbf{B}\,u(t_k)$$
$$\mathbf{x}(t_0) = \mathbf{x}_0 \tag{11.47}$$

where

$$\mathbf{x}(t_k) = \begin{bmatrix} x_1(t_k) \\ \\ x_2(t_k) \end{bmatrix} \tag{11.48}$$

is the state vector and $\mathbf{x}_0$ represents the initial state. The constant matrices $\mathbf{A}$ and $\mathbf{B}$ are given by

$$\mathbf{A} = \begin{bmatrix} 0 & 1 \\ \\ -a_0 & -a_1 \end{bmatrix} ; \quad \mathbf{B} = \begin{bmatrix} 0 \\ \\ 1 \end{bmatrix} \tag{11.49}$$

The solution of the linear difference Eq. (11.47) is analogous to that of the continuous case. Taking z transforms of Eq. (11.47) we obtain

$$z\,\mathbf{X}(z) = \mathbf{A}\,\mathbf{X}(z) + \mathbf{B}\,\mathbf{U}(z) + z\,\mathbf{x}(t_0) \tag{11.50}$$

where the transform of each term of $\mathbf{x}(t_k)$ is obtained. Solving for $X(z)$

$$\mathbf{X}(z) = (z\mathbf{I} - \mathbf{A})^{-1} z\,\mathbf{x}(t_0) + (z\mathbf{I} - \mathbf{A})^{-1}\mathbf{B}\,\mathbf{U}(z) \tag{11.51}$$

Since $\mathbf{A}$ is a constant matrix, the inverse transform of Eq. (11.51) can be evaluated to yield

$$\mathbf{x}(t_k) = \mathbf{\Phi}(t_k)\,\mathbf{x}(t_0) + \sum_{n=0}^{k-1} \mathbf{\Phi}(t_{k-1} - t_n)\,\mathbf{B}\,\mathbf{u}(t_n) \tag{11.52}$$

where the state transition matrix is given by

$$\mathbf{\Phi}(t_k) = Z^{-1}\left\{(zI - A)^{-1} z\right\} \tag{11.53}$$

and the second term on the right-hand side is the matrix form of the convolution summation.

The state transition matrix can also be obtained by considering the unforced system, and this approach yields considerable insight into the meaning of the state transitions. Consider the system of Eq. (11.47) without a forcing term, i.e.

$$\mathbf{x}(t_{k+1}) = \mathbf{A}\,\mathbf{x}(t_k) \tag{11.54}$$

If the initial state of the system is denoted by $\mathbf{x}(t_0)$, then at the first sampling instant $(t_1 = t_0 + T)$

$$\mathbf{x}(t_1) = \mathbf{A}\,x(t_0) \tag{11.55}$$

which is a transformation of the initial state. At the second sampling instant

$$\mathbf{x}(t_2) = \mathbf{A}\,\mathbf{x}(t_1) = \mathbf{A}^2\,\mathbf{x}(t_0) \tag{11.56}$$

and at the kth sampling instant

$$\mathbf{x}(t_k) = \mathbf{A}^k\,\mathbf{x}(t_0) \tag{11.57}$$

consequently, this vector difference equation represents a successive series of transformations of the initial state. Comparison of Eq. (11.57) with Eq. (11.52) shows that the state transition matrix can also be written as

$$\mathbf{\Phi}(t_k) = \mathbf{A}^k \tag{11.58}$$

for the linear sampled-data system with conventional sampling.

The vector difference equation (11.47) can be viewed as a recurrence relation which describes the state at time t_{k+1} given the state and the input at time t_k. Such expressions are conveniently solved on digital computers.

Example 11.5. As a detailed example of the formulation and solution of the vector difference equation for a linear discrete system, consider the block diagram of Fig. 11-3. Let $h(t)$ represent the output of the hold circuit, and let

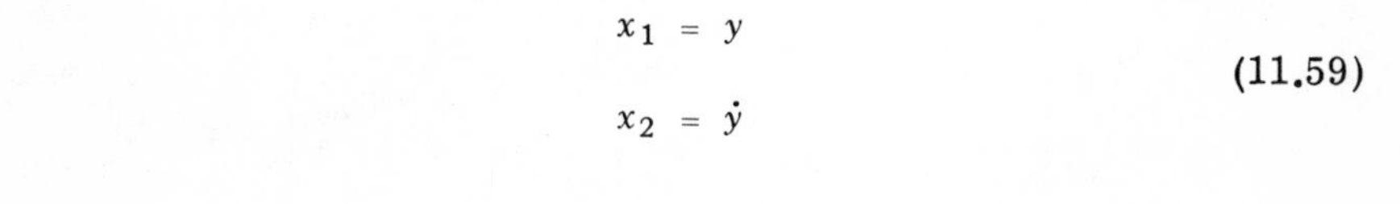

$$\begin{aligned} x_1 &= y \\ x_2 &= \dot{y} \end{aligned} \tag{11.59}$$

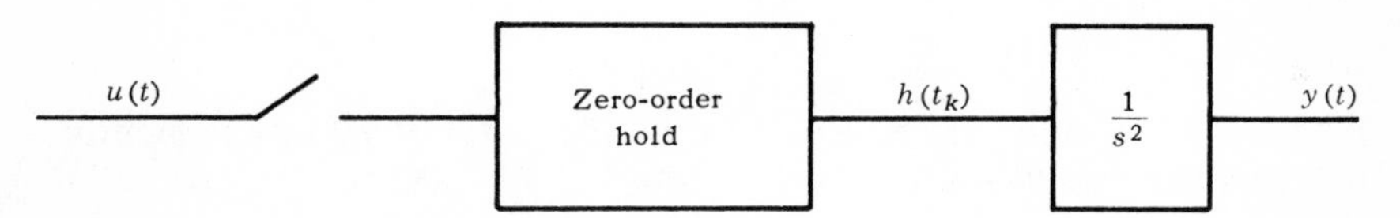

FIG. 11-3. Open-loop linear discrete system.

Then we can write directly

$$\begin{aligned} h(t_k) &= u(t_k) \\ x_2(t_{k+1}) &= x_2(t_k) + T\,h(t_k) \\ x_1(t_{k+1}) &= x_1(t_k) + T\,x_2(t_k) + T^2/2\,h(t_k) \end{aligned} \tag{11.60}$$

Making appropriate substitutions, these equations can be written

$$\mathbf{x}(t_{k+1}) = \mathbf{A}\,\mathbf{x}(t_k) + \mathbf{B}\,u(t_k)$$
$$\mathbf{x}(t_0) = C \tag{11.61}$$

where $\mathbf{x}(t_k)$ is the state vector at time t_k,

$$\mathbf{x}(t_k) = \begin{bmatrix} x_1(t_k) \\ x_2(t_k) \end{bmatrix} \tag{11.62}$$

and

$$\mathbf{A} = \begin{bmatrix} 1 & T \\ 0 & 1 \end{bmatrix}; \quad \mathbf{B} = \begin{bmatrix} T^2/2 \\ T \end{bmatrix}; \quad \mathbf{C} = \begin{bmatrix} c_1 \\ c_2 \end{bmatrix} \tag{11.63}$$

(c_1 and c_2 are constants.) The complete solution of this equation can be obtained from Eq. (11.52). We first obtain the state transition matrix using Eq. (11.53)

$$[z\mathbf{I} - \mathbf{A}]^{-1} = \begin{bmatrix} \left(\dfrac{1}{z-1}\right) & \dfrac{T}{(z-1)^2} \\ 0 & \left(\dfrac{1}{z-1}\right) \end{bmatrix}$$

$$\Phi(t_k) = Z^{-1}\left\{[z\mathbf{I} - \mathbf{A}]^{-1} z\right\} = \begin{bmatrix} 1 & kT \\ 0 & 1 \end{bmatrix}$$

It is easy to verify, using Eq. (11.53), that $\Phi(t_k) = \mathbf{A}^k$. Consequently, the complete solution may be written

$$\mathbf{x}(t_k) = \begin{pmatrix} 1 & kT \\ 0 & 1 \end{pmatrix} \mathbf{x}(t_0) + \sum_{n=0}^{k-1} \begin{pmatrix} T^2/2 + (k-1-n)T^2 \\ T \end{pmatrix} u(t_n) \tag{11.66}$$

This expression describes the state of the system at time $t_k = kT$ in terms of the control input $u(t)$ and the initial state $\mathbf{x}(t_0)$.

Closed-loop systems. The techniques described above apply directly to the closed-loop case. If the control input is the closed-loop error, as in Fig. 11-4, it is clear that

$$u(t_k) = r(t_k) - x_1(t_k) \tag{11.67}$$

where $r(t)$ is the reference input. With this additional relationship, to supplement the open-loop difference equations of the previous paragraphs, the closed-loop system can be described completely.

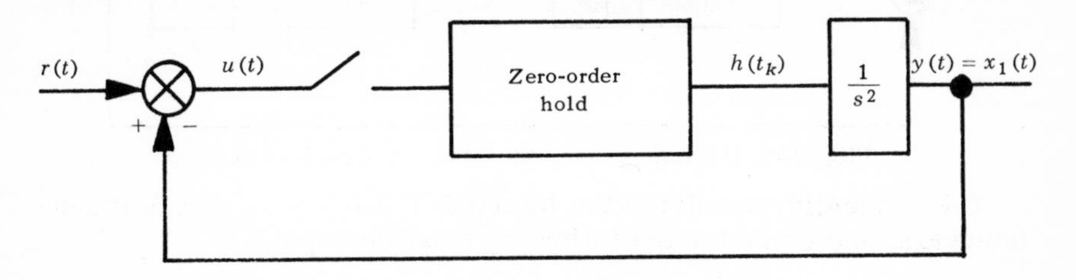

FIG. 11-4. Closed-loop linear discrete system

Discrete systems with nonconventional sampling. One of the advantages of the state-space formulation is that it provides a unified approach to the study of discrete-time problems [12], even when the sampling is non-conventional, in cases where:

(a) response between sampling instants is desired,
(b) the sampling period T_k is not constant but is a periodic function of k,
(c) sampling operations in the system are not synchronized,
(d) multirate sampling is present,
(e) finite pulse width or non-instantaneous sampling is present.

If the plant is linear, the vector difference equation obtained in these cases will also be linear, but, in general, will have time-varying coefficients. If the plant is nonlinear or when the sampling depends on the state of the system (such as pulse-width modulated discrete systems) the resulting nonlinear difference equations may be written

$$\mathbf{x}(t_{k+1}) = \mathbf{f}\big(\mathbf{x}(t_k), \mathbf{u}(t_k)\big) \tag{11.68}$$

where $\mathbf{f}$ is a vector-valued vector function describing the functional relationship. To illustrate the applicability of the state-space formulation, three examples will be used, two linear and one nonlinear.

Example 11.6. Discrete system with nonsynchronized sampling. Consider the system of Fig. 11-5 where both samplers are synchronized but Sampler S_2 lags behind Sampler S_1 by τ seconds. (This example is based on one published by Bertram [14].) The formulation of the difference equation is facilitated by careful selection of state variables and division of the sampling interval into subintervals. The output can always be selected as one of the state variables, as indicated in Fig. 11-5. It should be noted that two kinds of state transitions take place in this system:

(a) transitions of sample-and-hold elements, which occur at the sampling instants, and

(b) transitions of the continuous elements which occur during the intervals between samples.

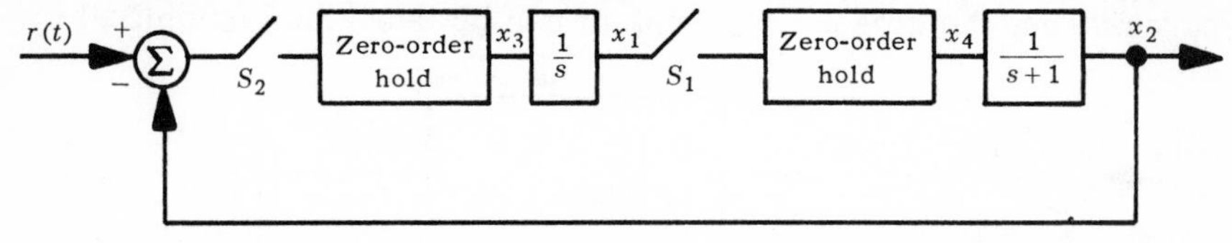

FIG. 11-5. Discrete system with non-synchronized sampling.

Consequently, we divide the interval $T = t_{k+1} - t_k$ into four subintervals, and consider the following relationships:

$$\begin{aligned}
&\text{I} \quad & \mathbf{x}(t_k + \tau^-) &= \mathbf{\Phi}_1 \mathbf{x}(t_k^+) \\
&\text{II} \quad & \mathbf{x}(t_k + \tau^+) &= \mathbf{D}_2 \mathbf{x}(t_k + \tau^-) \\
&\text{III} \quad & \mathbf{x}(t_{k+1}^-) &= \mathbf{\Phi}_2 \mathbf{x}(t_k + \tau^+) \\
&\text{IV} \quad & \mathbf{x}(t_{k+1}^+) &= \mathbf{D}_1 \mathbf{x}(t_{k+1}^-)
\end{aligned} \tag{11.69}$$

Relations I and III represent the transitions of the continuous elements, and relations II and IV those of the discrete (sample-and-hold) elements. The matrices $\mathbf{D}_1$ and $\mathbf{D}_2$ represent the effects of sampling of samplers S_1 and S_2 respectively and are given by

$$\mathbf{D}_1 = \begin{bmatrix} 1 & 0 & 0 & 0 \\ 0 & 1 & 0 & 0 \\ 0 & 0 & 1 & 0 \\ 1 & 0 & 0 & 0 \end{bmatrix}, \quad \mathbf{D}_2 = \begin{bmatrix} 1 & 0 & 0 & 0 \\ 0 & 1 & 0 & 0 \\ 0 & -1 & 0 & 0 \\ 0 & 0 & 0 & 1 \end{bmatrix} \tag{11.70}$$

(The input $r(t)$ is assumed equal to zero.) The transition of the system during the interval $(t_k^+, t_k + \tau^-)$ is given by

$$\mathbf{\Phi}_1(\tau) = \begin{bmatrix} 1 & 0 & \tau & 0 \\ 0 & e^{-\tau} & 0 & (1 - e^{-\tau}) \\ 0 & 0 & 1 & 0 \\ 0 & 0 & 0 & 1 \end{bmatrix} \tag{11.71}$$

and it can be seen that

$$\mathbf{\Phi}_2 = \mathbf{\Phi}_1(T - \tau) \tag{11.72}$$

Expressions (11.69) can be combined to yield the vector difference equation

$$\mathbf{x}(t_{k+1}) = \mathbf{A}(\tau)\, \mathbf{x}(t_k) \tag{11.73}$$

where the state transition matrix is given by

$$\mathbf{A}(\tau) = \mathbf{D}_1 \boldsymbol{\Phi}_1 (T - \tau) \mathbf{D}_2 \boldsymbol{\Phi}_1(\tau) \tag{11.74}$$

As is customary, the sampling instant t_k in (11.73) is understood as t_k^+.

Example 11.7. Multirate sampled-data system. The analysis of systems where two or more samplers operate at different sampling frequencies can be carried out in the frequency domain by means of the modified z transform [1-3, 23]. The state-space formulation of such problems proceeds in a manner analogous to that of the previous example, where the largest sampling interval is subdivided into several subintervals.

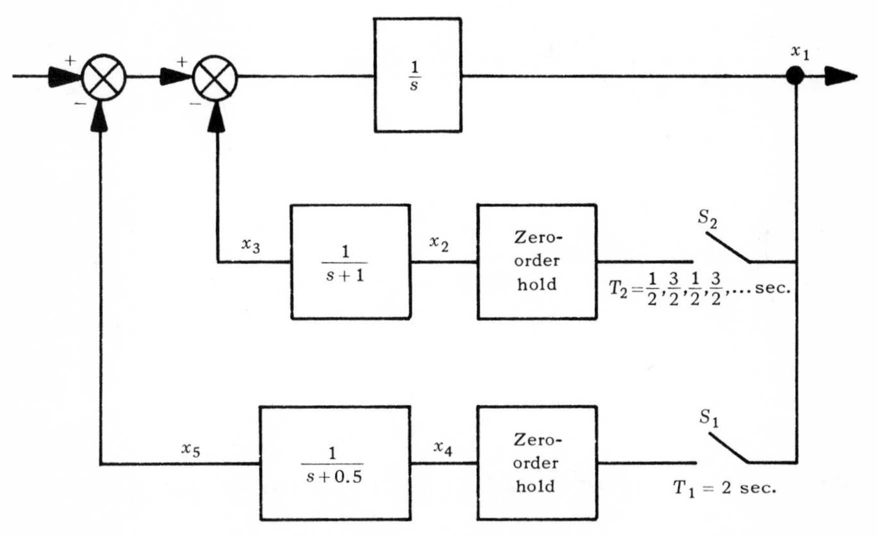

FIG. 11-6. Multirate sampled-data system.

Consider the system illustrated in Fig. 11-6 where we again assume $r(t) = 0$. Sampler S_1 is periodic with a period of 2 seconds. Sampler S_2 is aperiodic having a period which is alternately 0.5 and 1.5 seconds. Therefore, the sequence of sampling operations is: at time t_k both S_1 and S_2 sample; at time $t_k + 0.5$ only S_2 samples; at time $t_{k+1} = t_k + 2.0$ sec both S_1 and S_2 sample. We describe the system by the following state transitions:

$$\begin{aligned}
&\text{I.} \quad \mathbf{x}(t_k + 0.5^-) = \boldsymbol{\Phi}_1 \mathbf{x}(t_k^+) \\
&\text{II.} \quad \mathbf{x}(t_k + 0.5^+) = \mathbf{E}_1 \mathbf{x}(t_k + 0.5) \\
&\text{III.} \quad \mathbf{x}(t_{k+1}^-) = \boldsymbol{\Phi}_2 \mathbf{x}(t_k + 0.5^+) \\
&\text{IV.} \quad \mathbf{x}(t_{k+1}^+) = \mathbf{E}_2 \mathbf{x}(t_{k+1}^-)
\end{aligned} \tag{11.75}$$

Again, relations I and III represent continuous transitions while II and IV represent discrete transitions. The discrete (sample-and-hold) transitions are given by

$$\mathbf{E}_1 = \begin{bmatrix} 1 & 0 & 0 & 0 & 0 \\ 1 & 0 & 0 & 0 & 0 \\ 0 & 0 & 1 & 0 & 0 \\ 0 & 0 & 0 & 1 & 0 \\ 0 & 0 & 0 & 0 & 1 \end{bmatrix}, \quad \mathbf{E}_2 = \begin{bmatrix} 1 & 0 & 0 & 0 & 0 \\ 1 & 0 & 0 & 0 & 0 \\ 0 & 0 & 1 & 0 & 0 \\ 1 & 0 & 0 & 0 & 0 \\ 0 & 0 & 0 & 0 & 1 \end{bmatrix} \tag{11.76}$$

The behavior of the system during the interval $(t_k^+, t_k + 0.5^-)$ is described by relation (11.75-I) where

$$\mathbf{\Phi}_1 = \begin{bmatrix} 1 & \left(-\frac{1}{2} + e^{-\frac{1}{2}}\right) & \left(1 - e^{-\frac{1}{2}}\right) & \left(-\frac{3}{2} + 2e^{-\frac{1}{4}}\right) & 2\left(1 - e^{-\frac{1}{4}}\right) \\ 0 & 1 & 0 & 0 & 0 \\ 0 & \left(1 - e^{-\frac{1}{2}}\right) & \left(e^{-\frac{1}{2}}\right) & 0 & 0 \\ 0 & 0 & 0 & 1 & 0 \\ 0 & 0 & 0 & \left(1 - e^{-\frac{1}{4}}\right) & \left(e^{-\frac{1}{4}}\right) \end{bmatrix} \tag{11.77}$$

and during the interval $(t_k + 0.5^+, t_{k+1}^-)$ by the relation (11.75-III) where

$$\mathbf{\Phi}_2 = \begin{bmatrix} 1 & \left(\frac{1}{2} + e^{-\frac{3}{2}}\right) & \left(1 - e^{-\frac{3}{2}}\right) & \left(-\frac{1}{2} + 2e^{-\frac{3}{4}}\right) & 2\left(1 - e^{-\frac{3}{4}}\right) \\ 0 & 1 & 0 & 0 & 0 \\ 0 & \left(1 - e^{-\frac{3}{2}}\right) & \left(e^{-\frac{3}{2}}\right) & 0 & 0 \\ 0 & 0 & 0 & \left(1 - e^{-\frac{3}{4}}\right) & \left(e^{-\frac{3}{4}}\right) \end{bmatrix} \tag{11.78}$$

Consequently, the vector difference equation describing the system at the sampling instants of the slower sampler is

$$\mathbf{x}(t_{k+1}) = \mathbf{\Phi}_T\, \mathbf{x}(t_k) \tag{11.79}$$

where the overall state stransition matrix is defined by

$$\Phi_T = E_2 \Phi_2 E_1 \Phi_1 \tag{11.80}$$

Example 11.8. Discrete system with nonlinear gain. Consider the system of Fig. 11-7 which adds a nonlinear gain (e.g. a saturating

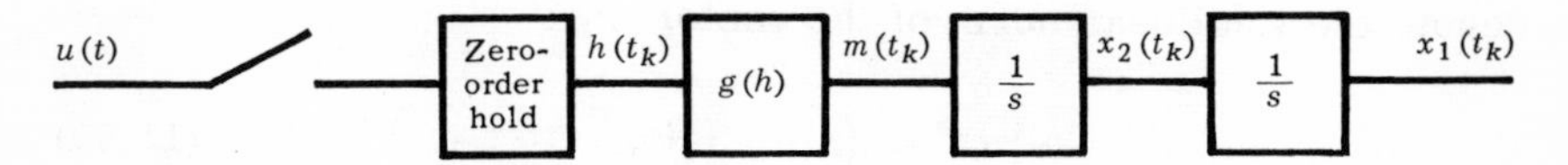

FIG. 11-7. Nonlinear discrete system.

amplifier) to the system previously considered in Example 11.3. Now transform techniques fail, but the recurrence relations can be formulated in a straightforward manner. For this simple case they can be written by inspection as:

$$\begin{aligned} x_1(t_{k+1}) &= x_1(t_k) + T x_2(t_k) + \frac{T^2}{2} g\big(u(t_k)\big) \\ x_2(t_{k+1}) &= x_2(t_k) + T g\big(u(t_k)\big) \end{aligned} \tag{11.81}$$

or in vector form

$$\mathrm{x}(t_{k+1}) = \mathrm{f}[\mathrm{x}(t_k),\ u(t_k)] \tag{11.82}$$

11.6 Stability of Discrete Time Systems

Definitions of Stability. We concentrate in this section on the unforced system (i.e., $u(t_k) \equiv 0$ for all t_k) represented by the equation

$$\mathrm{x}(t_{k+1}) = \mathrm{f}[\mathrm{x}(t_k)] \tag{11.83}$$

Let us assume that the system has an equilibrium state, denoting it by x_e. Then, mathematically, the system is in its equilibrium state if

$$\mathrm{x}_e = \mathrm{f}(\mathrm{x}_e) \tag{11.84}$$

That is, the equilibrium state has the property that if the initial state $\mathrm{x}(t_0) = \mathrm{x}_e$, then repeated iterations of the transformation (11.83) do not result in any change of state. The question of engineering interest, however, is whether the system will return to the equilibrium state if disturbed. Intuitively, if the system remains near the equilibrium state it is called *stable*. If it is stable and tends to the equilibrium state as $k \to \infty$, it is called *asymptotically stable.* If the system is asymptotically stable regardless of the magnitude of the disturbance from equilibrium, it is called *asymptotically stable in the large,* or *globally asymptotically stable.*

Let us now formulate these definitions more precisely [11, 13]. Assume that equilibrium is at the origin (this can always be accomplished by a translation of coordinates), i.e., $\mathbf{x}_e = 0$. Let $||\mathbf{x}||$ denote the Euclidean norm of the vector $\mathbf{x}$:

$$||\mathbf{x}|| = (\mathbf{x}^1 \mathbf{x})^{1/2} = \left(x_1^2 + x_2^2 + \cdots + x_n^2\right)^{1/2} \tag{11.85}$$

Definition. The equilibrium solution $\mathbf{x}_e$ is *stable* if given any $\epsilon > 0$, there exists a $\delta(\epsilon) > 0$ such that for all initial states $\mathbf{x}_0$ in the sphere of radius δ

$$||\mathbf{x}_0 - \mathbf{x}_e|| \leq \delta$$

the solution for all k, $\mathbf{x}(t_k)$, remains within a sphere of radius

$$||\mathbf{x}(t_k) - \mathbf{x}_e|| < \epsilon$$

If the equilibrium solution is stable and if, in addition,

$$\lim_{k \to \infty} ||\mathbf{x}(t_k) - \mathbf{x}_e|| = 0$$

the solution is *asymptotically* stable. If, in addition, the radius δ of the initial disturbance can be arbitrarily large, the solution is *asymptotically stable in the large.*

It is important to note that stability is defined here in terms of the motion of the state of the system in state space, not in terms of the system output.

Stability of Linear Sampled-Data Systems. Unforced linear sampled data systems can be represented by

$$\mathbf{x}(t_{k+1}) = \mathbf{\Lambda}\, \mathbf{x}(t_k) \tag{11.86}$$

where $\mathbf{\Lambda}$ is a constant (time-invariant) matrix (see example 11.5 above). The solution is of the form

$$\mathbf{x}(t_k) = \mathbf{\Lambda}^k\, \mathbf{x}(t_0) \tag{11.87}$$

i.e., it represents n successive transformations of the initial state. The null solution $\mathbf{x}_e = 0$ is asymptotically stable in the large if and only if every element of $\mathbf{\Lambda}^k$ tends to zero uniformly with k as $k \to \infty$. It can be shown [12, 17, 24] that this statement implies that the system is globally asymptotically stable if and only if the roots

$\lambda_1, \lambda_2, \ldots \lambda_k$ of the characteristic equation

$$\det(\mathbf{A} - \lambda\mathbf{I}) = 0$$

(the eigenvalues of $\mathbf{A}$) satisfy the condition

$$|\lambda_i| < 1, \quad i = 1, 2, \ldots k$$

This statement is equivalent to the statement that the poles of the closed-loop pulse transfer function of the system must lie inside the unit circle in the z plane.

Example 11.9. Consider the simple system of Fig. 11-4 again, where we assume $T = 1$ for simplicity. The pulse transfer function is given by

$$G(z) = \frac{Y(z)}{U(z)} = Z\left[\frac{1 - e^{-Ts}}{s^3}\right] = \frac{T^2(z - 1)}{2(z - 1)^2} \tag{11.88}$$

The denominator of the closed-loop transfer function (characteristic equation) is (for $T = 1$)

$$1 + G(z) = z^2 - 3/2\,z + 3/2 = 0 \tag{11.89}$$

and since for the roots z_i, $|z_i| > i = 1, 2,$ the system is unstable.

The difference equations are written, as in Example 11.5, with the additional feature that the system is closed-loop, and consequently (11.67) applies. For $r(t) = 0$ the vector difference equation is (11.86), where

$$\mathbf{A} = \begin{bmatrix} 1/2 & 1 \\ -1 & 1 \end{bmatrix} \tag{11.90}$$

from which the characteristic equation is obtained as

$$|\mathbf{A} - \lambda\mathbf{I}| = \begin{vmatrix} (1/2 - \lambda) & 1 \\ -1 & (1 - \lambda) \end{vmatrix} = 0 \tag{11.91}$$

which is identical with Eq. (11.89) if λ is substituted for z.

Stability of linear systems with inputs. The definitions of asymptotic stability given above are based on the free or unforced behavior of the state in state space. If bounded input vectors are applied to an asymptotically stable system, the state vector remains bounded. This statement can be made more precise in the following theorem [24], given here without proof:

Theorem 11.1. If a linear discrete system is described by

$$\mathbf{x}(t_{k+1}) = \mathbf{A}\,\mathbf{x}(t_k) + \mathbf{B}\,\mathbf{u}(t_k) \tag{11.92}$$

and the eigenvalues of $\mathbf{A}$ are in the open disk $|\lambda| < 1$, then, for all initial states, any bounded input vector sequence $\{\mathbf{u}(t_k)\}$ produces a bounded state vector $\mathbf{x}(t_k)$.

In other words, if $\|\mathbf{u}(t_i)\| < M$, $i = 0,1,2, \cdots$ where M is a real positive number, then it is possible to find a real positive number C such that $\|\mathbf{x}(t_k)\| < C\|\mathbf{x}(t_0)\|$ for all k, where $\mathbf{x}(t_0)$ is the initial state.

Applications of the Second Method of Lyapunov to Discrete Systems. The so-called "direct" or "second method" of Lyapunov for determining asymptotic stability of nonlinear differential equations has become quite important in recent years. The method is based on finding a scalar function of the state variables of the system which satisfies certain conditions. If such a function, called a *Lyapunov function,* does indeed exist, then the null solution of the differential equation is asymptotically stable in the large. The importance of the method is based on the fact that the stability information is obtained without having to solve the differential equation. A detailed discussion of the Second Method is given in Chapter 5.

Much less literature is available on the use of the Lyapunov method for determining asymptotic stability (either global or local) of difference equations. The basic references are those of Hahn [18] and Kalman and Bertram [11, 13]. From these references the following stability theorem can be stated (for proof, see ref. 11):

Theorem 11.2. If, for the vector difference equation

$$\mathbf{x}(t_{n+1}) = \mathbf{f}\big(\mathbf{x}(t_n)\big)$$

there exists a scalar function of the state variables $V(\mathbf{x})$, such that $V(\mathbf{0}) = 0$ and

(i) $V(\mathbf{x}) > 0$ when $\mathbf{x} \neq 0$
(ii) $V[\mathbf{x}(t_{k+1})] < V[\mathbf{x}(t_k)]$ for $k > K$, K finite
(iii) $V(\mathbf{x})$ is continuous in $\mathbf{x}$
(iv) $V(\mathbf{x}) \to \infty$ when $\|\mathbf{x}\| \to \infty$

then the equilibrium solution $\mathbf{x} = 0$ is *asymptotically stable in the large* and $V(\mathbf{x})$ is a Lyapunov function for this system. (It should be noted that this is only a sufficient condition.)

As with continuous systems, considerable ingenuity is required to find appropriate Lyapunov functions. For example, Bertram [13] discusses the application of functions of the form

$$V_1(\mathbf{x}) = \sum_{i=1}^{n} c_i |x_i| \tag{11.93}$$

(where the c_i are constants) to the study of stability of sampled systems with nonlinear gain elements. For similar systems with finite pulse width, Kadota [19] uses functions of the form

$$V_2(\mathbf{x}) = \sum_{i=1}^{n} e^{c_i T} x_i^2 \tag{11.94}$$

Clearly, both functions are positive definite in the whole space.

Most of the theorems of Chapter 5 have corresponding statements for the discrete time case. Consequently, the theory of the Second Method will not be discussed further here. Examples of the use of the Second Method in the study of asymptotic stability of nonlinear discrete-time systems will be found in the following sections.

11.7 Pulse Frequency Modulated Sampled Data Systems

In all the systems previously considered, information was transmitted using pulse-amplitude modulation, and the sampling intervals were assumed either fixed or periodically time-varying. If, however, the sampling intervals are functions of the state variables, the system becomes nonlinear. An illustration of such a system is presented in this section [20].

Difference Equations of the System. Consider the example illustrated in Fig. 11-8. The nth sampling interval is defined as

$$T_n = t_{n+1} - t_n \tag{11.95}$$

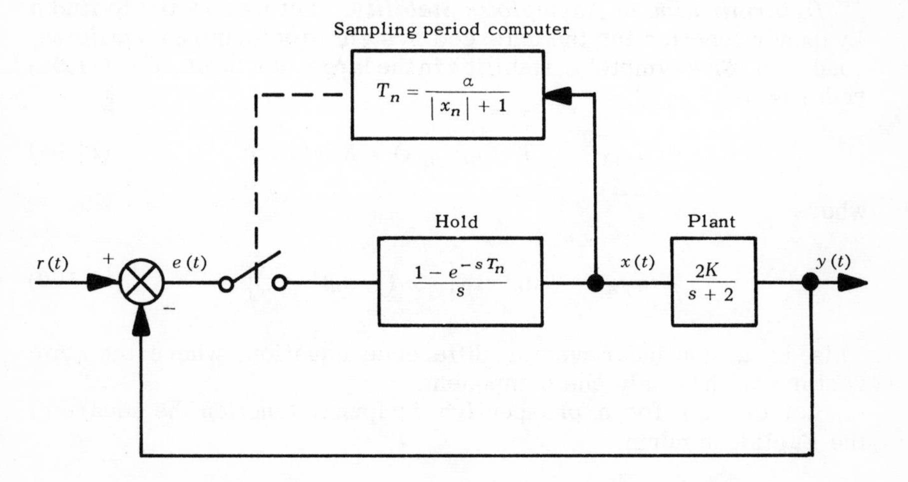

FIG. 11-8. Sampled-data system with variable frequency sampler.

Note that the hold periods will be variable, as well as the sampling periods. Let a control law governing the variable sample and hold device be given as

$$T_n = \frac{a}{|e(t_n)| + 1} \tag{11.96}$$

with the result that a large error results in an increase in the sampling frequency.

By using the techniques of previous sections, the system equations at the sampling instants may be written as

$$\left.\begin{aligned} x(t_n) &= e(t_n) \\ y(t_{n+1}) &= y(t_n)\exp(-2T_n) + K\,e(t_n)\bigl(1 - \exp(-2T_n)\bigr) \\ e(t_n) &= r(t_n) - y(t_n) \end{aligned}\right\} \tag{11.97}$$

where T_n is given by Eq. (11.96). Equations (11.97) can be combined to yield a single expression

$$\begin{aligned} y(t_{n+1}) = {} & -y(t_n)\Bigl\{K - (1+K)\exp\Bigl[-2a/\bigl(|r(t_n) - y(t_n)| + 1\bigr)\Bigr]\Bigr\} \\ & + Kr(t_n)\Bigl\{1 - \exp\Bigl[-2a/\bigl(|r(t_n) - y(t_n)| + 1\bigr)\Bigr]\Bigr\} \end{aligned} \tag{11.98}$$

This is a nonlinear difference equation which can be solved sample by sample if the initial state $y(t_0)$ is known and the input $r(t_n)$ is specified for $t_n > t_0$.

Determination of Asymptotic Stability. Let us now try to find a Lyapunov function for the unforced system to determine a sufficient condition for asymptotic stability in the large. If $r(t) \equiv 0$, Eq. (11.98) reduces to

$$y(t_{n+1}) = -K\,y(t_n) + (1+K)\,g[y(t_n)] \tag{11.99}$$

where

$$g(t_n) = y(t_n)\exp\Bigl[-2a/\bigl(|y(t_n)| + 1\bigr)\Bigr] \tag{11.100}$$

This is a nonlinear vector difference equation, where the state vector $\mathrm{y}(t_n)$ has only one component.

Let us pick for a prospective Lyapunov function the square of the Euclidean norm

$$V_1\bigl(\mathrm{y}(t_n)\bigr) = \|\mathrm{y}(t_n)\|^2 = y(t_n)^2 \tag{11.101}$$

The first difference of V_1 is

$$\Delta V_1(y) = y(t_{n+1})^2 - y(t_n)^2 \tag{11.102}$$

Since $V_1(y_n)$ is by inspection positive-definite, continuous in y_n, and tends to ∞ as $\|y_n\| \to \infty$, all that remains to be shown is that $\Delta V_1(y_n)$ is negative-definite. This implies the existence of a region in the a, K parameter space in which

$$y_n^2 > y_{n+1}^2 \quad \text{for all } n \tag{11.103}$$

If Eq. (11.96) is combined with Eq. (11.100) to obtain g as a function of T_n, and the result substituted in Eq. (11.99), inequality (11.103) becomes

$$y_n^2 \left[(1 + K)^2 \exp(-4T_n) - 2K(1 + K)\exp(-2T_n) + K^2\right] < y_n^2 \tag{11.104}$$

where

$$T_n = T_n(a, y_n)$$

according to Eq. (11.96). Note that $T_n > 0$ and $K > 0$ from physical considerations. Since $y_n^2 \geq 0$, inequality (11.104) is equivalent to

$$(1 + K)^2 \exp(-4T_n) - 2K(1 + K)\exp(-2T_n) + K^2 < 1 \tag{11.105}$$

This inequality can be used to determine bounds on the parameters a and K. If such bounds can be found, $V_1(y_n)$ will qualify as a Lyapunov function, and the system is asymptotically stable in the large.

As an illustration, Let $K = 2$ to simplify the arithmetic. Then Eq. (11.105) becomes

$$e^{-2T_n}(9e^{-2T_n} - 12) < -3 \tag{11.106}$$

or, since e^{-2T_n} is greater than zero for all $T_n > 0$, we must have

$$3e^{-2T_n} - 4 < -\frac{1}{e^{-2T_n}} \tag{11.107}$$

If we interpret the two sides of this inequality as equations defining two functions $f_1\left(e^{-2T_n}\right)$ and $f_2\left(e^{-2T_n}\right)$, we can obtain a graphical interpretation of the stability requirement by plotting f_1 and f_2 vs. e^{-2T_n}, as in Fig. 11-9.

The points of intersection are obtained from solution of the equation $f_1 = f_2$. Therefore, $\Delta V[y(t_n)] < 0$ for

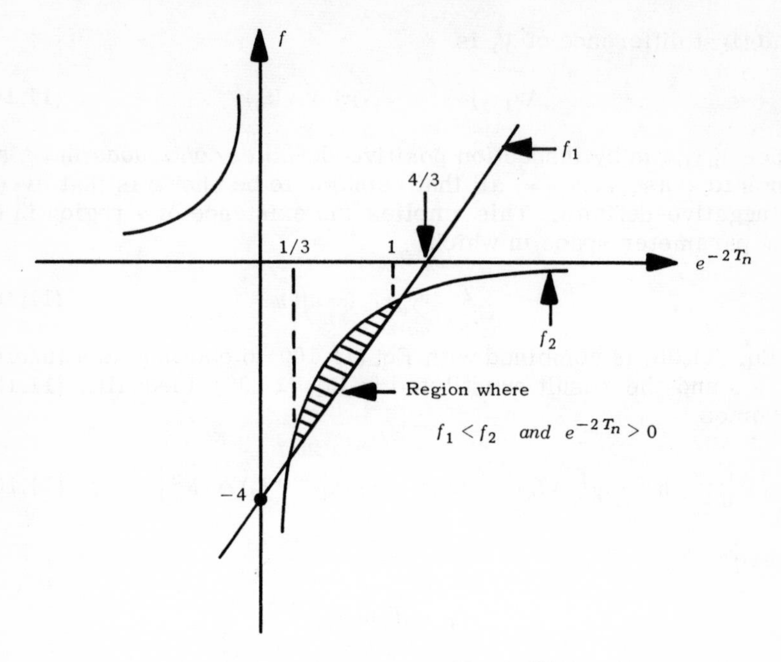

FIG. 11-9. Plot of f_1 and f_2 vs. e^{-2T_n}

$$\frac{1}{3} < e^{-2T_n} < 1 \tag{11.108}$$

The upper limit clearly cannot be exceeded, since for all

$$T_n > 0, \ e^{-2T_n} < 1$$

The lower limit means that

$$-2T_n > ln \frac{1}{3}; \quad T_n < \frac{1}{2} ln 3 \tag{11.109}$$

Using the "control law" for T_n as given by Eq. (11.96), we have

$$T_n = \frac{a}{|y(t_n)| + 1} < \frac{ln 3}{2} \tag{11.110}$$

This expression assumes its maximum value for $|y(t_n)| = 0$ and consequently

$$a = (ln 3)/2 \tag{11.111}$$

Therefore, provided that a stays below the limit of Eq. (11.111), the null solution of the nonlinear system is asymptotically stable in the large.

Integral pulse-frequency modulation. The system discussed above employed pulse amplitude modulation with an additional degree of freedom introduced by the adjustment of sampling frequency. A pure pulse-frequency modulator would produce a series of equal pulses the frequency of which is dependent on the modulator input, as illustrated in Fig. 11-10. If the modulator input is denoted

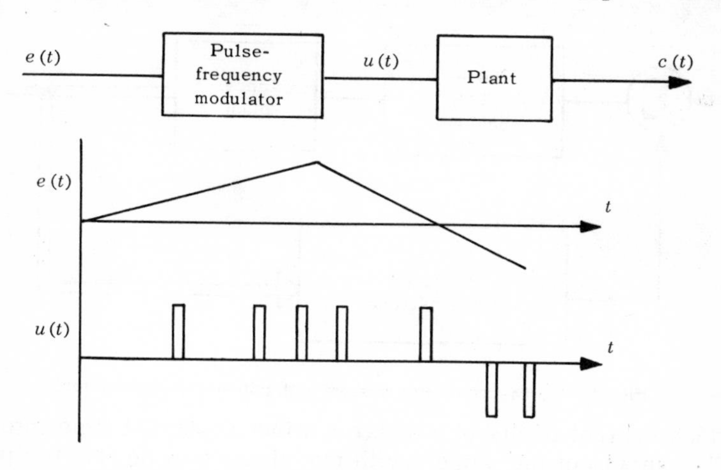

FIG. 11-10. Integral pulse-frequency modulator and waveforms.

by $e(t)$, then the nth interval between pulses $T_n = t_{n+1} - t_n$ can be obtained from the equation

$$\int_{t_n}^{t_n + T_n} e(t)\,dt = \pm K \tag{11.112}$$

where K is a design parameter. This equation indicates that if a pulse is produced at time t_n, the next pulse occurs when the magnitude of the integral reaches K. The pulse then carries the sign of the integral at that time. This type of modulation is known as "integral pulse-frequency modulation" and appears to be similar to a type of modulation occurring in the nervous system [25].

A detailed analysis of an integral pulse-frequency (IPF) modulation attitude control system has been made [26]. This analysis shows, by an extension of the second method of Lyapunov, that the ultimate state of an IPF-controlled second-order plant is a limit-cycle oscillation to which the system converges asymptotically.

11.8 Pulse-Width Modulated Discrete-Time Systems

One of the most interesting areas of study in discrete systems involves with the analysis and design of PWM systems. These systems are inherently nonlinear. In this section we shall review the work on PWM systems by Kadota and Bourne [21, 27], Nelson [22], since they illustrate the usefullness of the concepts discussed above.

Formulation of the difference equations. The system we consider is illustrated in Fig. 11-11, using a pulse-width modulator which

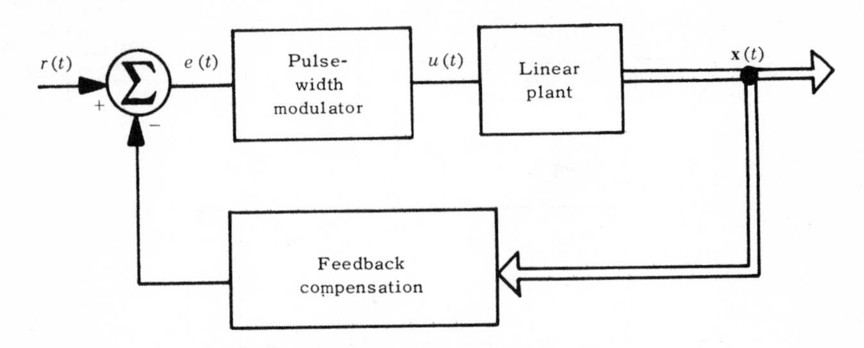

FIG. 11-11. Discrete-time system with pulse-width modulator.

provides control inputs to a linear continuous plant at time intervals T. The outputs of the pulse-width modulator will be flat-top pulses of constant amplitude M and variable width, given by

$$u(t) = \begin{cases} M \operatorname{sgn} e(kT), & kT \le t < kT + h(kT) \\ 0, & kT + h(kT) \le t < (k+1)T \end{cases} \tag{11.113}$$

where sgn is the signum function and $h(kT)$ (to be denoted as $h(k)$ henceforth for simplicity) is the width of the kth pulse,

$$h(k) = T \operatorname{sat} \frac{|e(kT)|}{\beta} \tag{11.114}$$

where β is a positive constant and the saturation function $\operatorname{sat} x$ is defined as

$$\operatorname{sat} x = \begin{cases} +1, & x > 1 \\ x, & |x| \leq 1 \\ -1, & x < -1 \end{cases} \tag{11.115}$$

Then $u(t)$ represents the control input to the plant.

The plant can be described by the vector differential equation

$$\dot{\mathbf{x}}(t) = \mathbf{A}\,\mathbf{x}(t) + \mathbf{B}\,\mathbf{u}(t) \tag{11.116}$$

where $\mathbf{x}(t)$ is the state vector of the plant at time t.

Since the plant is assumed linear and invariant, $\mathbf{A}$ is an $n \times n$ matrix with constant elements, and $\mathbf{B}$ is an n vector with constant elements. From Eq. (11.113), the input to the plant is either zero or constant at the value $\pm M$.

Then, following the techniques of Sec. 11.5, for the interval of time that the input is equal to M, $t_0 \leq t < t_1$, the solution of Eq. (11.116) is given by Eq. (11.34)

$$\mathbf{x}(t) = \mathbf{\Phi}(t - t_0)\mathbf{x}(t_0) + M\,\mathbf{g}(t - t_0), \quad t_0 \leq t \leq t_1 \tag{11.117}$$

where $\mathbf{\Phi}(t)$ is the *fundamental matrix* or *state transition matrix* of the differential equation (11.116)

$$\mathbf{\Phi}(t) = \exp \mathbf{A}t = \sum_{k=0}^{\infty} \mathbf{A}^k \frac{t^k}{k!} \tag{11.118}$$

and $\mathbf{g}(t)$ is the forcing vector given by

$$\mathbf{g}(t) = \int_0^t \mathbf{\Phi}(u)\,\mathbf{B}\,du \tag{11.119}$$

It can be seen that, to satisfy Eq. (11.117), we must have

$$\mathbf{\Phi}(0) = \mathbf{I} \quad \text{and} \quad \mathbf{g}(0) = \mathbf{0}$$

where $\mathbf{I}$ is the unit matrix of order n and $\mathbf{0}$ is the null matrix. Other useful properties of $\mathbf{\Phi}(t)$ and $\mathbf{g}(t)$ are [17, 22]

$$\begin{aligned} \mathbf{\Phi}(u + v) &= \mathbf{\Phi}(u)\,\mathbf{\Phi}(v) \\ \mathbf{g}(-v) &= -\mathbf{\Phi}(-v)\,\mathbf{g}(v) \end{aligned} \tag{11.120}$$

which arise from the properties of linear systems.

Applying the above general results to the PWM problem, with $u(t)$ given by Eq. (11.113), we can write the equation

$$\mathbf{x}[kT + h(k)] = \mathbf{\Phi}[h(k)]\,\mathbf{x}(kT) + u(k)\,\mathbf{g}[h(k)] \tag{11.121}$$

for the "input-on" time, where in Eq. (11.117) we let

$$t_0 = kT, \quad t = kT + h(k)$$

and similarly

$$\mathbf{x}[(k + 1)T] = \mathbf{\Phi}[T - h(k)]\,\mathbf{x}[kT + h(k)] + 0 \tag{11.122}$$

for the "input-off" time, where

$$t_0 = kT + h(k), \quad t = (k + 1)T$$

Equations (11.121) and (11.122) can be combined to give

$$\mathbf{x}[(k + 1)T] = \mathbf{\Phi}[T - h(k)]\left\{\mathbf{\Phi}[h(k)]\,\mathbf{x}(kT) + u(k)\,\mathbf{g}[h(k)]\right\} \tag{11.123}$$

Using property (11.120) of the state transition matrix, we can write

$$\Phi[T - h(k)]\,\Phi[h(k)] = \Phi(T) \tag{11.124}$$

and consequently obtain a single vector difference equation for the system

$$\mathbf{x}[(k+1)T] = \Phi(T)\,\mathbf{x}(kT) + u(k)\,\Phi[T - h(k)]\,\mathbf{g}[h(k)] \tag{11.125}$$

This equation can be used to study the time behavior of the system.

Stability of the PWM System. In order to apply the Second Method of Lyapunov to the system of Eq. (11.125), let us first diagonalize the matrix A.

We consider the class of systems where (for simplicity) the poles a_i of the plant transfer function $G(s)$ are all real and distinct. Then there exists a real nonsingular matrix P such that

$$\mathbf{P}^{-1}\mathbf{A}\mathbf{P} = \mathbf{J} \tag{11.126}$$

is a diagonal matrix with the poles a_i along its diagonal, i.e.

$$\mathbf{J} = \begin{bmatrix} a_1 & 0 & \cdot & \cdot & \cdot & 0 \\ 0 & a_2 & \cdot & \cdot & \cdot & 0 \\ \cdot & & & & & \cdot \\ \cdot & & & & & \cdot \\ 0 & \cdot & \cdot & \cdot & \cdot & a_n \end{bmatrix} \tag{11.127}$$

We then perform the transformation

$$\mathbf{x} = \mathbf{P}\mathbf{y}$$

which maps the state space X onto the state space Y. In terms of the new state vector $\mathbf{y}(kT)$, the difference Eq. (11.125) can now be written as

$$\mathbf{y}[(K+1)T] = \mathbf{E}(T)\,\mathbf{y}(kT) + u(k)\,\mathbf{E}[T - h(k)]\,\mathbf{f}[h(k)] \tag{11.128}$$

where

$$\mathbf{E}(T) = \mathbf{P}^{-1}\Phi(T)\mathbf{P} = \begin{bmatrix} e^{a_1T} & 0 & \cdot & \cdot & \cdot & \cdot & 0 \\ 0 & e^{a_2T} & \cdot & \cdot & \cdot & \cdot & 0 \\ \cdot & \cdot & \cdot & \cdot & \cdot & \cdot & \cdot \\ \cdot & \cdot & \cdot & \cdot & \cdot & \cdot & \cdot \\ \cdot & \cdot & \cdot & \cdot & \cdot & \cdot & 0 \\ 0 & \cdot & \cdot & \cdot & \cdot & \cdot & e^{a_nT} \end{bmatrix} \tag{11.129}$$

and

$$\mathbf{f}[h(k)] = \mathbf{P}^{-1}\mathbf{g}[h(k)] = \begin{bmatrix} \frac{1}{a_1}\left(1 - e^{a_1 h_k}\right) \\ \cdot \\ \cdot \\ \cdot \\ \cdot \\ \frac{1}{a_n}\left(1 - e^{a_n h_k}\right) \end{bmatrix} \tag{11.130}$$

By taking advantage of the properties (11.120), expression (11.128) can be further simplified to the form

$$\mathbf{y}[(k+1)T] = \mathbf{E}(T)\left\{\mathbf{y}(kT) - u(k)\,\mathbf{g}[-h(k)]\right\} \tag{11.131}$$

We now choose as a Lyapunov function the square of the generalized Euclidean norm* of the state vector $\mathbf{y}$

$$V = \|\mathbf{y}\|_p^2 = \sum_{i=1}^{n} c_i y_i^2 \tag{11.132}$$

where the c_i are positive constants. V is obviously positive-definite. Then, a sufficient condition for asymptotic stability in the large of the equilibrium solution $\mathbf{y}_e = 0$ is that $\Delta V(\mathbf{y}) < 0$ for all $\mathbf{y}$. To prove this, we follow Kadota [21] and write $\Delta V[\mathbf{y}(k)]$ in explicit form using Eq. (11.131) and

$$\Delta V[\mathbf{y}(kT)] = V[\mathbf{y}(k+1)T] - V[\mathbf{y}(kT)] < 0 \tag{11.133}$$

so that

$$\Delta V = \sum_{i=1}^{n} c_i\left[e^{2a_i T}(y_i + w_i)^2 - y_i^2\right] \tag{11.134}$$

where the y_i are the state variables (components of the state vector) and the w_i are defined as

$$w_i(k) = \frac{u(k)\left(e^{-a_i h_k} - 1\right)}{a_i} \tag{11.135}$$

By manipulation of Eq. (11.134), it can be shown that condition (11.133) reduces to finding constants c_i such that certain matrices (whose elements are determined by the c_i and the system parameters) have negative eignevalues. Consequently it is possible to

*For a discussion of the suitability of various norms of the state vector for Lyapunov functions, see ref. 13.

find conditions on the system parameters such that the pulse-width modulated system of Fig. 11-11 is asymptotically stable in the large. The conditions are then used to instrument the feedback control function $e(t)$ to insure stability. These concepts are best illustrated by means of a simple example based on the work of Kadota [21, 27].

Example 11.10. Let the linear plant be defined by the transfer function

$$G(s) = \frac{1}{s - a} \tag{11.136}$$

The differential equation for the system during one sampling interval (pulse-on) is

$$\dot{x} = ax + u(t) \tag{11.137}$$

where $u(t)$ is defined by Eq. (11.113). The general solution of Eq. (11.137) is

$$x(t) = e^{a(t - t_0)} x(t_0) + \int_{t_0}^{t} e^{a(t - \tau)} u(\tau) d\tau \tag{11.138}$$

Let the width of the kth pulse $h(k)$ be denoted by h_k. Then, for $t_0 = t_k$ and $t = t_k + h_k$

$$x(t_k + h_k) = e^{ah_k} x(t_k) + \int_{t_k}^{t_k + h_k} e^{a(t_k + h_k - \tau)} u(\tau) d\tau \tag{11.139}$$

and, for $t_0 = t_k + h_k$, $t = (k + 1)T = t_{k+1}$

$$x(t_{k+1}) = e^{a(T - h_k)} x(t_k + h_k) \tag{11.140}$$

Combining Eq. (11.139) and (11.140), we obtain the difference equation which describes the system

$$x(t_{k+1}) = e^{aT} \left[x(t_k) + \int_{t_k}^{t_k + h_k} e^{a(t_k - \tau)} u(\tau) d\tau \right] \tag{11.141}$$

Now, consider the regulator problem $[r(t) = 0]$ and assume that the feedback function in Fig. 11-11 is only a constant mulitplier a_{fb}. Then $\operatorname{sgn} e(t_k) = -\operatorname{sgn} x(t_k)$, and for the interval $(t_k, t_k + h_k)$ we can write, from Eq. (11.113)

$$u(t) = M \operatorname{sgn} e(t_k) = -M \operatorname{sgn} x(t_k) \tag{11.142}$$

Substituting in Eq. (11.141) and carrying out the integration

$$x(t_{k+1}) = e^{aT}\left[x(t_k) + (M/a)\left(e^{-ah_k} - 1\right)\operatorname{sgn} x(t_k)\right] \qquad (11.143)$$

or, since $\operatorname{sgn} x(t_k) = x(t_k)/|x(t_k)|$

$$x(t_{k+1}) = e^{aT}\left[1 + \frac{M\left(e^{-ah_k} - 1\right)}{a|x(t_k)|}\right] x(t_k) \qquad (11.144)$$

Now, we choose for the Lyapunov function the square of the Euclidean norm which, for this trivial case, is simply

$$V(x) = x^2(t_k) \qquad (11.145)$$

Then the system is asymptotically stable in the large if

$$\Delta V(x) = x^2(t_{k+1}) - x^2(t_k) < 0 \quad \text{for all} \quad x(t_k) \neq 0 \qquad (11.146)$$

Substituting Eq. (11.144) in Eq. (11.146)

$$\Delta V(x) = e^{2aT}\left[1 + \frac{M\left(e^{-a\,\mathrm{sat}|x(t_k)|/\beta} - 1\right)}{a|x(t_k)|}\right]^2 - 1 < 0 \qquad (11.147)$$

Now, this expression is piecewise monotonic in $|x(t_k)|$, due to the nature of the $\mathrm{sat}\, x$ function, and therefore its maxima and minima occur at the boundaries of the intervals $[0, \beta]$ and $[\beta, \infty]$. Therefore, we must find values for a, M, β, and T such that $\Delta V < 0$ for $|x| = 0$, $|x| = \beta$ and $|x| = \infty$. These three conditions are

$$\text{I} \quad e^{aT}\left|1 - \frac{TM}{\beta}\right| < 1$$

$$\text{II} \quad e^{aT}\left|1 + \frac{TM}{\beta}\,\frac{e^{-aT} - 1}{aT}\right| < 1 \qquad (11.148)$$

$$\text{III} \quad a < 0$$

Note that the second and third inequality imply the first since

$$\frac{e^{-aT} - 1}{aT} < 1 \quad \text{for} \quad a < 0$$

If, for example, $a = -1$, the condition for stability reduces to

$$e^{-T}\left|1 + \frac{TM}{\beta}\left(\frac{e^{T} - 1}{-T}\right)\right| < 1 \qquad (11.149)$$

The parameters T, M, and β must now be selected in order to satisfy this relationship, and the system is asymptotically stable in the large.

11.9 Synthesis of Discrete-Time Systems

The previous sections of this chapter have been devoted to the analysis of linear and nonlinear sampled-data systems. The purpose of this section is to introduce the synthesis problem, with major emphasis on linear systems capable of reaching equilibrium states in the minimum possible number of sampling periods. We shall begin by examining the use of z-transform techniques for synthesizing discrete controllers which achieve minimum settling time for particular classes of input signals. The state-space approach to synthesis of optimal discrete systems is discussed briefly.

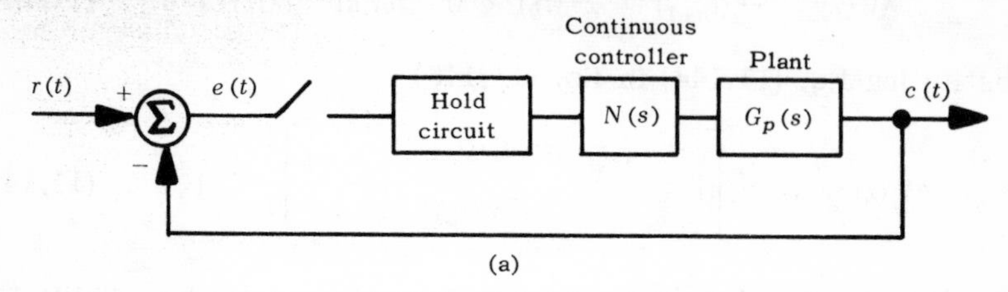

(a)

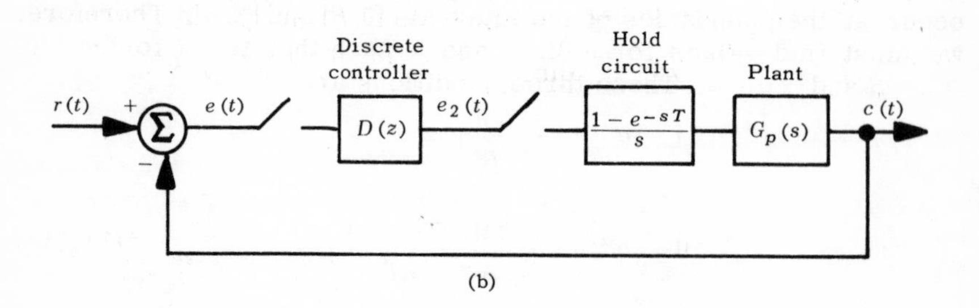

(b)

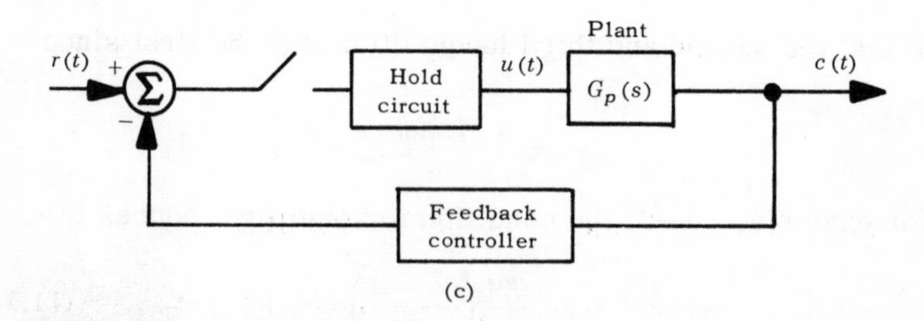

(c)

FIG. 11-12. Sampled-data systems with cascade and feedback controllers.

Synthesis of Discrete Controllers Using z Transform Techniques. Consider the unity feedback error-sampled systems of Fig. 11-12, which illustrate two common approaches to compensation of sampled-data systems. Frequency-domain synthesis techniques used with linear continuous systems are based on the following steps:

(1) Performance criteria are fomulated as a goal for the synthesis procedure. These may be such conventional factors as peak time or maximum overshoot to step inputs, minimum mean-squared error or the time required for the system to reach an equilibrium state from arbitrary initial conditions.

(2) The closed-loop transfer function is determined from the specifications of (1) and the plant transfer function.

(3) The transfer function of a physically realizable controller is computed from (2).

(4) The controller is synthesized exactly or approximately.

The steps listed above cannot be applied directly to the synthesis of the controller $N(s)$ in Fig. 11-12a. The closed-loop pulse transfer function of the system of Fig. 11-12a is given by

$$K(z) = \frac{C(z)}{R(z)} = \frac{G_h NG_p(z)}{1 + G_h NG_p(z)} \tag{11.150}$$

where

$$G_h NG_p(z) = Z\left\{G_h(s)N(s)G_p(s)\right\} \tag{11.151}$$

and $G_h(s)$ is the transfer function of the hold circuit. It can be seen that the controller characteristic $N(s)$ cannot be isolated in Eq. (11.150). Consequently, continuous compensation of discrete-time systems generally requires the use of approximations or cut-and-dry techniques [1-3, 23]. When a discrete compensator is used, the closed-loop pulse transfer function is

$$K(z) = \frac{D(z)G(z)}{1 + D(z)G(z)} \tag{11.152}$$

where

$$G(z) = Z\left\{G_h(s)G_p(s)\right\} \tag{11.153}$$

From Eq. (11.152) it is possible to write directly

$$D(z) = \frac{1}{G(z)} \frac{K(z)}{1 - K(z)} \tag{11.154}$$

Consequently, if the desired performance specifications can be incorporated into a closed-loop pulse transfer function $K(z)$, the discrete controller pulse transfer function can be obtained from Eq. (11.154). Such controllers are commonly referred to as "digital controllers" in the literature, since the difference equation represented by $D(z)$ can be implemented on a digital computer. (The phrase "discrete controller" is more accurate, however, since $D(z)$ can also be implemented using analog elements and sample-hold circuits without the amplitude quantization present in a digital computer. In fact, the nonlinear effect of quantization may have to be considered in the design of strictly digital controllers.)

It remains to be shown how performance criteria can be incorporated in the selection of an overall pulse transfer function $K(z)$. We shall concentrate on the following performance criteria:

(1) Zero steady-state error at the sampling instants.

(2) Minimum settling time (i.e., the system error must become zero in the minimum possible number of sampling periods). Systems of this type are known as "minimal systems." To incorporate these performance specifications in $K(z)$ we begin by writing the error signal transform

$$E(z) = R(z)[1 - K(z)] \tag{11.155}$$

where $R(z)$ is the z transform of the input signal $r(t)$. If the steady-state error is to be zero, the final value theorem of z transforms [1-3] is used

$$\lim_{n \to \infty} e(nT) = \lim_{z \to 1} \left(1 - z^{-1}\right) R(z)\,[1 - K(z)] = 0 \tag{11.156}$$

Consider now polynomial inputs of the form $r(t) = t^m$, for which

$$R(z) = \frac{A(z)}{\left(1 - z^{-1}\right)^n} \tag{11.157}$$

where $A(z)$ is a polynomial in z with no roots at $z = 1$ and $n = m + 1$, Substituting in Eq. (11.156), it can be seen that a necessary condition for zero steady-state error is that $1 - K(z)$ contain the factor $\left(1 - z^{-1}\right)^n$. Consequently, we must have

$$1 - K(z) = \left(1 - z^{-1}\right)^n F(z) \tag{11.158}$$

where $F(z)$ is a polynomial in z^{-1} with no roots at $z = 1$. The desired closed-loop pulse transfer function is given by

$$K(z) = 1 - \left(1 - z^{-1}\right)^n F(z) \tag{11.159}$$

which guarantees zero steady-state error. To obtain minimum settling time, we substitute Eqs. (11.158) and (11.157) in (11.55), and note that

$$E(z) = A(z)F(z) \tag{11.160}$$

For the error to settle in the minimum number of sampling periods, $F(z)$ must be a polynomial of the lowest order possible. If $G(z)$ has no poles at the origin and no zeros on or outside the unit circle, it is possible to choose $F(z) = 1$ and still meet physical realizability considerations. With these conditions, it can be seen that $K(z)$ takes the following form for the "minimal system."

Step input $\quad K(z) = z^{-1}$

Ramp input $\quad K(z) = 2z^{-1} - z^{-2}$

Parabolic input $\quad K(z) = 3z^{-1} - 3z^{-2} + z^{-3}$

and so forth. In general, if the input is of the form $r(t) = t^m$, the minimal response system settles in $(m + 1)$ sampling periods.

The following assumptions have been tacitly made in the above development: (1) The initial conditions are equal to zero; (2) the plant has no poles on or outside the unit circle; (3) the plant contains enough integrations to make zero steady-state error possible for the particular input selected

Comments on the Minimal Synthesis. Unfortunately, the method of synthesis outlined above is completely impractical for the following reasons:

(1) Minimum settling time at the sampling instants does not guarantee zero ripple *between* the sampling instants.

(2) The synthesis depends on cancellation of poles and zeros of $G(z)$ in order to obtain a closed-loop transfer function $K(z)$ with poles only at the origin. Such cancellation is not possible if $G(z)$ has poles on or outside the unit circle.

(3) Multiple poles of $K(z)$ at the origin are undesirable from the standpoint of sensitivity.

(4) Minimal systems are optimum only for the specific input for which they are designed, and in general are not satisfactory for other inputs.

It can be shown that in order to overcome difficulties (1) and (2), it is necessary to modify the specifications on $K(z)$ as follows:

(a) $K(z)$ must be a finite polynomial in z^{-1},

(b) The zeros of $K(z)$ must include all the zeros of $G(z)$,

(c) The zeros of $[1 - K(z)]$ must include all the poles of $G(z)$ on or outside the unit circle.

These synthesis concepts are best illustrated by means of a simple example.

Example 11.11. Consider the system of Fig. 11-12b with

$$G_p(s) = \frac{1}{s^2} \tag{11.160}$$

and a zero-order hold. Minimal response to a ramp input is derived by synthesizing an appropriate discrete compensator. Then

$$G(z) = Z\left\{\frac{1 - e^{-sT}}{s^3}\right\} = \frac{T^2}{2}\,\frac{z^{-1}\left(1 + z^{-1}\right)}{\left(1 - z^{-1}\right)^2} \tag{11.161}$$

For a ramp input we choose

$$1 - K(z) = \left(1 - z^{-1}\right)^2 F(z) \tag{11.162}$$

and begin by letting $F(z) = 1$. Then

$$K(z) = 2z^{-1} - z^{-2} \tag{11.163}$$

and the compensator becomes

$$D(z) = \frac{1}{G(z)}\,\frac{K(z)}{1 - K(z)} = \frac{2\left(2 - z^{-1}\right)}{T^2\left(1 + z^{-1}\right)} \tag{11.164}$$

It can be seen that the compensator attempts to cancel a zero of $G(t)$ on the unit circle. As is well known, imperfect cancellation results in instability. Furthermore, even with perfect cancellation, ripple is present between sampling instants. Consider the system error

$$E(z) = R(z)[1 - K(z)] = Tz^{-1} \tag{11.165}$$

so that the system settles in two sampling instants. The output of the compensator is obtained from

$$\begin{aligned} E_2(z) = E(z)D(z) &= \frac{2}{T}\,\frac{\left(2 - z^{-1}\right)}{\left(1 + z^{-1}\right)} \\ &= 2z^{-1} - 3z^{-2} + 3z^{-3} - 3z^{-4} + \cdots \end{aligned} \tag{11.166}$$

Thus, the minimal system not only requires perfect zero cancellation but also produces ripple. If we modify the design by including the zeros of $G(z)$ in $K(z)$, we obtain

$$1 - K(z) = \left(1 - z^{-1}\right)^2 F_1(z) \tag{11.167}$$

$$K(z) = \left(1 + z^{-1}\right) F_2(z) \tag{11.168}$$

where the simplest polynomials which will satisfy these relationships become

$$F_1(z) = 1 + 4/5z^{-1} \tag{11.169}$$

$$F_2(z) = \left(6/5z^{-1} - 4/5z^{-2}\right) \tag{11.170}$$

and the resulting discrete compensator is

$$D(z) = \frac{2\left(6/5 - 4/5z^{-1}\right)}{T^2\left(1 + 4/5z^{-1}\right)} \tag{11.171}$$

It is easy to show that this improved system requires three sampling periods to settle. Furthermore, $E_2(z)$ is now a finite polynomial and consequently the system settles with zero ripple.

Introduction to Synthesis of Time-Optimal Systems. The synthesis of optimal sampled-data systems using the state-space formulation has been the subject of intensive research in recent years [28-37]. While a detailed discussion of time-optimal synthesis is beyond the scope of this chapter, a simple formulation of the problem is presented here. Assume that the system is linear and time-invariant, and described by the following vector difference equation (state transition equation).

$$\mathbf{x}(t_{k+1}) = \mathbf{A}\,\mathbf{x}(t_k) + \mathbf{h}\,u(t_k) \tag{11.172}$$

where $\mathbf{x}(t_k)$ is the n-dimensional state vector, $u(t_k)$ is a scalar input, $\mathbf{A}$ is an $n \times n$ matrix and $\mathbf{h}$ is an n-dimensional vector. The state of the system at the first two sampling instants is given by

$$\begin{aligned} \mathbf{x}(t_1) &= \mathbf{A}\,\mathbf{x}(t_0) + \mathbf{h}\,u(t_0) \\ \mathbf{x}(t_2) &= \mathbf{A}^2\,\mathbf{x}(t_0) + \mathbf{A}\,\mathbf{h}\,u(t_0) + \mathbf{h}\,u(t_1) \end{aligned} \tag{11.173}$$

Successive iteration results in

$$\mathbf{x}(t_N) = \mathbf{A}^N\,\mathbf{x}(t_0) + \mathbf{A}^{N-1}\,\mathbf{h}\,u(t_0) + \cdots + \mathbf{h}\,u(t_{N-1}) \tag{11.174}$$

The time-optimal regulator problem is concerned with bringing the state of the system to the origin from an arbitrary initial $\mathbf{x}(t_0)$. Consequently, if we desire $\mathbf{x}(t_N) = 0$, the above equation can be premultiplied by $-\mathbf{A}^{-N}$ and written

$$\mathbf{x}(t_0) = -\mathbf{A}^{-1}\mathbf{h}u(t_0) - \mathbf{A}^{-2}\mathbf{h}u(t_1) - \cdots - \mathbf{A}^{-N}\mathbf{h}u(t_{N-1}) \tag{11.175}$$

If we collect $u(t_0), u(t_1), \cdots u(t_{N-1})$ into a vector $\mathbf{U}(t_N)$, where

$$U_i(t_N) = u(t_{i-1}) \tag{11.176}$$

we can state the objective of time-optimal synthesis as finding the minimum N and the corresponding $\mathbf{U}(t_N)$ which satisfy Eq. (11.175), i.e., which bring the state of the system to the origin of the state space. In general, $\mathbf{U}(t_N)$ is constrained in magnitude, so that only those control vectors which satisfy the constraint are considered admissible.

To illustrate one approach to the problem, consider the following simple example [28]

Example 11.12. Let the system to be controlled be given by

$$G_p(s) = \frac{1}{s(s+1)} \tag{11.177}$$

and the input applied through a hold circuit, as in Fig. 11-12c. The vector difference equation describing the system can be formulated by the techniques of Sec. 11.5. If we let the output $c(t) = x_1(t)$ and $\dot{c}(t) = x_2(t)$, the system is described by Eq. (11.172) where

$$\mathbf{A}(T) = \begin{bmatrix} 1 & (1 - e^{-T}) \\ 0 & e^{-T} \end{bmatrix}, \quad \mathbf{h} = \begin{bmatrix} T - 1 + e^{-T} \\ 1 - e^{-T} \end{bmatrix} \tag{11.178}$$

Now let $\mathbf{x}^{(1)}(t_0)$ be an initial state from which it is possible to reach the origin in exactly one sampling interval. Then, from Eq. (11.175)

$$\mathbf{x}^{(1)}(t_0) = -u(t_0)\mathbf{A}^{-1}(T)\mathbf{h}(T) \tag{11.179}$$

If we let $\mathbf{A}^{-1}(T)\mathbf{h}(T) = \mathbf{v}_1$, we have

$$\mathbf{x}^{(1)}(t_0) = -u(t_0)\mathbf{v}_1 \tag{11.180}$$

Consider now the state $\mathbf{x}^{(2)}(t_0)$ from which the origin can be reached in two sampling intervals. Clearly, this is equivalent to being able to reach the state of Eq. (11.180) in one step. Then, applying Eqs. (11.180) and (11.175) we have

$$\begin{aligned} \mathbf{x}^{(2)}(t_0) &= \mathbf{A}^{-1}(T)\left[-u(t_1)\mathbf{v}_1 - u(t_0)\mathbf{h}(T)\right] \\ &= -u(t_1)\mathbf{A}^{-2}(T)\mathbf{h}(T) - u(t_0)\mathbf{A}^{-T}(T)\mathbf{h}(T) \end{aligned} \tag{11.181}$$

Now letting $\mathbf{A}^{-2}\mathbf{h} = \mathbf{v}_2$, Eq. (11.181) becomes

$$\mathbf{x}^{(2)}(t_0) = u(t_1)\mathbf{v}_2 - u(t_0)\mathbf{v}_1 \tag{11.182}$$

Now it can be shown [28] that the two vectors $\mathbf{v}_1$ and $\mathbf{v}_2$ are linearly independent for any value of T. Since any n-dimensional vector can be represented by a linear combination of n linearly independent vectors, it follows that since $x(t_k)$ in the example is two-dimensional, the origin can be achieved in two steps. Consequently, Eq. (11.181) can be solved for $u(t_1)$ and $u(t_0)$

$$\begin{aligned} u(t_0) &= a_{11}x_1(t_0) + a_{12}x_2(t_0) \\ u(t_1) &= a_{21}x_1(t_0) + a_{22}x_2(t_0) \end{aligned} \tag{11.183}$$

This is a sequence of values which depends only on the initial state $x(t_0)$. The a's in Eq. (11.183) represent terms obtained from the solution of Eq. (11.182). But if one considers the system at time t_1 from which it is possible to reach the origin in one sampling interval, it is clear that the second of Eqs. (11.183) can be written as

$$u(t_1) = a_{11}x_1(t_1) + a_{12}x_2(t_1) \tag{11.184}$$

The significance of this result is that the optimal controller can be instrumented using linear time invariant feedback. In fact, the feedback controller for this case is simply

$$F(s) = a_{11} + a_{12}s \tag{11.185}$$

To obtain the controller coefficients a_{11} and a_{12} in terms of the system parameters, we substitute in Eq. (11.183). This can be facilitated by noting that

$$\Lambda^{-j}(T) = \Lambda(-jT) \tag{11.186}$$

and one obtains

$$a_{11} = \frac{-e^T}{T(e^T - 1)} \; ; \quad a_{12} = \frac{-(e^{2T} - e^T - T)}{T(e^T - 1)^2} \tag{11.187}$$

The development of a linear discrete time-optimal controller for the same type of linear system is discussed by Mullin [37].

When restrictions are placed on the class of admissible control vectors $U(t_k)$, a possible approach to the problem is to divide the state space into regions, such that an optimum solution is known if the system is in one of these regions [28–32]. A similar approach can be extended to the synthesis of optimal pulse-width modulated discrete systems [38–40], where regions from which convergence to the origin in a minimum number of sampling instants are found. Another approach is based on finding a linear functional which completely specifies the solution [30, 35, 36]. It can also be shown [30] that under certain conditions, time-optimal discrete controls approach continuous time optimal control if the sampling period is allowed to approach zero.

REFERENCES

1. Ragazzini, J. R. and Franklin, G. *Sampled-Data Control Systems*, McGraw-Hill, 1958.
2. Tou, J. T. *Digital and Sampled-Data Control Systems*, McGraw-Hill, 1959.
3. Jury, E. I. *Sampled-Data Control Systems*, Wiley, 1958.
4. Tsypkin, Ya. Z. *Theory of Sampled-Data Systems* (in Russian) Moscow: Fizmatgiz, 1958.
5. Barker, R. H. "The Pulse Transfer Function and its Application to Sampling Servo Systems," *Proc. Instn. of Elec. Engrs.* v. 99, pt IV, pp 302f, 1952.
6. Hurewicz, W. "Filters and Servo Systems with Pulsed Data" (Chapter 5 of *Theory of Servomechanisms*, by H. M. James, N. B. Nichols, and R. S. Phillips,) McGraw-Hill, 1947.
7. Lighthill, M. J. *Fourier Analysis and Generalized Functions*, Cambridge University Press, 1958.
8. Friedland, B. "Sampled-Data Systems Containing Periodically Varying Members" *Proc. First IFAC Congress*, Moscow, 1960.
9. Bekey, G. A. "A Survey of Techniques for the Analysis of Sampled-Data Systems with a Variable Sampling Rate," *ASD Technical Report ASD-TDR 62-35*, February, 1962.
10. Zadeh, L. A. "An Introduction to State-Space Concepts" AIEE Paper 10-1, presented at the *1962 Joint Automatic Control Conference*, June, 1962.
11. Kalman, R. E. and Bertram, J. E. "Control System Analysis and Design via the Second Method of Lyapunov, II. Discrete-Time Systems" *Jour. Basic Engrg.*, June 1960, pp. 394-399.
12. Kalman, R. E. and Bertram, J. E. "A Unified Approach to the Theory of Sampling Systems," *Jour. Franklin Inst.* v. 267, pp. 405-436, May, 1959.
13. Bertram, J. E. "The Direct Method of Lyapunov in the Analysis and Design of Discrete-Time Control Systems," pp. 79-104 *Work Session on Lyapunov's Second Method*, ed. by L. F. Kazda, University of Michigan, Ann Arbor, September, 1960.
14. Bertram, J. E. "The Concept of State in the Analysis of Discrete-Time Control Systems," AIEE Paper 11-1, Presented at *1962 Joint Automatic Control Conference,* June 1962.
15. Bellman, R. *Adaptive Control Processes: A Guided Tour*, Princeton University Press, 1961.
16. Desoer, C. A. "An Introduction to State-Space Techniques in Linear Systems," AIEE Paper 10-2 presented at *1962 Joint Automatic Control Conference*, June, 1962.
17. Coddington, E. A. and Levinson, N. *Theory of Ordinary Differential Equations*, McGraw-Hill, 1955.
18. Hahn, W. "On the Application of the Method of Lyapunov to Difference Equations" (in German) *Mathematische Annalen*, v. 136, pp. 402-441 (1958). Also translated into English by G. A.

Bekey and available from Space Technology Laboratories, Inc. as STL, Translation TR-61-5110-16, April, 1961.

19. Kadota, T. T. "Asymptotic Stability of Some Nonlinear Feedback Systems," University of California, Berkeley, Department of Electrical Engineering Report, Issue No. 264, Series No. 60, January, 1960.
20. Bekey, G. A. "Sampled-Data Models of the Human Operator in a Control System," pp. 178-186, *ASD Technical Report ASD-TDR-62-36,* February, 1962.
21. Kadota, T. T. and Bourne, H. C. "Stability Conditions of Pulse-Width Modulated Systems through the Second Method of Lyapunov," *IRE Transactions,* v. AC-6, pp. 266-275, September, 1961.
22. Nelson, W. L. "Pulse-Width Relay Control in Sampling Systems," *Trans. ASME*, series D, v. 83, pp. 65-76, 1961.
23. Kuo, Benjamin C. *Analysis and Synthesis of Sampled-Data Control Systems,* Prentice-Hall, 1963.
24. Zadeh, L. A. and Desoer, C. A. *Linear System Theory: The State Space Approach,* McGraw-Hill, 1963.
25. Li, C. C. and Jones, R. W. "Integral Pulse Frequency Modulated Control Systems," presented at the 2nd Congress of the International Federation of Automatic Control (IFAC), Basle, 1963. (To be published in the *Proceedings of the 2nd IFAC Congress,* by Butterworths, 1964.)
26. Farrenkopf, R. L., Sabroff, A. E., and Wheeler, P. C. "An Integral Pulse Frequency On-Off Attitude Control System," Space Technology Laboratories, Inc. *Report No. 8637-6117-RJ000,* Redondo Beach, Calif., March, 1963.
27. Kadota, T. T. "Analysis of Nonlinear Sampled-Data Systems with Pulse-Width Modulators," University of California, Berkeley, Department of Electrical Engineering Report, Series No. 60, Issue No. 290, June 30, 1960.
28. Kalman, R. E. and Bertram, J. E. "General Synthesis Procedure for Computer Control of Linear Systems—An Optimal Sampling System," *Trans. AIEE*, v. 77, pt II, pp. 602-609, 1958.
29. Kalman, R. E. and Koepcke, R. W. "Optimal Synthesis of Linear Sampling Control Systems Using Generalized Performance Indexes," *Trans. ASME*, v. 80, pp. 1820-1826, November 1958.
30. Neustadt, L. W. "Discrete Time-Optimal Control Systems," *International Symposium on Nonlinear Differential Equations and Applied Mechanics,* ed. by J. P. LaSalle and S. Lefschetz, Academic Press, 1963, pp. 267-283.
31. Desoer, C. A. and Wing, J. "A Minimal Time Discrete System," *IRE Trans. on Auto. Control*, v. AC-6, pp. 111-125, 1961.
32. Desoer, C. A., E. Polak, and J. Wing, "Theory of Minimum-Time Discrete Regulators," presented at 2nd IFAC Congress, Basle, 1963 (To be published by Butterworths, 1964).

33. Tsypkin, Ya. Z. "Optimal Processes in Sampled-Data Control systems," *Energetika i Avtomatika*, No. 4, 1960; translated by Z. Jakubski in Space Technology Laboratories Technical Translation No. STL-TR-61-5110-27, July, 1961.
34. Tou, J. T. and Vadhanaphuti, B. "Optimum Control of Nonlinear Discrete-Data Systems," *Trans. AIEE*, pt II, pp. 166-171, September, 1961.
35. Krasovskiy, N. N. "On an Optimal Control Problem," *Prikl. Matematika i Mekhanika*, v. 21, pp. 670-677, 1956; translated by L. W. Neustadt in Space Technology Laboratories Document AR 61-0007, July, 1961.
36. Koepcke, R. W. "A Solution to the Sampled, Minimum Time Problem," Preprints of 4th Joint Automatic Control Conf., pp. 94-100, 1963.
37. Mullin, F. J. and deBarbyrac, J. "Linear Digital Control," *Preprints of 4th Joint Automatic Control Conference*, pp. 582-588, 1963.
38. Pyshkin, I. V. "Processes of Finite Duration in Pulse-Width Systems," *Avtomatika i Telemekhanika*, v. 21, pp. 201-208, February, 1960.
39. Nelson, W. L. "Optimal Control Methods for On-Off Sampling Systems," *Trans. ASME (J. Basic Engr.)* v. 84, Ser. D. pp. 91-101, 1962.
40. Polak, E. "Minimum Time Control of Second-Order Pulse-Width-Modulated Sampled-Data Systems," *Trans. ASME (J. Basic Engr.)* v. 84, Ser. D., pp. 101-110, 1962.
41. Smith, F. T. "A Discussion of Several Concepts used in the Optimization of Control Systems by Dynamic Programming," RAND Corp. Report P-1665, April 9, 1959.
42. Monroe, A. J. *Digital Processes for Sampled-Data Systems*, New York: Wiley, 1962.

12

Description of the Human Operator in Control Systems

GEORGE A. BEKEY
ASSOCIATE PROFESSOR
ELECTRICAL ENGINEERING DEPARTMENT
UNIVERSITY OF SOUTHERN CALIFORNIA
LOS ANGELES, CALIFORNIA

12.1 Introduction

This chapter is concerned with the mathematical description of the input-output behavior of the human operator when he acts as an element in a closed-loop control system. Consider the simplified block diagram of the system shown in Fig. 12-1. This diagram could represent, for example, the manual control of one axis of an aerospace vehicle. The operator is required to detect a system error visually and to perform a manual movement such that the error is reduced. His response must be consistent with both stability and dynamic performance requirements.

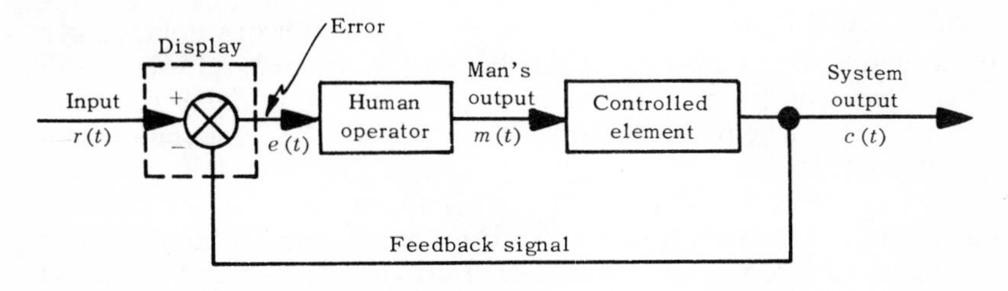

FIG. 12-1. Simple control system with human operator.

From the standpoint of the control system designer, it is extremely desirable that the input-output characteristics of the human operator be expressed mathematically to predict system performance. However, the construction of an adequate mathematical model of a human operator's behavior, even for a particular task,* is extremely difficult. Some of the sources of this difficulty are discussed in the following section.

The development of human operator models in manual tracking includes a series of attempts to apply techniques used for the identification of other physical systems. The two major approaches that have been used may be called the "spectral analysis method" and the "parameter tracking method." Spectral analysis techniques have been applied to the development of both continuous and sampled-data models. The parameter tracking method is based on assuming a form for the human operator model and adjusting the parameters of this model until an appropriate error criterion is minimized. In the following sections of this chapter, both methods are reviewed and their advantages and limitations are noted briefly.

Due to the difficulty of obtaining adequate descriptions of the human operator's behavior in a complex system, nearly all of the work in the past has been done with control systems as simple as that of Fig. 12-1. That is, the operator is assumed to have one or at the most two inputs. Extraneous inputs and feedback loops are generally neglected in the formulation of models. If the display of Fig. 12-1 is a subtraction device which presents to the man only the difference between the input r and the controlled variable c, the task is known as "compensatory tracking." The operator's function is to reduce the error e to zero. If both the reference input r and the output c are displayed, the task is called "pursuit tracking."

12.2 Characteristics of the Human Operator in a Control System

The input-output behavior of a human operator in a control system is characterized by the following major features [1-4]:

(a) *Reaction time:* the presence of a pure time delay or transport lag, which can be clearly observed in the response to step function inputs.

(b) *Low-pass behavior*: the tracker tends to attenuate high frequencies, the amount of attenuation increasing as the frequency increases, as indicated by visual examination and Fourier analysis of tracking records.

(c) *Task dependence*: ability of the operator to adjust his input-output characteristics to perform his control function with a wide range of controlled element dynamics.

(d) *Time dependence* : the dependence of the operator's characteristics on time, as seen in two forms: first, his performance

*The word *task* refers to the combination of a particular input function, controlled element, and display-control configuration in a manual control system.

changes with time as he learns; second, he is capable of sensing changes in environmental parameters and controlled system parameters, and adjusting his characteristics accordingly.

(e) *Prediction*: the well-known ability of the human operator to predict the course of a target based on past performance [5]. This ability to extrapolate is important in tracking since it means that tracking behavior is different with "predictable inputs" (such as sine waves or constant-frequency square waves) than it is with random or random-appearing inputs. Tracking with a predictable input has been called "precognitive" tracking [3, 4].

(f) *Nonlinearity*: for certain tasks the operator's behavior appears to be approximately linear while for other tasks his behavior is nonlinear.

(g) *Determinacy*: a human operator is a nondeterministic system, since his performance is different in successive trials of the same experiment. However, his variability is small in situations where training time is adequate and the task is not considered difficult. Consequently, a deterministic model may be used to describe his performance in a statistical sense.

(h) *Intermittency*: there is a considerable body of evidence to indicate that the human operator behaves as a discrete or sampling system in certain tracking operations.

The various human operator models appearing in the literature represent attempts to insert the major characteristics listed above into a mathematical description.

12.3 Quasi-Linear Continuous Models

The earliest models constructed from the point of view of control engineering were those of Tustin [10] and Ragazzini [11]. These models were postulated as linear and continuous with an additional disturbance of unknown origin. Tustin called the additional term which did not result from linear operations on the input the "remnant" (i.e., those components of the response at frequencies other than the input frequencies in this case). The name "remnant" is still used in the quasi-linear models to be discussed below. In block-diagram form, the human operator model may be indicated as shown in Fig. 12-2 below.

The Tustin model is based on Fourier analysis of the outputs when the input is a sum of 3 sinusoids. The transfer function of the resulting model is of the form

$$G_H(s) = K\left(\frac{K_1}{s} + K_2\right)e^{-Ds} \tag{12.1}$$

where s is the complex frequency variable, D is the time delay, and K, K_1, and K_2 represent parameters which depend on the choice of controlled element dynamics.

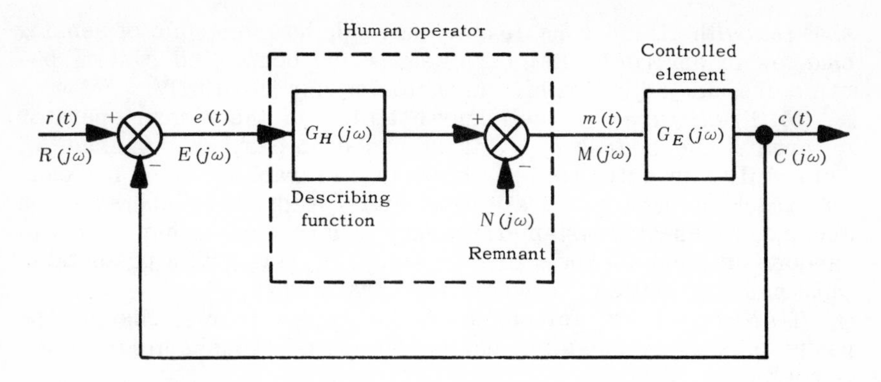

FIG. 12-2. Quasi-linear model of human operator.

J. R. Ragazzini reported in 1948, in an unpublished paper delivered at the American Psychological Association, that the transfer relationship may be of the form:

$$G_H(s) = K\left(K_1 s + K_2 + \frac{K_3}{s}\right)e^{-Ds} \tag{12.2}$$

The parameters of the transfer functions (12.1) and (12.2) depend upon the particular task. Transfer functions of this type will be termed quasi-linear. By quasi-linear in the present discussion we mean that the system has the following properties:

(a) The system is describable by a linear differential equation with coefficients which are dependent on the system configuration and the input signal bandwidth, but remain constant for a particular system, and

(b) The linear relationship determines only a portion of the system's output (the linear part). In addition, a random or uncorrelated component may exist. The validity of the quasi-linear model is dependent on the proportion of the system output which it specifies.

The definition of quasi-linearity given above was assumed by Tustin in his studies, even if not stated explicitly. It is interesting to review the methods used to arrive at quasi-linear transfer relationships [3, 12].

Phillips [13] described a model based on a step-function response of the form

$$G_H(s) = \frac{K(1 + T_2 s)e^{-Ds}}{s} \tag{12.3}$$

Mayne [14] recorded responses to step inputs and obtained human operator models from Fourier analysis of the responses.

Step responses have been used by Hyndman and Beach [15] to study open-loop responses. Their model was

$$G_{OL}(s) = \frac{K}{s^2 + 2\zeta\omega_n s + \omega_n^2} e^{-Ds} \qquad (12.4)$$

Open-loop relationships are not directly applicable to closed-loop systems with human operators.

Response of the human operator to simple sinusoids has been used as well, but, as previously noted, the response is expected to be quantitatively different from that obtained when tracking more complex signals. A number of interesting studies have been performed with sine-wave inputs [16-18]. Two major conclusions may be drawn from these studies: (1) if the frequencies are sufficiently low, a linear invariant model adequately represents the major portion of the operator's output; (2) as the frequency increases, the human tracker's performance gradually deviates from that expected with a linear invariant model.

Most of the significant work has been performed recently with complex signals consisting either of sums of sine waves or random signals having appropriate spectral characteristics. Tustin used a sum of 3 sine waves. Russell [19] used a sum of 4 sine waves and also measured the output components at the same frequencies as those present in the input. Elkind [6] performed an extremely thorough study of compensatory and pursuit tracking using input signals which consisted of the sum of a large number of sinusoids (as many as 144) of random phase. The method used by Elkind for the synthesis of the quasi-linear model will be described in the next section. Krendel [7] used random input signals to determine linear relationships. The best quasi-linear results available to date are those of Elkind and Krendel.

12.4 Synthesis of Quasi-Linear Continuous Models

The models described by Elkind are obtained by using a technique suggested by Booton [20] for the analysis of nonlinear systems with Gaussian random inputs. The theory of the resulting random-input describing function has been discussed in detail by McRuer and Krendel [3] as well as by Booton and will be briefly outlined here. Basically the method consists in assuming that the output of the system is the sum of two signals, one linearly correlated with the input shown in the block diagram of Fig. 12-2, the other not.

The input $r(t)$ is assumed to be a Gaussian random process. We seek to minimize the error between the output $m'(t)$ and the actual output $m(t)$. We can describe the linear process in the time domain by a weighting function $h(t)$ such that

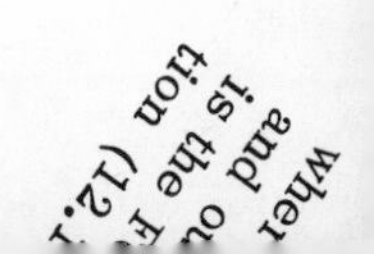

$$m'(t) = \int_0^\infty h(\tau)\, e(t - \tau)\, d\tau \tag{12.5}$$

The error of approximation is then given by

$$\begin{aligned} \epsilon(\tau) &= m(t) - m'(t) \\ &= m(t) - \int_0^\infty h(\tau)\, e(t - \tau)\, d\tau, \quad \tau > 0 \end{aligned} \tag{12.6}$$

We desire to select $h(t)$ such that it minimizes this mean-square error, i.e.,

$$\overline{\epsilon^2(t)} = \overline{[m(t) - m'(t)]^2} \tag{12.7}$$

The $h(t)$ which minimizes (12.7) is the solution of the Wiener-Hopf integral equation:

$$R_{\text{em}}(\tau) = \int_0^\infty h(u)\, R_{\text{ee}}(\tau - u)\, du \tag{12.8}$$

where $R_{\text{em}}(\tau)$ is the cross-correlation function between the error and output
and $R_{\text{ee}}(\tau)$ is the auto-correlation function of the error.

To make use of these relations for the synthesis of quasi-linear models, it is more convenient to express them in the frequency domain. From the block diagram of Fig. 12-2, the Fourier transforms of the system error and operator's output respectively are given by

$$E = R\left[\frac{1}{1 + G_H G_E}\right] - N\left[\frac{G_E}{1 + G_H G_E}\right] \tag{12.9}$$

$$M = R\left[\frac{G_H}{1 + G_H G_E}\right] + N\left[\frac{1}{1 + G_H G_E}\right] \tag{12.10}$$

The frequency domain expression corresponding to (12.8) is

$$S_{\text{em}}(j\omega) = G_H(j\omega)\, S_{\text{ee}}(j\omega) \tag{12.11}$$

$_{\text{em}}(j\omega)$ is the cross-power spectral density between error 't, $S_{\text{ee}}(j\omega)$ is the error-power spectral density and $G_H(j\omega)$ 'ier transform of the weighting function $h(t)$. Using equa- the cross-spectral density $S_{\text{re}}(j\omega)$ between input and error, and $S_{rm}(j\omega)$ between input and operator output are given, respectively, by

$$S_{re}(j\omega) = \frac{1}{1 + G_H(j\omega)G_E(j\omega)} S_{rr}(\omega) \tag{12.12}$$

and

$$S_{rm}(j\omega) = \frac{G_H(j\omega)}{1 + G_H(j\omega)G_E(j\omega)} S_{rr}(\omega) \tag{12.13}$$

The noise spectrum $S_{nn}(\omega)$ does not appear in these relations since it is uncorrelated with the input, by hypothesis. The cross-spectral densities of (12.12) and (12.13) can be measured experimentally. The describing function or quasi-linear transfer relationship is then obtained as the ratio of the two experimental quantities above, i.e.,

$$G_H(j\omega) = \frac{S_{rm}(j\omega)}{S_{re}(j\omega)} \tag{12.14}$$

Relation (12.14) represents a set of experimental points which define the amplitude and phase of the complex number $G_H(j\omega)$. When these quantities are plotted as a function of frequency, they can be fitted closely with a linear analytic relationship of the form:

$$G_H(j\omega) = \frac{Ke^{-j\omega D}(1 + j\omega T_L)}{(1 + j\omega T_N)(1 + j\omega T_I)} \tag{12.15}$$

When the controlled element dynamics are negligible [i.e., $G_E(j\omega) \cong 1$] then T_L and T_I are approximately zero, and (12.15) reduces to

$$G_H(j\omega) = \frac{Ke^{-j\omega D}}{1 + j\omega T_N} \tag{12.16}$$

The "remnant" term is now represented by the noise injected at the operator's output in Fig. 12-2. If we define a "closed-loop quasi-linear describing function" $H(j\omega)$ as

$$H(j\omega) = \frac{G_H(j\omega)}{1 + G_H(j\omega)} \tag{12.17}$$

and a closed-loop remnant spectrum as S'_{nn}:

$$S'_{nn}(\omega) = \left| \frac{1}{1 + G_H(j\omega)G_E(j\omega)} \right|^2 S_{nn}(\omega) \tag{12.18}$$

then the operator's output-power spectrum can be described by the relation:

$$S_{mm}(\omega) = |H(j\omega)|^2 S_{rr}(\omega) + S'_{nn}(\omega) \tag{12.19}$$

Equation (12.19) shows that the trained operator's output spectrum is composed of two portions, one resulting from a linear operation on the input spectrum S_{rr}, the other an additional "noise" term. The degree of validity of the quasi-linear model can now be related to the proportion of total operator output power at each frequency predicted by the model. This can be done by defining the linear correlation between m and r (sometimes called the "coherence function") as

$$\rho^2 = \frac{|S_{rm}(j\omega)|}{\sqrt{S_{rr}(\omega) S_{mm}(\omega)}} \tag{12.20}$$

Elkind [6] reported linear correlations of 0.9 or better for his studies when the input spectrum was flat with a sharp cutoff at or below 0.75 cps, approximately. The linear correlation drops off rapidly with increasing frequency content of the input signal. With a flat input spectrum with a sharp cutoff at 1.6 cps, ρ is approximately 0.75, and with a cutoff at 2.4 cps, ρ is about 0.6. Consequently, the quasi-linear model can be considered a valid representation of human tracking behavior only for the situation where the input signal contains no appreciable energy above 0.75 cps.

12.5 Dependence of Model Parameters on the Forcing Function

When the controlled element dynamics are negligible, the quasi-linear model of Eq. (12.15) reduces to

$$G_H(j\omega) = \frac{Ke^{-j\omega D}}{1 + j\omega T_N} = \frac{Ke^{-j2\pi fD}}{1 + jf/f_N} \tag{12.21}$$

Elkind [6] was able to show that the parameters K and f_N are related for a wide range of input signal bandwidths. For all the input conditions when $\rho > .9$, the data show that

$$Kf_N \cong 1.5 \tag{12.22}$$

Furthermore, when the input spectra were flat, the model gain K depended on the input function cutoff frequency, f_{co}, as follows:

$$K \cong \frac{2.2}{f_{co}{}^2} \tag{12.23}$$

Equations (12.22) and (12.23) are strictly empirical and have not been derived from theoretical considerations. They do, however, indicate the dependence of the operator's performance upon tasks of varying difficulty. Thus, as f_{co} increases (and the task becomes more difficult) the model gain K decreases and the lag frequency f_N increases.

For all the tracking runs where (12.22) and (12.23) were approximately valid, the model delay was given by

$$D \cong 0.13 \text{ seconds} \tag{12.24}$$

In open-loop step responses, the reaction-time delay is approximately 0.25 seconds. Since, in closed-loop tracking of continuous signals, a certain amount of prediction is possible, the lower value of D given by (12.24) is reasonable.

Similar work has been done with pursuit tracking. To a lesser extent, quasi-linear models have been obtained for both compensatory and pursuit tracking systems with controlled element dynamics. Where quasi-linear models were appropriate, they were of the form of Eq. (12.15).

12.6 Difficulties with the Quasi-Linear Continuous Model

The quasi-linear model gives impressive evidence of the nearly linear behavior of the human operator when tracking signals of low frequency. However, the model suffers from a number of drawbacks in addition to the frequency limitations. Among these are the following:

(a) Being linear and continuous, the output of the model cannot contain frequencies not present in the input signal (which are known to exist in human operator outputs).

(b) The model cannot account for a substantial body of experimental evidence (to be discussed below) which suggests intermittent behavior of the tracker.

(c) The model does not account for the predictive ability of the human operator.

12.7 Discrete or Sampled-Data Models of the Human Operator

There is considerable evidence which suggests that the human operator's response may be intermittent rather than continuous, i.e., he acts upon discrete samples of input information rather than upon a continuous input [2, 8, 9]. Several mathematical models have been based on the hypothesis that the human operator behaves as a discrete system [9, 3].

The Difference Equation Approach of North

The first discrete mathematical model was proposed by J. D. North [22]. North examined the behavior of a typical quasi-linear differential equation model, of the type proposed by Tustin. Then he replaced the derivatives in the equation by finite differences and examined the resulting linear difference equation. His final equations included not only the deterministic components of the tracker's response but also nondeterministic components which in the discrete case represented samples from "white noise." The approach of North is quite sophisticated and attempts to include many effects. However, his difference-equation approach suffers from the fact that he concentrated on the system behavior at the "sampling instants" while the human operator's output is clearly continuous. That is, any discrete processes which may occur should be internal, and a data-reconstruction element must appear in the model for the output to be continuous. North evaluated the power spectral density of his model output, in both the continuous and the discrete case. However, since the time-domain relation was valid only for integral values of sampling intervals, he could say no more than that the resulting spectral densities approach each other as the sampling interval approaches zero.

The Analog Computer Study of Ward

In a thesis submitted in Australia, Ward [23] has discussed a study of a sampled-data model using an analog computer. Ward's model is shown in Fig. 11-3. The results of Ward's study are given in the form of computer output traces, which were compared with human operator output recordings to detect similarities in appearance. The model parameters were adjusted manually to yield a good visual fit to tracking data. The input functions consisted of the sum of 3 sine waves, the highest frequency being 0.2 cps and the lowest 0.01 cps. The apparent amplitude of the "noise" in both man and model is of the same order of magnitude. The study is severely limited for several reasons: (1) the lack of analytical work (apparently Ward considered the system to be nonlinear merely due to the presence of the sampler); (2) the poor choice of sampling and data reconstruction circuits for the analog simulation. As can be noted in Fig. 12-3, the model includes a sampler and zero-order hold, followed by a delayed sampler (to account for reaction time), and then an integrator. However, the integrator is a nonresetting hold circuit, and consequently the reconstructed error signal e_R has no resemblance to the continuous input.

However, Ward's study, while limited, is useful in that it provides additional evidence that at least a portion of the "remnant" noise term required in continuous models may be accounted for by harmonics due to sampling.

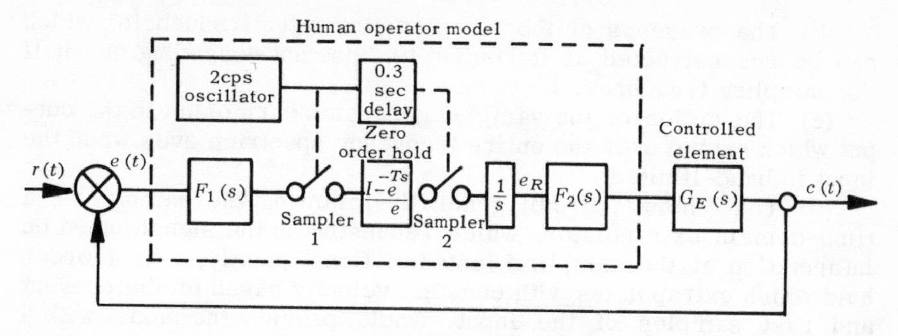

FIG. 12-3. Sampled-data operator model due to Ward.

The transfer functions $F_1(s)$ and $F_2(s)$ in Ward's model were also obtained empirically, and are given as

$$F_1(s) = A + Bs + D\frac{s^2 + 2\zeta_1\omega_0 s + \omega_0^2}{s^2 + 2(4\zeta_1)\omega_0 s + \omega_0^2} \cdot \frac{1}{1 + 0.5s} \tag{12.25}$$

$$F_2(s) = \frac{1}{1 + C\dfrac{s^2 + 2\zeta_1\omega_0 s + \omega_0^2}{s^2 + 2(4\zeta_1)\omega_0 s + \omega_0^2} \cdot \dfrac{1}{1 + 0.5s}} \tag{12.26}$$

A Quasi-Linear Sampled-Data Model

Spectral analysis techniques have also been used for the synthesis of a discrete model [9]. In its simplest form, this model is shown in Fig. 12-4 for the case where controlled element dynamics are negligible. The human operator is represented by a periodic sampler, a hold circuit, and a continuous element represented by Eq. (12.21). The ability of the sampled-data model of Fig. 12-4 to

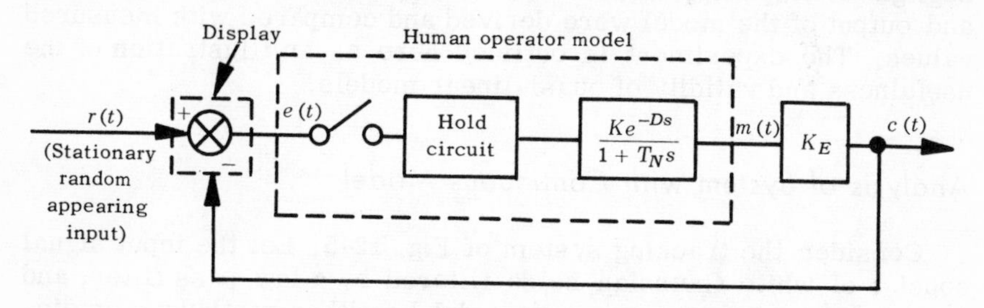

FIG. 12-4. Discrete model of the human operator.

meet some of the problems facing the continuous model can be seen intuitively by considering the following characteristics of sampled systems:

(a) Changes in the input cannot have any effect until the next sampling instant occurs.

(b) The presence of the sampler limits the frequencies which can be reconstructed at its output to those not exceeding one-half the sampling frequency.

(c) The action of the sampler generates harmonics in the output which extend over the entire frequency spectrum even when the input is band-limited.

(d) The "hold" circuit generally following the sampler is a time-domain extrapolator, which reconstructs the signal based on information at the sampling instants. Consequently, a first-order hold which extrapolates with constant velocity based on the present and past samples of the input, would provide the model with a characteristic known to exist in human tracking.

(e) In the limit as the input frequency approaches zero, the sampled system output approaches the continuous system output. This is desirable since the continuous model is quite adequate for low-frequency inputs.

To examine the implications of sampling for the behavior of the system, the model of Fig. 12-4 has been analyzed with several types of hold circuits, and expressions for spectral density functions of the error $e(t)$ and output $c(t)$ have been obtained [9].

12.8 Analysis of Continuous and Sampled Models with Random Inputs

The parameters of the models discussed above may be obtained experimentally. This section describes a particular experiment, shows how the parameters were obtained, and compares the performances of the models and the human operator. For this particular experiment, the controlled-element dynamics were assumed negligible. Expressions for the power spectral density of the error and output of the model were derived and compared with measured values. The experiment is outlined here as an illustration of the usefulness and validity of quasi-linear models.

Analysis of System with Continuous Model

Consider the tracking system of Fig. 12-5. Let the input signal consist of white Gaussian noise filtered by a low-pass filter, and let the linear human operator model be either continuous or discrete.

If the input to a linear invariant system is a stationary process $x(t)$ characterized by a power spectral density $S_{xx}(\omega)$, then the power spectral density of the output $S_{yy}(\omega)$ is

$$S_{yy}(\omega) = |F(j\omega)|^2 S_{xx}(\omega) \tag{12.27}$$

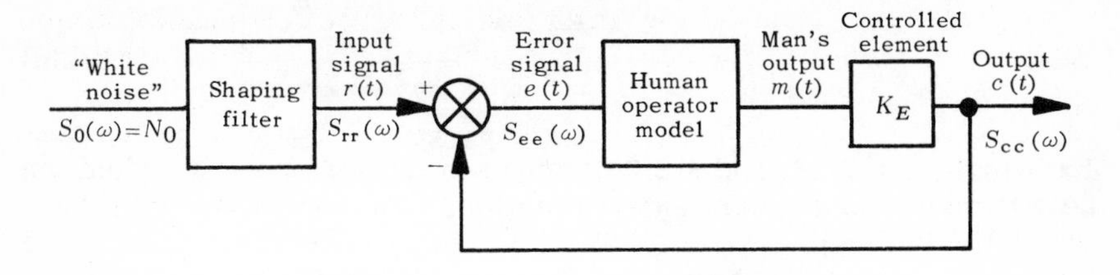

FIG. 12-5. Compensatory tracking system.

where $F(j\omega)$ is the frequency-response function of the system. If the noise source is assumed white then

$$S_0(\omega) = N_0 = \text{constant} \tag{12.28}$$

The input to the tracking system then has the spectral density

$$S_{rr}(\omega) = |F(j\omega)|^2 N_0 \tag{12.29}$$

where the frequency-response function of the low-pass filter is given by

$$F(j\omega) = \frac{1}{1 + j\omega/\omega_B} \tag{12.30}$$

We are interested in obtaining an expression for the error spectral density and the output spectral density when the transfer function of the continuous operator model is

$$G_c(j\omega) = \frac{K_{HE}\, e^{-j\omega D_c}}{1 + j\tau_c\omega} \tag{12.31}$$

where K_{HE} is a gain which includes the model gain K_H and the controlled element gain K_E. The input spectral density is given by

$$S_{rr}(\omega) = \frac{N_0\,\omega_B^{\,2}}{\omega^2 + \omega_B^{\,2}} \tag{12.32}$$

Using Eqs. (12.31) and (12.32), the error power spectral density is given by:

$$S_{ee}(\omega) = \frac{\left(1 + \tau^2\omega^2\right)\omega_B^{\,2} N_0}{\left(\omega^2 + \omega_B^{\,2}\right)\left[\left(1 + K^2\right) + \tau^2\omega^2 + 2K(\cos\omega D - \tau\omega \sin\omega D)\right]} \tag{12.33}$$

where the subscripts on τ, K, and D have been dropped for convenience. Similarly, the output power spectral density is

$$S_{cc}(\omega) = \frac{K^2 \omega_B^2 N_0}{\left(\omega^2 + \omega_B^2\right)\left[\left(1 + K^2\right) + \tau^2 \omega^2 + 2K(\cos \omega D - \tau\omega \sin \omega D)\right]} \qquad (12.34)$$

Expressions (12.33) and (12.34) represent power spectra which can be estimated in experimental situations and compared with their theoretical values.

Analysis of the Sampled Model

Analyses similar to the above have been carried out using zero-order and first-order hold circuits in the discrete models [9]. However, since much more significant results are obtained with first-order hold circuits, only the results of first-order hold models are reported here.

Signals in sampled-data systems are in general nonstationary, even with stationary inputs. Consequently, statistical functions computed from ensemble averages do not equal those computed from time averages. In this study, the spectral density functions were obtained from time-averaged autocorrelation functions.

Consider now the sampled form of the operator model given in Fig. 12-4. The derivation of an expression for the output power spectral density follows a procedure analogous to that for the continuous system, with complications introduced by sampled signals which have repeated spectra along the entire frequency axis. The power spectral density of the continuous error signal $e(t)$ cannot be obtained by a single relation analogous to (12.33) since there is no transfer relationship which explicitly relates $E(j\omega)$ to $R(j\omega)$ in an error-sampled system. Thus a more complex procedure is required. Let $G_{s1}(s)$ represent the transfer function of the combination of first-order hold and continuous element in Fig. 12-4. Then it can be shown that since

$$\frac{C(j\omega)}{R^*(j\omega)} = \frac{G_{s1}(j\omega)}{1 + G_{s1}^*(j\omega)} \qquad (12.35)$$

the power spectral density of the output is given by

$$S_{cc}(\omega) = \left|\frac{G_{s1}(j\omega)}{1 + G_{s1}^*(j\omega)}\right|^2 S_{rr}^*(\omega) \qquad (12.36)$$

where, as usual, the asterisk represents sampled quantities. The sampled spectral density is defined in terms of the $\mathcal{Z}$ transform of $S_{rr}(s)$:

$$S_{rr}^*(s) = \frac{1}{T}\,\mathcal{Z}[S_{rr}(s)]\Big|_{z=e^{sT}} = \frac{1}{T^2}\sum_{n=-\infty}^{+\infty} S_{rr}(s + jn\omega_s) \qquad (12.37)$$

The additional factor of $1/T$ arises due to time-averaging over a sampling period; $S_{rr}(s)$ is the bilateral Laplace transform representation of the power spectral density.

It can be shown that the expression for the power spectral density of the continuous error $e(t)$ is

$$S_{ee}(\omega) = S_{rr}(\omega) - \frac{2}{T} S_{rr}(\omega) \, Re\left[\frac{G_{s1}(j\omega)}{1 + G^*_{s1}(j\omega)}\right] + S^*_{rr}(\omega)\left|\frac{G_{s1}(j\omega)}{1 + G^*_{s1}(j\omega)}\right|^2 \tag{12.38}$$

where, as before, $S_{rr}(j\omega)$ is the power spectral density of the continuous input, and $S^*_{rr}(j\omega)$ is the corresponding sampled spectral density. The third term in Eq. (12.38) will be recognized from Eq. (12.36) as the power spectral density of the output signal $c(t)$. Consequently, the only unknown in Eq. (12.38) is the second term, which is twice the real part of the cross-spectral density between $r(t)$ and $c(t)$, $S_{rc}(j\omega)$.

To evaluate the spectral densities $S_{cc}(\omega)$ in Eq. (12.36) and $S_{ee}(\omega)$ in Eq. (12.38), the z transform of the open-loop transfer function must be determined.

$$G_{s1}(s) = \left(\frac{1 + Ts}{T}\right)\left(\frac{1 - e^{-Ts}}{s}\right)^2 \left(\frac{Ke^{-D_1 s}}{1 + \tau s}\right) \tag{12.39}$$

The z transform of (12.39) is

$$G_1(z) = \frac{K\left(Q_1 z^2 + Q_2 z + Q_3\right)}{Tz^2\left(z - e^{-aT}\right)} \tag{12.40}$$

where

$$\begin{aligned} Q_1 &= (2T - D_1 - \tau) + (\tau - T)e^{-a(T-D_1)} \\ Q_2 &= T - \left(1 + e^{-aT}\right)(2T - D_1 - \tau) - 2(\tau - T)e^{-a(T-D)} \\ Q_3 &= -Te^{-aT} + (2T - D_1 - \tau)e^{-aT} + (\tau - T)e^{-a(T-D)} \end{aligned} \tag{12.41}$$

Substitution of Eq. (12.40) in Eq. (12.36) gives the output power spectral density:

$$S_{cc}(\omega) = \frac{\{4K^2(1+T^2\omega^2)(1+e^{-2aT} - 2e^{-aT}\cos\omega T)(1-\cos\omega T)^2\} S_{rr}^*(\omega)}{T^2\omega^2(1+\tau^2\omega^2)\left[(1+P_1^2+P_2^2+P_3^2)+2(P_1+P_1P_2+P_2P_3)\cos\omega T+2(P_2+P_1P_3)\cos 2\omega T+2P_3\cos 3\omega T\right]} \tag{12.42}$$

where, for the filtered noise input, $S_{rr}^*(\omega)$ is

$$S_{rr}^*(\omega) = \frac{N_0 \omega_B \sinh \omega_B T}{2T(\cosh \omega_B T - \cos \omega T)} \tag{12.43}$$

Similarly, from Eq. (12.38) the continuous error power spectral density is

$$S_{ee}(\omega) = S_{rr}(\omega) + S_{cc}(\omega) - \frac{2}{T} S_{rr} Re \left[\frac{G_{s1}(j\omega)}{1 + G_{s1}^*(j\omega)} \right] \tag{12.44}$$

where

$$Re \frac{C(j\omega)}{R^*(j\omega)} = \frac{-KF_R(P,\omega)\left[\left(1 + \tau T\omega^2\right)\cos \omega D + \omega(T - \tau)\sin \omega D\right]}{T\omega^2\left(1 + \tau^2\omega^2\right)(P_{10} + P_{11}\cos \omega T + P_{12}\cos 2\omega T + P_{13}\cos 3\omega T)}$$
$$+ \frac{K\left[\omega(T - \tau)\cos \omega D - \left(1 + \tau T\omega^2\right)\sin \omega D\right]F_I(P,\omega)}{T\omega^2\left(1 + \tau^2\omega^2\right)(P_{10} + P_{11}\cos \omega T + P_{12}\cos 2\omega T + P_{13}\cos 3\omega T)} \tag{12.45}$$

and the coefficients are defined as follows:

$$\text{and} \begin{cases} F_R(P,\omega) = P_4 + (P_5 + P_7)\cos \omega T + (P_6 + P_8)\cos 2\omega T + (P_3 + P_9)\cos 3\omega T \\ F_I(P,\omega) = (P_5 - P_7)\sin \omega T + (P_6 - P_8)\sin 2\omega T + (P_3 - P_9)\sin 3\omega T \end{cases}$$

$$P_1 \triangleq \left(KQ_1 - Te^{-aT}\right)/T$$

$$P_2 \triangleq KQ_2/T$$

$$P_3 \triangleq KQ_3/T$$

$$P_4 \triangleq 1 - \left(2 + e^{-aT}\right)P_2 - e^{-aT} P_3$$

$$P_5 \triangleq P_1 - \left(2 + e^{-aT}\right)P_2 + \left(1 + 2e^{-aT}\right)P_3$$

$$P_6 \triangleq P_2 - \left(2 + e^{-aT}\right)P_3$$

$$P_7 \triangleq -\left(2 + e^{-aT}\right) + \left(1 + 2e^{-aT}\right)P_1 - e^{-aT} P_2$$

$$P_8 \triangleq \left(1 + 2e^{-aT}\right) - e^{-aT} P_1$$

$$P_9 \triangleq -e^{-aT}$$

$$P_{10} \triangleq 1 + P_1^2 + P_2^2 + P_3^2 \tag{12.46}$$

$$P_{11} \triangleq 2(P_1 + P_1P_2 + P_2P_3)$$

$$P_{12} \triangleq 2(P_2 + P_1P_3) \qquad \text{(12.46) (Cont'd)}$$

$$P_{13} \triangleq 2P_3$$

Thus, expressions are available for computation of the power spectral density of the error and output signals of the sampled models in terms of the basic parameters of the continuous model (K, D, and $\tau = 1/a$) and the additional parameters D_1 and T.

Experiments and Selection of Parameter Values

To verify the feasibility of the proposed discrete model, an experimental program was devised [9, 33]. Measurements were made of the power spectral density of output and error signals from a number of human operators tracking random-appearing inputs. The experimental situation was based on compensatory tracking in one dimension using a hand controller with negligible dynamics. An oscilloscope was used for the display. The input signal $r(t)$ was the output of a low-pass filter, the input to which consisted of a sum of 10 sine waves of equal amplitude and nonharmonic frequencies. This experimental arrangement approximates the idealized situation of Fig. 12-4. Approximately 100 runs were made with 8 operators, after a period of training.

The experiments were also used to provide the values of parameters to be used in the numerical evaluation of the power spectral densities obtained from the model. The parameters were selected as follows:

(a) From cross-spectral measurements of the operator's input, error, and output, the data for the quasi-linear continuous model were obtained by the method outlined in Section 12.4. The continuous model parameters, K, D_c, and τ were obtained by fitting the transfer function of Eq. (12.6) to the experimental data.

(b) The sampling period T was obtained by examining the recorded spectra for a pronounced peak in the vicinity of 1 to 1.5 cps. The frequency at which the peak occurred was assumed to correspond to one-half the sampling frequency.

(c) The time delay D_s of the sampled model was obtained by evaluating the effective time delay due to the hold circuit (D_h) and then letting $D_s = D_c - D_h$. This method was used so that the total open-loop phase shift remains approximately equal for the sampled and continuous models in the frequency range of interest.

The expressions for $S_{ee}(\omega)$ and $S_{cc}(\omega)$ derived above were programmed for solution on a digital computer. Typical results are shown in Fig. 12-6. The plotted points represent the power spectral density measured at the 10 component frequencies of the input signal.

The results of Fig. 12-6 illustrate the peaking in the output spectrum, characteristic of both the operator's output and the output of the discrete model. The continuous model, while providing an excellent fit at low frequencies, attenuates too rapidly at high frequencies. The output power spectral density from the discrete models shows remarkable agreement with the experimental data, in spite of the idealizing assumptions made.

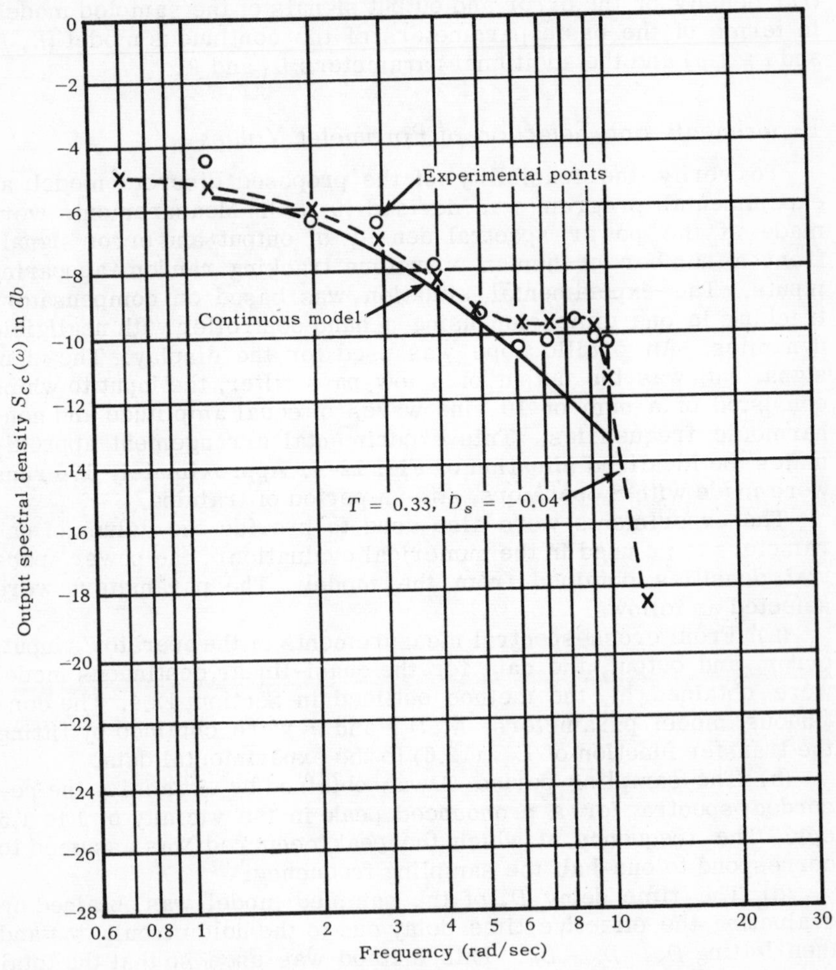

FIG. 12-6. Comparison of experimental and analytical values of output power spectral density, from ref. 9.

It should be noted that the sampled model delay D_s shown on Fig. 12-6 is negative, i.e., the model incorporates a predictor. It is likely that the use of partial velocity hold circuits, which result in less phase shift, wouldyield models with positive values of delay.

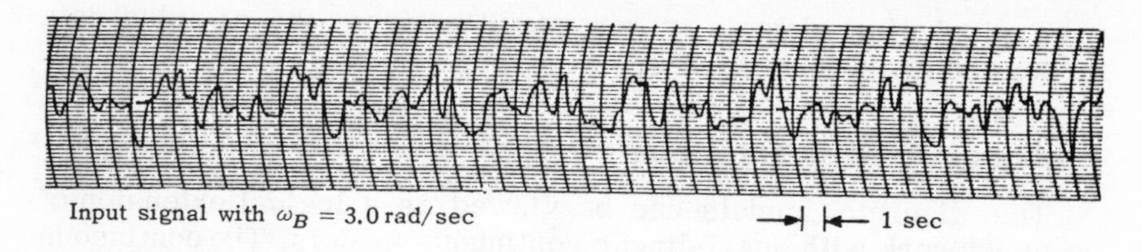

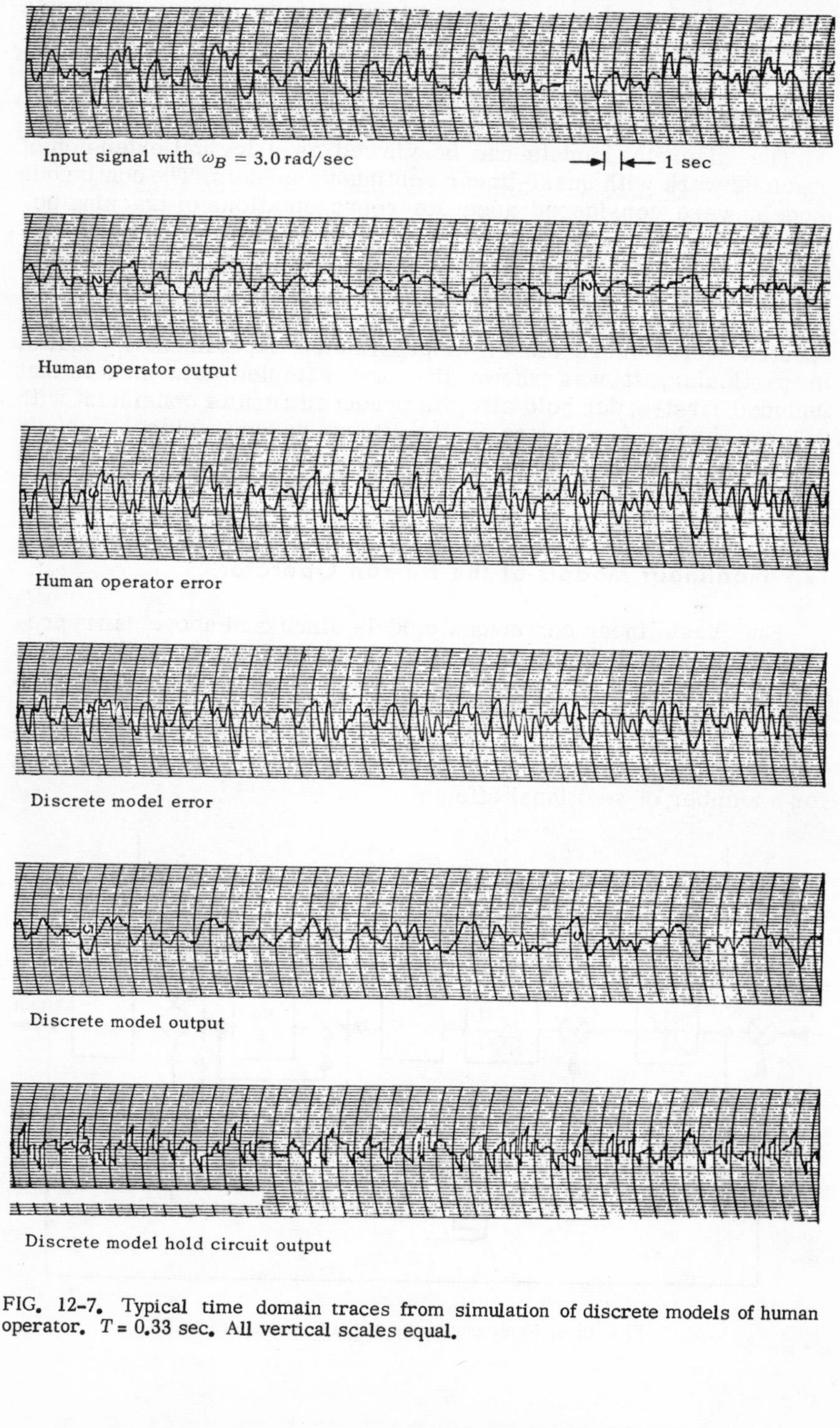

Input signal with $\omega_B = 3.0$ rad/sec

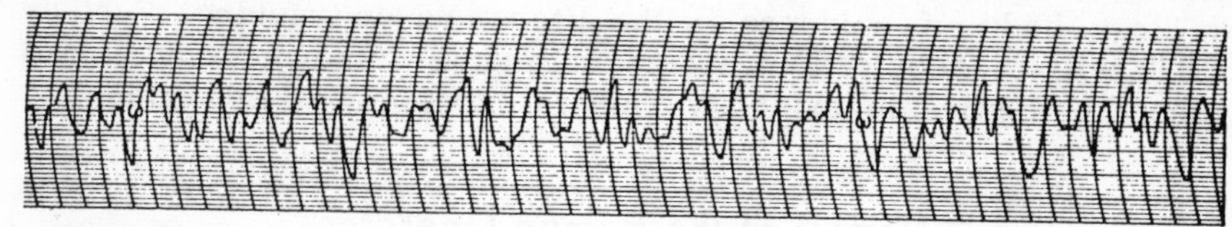
Human operator output

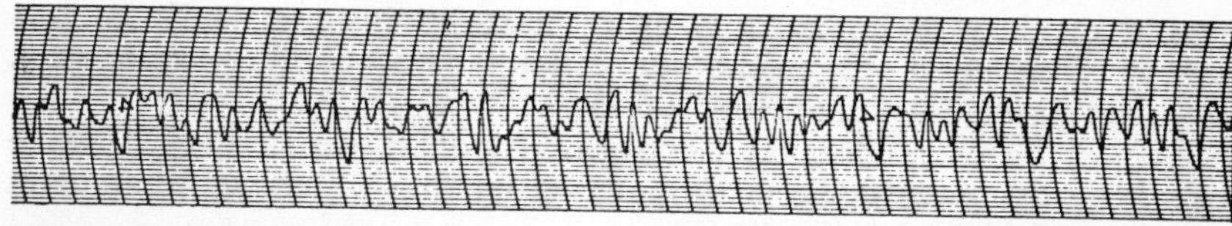
Human operator error

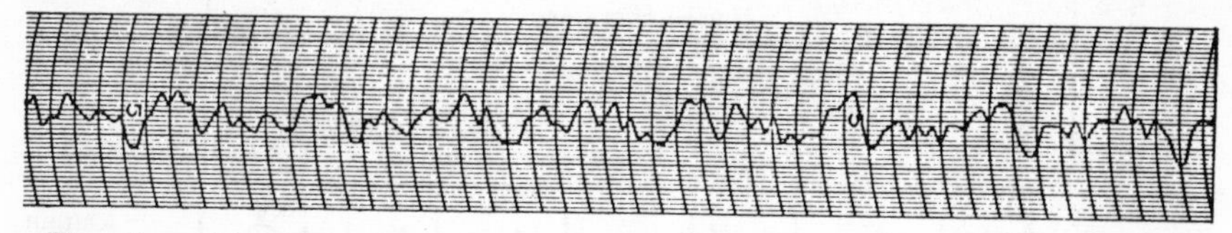
Discrete model error

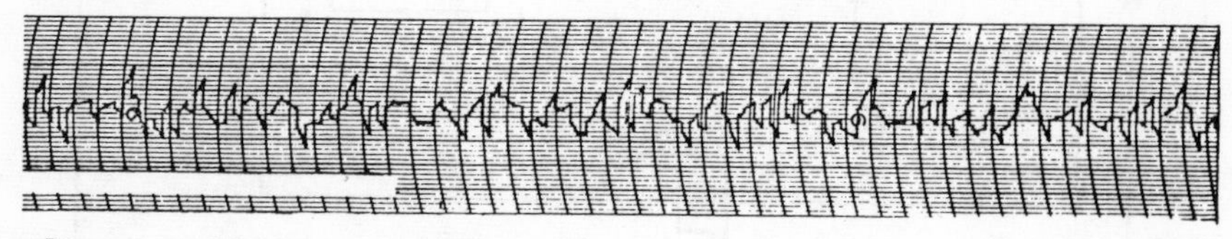
Discrete model output

Discrete model hold circuit output

FIG. 12-7. Typical time domain traces from simulation of discrete models of human operator. $T = 0.33$ sec. All vertical scales equal.

Typical time domain traces resulting from the sampled data model are given in Fig. 12-7.

Evaluation of Discrete Models

The discrete models can be viewed as a logical extension of previous work with quasi-linear continuous models. The continuous models were considered adequate representations of tracking behavior when the input function bandwidth did not exceed approximately 3/4 cps. In the present study, the bandwidth extended to 1.6 cps. The results showed that for the particular tracking system considered, use of the discrete models resulted in output spectra which more closely approximated experimental results. In particular, it was shown that the sampled-data models that included first-order hold circuits produced results consistent with a large body of evidence in the literature on tracking. Further study will be required to ascertain whether discrete models are generally applicable under a variety of experimental conditions.

12.9 Nonlinear Models of the Human Operator

The quasi-linear continuous models discussed above can represent a number of human operator characteristics. Nonlinear behavior is included only in the sense that it is a possible explanation of a portion of the "remnant" term.

A series of studies at Goodyear Aircraft Corp. [24-26] were aimed at the development of operator models which would account for a number of additional effects.

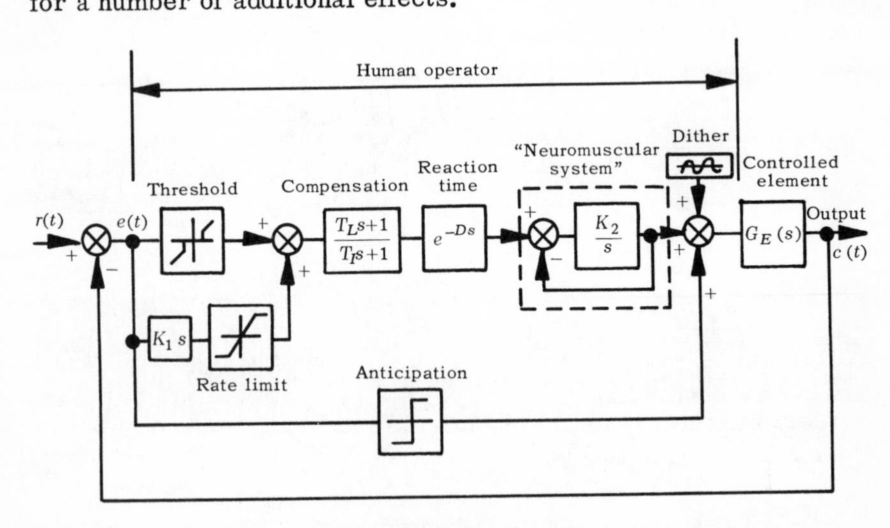

FIG. 12-8. Nonlinear model of the human operator.

The Goodyear model is shown in Fig. 12-8. The nonlinearities included in this model are:

(a) A dead zone to represent the operator's sensory threshold.

(b) Input-rate saturation.

(c) A relay or bang-bang function in parallel to represent the operator's "anticipation."

The output of the model was compared with that of a human pilot responding to the same input signal. The model parameters were adjusted manually until, on the basis of visual observation, the output of the model resembled that of the human pilot. This technique can be considered a manual version of the automatic parameter-tracking method discussed in Section 12.10 below.

The outputs of the Goodyear model, after adjustment of the parameters, showed excellent agreement with time-domain tracking data. The model is of historical interest but limited in usefulness because of its complexity and the requirement for computer simulation.

12.10 A Time Varying Model of the Human Operator

The human operator models discussed above were applicable only to specific tasks. The parameters of the models were dependent upon the controlled element dynamics and upon the spectral characteristics of the forcing function. Thus, by hypothesis, the models were not applicable during a transient period, e.g., the period following a sudden change in either the forcing function or the controlled element dynamics. Sheridan [27] attempted to investigate the operator's performance during this transient period.

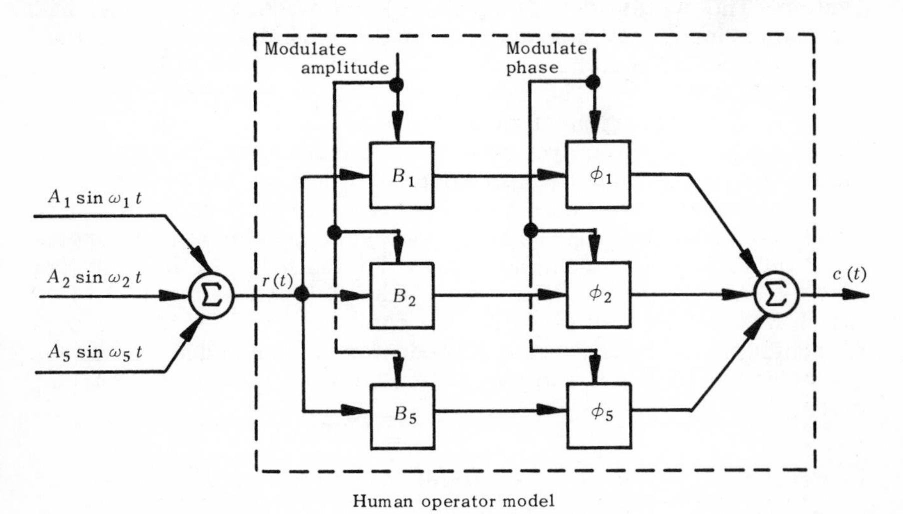

FIG. 12-9. Time-varying linear model.

Sheridan's model is shown in Fig. 12-9. The reference input was given by the sum of 5 sinusoids,

$$r(t) = \sum_{i=1}^{5} A_i \sin \omega_i t \tag{12.47}$$

and the operator's output was represented by

$$c(t) = \sum_{i=1}^{5} B_i(t) \sin [\omega_i t + \phi_i(t)] + n(t) \tag{12.48}$$

The operator is represented as a modulation device which adjusts the amplitude and phase of each of the component frequencies of the input.

Most of Sheridan's work was concerned with the theoretical and practical problems of estimating the time-varying amplitudes $B_i(t)$ and phase angles $\phi_i(t)$ by harmonic analysis techniques. The results are presented as polar plots of $B_i(t_j)$ and $\phi_i(t_j)$ for 5 values of time t_j during a period of 60 seconds after a sudden change in the task.

12.11 Automatic Model-Matching (Parameter-Tracking) Techniques for the Description of Human Operator Characteristics [34]

The previous paragraphs have emphasized the description of human operator characteristics using Fourier analysis and spectral analysis techniques. Model-matching techniques are based on the formulation of an assumed mathematical model of the unknown system. The system (human operator) and model are supplied with the same input function and the model parameters are adjusted until an adequate match between the behaviors of the model and the system is achieved. The use of this technique for the tracking of unknown but varying plant parameters in connection with a model-reference adaptive control system is discussed in Chapter 6. The emphasis in this and the following sections of this chapter is on the application of the method for determination of parameter values which minimize an appropriate function of the "matching error," i.e., the difference in outputs of model and man. In the Goodyear studies (Section 12.9) and in the study of Ward (Section 12.7) the match was determined visually. We shall now consider two methods of achieving a match between operator performance and model performance by automatic (computer) adjustment of model parameters.

General Formulation of the Model Matching Problem

Consider the diagram of Fig. 12-10. The controlled element G_E is assumed to be a known deterministic system having an output

$c(t)$ which is uniquely determined for $t > t_0$ by the state of G_E at $t = t_0$ and by $m(t)$ for $t > t_0$. The display represented by the summation symbol, forms the error signal

$$e(t) = r(t) - c(t) \tag{12.49}$$

The human operator, H, has an input $e(t)$ and an output $m(t)$. In addition to the input $e(t)$, the operator may be assumed to respond to additional inputs which are unknown. Consequently, he is considered to be a nondeterministic-element.

The basic problem is that of determining the parameters of M, an assumed mathematical model of H. (Since H is nondeterministic, a deterministic model can only describe his performance in a statistical sense. However, the background of previous studies assures us that human operator variability is small in situations where training time is adequate and the task is not considered difficult.) Both M and H receive $e(t)$ as an input signal in Fig. 12-10. The output of the model is denoted by $u(t)$ and the matching error is

$$\epsilon(t) = m(t) - u(t) \tag{12.50}$$

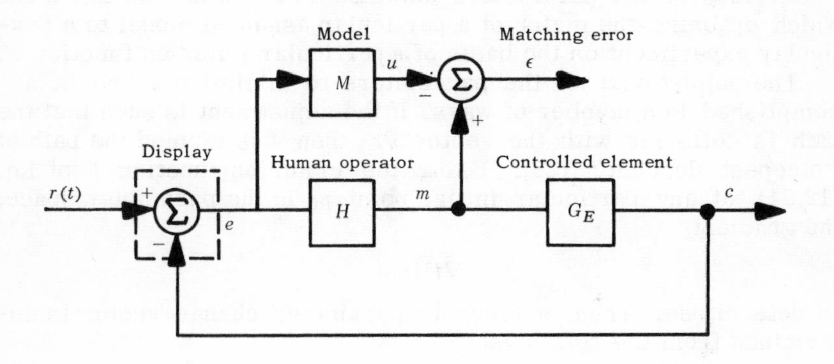

FIG. 12-10. Model-matching situation.

An ideal model is one which makes $\epsilon(t)$ zero for all t. We assume that the model can be represented by the system of nonlinear differential equations:

$$u_i = M_i(u_1, u_2, \ldots u_n;\ p_1, p_2 \ldots p_m;\ e, t), \quad i = 1, 2, \ldots n \tag{12.51}$$

over the interval $t_0 \leq t \leq t_1$. The initial state of the model is

$$u_i(t_0) = u_{i0} \quad i = 1, 2, \ldots n \tag{12.52}$$

and is assumed specified in advance. Equation (12.3) can be expressed in vector form as

$$\dot{\bar{u}} = \bar{M}(\bar{u}, \bar{p}, c, t) \tag{12.53}$$

where $\bar{u}$ represents the state of the model, $\bar{p}$ is a vector whose m components are the parameters of the model, and $\bar{M}$ specifies the form of the model.

To make $\epsilon(t)$ as nearly zero as possible, an appropriate function $F(\epsilon)$ is minimized. To convert the minimization problem to one of ordinary calculus, it is necessary to choose a criterion function which is an ordinary function of the parameters, such as

$$F = \int_{t_0}^{t_0+T} \epsilon^2(t)\,dt = \int_{t_0}^{t_0+T} [m(t) - u(\bar{p}, t)]^2\,dt \qquad (12.54)$$

where $\bar{p}$ is assumed to be constant during the integration interval T. Now, if F is a continuous function of $p_1, p_2, \ldots, p_m$, and has continuous first partial derivatives with respect to $p_1, p_2, \ldots, p_m$, then a necessary condition for F to be a minimum is

$$\overline{\nabla F} = \left(\frac{\partial F}{\partial p_1}, \frac{\partial F}{\partial p_2}, \ldots \frac{\partial F}{\partial p_m}\right) = \bar{0} \qquad (12.55)$$

The values of the parameters obtained at the minimum are those which optimize the match of a particular assumed model to a particular experiment on the basis of a particular criterion function F.

The adjustment of the parameters to minimize F can be accomplished in a number of ways. If the adjustment is such that the path is collinear with the vector ∇F, then it is termed the path of "steepest descent" [28]. Using the criterion function F of Eq. (12.54), at any particular initial point $\bar{p}_0$ in the parameter space, the gradient

$$\overline{\nabla F}(\bar{p}_0)$$

is determined. Then, a discrete parameter-change vector is determined from the relationship

$$\overline{\Delta p_0} = -K\,\overline{\nabla F}(\bar{p}_0) \qquad (12.56)$$

where K is a positive constant. The iterative implementation of Eq. (12.56) results in a discrete steepest-descent trajectory in the parameter space. The method has been applied to the determination of models of human operators performing one-dimensional compensatory tracking.

2.12 Continuous Parameter-Tracking Methods

If the computation interval T in Eq. (12.54) is allowed to approach zero, one obtains a time-dependent criterion function:

$$f = \epsilon^2(t) = [m(t) - u(\bar{p}, t)]^2 \qquad (12.57)$$

In contrast to Eq. (12.54), f is a functional which depends on the entire time history of the parameter vector $\bar{p}$ (see Chapter 6). Consequently, if the model is a dynamic system, it is not possible to define a meaningful gradient vector unless the parameters are fixed. An "approximate gradient" or "approximate steepest-descent" technique may be obtained in the continuous case if one assumes that the rate of adjustment of the parameters is sufficiently slow compared to the basic time constants of the model. Then, a continuous version of Eq. (12.56) may be written as

$$\frac{d\bar{p}}{dt} = -K\,\overline{\nabla F} \tag{12.58}$$

and implemented on an analog computer. The method has been applied to the determination of parameters of a "learning model" of a physical system by Margolis and Leondes [29]. Stability and convergence of continuous parameter-tracking methods have been demonstrated in certain cases by Donalson [30]. The analog implementation may lead to serious stability difficulties if a rapid rate of adjustment is attempted.

Approximate gradient methods, despite their difficulty, have been applied to the study of human operator models by Ornstein [31] and Wertz [32]. Ornstein instrumented equations of the form (12.58) on an analog computer and adjusted K until the system was stable. He used a model of the form

$$bu(t) + c\dot{u}(t) + d\ddot{u}(t) = k[e(t - 0.2) + a\,\dot{e}(t - 0.2)] \tag{12.59}$$

where k, a, b, c, and d were the parameters to be determined. The reaction time delay was assumed equal to 0.2 sec. The error criterion $f = |\epsilon|$ was used.

To improve the convergence of the parameter adjustment procedure, Ornstein performed an iterative adjustment of the model initial conditions, using the final values of the parameters p_i as initial conditions for the next run. The variability of his results was extremely large and the work suffers from the lack of analytical effort. However, it is significant since it represents the first attempt to apply approximate gradient methods to human operator studies.

Wertz [32] used steepest descent with the criterion $f(\epsilon) = \epsilon^2$, and adjusted only the parameters k and a in Eq. (12.59). Wertz was able to prove convergence and stability analytically.

Continuous model-matching techniques have also been applied to parameter determination in nonlinear models of the human operator. As an interesting illustration of the method, consider the problem of constructing a model of a human operator performing one-dimensional tracking with a bang-bang controller [35], as illustrated in the diagram of Fig. 12-11.

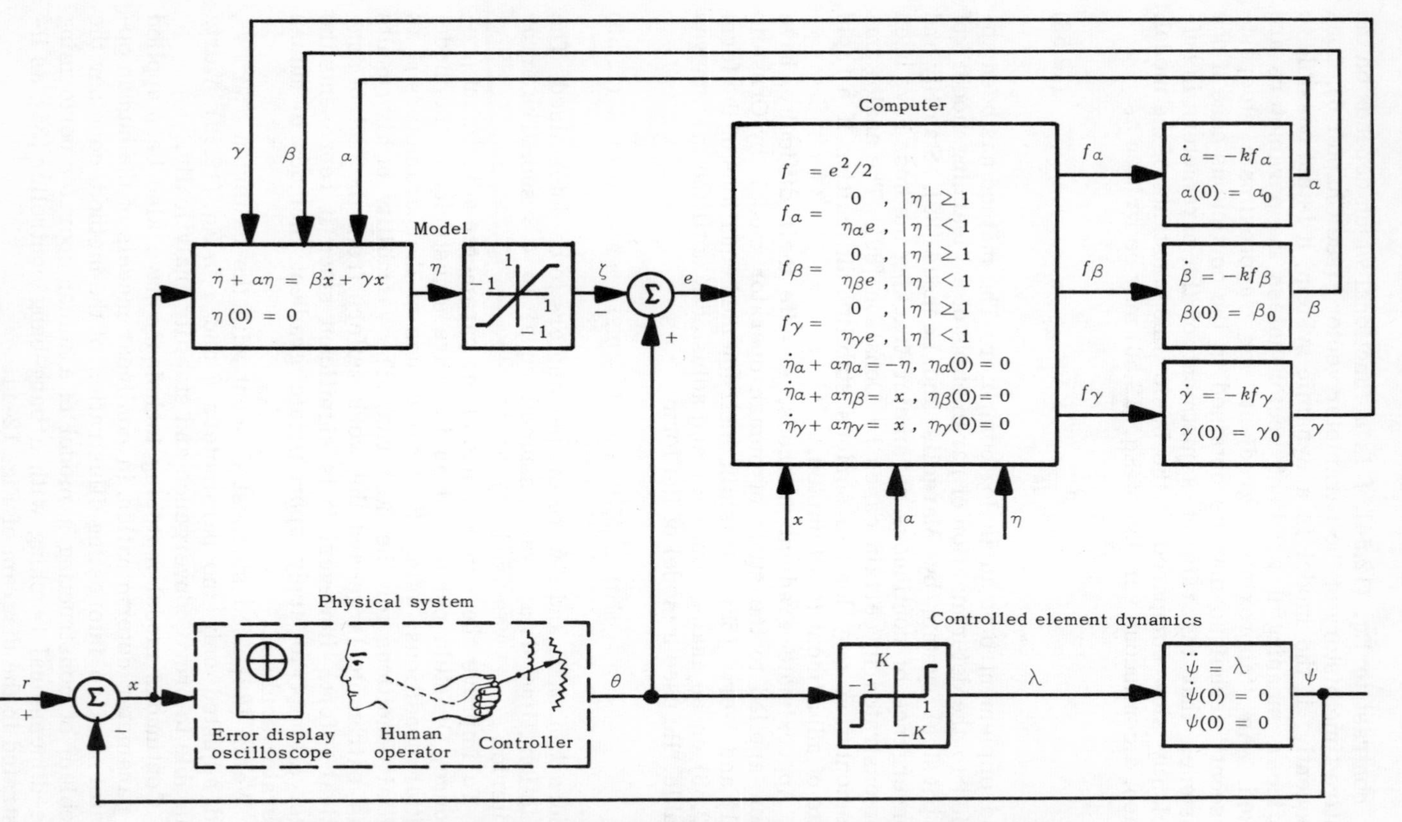

FIG. 12–11. Block diagram of parameter identification system with human operator.

The human operator indicated in the block diagram detects an error between the desired and actual system outputs and applies a correcting torque through an on-off controller to a vehicle defined simply by its inertia. No torque is exerted on the vehicle until the hand controller reaches the limit of its travel, at which point a fixed level of torque is applied. It is desired to construct a simple mathematical model representing the dynamic behavior relating the man's visual input to the controller output, and consequently to include the fact that the controller output is limited. The mathematical model is also indicated in the figure as a first-order linear differential equation followed by a limiter. The objective of the optimization problem is to adjust the parameters α, β, and γ continuously by means of an approximate gradient method. The criterion function $f = e^2$ is used, and the required computations are indicated in the figure. Following the computation of the approximate components of the gradient vector, the desired derivatives of the parameters with respect to time are obtained. They are integrated and used as inputs to the mathematical model. The typical results of this optimization problem, starting from arbitrary initial values for α, β, and γ are shown in Fig. 12-12.

The first channel in this figure represents the visual input the operator sees on the oscilloscope. The second trace is his manual controller output, and the third trace is the output of the mathematical model. It is desired to make trace no. 3 as close to trace no. 2 as possible with the preselected mathematical model and by a continuous parameter adjustment technique. The error or difference between the desired (operator's) output and the model output is shown in the fourth trace. Channel 5 represents the rms value of the error as indicated by a highly damped meter having a 10-second time constant. Channels 6, 7, and 8 show the recorded values of α, β, and γ. With reference to these tracking records, it can be noted that the outputs of the model and the operator are quite different during the first 100 seconds of the run, but become quite similar during the last 100 seconds. It can also be seen that approximately 100 seconds are required for the parameters to attain approximately steady values. The presence of high frequencies in the error after 100 seconds indicates the desirability of choosing a higher order model if possible, and illustrates the fact that this type of parameter optimization can only select the optimum value of parameters for the particular criterion and for the particular model chosen in the experiment. Attempts to obtain faster convergence resulted in instability.

12.13 Model-Matching with Orthogonal Filters

Elkind and Green [36] have represented the human operator by means of a linear model composed of a set of filters whose impulse

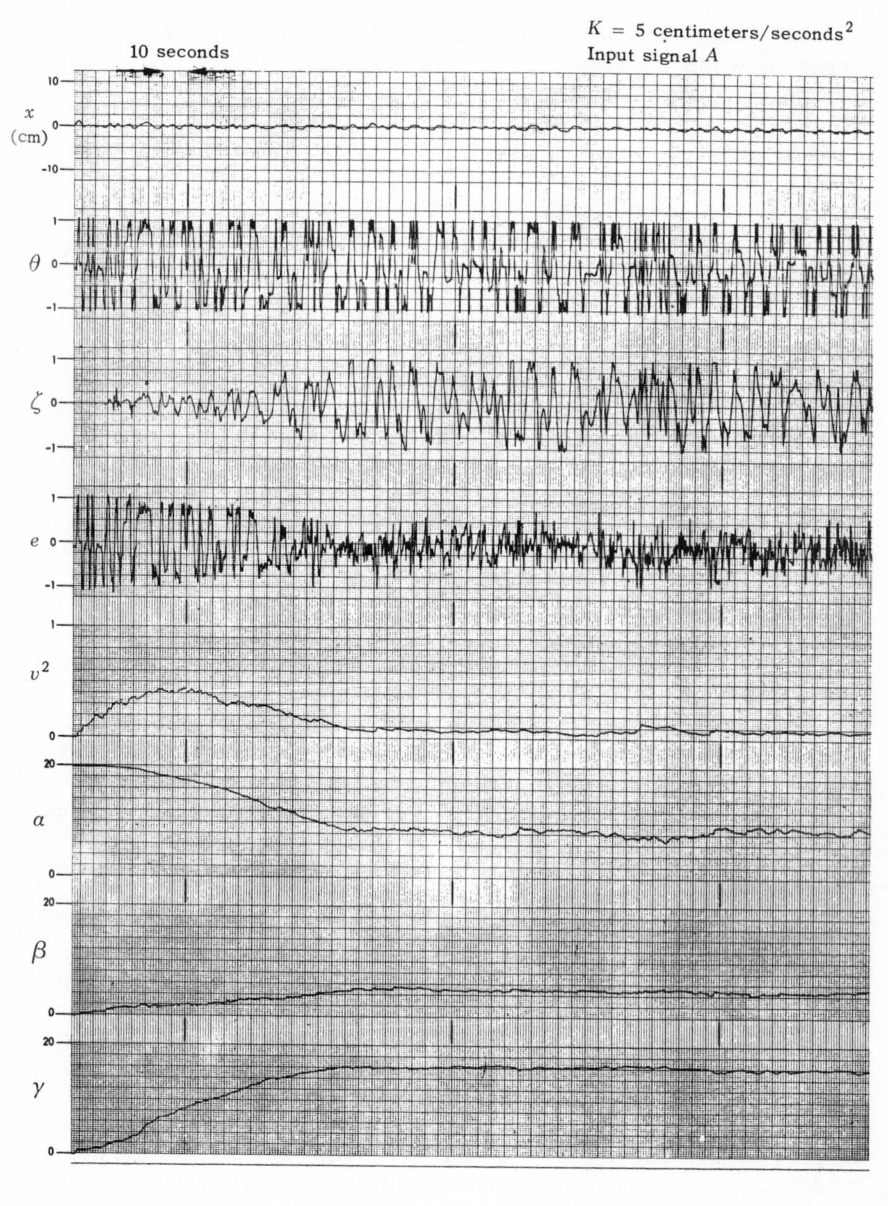

θ = Unit controller angle

ζ = Unit controller angle

e = Unit controller angle

v^2 = (Unit controller angle)^2

α = (Seconds)$^{-1}$

$\beta = \dfrac{\text{Unit controller angle}}{\text{Centimeters}}$

$\gamma = \dfrac{\text{Unit controller angle}}{\text{(Centimeter)(Seconds)}}$

FIG. 12-12. Tracking record.

responses are orthonormal. The filter outputs are weighted and summed to yield the model output. The weights are chosen such that the mean-square matching error, $F(\epsilon) = \overline{\epsilon^2}$ (over some particular time interval T) is minimized.

In terms of the notation of Fig. 12-10, the matching error is

$$\begin{aligned} \epsilon(t) &= m(t) - u(t) \\ &= m(t) - \sum_{j=1}^{k} b_j z_j(t) \end{aligned} \qquad (12.60)$$

if k filters are used. The mean square is

$$\frac{1}{T}\int_0^T \epsilon(t)^2 dt = \overline{\epsilon}^2 = \overline{[m(t) - u(t)]^2} \qquad (12.61)$$

The values of the coefficients b_j which minimize Eq. (12.61) are obtained by differentiating Eq. (12.61) with respect to each b_j and setting the result equal to zero. A set of k equations is obtained which can be solved for the b_j.

When a linear invariant model is assumed, a finite number of filters is selected from a complete orthogonal set and used to approximate the weighting function. With appropriate restrictions, the method can be extended to time-varying linear systems. Usually, the restrictions require that system parameters remain approximately constant during the measurement interval T.

Elkind and Green also suggest that, in the techniques of Wiener [37] and Bose [38], sets of orthogonal filters can be used to represent the outputs of nonlinear systems. However, no work has yet been reported on attempts to use the nonlinear theory for human operator models.

12.14 Conclusions

The human operator's response can be approximated by a linear model for many simple tasks. The parameters of this linear (continuous or discrete) model will depend on the particular task. Determination of parameters of quasi-linear models has been made by cross-correlation and model-matching techniques.

Future developments may require models of the human operator for tasks in which his operation is nonlinear and may also be time-varying. None of the methods outlined above are at present adequate for the synthesis of models and identification of their parameters in the time-varying, nonlinear case. Current studies in model-matching techniques using approximations to gradient methods offer some hope that there will be progress in this area in the next few years.

REFERENCES

1. Bates, J. A. V., "Some Characteristics of a Human Operator," *Journal of the I.E.E.* (England), v. 94, pt. IIA, pp. 298-304, 1947.
2. Craik, J., "Theory of the Human Operator in Control Systems," *British Jour. of Psych.*, December, 1947 and March, 1948.
3. McRuer, D. T. and Krendel, E. S., "Dynamic Response of Human Operators," *WADC Technical Report 56-524*, October, 1957.
4. McRuer, D. T. and Krendel, E. S., "The Human Operator as a Servo System Element," *Jour. Franklin Institute* 267:381-403, May, 1959 and 267:511-536, June, 1959.
5. Booton, R. C., "The Analysis of Nonlinear Control Systems with Random Inputs," *Proc. Symposium on Nonlinear Circuit Analysis*, 2:369-391, April, 1953.
6. Elkind, J. I., "Characteristics of Simple Manual Control Systems," *Lincoln Laboratory Report No. 111*, M.I.T. Lincoln Laboratory, Lexington, Mass., April, 1956.
7. Krendel, E. and Barnes, G. H., "Interim Report on Human Frequency Response Studies," *WADC Technical Report TR 54-370*, June, 1954.
8. Broadbent, D. E., *Perception and Communication* pp. 268-296, New York: Pergamon Press, 1958.
9. Bekey, G. A., *Sampled Data Models of the Human Operator in a Control System*, Ph.D. Dissertation, Department of Engineering, University of California, Los Angeles, January, 1962 (also ASD Technical Report 62-36, February, -962).
10. Tustin, A., "The Nature of the Operator's Response in Manual Control and its Implications for Controller Design," *Jour. Institution Elec. Engrs.* (London) 94 (IIA):190-202, 1947.
11. Ragazzini, J. R., "Engineering Aspects of the Human Being as a Servo-mechanism," unpublished paper presented at the American Psychological Association Meeting, 1948.
12. Licklider, J. C. R., "Quasi-Linear Operator Models in the Study of Manual Tracking," pp. 171-279 in *Developments in Mathematical Psychology*, ed. by R. D. Luce, Glencoe, Illinois: The Free Press, 1960.
13. Phillips, R. S., "Manual Tracking," in *Theory of Servomechanisms*, ed. by H. M. James, N. B. Nichols and R. S. Phillips, pp. 360-368, New York: McGraw-Hill, 1947.
14. Mayne, R., "Some Engineering Aspects of the Mechanism of Body Control," *Electrical Engineering*, 70:207-212, March, 1951.
15. Hyndman, R. W. and Beach, R, L., "The Transient Response of the Human Operator," *IRE Trans. on Medical Electronics*, PGME-12:67-71, December, 1959.

16. Ellson, D. G., Gray, F., et al., "Wavelength and Amplitude Characteristics of Tracking Error Curves," *AAF-AMC Engr. Report TSEAA-694-2D*, Wright Field, Ohio, April, 1947.
17. Walston, C. E. and Warren, C. E., "Analysis of the Human Operator in a Closed-Loop System," *Research Bulletin* 53-32, Human Resources Center, ARDC, Ohio State University, August, 1953.
18. Noble, M., Fitts, P. M. and Warren, C. E., "The Frequency Response of Skilled Subjects in a Pursuit Tracking Task," *Jour. Exp. Psych.* 49, 249-56, 1955.
19. Russell, L., *Characteristics of the Human as a Linear Servo Element*, M. S. Thesis, Massachusetts Institute of Technology, May, 1951.
20. Booton, R. C., "The Analysis of Nonlinear Control Systems with Random Inputs," *Proc. Symposium on Nonlinear Circuit Analysis*, 2:369-391, April, 1953.
21. Jackson, A. S., "Synthesis of a Linear Quasi-Transfer Function for the Operator in Man-Machine Systems," *WESCON Convention Record*, pt. 4, 2:263-271, 1958.
22. North, J. D., "The Human Transfer Function in Servo Systems," in *Automatic and Manual Control*, ed. by A. Tustin, 473-502, London: Butterworths, 1958.
23. Ward, J. R., *The Dynamics of a Human Operator in a Control System: A Study Based on the Hypothesis of Intermittency*, Ph.D. Dissertation, Department of Aeronautical Engineering, University of Sydney, Australia, May, 1958.
24. Diamantides, N. D., "Informative Feedback in Jet-Pilot Control Stick Motion," *AIEE Applications and Industry*, 243-39, November, 1957.
25. Diamantides, N. D., "A Pilot Analog for Airplane Pitch Control," *J. Aero. Sciences*, 25:361-71, June, 1958.
26. Goodyear Aircraft Corp., "Final Report: Human Dynamics Study," GAC Report GER-4750, April, 1952.
27. Sheridan, T. B., "Time Variable Dynamics of Human Operator Systems," *MIT Dynamic Analysis and Control Lab Report* AFCRC-TN-60-169, Cambridge, Massachusetts, March, 1960.
28. Tompkins, C. B., "Methods of Steep Descent," pp. 448-79 in *Modern Mathematic for the Engineer*, ed. by E. F. Beckenbach, McGraw-Hill Book Co., 1956.
29. Margolis, M. and Leondes, C. T., "A Parameter-Tracking Servo for Automatic Control Systems," *IRE Trans. on Automatic Control*, AC-4:100,111, 1959.
30. Donalson, D. D., "The Theory and Stability Analysis of a Model Reference Parameter Tracking Technique for Adaptive Automatic Control Systems," *Ph.D. Dissertation*, University of California, Los Angeles, May, 1961.
31. Ornstein, G. N., "Applications of a Technique for the Automatic Analog Determination of Human Response Equation

Parameters," *Report NA61H-1,* North American Aviation, Columbus, Ohio, January, 1961 (also available as Ph.D. Thesis, Ohio State University, Department of Psychology, 1960).

32. Wertz, H. J., "A Learning Model to Evaluate and Aid Human Operator Adaptation," paper presented at IRE International Congress on Human Factors in Electronics, Long Beach, California, May, 1962.
33. Bekey, G. A., "The Human Operator as a Sampled-Data System," *IRE Trans. on Human Factors in Electronics*, HFE-3: 43-51, September, 1962.
34. Bekey, G. A. and Humphrey, R. E., "Review of Model Matching Techniques for the Determination of Parameters in Human Pilot Models," Space Technology Laboratories Technical Report 9865-6002-MU000, Redondo Beach, Calif., Nov., 1962.
35. Humphrey, R. E. and Bekey, G. A., "A Technique for the Determination of Parameters in a Nonlinear Model of a Human Operator," Space Technology Laboratories Technical Report 9865-6003-MU000, March, 1963.
36. Elkind, J. I. and Green, D. M., "Measurement of Time-Varying and Nonlinear Dynamic Characteristics of Human Pilots," *ASD Technical Report 61-225*, December, 1961.
37. Wiener, N., *Nonlinear Problems in Random Theory,* John Wiley and Sons, New York, 1958.
38. Bose, A. G., "A Theory of Nonlinear Systems," Tech. Report No. 309, Res. Lab. of Electronics, Mass. Inst. of Tech., Cambridge, Massachusetts, 1956.

13

Applications of Modern Methods to Aerospace Vehicle Control Systems

R. A. NESBIT, SCIENTIST
BECKMAN INSTRUMENTS, COMPUTER OPERATIONS
SANTA MONICA, CALIFORNIA

13.1 Introduction

There are many significant control problems associated with the design and operation of the class of vehicles referred to as aerospace vehicles. These control problems have some special characteristics which may be utilized in the design of the controllers. The utilization of the "special facts" about aerospace vehicles is very important to the technology of these controllers. Of course, there are also other problems common to more general classes of control systems. It is the purpose of this discussion to list some of the problems and indicate the approaches to these problems which have been found profitable. A complete derivation of the special characteristics of these vehicles is not attempted here, but rather a description of their general significance in controller design.

Each time a new type of control problem is approached, certain assumptions are made in order to apply a mathematical or logical analysis. To obtain the "right" conclusions, the premises must be right and the logical analysis valid. This is not to say that it is not profitable to make simplifying assumptions, but rather to place these assumptions in proper perspective. There are certain ideas concerning the control of aerospace vehicles which can properly be called theories. The most important of these is the use of Newtonian or post-Newtonian mechanics for the description of the vehicle motion. The fact that the motion of all the vehicles in the class being discussed can be predicted through the use of Newtonian mechanics is the main unifying feature. Such vehicles include aircraft, airships, rockets, gliders, satellites, ballistic missiles, and lifting reentry vehicles, as well as others with similar equations of motion and many common characteristics, such as submarines and hydrofoils.

The following discussion is an attempt to identify the important technical problems and ways in which they have been solved. In the design of any control system it is necessary to make some assumptions about the process being controlled. The first assumption is that the process is controllable; the system must be well enough built to perform the tasks required of it. The obvious requirement is often at the heart of technological limitations on a system, and places an upper bound on the value of all controllers including "adaptive" controllers. It is also important to build controllers which utilize the performance capability of the vehicle.

The controllability of a system is closely linked with the available instrumentation and data-processing components and actuators. Another important problem which prevades all aspects of vehicle design is obtaining sufficient reliability. The problem is not simple, as the following example from the design of a rocket booster illustrates. Without aerodynamic fins on the vehicle, an autopilot failure would be catstrophic. But would the addition of fins improve the chances of successful flight? Such questions are typical of reliability problems.

13.2 General Properties of the Dynamic Equations

One unifying feature of the control of vehicles is the fact that the classical six-degrees-of-freedom equations of motion [1] describe part of the dynamic process. The gross problem of control is the appropriate application of forces and torques to perform the required maneuvers [2, 3, 4].

The forces and torques are supplied either by interaction with the surrounding media and potential fields, or by reaction devices. These forces are often related to the geometric variables of control-surface shape and position through the complicated dynamics of fluid flow; the analytical determination of the complete dynamics of these systems is quite difficult. Wind tunnels and other experimental apparatus are often used to determine the torques and forces resulting from interaction with surrounding media. (It is noteworthy that the Wright Brothers constructed a wind tunnel in their bicycle shop.) Another method of analyzing this type of problem in the preliminary design stage is to solve for the aerodynamic pressure distribution on a general rather than a special purpose computer (i.e., the wind tunnel). This method is profitable when sufficient accuracy is obtainable at less cost than an experimental program. The efficient solution of the Navier-Stokes equation on a general purpose computer is a formidable task. One feature of a properly programmed computational method is the ability to make modifications of the vehicle geometry easily.

Outside the atmosphere, forces on vehicles are very small, and torques which are usually neglected (such as gravity gradient and solar radiation pressure) become important. Because of this dynamic pressure change, it is reasonable to subdivide the class of aerospace vehicles into spacecraft and aircraft. Of course, some vehicles operate in both regimes, but even in these cases the mode of control is usually changed as the dynamic pressure changes.

The equations of motion are usually simplified for any analytical study. The first simplification is usually made by linearizing about a nominal trajectory, and the second by assuming that the linear equations thus obtained can be approximated by constant coefficient equations [2]. Another simplification is made possible because of the different time responses of the path and rotational equations. This difference in time response is clearly a characteristic of the actual system, not just of the linearized model, and is manifested by a difference in the natural frequency of the linearized model. For instance, for a rigid body pitching frequency of about 1 radian/second, the normal force root might be about 0.015 radian/second. This frequency separation gives rise to the separation of the linearized navigational control systems from the linearized attitude control systems. Considered separately, the navigation problem can include a reasonably simple dynamic representation of the autopilot-airframe combination. Most of the applications of control methods are made on the basis of such simplified models, and are then checked by computer simulation and in the actual system.

The aerodynamic forces on the vehicle are often described in terms of components along the velocity vector and normal to it. The component in the direction of minus velocity is called drag and the component normal to the velocity vector is called lift. The aerodynamic efficiency of a vehicle is often described in terms of the maximum lift-to-drag ratio possible with the vehicle. This parameter is one of the distinguishing features of aircraft designs. There are many other competing features of a vehicle design, and the most economical design does not always have the maximum lift-to-drag ratio. For reentry vehicles, another parameter is the specific weight of the heat protection system, and the cost of the heat protection system is also dependent on the vehicle geometry. The subsonic and hypersonic performances may require different configurations. The X-15 has a subsonic maximum lift-to-drag (L/D) of around 4, an airline transport around 10, and high performance gliders around 35. Power-off horizontal landing is possible with an L/D of 2 or somewhat less. Another important parameter of the dynamics is wing loading. Wing loading is defined as the weight of the vehicle divided by the reference area. The dynamic pressure $\left(\rho \frac{V^2}{2}\right)$ is used to scale the aerodynamic lift and drag. The range of dynamic pressures for which the aerodynamic forces and moments are important is three or four orders of magnitude wide. The magnitudes of forces and torques are roughly proportional to the dynamic pressure.* In some vehicles, the change in dynamic pressure is very rapid; in these cases, the constant coefficient approximation to the dynamics is of questionable value.

The aerodynamic phenomenon of stalling or flow separation is an important factor which limits the performance of aircraft. This phenomenon occurs when the angle of attack exceeds a particular value, and thus a reasonable way to avoid stalling is to limit the angle of attack. If one wing stalls while the other is unstalled, the aircraft executes a maneuver called a spin. Spins and stalls are usually accompanied by rapid loss of altitude. Recovery from either maneuver is usually possible if there is sufficient altitude, but satisfactory operation of most aircraft does not require performance of these maneuvers.

The aerodynamic torques on a vehicle also depend on the orientation of the vehicle with respect to the flow field. For a nonrotating vehicle, these torques must balance near the trim point, which depends upon the particular control-surface deflections. If small angular displacements are made from the trim point, the resulting moments may be in a direction to move the vehicle toward the trim point. This situation is called static stability. Dynamic stability means that the linearized equations of motion have stable roots near the trim point. Part of the aerodynamic design problem is to obtain a reasonable amount of static and dynamic stability. In some cases, such as boost rockets, the vehicle is statically unstable, and a properly functioning autopilot is required for safe flight. The partial derivatives of the moments and forces with respect to the angles and angular rates are known as stability derivatives [2].

The stability derivatives change with mach number. This is another way of saying that the forces and moments are not strictly proportional to dynamic pressure. This mach dependence is especially strong near mach 1.

The structural flexibility of the vehicle gives rise to another fairly common dynamic property of this class of systems [5, 6, 7]. This dynamic property has been neglected in the design of many autopilots, but even when neglected, its presence has prevented the indiscriminate use of high gain with its accompanying advantages. Structural modes and frequencies are approximated by numerical solutions of the beam equations. The structural damping of these

*Flow separation and Mach effect cause the functional relation to vary from proportionality.

modes is not easily predicted in structures with no intentional damping mechanism. In many cases the structural modes are lightly damped. The design of boost rocket autopilots involves prevention of instability in the structure. The structural design and controller design are very strongly related in this case. In analysis of satellite attitude control using passive stabilization, normal-mode damping is required.

Another important characteristic of rigid body dynamics which affects aircraft performance is "pitch up." This effect is easily visualized if one considers the rotation of a dumbbell about a nonprincipal axis; the term "pitch up" is used to describe this phenomenon in aircraft because it has the effect of raising the nose of the vehicle. It occurs during rapid banking maneuvers made about a nonprincipal axis (usually at high angles of attack). The pitch up torque is proportional to the square of the bank rate; the pitch up tendency must be balanced by control torque, and the limitations on the available pitch control torque imply bank rate limitations. It is possible to use nonlinear feedback to coordinate the maneuver.

On powered aircraft, the thrust force is used to augment the aerodynamic lift and alleviate the drag. The control of powered vehicles includes control of the power plant. Power plant control systems may have thrust-direction control, thrust-magnitude control, or both. The power plant may also have voltage, fuel flow, pressure, temperature, and speed control systems.

The control actuators and sensors also have dynamic characteristics which must be considered in the analysis and solution of the control problems.

In summary, the basic dynamics of aerospace vehicles are described by the six-degrees-of-freedom equations of a rigid body. The aerodynamic forces and moments are related to the atmospheric density, the vehicle orientation, the velocity, and the control surface deflections. In addition to the basic dynamics, the structural modes of vibration must be considered. The thrust generation, sensors, and actuators involve additional dynamics which are dependent upon the particular units utilized.

In terms of the general dynamic properties of these vehicles, some general control problems can be stated. The reader is referred to other texts for a more complete description of the dynamic characteristics [2, 3, 4].

13.3 General Problems of Control

The problems of dynamically controlling a vehicle, as described above, can be subdivided into attitude control, trajectory control, power plant control, and support system control. Attitude and trajectory controls can sometimes be considered independently. In these cases, independent consideration is possible because of the dynamic property, already mentioned, that the response times of these controls differ from each other by at least one order of magnitude. Sometimes, in satellites, the entire control problem is one of providing attitude control [8].

The attitude control system requires attitude sensors and torque actuators. The sensors can be provided by a human pilot or by suitable physical devices. In the case of either the human pilot or the physical device, one must account for the measurement and actuation dynamics.

It is important to realize that the human operator may have other managerial functions besides attitude sensing and servo actuation. Where the pilot must rely entirely on data supplied by primary sensors, it is of questionable value to include him and his dynamics in the loop. One of the important problems of vehicle design is providing an economic balance between automatic control, manual control by the pilot, and information display. This problem is just one example of the more general problem of efficient utilization of the man-machine combination.

The function of the attitude control system, manned or automatic, is to maintain the attitude of the vehicle with satisfactory accuracy in the presence of disturbances. In the case of aircraft, the system must operate over a wide range of dynamic pressures and over changes in aerodynamic characteristics. It is only due to the properties of feedback that the system can be made insensitive to these changes.

In boost-rocket attitude control, the main control variable is thrust direction. The aerodynamic configuration is usually unstable. The sensors are gyros which respond to structural bending as well as attitude. The control problem is one of obtaining satisfactory attitude control over a wide range of bending-mode frequencies and shapes without exciting the structure. The dynamics also change due to changes in quantity of propellant and dynamic pressure. The sloshing of propellant is also part of the attitude dynamics [7].

The satellite attitude control problem is to obtain the desired accuracy with as little weight and as long a life as possible. In the study of this problem the dynamics may be very simple as in the case of one relatively rigid body. However, if long wires and booms are attached, the model analysis of the orbiting structure is required. This appendages are sometimes added to perform the attitude control function "passively" using gravity gradient torques.

The trajectory controls include a wide variety of different maneuvers. These maneuvers include landing, homing, rendezvous, area navigation, reentry energy management, orbit transfer, station keeping, and orbital injection. The system used to perform the trajectory control function is sometimes called a guidance system. On boost vehicles, the guidance system may use an inertial platform and a special purpose digital computer. The sensors for the trajectory control may use radio signals, radar systems, radio beams, light beams, and maps to aid in the determination of position.

A safe, comfortable landing is the function of a landing control. During approach for landing, the wind gusts and wind shear disturb the trajectory. The touchdown must occur on the runway at a low sink rate and with a reasonable attitude. There is a competitive trade-off between trajectory and attitude errors, and the relative importance of the two errors changes with distance from touchdown [9, 10, 11, 12].

The reentry trajectory must satisfy heating and acceleration constraints [13]. It should also utilize the performance capability of the vehicle to obtain the range and crossrange potential. The density of the atmosphere is variable and the aerodynamic characteristics of the vehicle may depend on the amount of erosion of the heat-protecting surface. If the vehicle is to utilize its maneuverability, the trajectory control must have the capability of selecting or following a variety of paths.

Orbital injection requires specification of a trajectory to give efficient fuel utilization. The terminal or burnout position and velocity must be very accurately controlled. This control requires accurate measurement. Orbit transfer has many features similar to orbital injection; efficient fuel utilization and accuracy are still problems [14, 15].

In addition to providing a controller which can perform the dynamical maneuver, there are many problems common to all control designs. In some cases the cost of system failure is measured in human life. The analytical approach to the reliability problem has fundamental limitations, since meaningful conclusions must be based on meaningful data. The experimental approach to reliability has fundamental limitations because it is not always economically feasible to test a representative sample. The "optimization" of reliability is not simple to formulate. One of the difficulties is description of the alternative design choices; another is evaluation of the types of failure and their likelihood. One irrational difficulty which may face a system designer is a requirement for a one-in-ten-million failure rate, with no means to verify or determine this with reasonable confidence. Parts counts and other similar operations sometimes give reasonable estimates, but they do not tell the whole story.

The other obvious requirements of these controllers are minimum cost, weight and power dissipation with maximum performance, versatility, and accuracy. These conflicting requirements are typical of "real world" oriented problems which require creative engineering, managerial excellence, and capital investments for their solution.

The new problems of aerospace vehicle control derive from attempts to improve performance of both the vehicle and its control system, to perform new maneuvers such as landing with zero visibility and landing on the moon, and to improve the accuracy of the basic sensors. Each significant improvement in device technology reopens the question of system improvement, and the system requirements motivate component development.

13.4 Methods of Analysis and Solution of Problems

The solution of a control problem is that system which competes favorably for the accomplishment of the control task. The market provides the final competition and integrates the economic factors. The analysis of the problem is the logical procedure used to study a proposed solution. In the other chapters of this text, various methods of analysis are discussed. These methods and others may be applied to the analysis of controllers. This section will attempt to survey the methods of analysis which have been profitably applied to the solution of the aerospace vehicle control problems.

It has already been mentioned that linearization plays an important role in the dynamic analysis of the controllers. Linear analysis of very complicated systems is made feasible by intelligent use of automatic computations. These computations can provide rapid and inexpensive analysis of the system dynamics and of the effect of compensating feedback [2, 16].

One type of useful computer program calculates the coefficients of the linear differential equations from the vehicle parameters. With this program, it is possible to determine the vehicle characteristics efficiently at different flight conditions, and to study a variety of vehicles. Other useful programs perform analyses of sets of linear constant coefficient equations. The characteristic roots or poles of the equations are determined by factoring the characteristic determinate. Transfer functions are computed by applying Cramer's rule to the transformed equations. The frequency response of transfer functions is also easily computed. Inverse transform and root locus routines are also available. These computational aids compete with the spirule and desk calculator, and offer an advantage on certain problems [16].

The solution of the control problem is usually conceived on the basis of the linear approximation. There are many useful techniques for the design of linear controllers. The concept of sensitivity introduced by Bode [17] and extended by Horowitz [16] explains the purpose of feedback quantitatively. The disturbances are to be rejected while the input-output relation is constrained to satisfy the requirements of the controller. These requirements are often stated in terms of bandwidth, power gain, time response, overshoot, and error maximums. They are often related by assuming a pair of "dominate" roots; that is, the dynamics are approximated by a second-order system. This approach is sometimes referred to as the "dominant operator" approach to system design.

The closed-loop bandwidth required of the overall control channel depends upon the type of signal it is expected to follow. Some of the gain and bandwidth capability of the actuator is required for disturbance rejection. For further discussion of a linear design technique which explicitly considers transmission and sensitivity, the reader is referred to Horowitz [16]. Overshoot requirements on the linear model become more important when the nonlinearities are considered. For example, overshoot could initiate a stall in an otherwise safe maneuver.

Linear constant coefficient systems can be designed by minimizing quadratic cost functions. This technique may require estimates of the disturbance statistics or it may trade off control gain and system accuracy. Due to the high order of most vehicle equations, the sensitivity approach may be more expedient. However, in satellite attitude control or orbit correction problems, the dynamics are simple enough to allow optimization of error cost versus fuel cost. Even in these problems the solution usually involves constraints on the system mechanization, and the optimal controls are fairly obvious physically if not mathematically [18, 19].

To summarize, the ability to design complex linear systems to meet specifications on bandwidth, time response, and sensitivity *is* required for the solution of vehicle control problems. Competitive superiority should be based on this principle.

The foregoing discussion mentions the usefulness of automatic computation for rapid and economical linear analysis. General purpose computers are indispensible for simulation of the nonlinear models of controllers and processes. These simulations are often economical ways to experiment with concepts and components. In some cases, only special purpose experiments are adequate for verification of a hypothesis. If a concept is reduced in practice to the form of hardware, the final test occurs through actual performance of the control task. It is important to balance the theoretical analysis, the computational analysis, and experimental verification. In complicated systems, theoretical and computational analyses are often much less expensive than construction and test. Unfortunately, this fact is sometimes used to justify disembodied analysis.

It is often worthwhile to "check" the design by using a relatively complete simulation. The simulation may include dead zones, hysteresis, limiters, coulomb friction, or other nonlinear dynamics. And the simulation may also be used to determine the controllability and performance limitations of the system model. Nominal trajectories may also be obtained from the complete simulation of the equations of motion. These nominal trajectories may then be used for planning and for obtaining the linearized perturbation equations.

Modern controllers often contain computer elements and data processing equipment. The data processing is used to enhance the signal-to-noise ratio of the sensor output. By using digital signal transmission and computation, the sensitivity of the transmission to variation of amplifier parameters is reduced. The development of microcircuits and improved sensors makes it possible to implement small reliable information processing elements. In some cases general purpose computers are valuable as one component of the complete controller. Thus computers serve both to facilitate the analysis of controllers and to implement them. It may even be profitable to analyze proposed special-purpose digital controllers on general-purpose digital computers before they are built.

The efficient utilization of the computer is also a man-machine problem. The man-machine problem is old and ever important. The man's desires may or may not be rational, but in either case the purpose of the machine is to serve him. In some cases it is cheaper to find the right man than to change the machine, and in other cases a simple modification of the machine is effective. One example of the importance of "psychology" is the retention of control cables in very long, flexible aircraft. Even though these cables are troublesome, heavy, and expensive, they are necessary because many pilots think they are more reliable than any possible alternative.

Obtaining representative models of the human operator is a difficult task. For simple tracking tasks, the human operator has some time delay, and can provide some data processing. For more complex tasks such as system-monitoring and decision-making, the available representations are much less than adequate. Most tasks require practice, and if man is to be useful in machine operation, it must be assumed that he has some minimal skill. The history of aviation is possible only because of the superb skill of human operators facing superhuman tasks. However, man's capabilities are limited, and some tasks can be done better by machine. As with many competitive situations, much time is spent extolling the virtues of one side or the other instead of attempting to find the most profitable combination. Some of the "Man in the loop" advocates make the assumption at the outset that it is desirable to have the particular task performed by the human. Others who assume that the system must be completely automatic find that the solution may be expensive and less reliable. As new equipment is developed, the division of labor must be reevaluated. The main techniques for analyzing man-machine interaction involve experiments under actual and simulated conditions. Elaborate simulations which utilize computers and visual models provide realistic situations for pilot training and equipment test. Effective displays and controls simplify the man-machine interaction, and improved sensors extend the capability.

The question of improving the sensors is inherently coupled to the analysis of the statistical properties of the signal and noise. The most useful technique for this analysis has been propagation of variance or mean-square error analysis. In signal-processing applications, matched filters and Wiener filters have extended the capabilities of separating signal from noise. Improvement of signal processing may well result from the implementation of the time-varying filters derived by Kalman [19] or improvements thereof.

A typical method of combining two sensors with different frequency capabilities is to use complementary filters and combine the output. This has been done for altitude rate by combining the pressure rate pneumatic sensor output with the output of an inertial velocity meter.

The propagation of variance is accomplished for sensors with bias errors by integration along the nominal trajectory to find the influence functions. These bias errors are not presently included in the Kalman formulation in a useful way. Guidance system error analyses are usually made by mapping the variance of the accelerometer, gyro bias, and scale factor errors into the variance of the position and velocity error. Again, automatic computation facilitates the analysis.

In general, optimization requires some choices, some prediction of the consequences of these choices, and a way of ordering the consequences with regard to their relative preference. In some problems of vehicle dynamics, there is a wide choice of trajectories, a reasonable criterion of performance, and a known relation between the criterion and the trajectory. For these problems, it may be profitable to utilize the mathematical theory of optimization. Fuel minimization has been one performance criterion. Sometimes the trajectory is parameterized to allow a simple expression of its form. Variation of the parameters is then identified with the corresponding variation in performance. If there are sufficient possibilities in the class of parameterized trajectories, the performance can be made reasonably close to the optimum. This reduction of dimensionality, sometimes known as the Rayleigh-Ritz method, is often used for obtaining functional optimization. Through the calculus of variations, one can sometimes solve useful problems [20, 21]. The Pontryagin extension of the Weierstrass condition (the Maximum Principle) is sometimes useful for the solution of control problems [14]. In some cases, the optimum solution may be found by a direct consideration of the problem; this approach is called the direct method. For most control problems, the criterion cannot be directly related to the possible choices. Fuel minimization in a satellite may be unimportant compared with system simplicity, rendering unprofitable analysis or implementation of a fuel optimum system. It may be much more profitable to invest in an improved sensor or actuator rather than optimize the performance of those which are marginally adequate. Once again, the importance of considering the "real" problem is apparent.

One method of functional optimization proposed by several people is the functional steepest descent [22, 23]. This method uses the gradient of the functional to modify a given control choice. The sequence thus obtained may converge to a local minimum. If the cost is a nonnegative quadratic functional, then the convergence can be shown, and the minimum value will be obtained. The numerical difficulties of this approach should not be underestimated. However, in one aircraft climb problem, Bryson and Nesline reported almost a 45% reduction in the time required when a steepest descent computation of the trajectory program was made. The optimum solution contained a zoom which was novel to the usual program.

In many vehicle problems, the binary difference (it works or it does not work) makes analytical optimization less useful. If a heat protection system is adequate for the extremes of vehicle performance, there is very little payoff for optimizing the trajectory with respect to heating. It is much more profitable to simplify the control system.

To summarize, aerospace vehicle control problems are analyzed by linearized design, by computer simulation, by mathematical and logical analysis, and by experimentation. They are solved by a combination of innovation, capital investment, and entrepreneur management. The concepts of stability, optimization, random processes, and system modeling can be useful for systems analysis. If the total problem is considered, these concepts may provide a basis for profitable innovation and solution of the problem.

13.5 Specific Examples

a. Aircraft Attitude Control

As mentioned previously, the high performance capabilities of modern aircraft give rise to wide variations in flight conditions. The "gain" of the control elements are roughly proportional to the dynamic pressure, and this quantity may vary over a range from about 1000 lb/ft^2 to about 10 lb/ft^2. Many systems cannot operate satisfactorily over this range of gain, and some compensatory change in the autopilot gain is usually required. One method of implementation is to sense some variable proportional to the gain; this method requires extensive instrumentation. There are several ways of arranging system variables proportional to gain.

The advantages of high gain in feedback systems are well known but often overlooked in favor of "gain margins." A significant improvement in system performance can be obtained by using one of the "adaptive" gain adjustment systems. These systems maintain the gain as high as possible without runaway instabilities.

The assumption about the dynamics in the application of these systems is that there is some range of system gain for which satisfactory operation is possible. The adaptive part of the system attempts to keep the gains in this range. A certain amount of knowledge of the system is required to assure that the gain-change logic will not result in an incorrect adjustment.

Systems of this type have been flown on high performance aircraft with satisfying results. The systems used have been built by Minneapolis-Honeywell and General Electric. The Honeywell system maintains a small limit cycle in the autopilot and uses the limit cycle properties to adjust the gain [6, 24, 25, 26].

As an example of a system of this type consider the following:

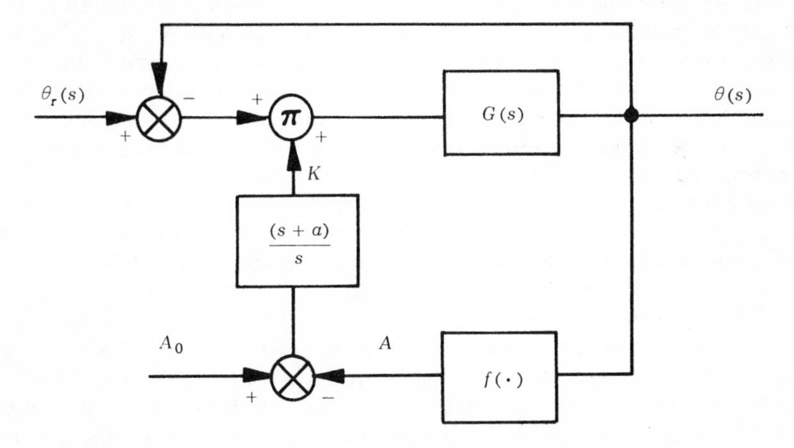

FIG. 13-1. Gain adaptive system.

The functional $A = f(\theta)$ is the amplitude or frequency of the limit cycle, and A_0 is the reference. The stability of the gain adjustment loop must be assured. The problem of designing the gain adjustment circuit is quite similar to the design of a stable oscillator. The basic feedback control system is designed as though K were a constant, and a computer check of the coupled system is made.

b. Automatic Landing

Automatic aircraft landing poses an interesting problem for the control engineer. Several systems have been proposed and developed. (The Autoland System built by Lear/Siegler, Inc. and Sud Aviation is being installed in the Italia Caravel fleet.) Several methods of realizing an automatic landing system are possible. Some of these methods require little or no additional equipment at the airport, and utilize the ILS systems now available. Another method adds a scanning glide slope beam which originates about 2000 feet down the runway.

For carrier-deck landing, one might consider a data link system. The control commands to the aircraft may be generated in a shipboard computer which processes radar data, shipboard inertial measurements, and shipboard wind velocity measurements. The commands would be transmitted to the aircraft autopilot. This system has an interesting mixture of control and information theory problems. There is a possible trade-off between the bandwidth required by the data link and the complexity of the autopilot.

Another obvious requirement of a landing system is reliability. The complexity of the carrier landing system mentioned above makes the reliability problem difficult, but it should be possible to make an automatic landing more reliable than a human pilot's.

Along with the usual noise in the sensors and variations in aircraft parameters, the system must contend with gust and turbulence along the flight path. These inputs are quite significant due to the low altitude of the aircraft. The usual approach speed is about 1.3 times the stall speed, and the maneuvers called for by the control system must be kept from stalling the aircraft or touching down too soon.

There are four types of flare control systems which the present discussion considers: the fixed-path control (FPC), and exponential flare controller (EFC), a final-value controller (FVC), and a terminal-recomputing flare controller (TRC). The application of the FVC to aircraft landing is discussed in a master's thesis by L. K. Mattingly [10], which reports that the system was successfully flight-tested. The EFC and TRC systems are compared with the performance of systems involving random gust and sensor inputs on the simulated dynamics of four quite different aircraft in a technical report published by the Air Force Aeronautical Systems Division [9]. The simulation used in this comparison includes a simulation of ground effect during landing. The FPC requires a longitudinal position measurement, and this added complexity has not been shown to be necessary. Further, some parallel translation of the touchdown point is even desirable to prevent unnecessary maneuvering during the flare.

c. *Theory of Final-Value Control*

The theory of the FVC will be presented in a slightly more general form than is actually applied in the landing system mentioned above. The study of the final-value system also provides an insight into the roll of the adjoint equations in optimization [12].

A final-value control system is designed to control the state of the system at one instant T. This instant is taken as the nominal touchdown time. The system equations are taken to be

$$\dot{\mathbf{x}} = A(t)\mathbf{x} + B(t)\mathbf{u}$$

$$\mathbf{x}(t_0) = \mathbf{c}$$

The state of the system $\mathbf{x}$ is controlled by the control vector $\mathbf{u}$. The equation

$$\dot{\mathbf{y}} = -A^T\mathbf{y}$$

is the formal adjoint to the homogeneous system equation. The solutions to the adjoint equation are used to make a "conditional response predictor." This "predictor" is just the analytical solution for the value of the state vector $\mathbf{x}(T)$ at the instant T under the assumptions of a specific control and no noise inputs to the system. Suppose $Y(t)$ is a matrix solution of the adjoint equation

$$\dot{Y} = -A^T Y$$

with $Y(T) = C$. Then this equation and the dynamics equation for $\dot{\mathbf{x}}$ can be combined to yield

$$\frac{d}{dt}(Y^T\mathbf{x}) = \dot{Y}^T\mathbf{x} + Y^T\dot{\mathbf{x}} = Y^T B\mathbf{u}$$

This equation is then integrated from t to T:

$$C^T\mathbf{x}(T) - Y^T(t)\mathbf{x}(t) = \int_t^T Y^T(\ )B(\tau)\mathbf{u}(\tau)\,d\tau$$

The quantities above are identified as follows: $C^T\mathbf{x}(T)$ is some vector describing the state of the system at time T, $Y^T(t)\mathbf{x}(t)$ is equal to $C^T\mathbf{x}(T)$ if the control $\mathbf{u}$ is zero. Suppose $\mathbf{u} = \mathbf{g} + \mathbf{v}$ where $\mathbf{g}$ is a preprogrammed input. Then

$$C^T\mathbf{x}(T) = Y^T(t)\mathbf{x}(t) + \int_t^T Y^T B\mathbf{g}\,d\tau + \int_t^T Y^T B\mathbf{v}\,d\tau$$

The quantity

$$S_{cp} = Y^T(t)\mathbf{x}(t) + \int_t^T Y^T B\mathbf{g}\,d\tau$$

is the conditional value of $C^T\mathbf{x}(t)$. That is, if $\mathbf{v} = 0$ for $t \le \tau \le T$, then

$$S = C^T\mathbf{x}(T) = Y^T(t)\mathbf{x}(t) + \int_t^T Y^T B\mathbf{g}\,d\tau$$

The difference between the value of S and the conditional prediction S_{cp} is the predicted error e_p.

The diagram of Fig. 13-2 shows this system.

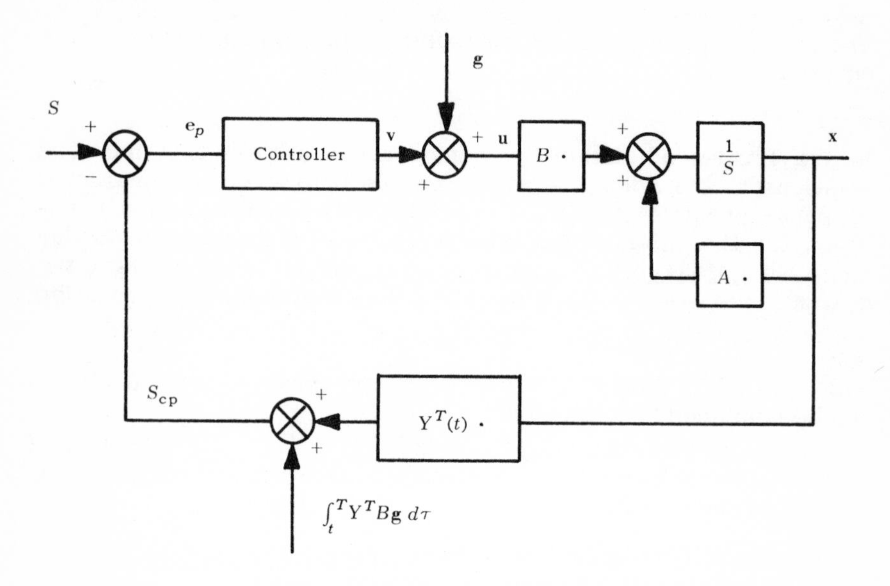

FIG. 13-2. Final value control system.

The controller design has not yet been specified, but a wide variety of controllers give suitable system performance. From the relation

$$e_p = \int_t^T Y^T B v(\tau)\, d\tau$$

one may obtain

$$\dot{e}_p = -Y^T B v$$

One may choose $v = f(e_p)$ and choose a Lyapunov function (dropping the subscript)

$$V = e^T P e > 0 \qquad e \neq 0 \quad P \cdot \text{constant}$$

$$\dot{V} = -\left[f^T B^T Y P e + e P Y^T B f\right]$$

From $\dot{V}$ it can be seen that if $e^T Y^T B f > 0$ for $e \neq 0$ and for some P, then the predicted error will continually decrease with respect to the norm $||e||^2 = e^T P e$; and if $e^{T_p} Y^T B f > h e^{T_p} e$, then $||e(T)|| \leq e^{-h(T-t)} ||e(t)||$.

For FVC application to flare, the nominal function g is selected so that the nominal flare is safe and comfortable. The controller may be chosen to be linear with saturation. The saturation limits prevent the command of unsafe maneuvers by the controller. The dynamics of the aircraft are included in the prediction.

The Terminal Recomputing Controller

This controller uses a family of nominal trajectories all of which satisfy the terminal conditions [9]. At each instant of time, the particular nominal trajectory passing through the aircraft trajectory is computed. The control command is then computed such that the aircraft will follow that nominal trajectory. The nominal trajectories are sometimes taken to be polynomials with coefficients which will satisfy the required conditions. For example, the function

$$h(\tau) = -\tau + a_2\tau^2 + a_3\tau^3 \qquad \tau = t - T$$

satisfies the terminal conditions $h(0) = 0, \dot{h}(0) = -1$ for all values of a_2 and a_3. The parameters are determined from the measured h_M and $\dot{h}_M$ by the conditions

$$\begin{aligned} h_M(\tau_0) + \tau_0 &= \\ \dot{h}_M(\tau_0) + 1 &= \end{aligned} \begin{bmatrix} \tau_0^2 & \tau_0^3 \\ 2\tau_0 & 3\tau_0^2 \end{bmatrix} \begin{bmatrix} a_2 \\ a_3 \end{bmatrix}$$

Solving these equations and substituting the coefficients a_2 and a_3 in the nominal function, one obtains the function

$$h_c(\tau) = -\tau + \left[3\left(\frac{\tau}{\tau_0}\right)^2 - 2\left(\frac{\tau}{\tau_0}\right)^3\right]\left[h_M(\tau_0) + \tau_0\right] + \tau\left[\left(\frac{\tau}{\tau_0}\right)^2 - \frac{\tau}{\tau_0}\right]\left[\dot{h}_M(\tau_0) + 1\right]$$

Problem: Plot the nominal trajectory for $h_M = 100$ ft, $\dot{h}_M = -20$ ft/sec, and $\tau_0 = -20$.

The proponents of this type of control system suggest that the altitude profile of the aircraft be commanded to realize the above nominal trajectory. It appears that a better class of nominals should be chosen.

The Exponential Path Controller

The EPC is quite simple conceptually [9]. The altitude rate command is proportional to the altitude above the reference (Fig. 13-3).

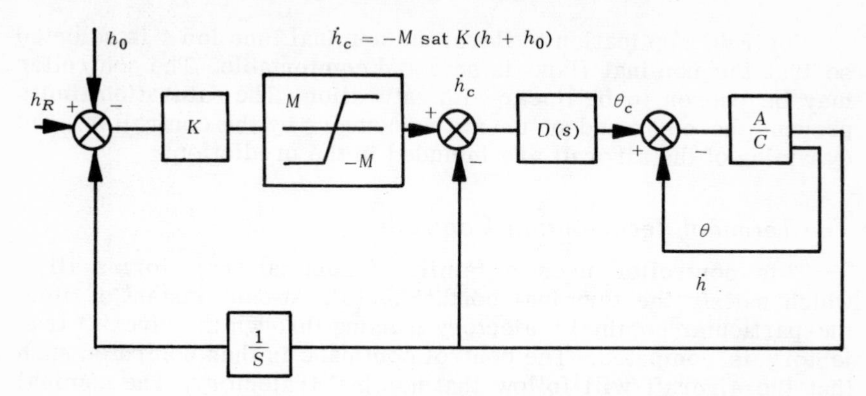

FIG. 13-3. Exponential path controller.

The constant h_0 is a bias to insure the desired sink rate at touchdown. The coefficient K determines the "time constant" of the exponential path. And the compensating dynamics $D(s)$ are chosen for stability and small errors in the vertical velocity. This system is reported to have a good response even in the presence of gust and noise inputs. The system does not have a predictor to compensate the commands for "ground effect." Ground effect influences the path because of the constant change of attitude in lift. In those aircraft for which ground effect produces a significant shift in the touchdown point, the rate command can be modified by a suitable nonlinearity to maintain a near-exponential path.

All the control systems discussed above require an altitude-rate measurement. Such measurements may be made in various ways; there is a trade-off between communication bandwidth and on-board instrumentation. A complete inertial measuring unit can be erected on the basis of a fairly simple message, or, in the absence of an on-board altimeter, the information must be transmitted at a fairly rapid rate.

From this brief discussion of the problems of automatic landing control, it is seen that various concepts can lead to the mechanization of a workable control system. The evaluation or comparison of the resulting systems must be made on the basis of economy, reliability, and performance.

ACKNOWLEDGMENTS

Many individuals have contributed to our understanding of aircraft and space vehicle control. To those who, throughout the history of aerospace vehicles, have contributed to the intellectural heritage which makes control of aerospace vehicles possible, I owe a debt of gratitide. To my immediate colleagues, including E. B. Stear, who have contributed to my understanding of these subjects, I am also grateful.

REFERENCES

1. Goldstein, H., *Classical Mechanics*, Addison-Wesley, Inc., Cambridge, Mass., 1950.
2. Etkin, B., *Dynamics of Flight*, John Wiley and Sons, Inc., New York, 1959.
3. Miele, A., *Flight Mechanics*, Addison-Wesley, Inc., Palo Alto, 1962.
4. Bodner, V. A. and Kozlov, M. S., *Stabilization of Flying Craft and Autopilots*, (Translation from Russian by Foreign Technology Division, AFSC, Ohio), Moskva, 1961.
5. Stear, E. B., "A Critical Look at Vehicle Control Techniques," *Astronautics and Aerospace Engineering*, August, 1963.
6. Prince, L. T., "Design, Development, and Flight Research of an Experimental Adaptive Control Technique for Advanced Booster Systems," *ASD-TDR-62-178*, February, 1962, I-II.
7. Zaborsky, J., Luedde, W., and Wendl, M., "New Flight Control Techniques for a Highly Elastic Booster," *ASD-TR-61-231*, September, 1961.
8. Roberson, R. E., "Attitude Control of Satellites and Space Vehicles," *Advances in Space Science*, Vol. II, 1960, pp. 351-436, Academic Press, New York.
9. Doniger, J. F. Belsky, Reynolds, G., Hanisch, B., and Flit, A., "Automatic Landing System Study, Part I. Results of Airborne Equipment Studies," *Report No. ASD-TR-61-114*, Wright-Patterson Air Force Base, Ohio, February, 1962.
10. Mattingly, L. K., "Application of Terminal Control Techniques for an Aircraft Landing Flare System," M. S. Thesis, University of California, Los Angeles, January, 1962.
11. Merriam, C. W. III, "Study of an Automatic Landing System for Aircraft," *MIT OACL Report No. 93*, May 10, 1955.
12. Matthews, M. V. and Steeg, C. W., "Terminal Controller Synthesis," *MIT Report No. 55-272*, p. 10, November 4, 1955.
13. Bryson, A. E., Denham, W. F., Carroll, F. J., and Mikami, K., "Determination of Lift or Drag Program to Minimize Reentry Heating," *Journal of Aerospace Sciences*, Vol. 29, No. 4, April, 1962.
14. Meditch, J. S., "Optimal Thrust Programming for Minimal Fuel Midcourse Guidance," presented at ASD Optimal System Synthesis Conference, Dayton, Ohio, September 11-13, 1962.
15. Kelley, J. H., Falco, M., and Ball, D. J., "Air Vehicle Trajectory Optimization," presented at SIAM Symposium on Multivariable System Theory, Cambridge, Mass., Nov. 1-3, 1962.
16. Horowitz, Isaac, *Synthesis of Feedback Systems*, Academic Press, New York, 1963.
17. Bode, H. W., *Network Analysis and Feedback Amplifier Design*, Van Nostrand, Princeton, New Jersey, 1945.

18. Ellert, F. J. and Merriam, C. W. III, "Synthesis of Feedback Controls Using Optimization Theory—An Example," presented at 1962 JACC in New York.
19. Bryson, A. E., Denham, W. F., "Guidance Scheme for Supercircular Reentry of a Lifting Vehicle," *Journal of the American Rocket Society*, Vol. 32, No. 6, June, 1962.
20. Hestenes, M. R., "A General Problem in the Calculus of Variations with Applications to Paths of Least Time," *Rand Report RM-100*, March 1, 1949.
21. Breakwell, J. V., "The Optimization of Trajectories," *J. Soc. Indust. Appl. Math.*, Vol. 7, No. 2, June, 1959.
22. Balakrishnan, A. V., "A Steepest Descent Method for a Class of Final Value Control Systems," *A-62-1730-211*, Aerospace Corporation, April 11, 1962.
23. Bryson, A. E. and Denham, W. F., "A Steepest Ascent Method for Solving Optimal Programming Problems," *Journal of Applied Mechanics* (Trans. ASME, Series E), Vol. 29, No. 2, pp. 247-257, June, 1962.
24. Stear, E. B. and Gregory, P. E., "Capabilities and Limitations of Some Adaptive Techniques," National Conference *Proceedings 1962 NAECON*, May, 1962.
25. Ostgaard, M. A. and Butsch, L. M., "Adaptive and Self-Organizing Flight Control System," *Aerospace Engineering*, p. 80, September, 1962.
26. Ostgaard, M. A., Stear, E. B., and Gregory, P. C., "The Case for Adaptive Controls," present to AGARD Flight Mechanics Panel, Paris, July, 1962.

Index